WITHDRAWN FROM
THE LIBRARY

UNIVERSITY OF
WINCHESTER

NOT TO BE REMOVED
FROM THE LIBRARY

D0301353

AN HISTORICAL GEOGRAPHY OF
ENGLAND BEFORE A.D. 1800

AN
HISTORICAL GEOGRAPHY
OF ENGLAND
BEFORE A.D. 1800

Fourteen Studies

edited by

H. C. DARBY, Litt.D.

Professor of Geography in the
University of Cambridge

CAMBRIDGE
AT THE UNIVERSITY PRESS
1969

Published by the Syndics of the Cambridge University Press
Bentley House, 200 Euston Road, London, N.W.1
American Branch: 32 East 57th Street, New York, N.Y. 10022

Standard Book Number: 521 04769 2

First published 1936
Reprinted with corrections 1948
Reprinted 1951   1961   1969

KING ALFRED'S COLLEGE
WINCHESTER.

REF   X4501259

911·42
DAR   41968

First printed in Great Britain at the University Press, Cambridge
Reprinted in Great Britain
by Bookprint Limited, Crawley, Sussex

CONTENTS

PREFACE

The term "Historical Geography" has been applied to a variety of subjects—to the story of geographical exploration and of geographical science, to the history of changing political frontiers, and to the study of the influences of geographical factors upon historical events. All these are most illuminating themes, and this is not the place to embark upon a discussion of terminology. Yet the fact remains that historical geography has been increasingly identified with another line of thought whose data are, of necessity, historical, but whose outlook is geographical. This study —to use Professor E. G. R. Taylor's words—"strictly speaking merely carries the geographer's studies into the past: his subject-matter remains the same." And this subject-matter is concerned with the reconstruction of past geographies, and aims to provide a sequence of cross-sections taken at successive periods in the development of a region. These surveys could hardly be the work of one individual, and this is the reason for a volume of co-operative studies.

The chapters have been written quite independently, and are not all upon the same pattern. Their differences reflect both the varying contents of their periods and the inequalities in the available materials. But although each chapter is quite separate, the arrangement of the book is designed to secure some continuity, and there are numerous cross-references. With one exception, the contributors are members of Departments of Geography in British Universities. We value very much the co-operation of so distinguished a place-name scholar as Professor Ekwall of the University of Lund.

It is not for me, or for us, to claim anything for this volume. Everyone who looks through its pages may find much to criticise. But I must point out that despite the increasing use of the term "historical geography" and despite the increasing importance of the subject in University studies, no substantial Historical Geography of England has yet appeared. This book therefore is, in a sense, experimental. As our text we might well take the words of Jean Brunhes, one of the many French geographers whose contributions are so outstanding:

L'homme entre en rapport avec le cadre naturel par les faits de travail, par la maison qu'il construit, par la route qu'il parcourt, par le champ qu'il cultive, par la carrière qu'il creuse, etc., . . . C'est, en effet, l'intermédiaire du travail et des conséquences directes de ce travail qui établit la vraie connexion entre la géographie et l'histoire.

These consequences, with their changing distributions, form much of the subject-matter of historical geography.

Quite deliberately, no attempt has been made to provide either a philosophical introduction or a comprehensive conclusion. Mr Bowen's chapter is by its nature an introductory one, and Mr Spate's account of London forms a logical full stop. Beyond 1800 we have not ventured to go, partly for practical reasons, partly because the subject-matter changes so much, and not least because the geography of the nineteenth century has been covered in certain chapters of Dr Clapham's *Economic History of Modern Britain*. Further, although Wales does not come within the scope of this book, we have not hesitated to make reference to it where this has seemed desirable.

This is an opportunity to express gratitude to those who have helped us. Some chapters contain individual acknowledgments; on p. xii is a list of the permissions to re-draw maps. We are also indebted to the many authors whose work we have quoted and used. Here, I can thank only those who have influenced the book as a whole. The kindness of Professor Debenham, and the facilities afforded by his Department of Geography at Cambridge, constitute the setting in which the idea of this book grew up; and in this connection I owe particular thanks to Mr B. L. Manning of Jesus College. Fostered by the encouragement of Dr Clapham, the idea became a concrete plan. Too often have I encroached upon the time alike of Dr Clapham, of Professor Debenham and of Mr Manning. I must also be allowed to give my warm thanks to the contributors. Without their courtesy and friendliness even a concrete plan could hardly have become a printed book! This is all too meagre an acknowledgment of much mutual co-operation spread over the last two years. To Professor Taylor, who contributes two chapters, I am indebted for advice from time to time. All the maps have been drawn by Mr L. D. Lambert and the index has been compiled by Mr W. H. Bromell; finally, only those who have had their books printed at the University Press know how helpful the Press officials can be. To their skill and care we owe a great deal.

H. C. DARBY

King's College
Cambridge
All Souls Night, 1935

MAPS AND DIAGRAMS

ACKNOWLEDGMENTS

We are much indebted for the following permissions:
For Figs. 1, 2, 4 and 7 (B and C) to Miss L. F. Chitty, Sir Cyril Fox and the National Museum of Wales; for Fig. 6 to Mr C. F. C. Hawkes, Mr G. C. Dunning and the Council of the Royal Archaeological Institute; for Fig. 7 to the Royal Irish Academy; for Fig. 12 to Mr W. F. Grimes and the Honourable Society of the Cymmrodorion; for Fig. 15 to the Controller of H.M. Stationery Office; for Fig. 21 to the editor of *The English Historical Review*; for Fig. 22 to the Royal Historical Society; for Fig. 23 to the Harvard University Press; for Fig. 24 to Mr F. W. Morgan; for Fig. 28 to Mr F. W. Morgan and the Royal Scottish Geographical Society; for Figs. 66 and 67 to Professor E. C. K. Gonner and Messrs Macmillan and Co., Ltd.; for Figs. 68 and 80 to the Clarendon Press, Oxford; for Fig. 69 to Professor J. U. Nef and Messrs George Routledge and Sons, Ltd.; for Fig. 70 to Mr L. Cossons and the Historical Association; for Fig. 76 to Dr Gilbert Slater.

Chapter I

INTRODUCTORY BACKGROUND:
PREHISTORIC SOUTH BRITAIN

E. G. BOWEN, M.A.[1]

The great advances in the study of prehistoric archaeology during recent years have made clear a sequence of well-defined cultures in Britain which stretch back in time over at least as many centuries before the Roman invasions as have elapsed since. The evidence is quite different from that of historical times, consisting, as it does, of the remains of earthworks, stone structures, and settlements, and of finds of pottery and implements of stone or metal. It is often difficult to be sure of the original purpose of many prehistoric sites and monuments, while many of the finds may have reached the place of their discovery by accident; many still remain undiscovered; many have not been recorded. Archaeological distributions, therefore, do not permit of absolutely certain correlations between prehistoric settlement and what can be reconstructed of the physiographical conditions of the time. For these reasons this must be an essay in the broadest of outlines, implicit with many reservations.

The nature of the material, and the limitations of the distributional method also affect considerably the scope of the enquiry. In order to delimit the boundaries of cultural provinces it is necessary to utilise the characteristics of articles that did not normally wander far from home (domestic and funerary pottery), or to select some immobile object like a dolmen, a barrow, or an earthwork. On the other hand, typological variation between monuments or objects is not an important consideration in demarcating the chief areas of settlement at any given period. For each of the phases—the Neolithic Age, the Bronze Age and the Iron Age—finds and sites of all descriptions may be mapped together on the assumption that their varying density provides some index of the differences in the intensity of human settlement at different epochs.

---

[1] I am greatly indebted to Sir Cyril Fox, Director of the National Museum of Wales, Cardiff, Prof. H. J. Fleure of the University of Manchester, Prof. C. Daryll Forde of the University of Wales, and Miss Lily F. Chitty for their kind assistance in reading the manuscript and in general discussion. To Mr O. G. S. Crawford also I owe many thanks for his comments upon the proof.

Finally, as many of the metal implements (certainly most of the gold ornaments) were imported objects, or objects made of imported material, it is evident that there must have existed an extensive long-distance trade in very early times. There were many sporadic losses along the trade routes, which are thereby indicated. Very often a chemical analysis of a metal find identifies its source. The indications of distant sources of raw material, and the fact that archaeological finds are dotted thickly across the necks of peninsulas, suggest that early commerce was to a large extent sea-borne, but that trans-peninsular routes were used to avoid the difficulty of doubling stormy headlands. Further, it is possible, by calculating the relative frequency of the occurrence of imported material in different regions at various times, to estimate the fluctuations in the volume and destination of prehistoric traffic along different routes.

Prehistoric material may thus be classified according to its value as evidence (1) in the delineation of cultural provinces, or (2) in the demarcating of the main settlement areas, or (3) in indicating the nature and extent of long-distance trade.[1] Among the many developments of prehistoric study at the present time these three lines of enquiry, fraught with difficulties though they may be, provide some indications of the changing geography of prehistoric Britain.

CULTURAL PROVINCES

It is usual to subdivide British prehistory into four ages, the Palaeolithic, the Neolithic, the Bronze and the Iron. The first and second periods, as their names imply, were marked by the use of stone implements. Dressed cores or worked-up flakes of flint characterise the Palaeolithic, while the advent of pottery and the beginning of agriculture mark the Neolithic. During the third and fourth periods there was widespread use of objects and implements of metal, first of bronze and later of iron. Besides this conventional fourfold division there was another and perhaps even more fundamental transition, that from a food-gathering to a food-producing stage, which took place between the Palaeolithic and the Neolithic. It is impossible to be precise about the date of such fundamental changes in economy, but they took place at different times in different areas in Britain between 2500 and 2000 B.C. During these centuries new influences emanating ultimately from the Ancient East reached Britain. The Neolithic Age

---

[1] It is obvious, however, that the same object might serve in more than one capacity. This general introduction is based on Mr O. G. S. Crawford's excellent critical study of archaeological method in *Man and his Past* (1921), especially Chapters VII, VIII and XIII.

is now considered to be in itself a comparatively brief transition period between the hunting and collecting of Palaeolithic times and the fully developed pastoral and agricultural economies of the Ages of Metal.

Little is known of the Palaeolithic and the Mesolithic food-gatherers of Britain. Their numbers must have been very small, and their material equipment was but slight. Their habitations were apparently confined to the coastal caves or to the caverns of calcareous uplands where supplies of flint could be obtained.[1] More detail is available concerning the later ages, and the two millenia before Christ can be further subdivided into a number of well-defined phases whose sequence and duration are very roughly indicated in the following table.[2]

Early Iron Age	c. 100 B.C.– 100 A.D.	Iron Age C (The Belgic Period).
	c. 250 B.C.– 100 B.C.	Iron Age B (Glastonbury and major Hill-fort Period).
	c. 550 B.C.– 250 B.C.	Iron Age A (All Cannings Cross).
Bronze Age	c. 1000 B.C.– 500 B.C.	Late Bronze Age.
	c. 1700 B.C.–1000 B.C.	Middle Bronze Age.
	c. 2000 B.C.–1700 B.C.	Early Bronze Age (Beakers).
"Neolithic Age"	c. 2500 B.C.–2000 B.C.	Neolithic camps and earlier Megaliths.

The establishment of the fundamental change in the economy of the prehistoric inhabitants of Britain from a life of food gathering to one of food production was also roughly contemporaneous with gradual, but nevertheless fundamental, changes in the physical geography of the area. Britain was becoming a group of islands. During Mesolithic times, its coastlands stood between 70 and 100 feet above their present level, thereby making

[1] For further information, see J. Reid Moir, *The Antiquity of Man in East Anglia* (1927); D. A. E. Garrod, *The Upper Palaeolithic in Britain* (1926); J. G. D. Clark, *The Mesolithic Age* (1934); together with the past volumes of the *Antiquaries Journal, Antiquity*, and the *Proceedings of the Prehistoric Society of East Anglia*.

[2] This table represents a rather maximum dating. It is based with some modifications on that printed in C. Fox, *The Personality of Britain* (1932). The chief changes concern the Iron Age, where a scheme, first suggested by Hawkes in *Antiquity*, v, 60 (1931), and subsequently modified by others, is thought to be more applicable to southern Britain in the Iron Age, than the following continental chronology based on Déchelette—c. 500 B.C.–250 B.C., Hallstatt and La Tène I; c. 250 B.C.–100 B.C., La Tène II; c. 100 B.C.–Roman Conquests, La Tène III. Hawkes has shown that in Britain the main divisions of Iron Age culture are as much regionally as chronologically distinct, and that they may overlap in time. Further comments on the scope of the table will be found on p. 9. Most of the technical archaeological terms are explained by implication later in the text, but for a complete survey see Miles Burkitt and V. Gordon Childe, *A Chronological Table of Prehistory*, reprinted from *Antiquity*, VI, 185 (1932).

the Straits of Dover either very narrow or non-existent, and adding to the present outline of the land a margin of coastal lowland which was most prominent in the south and east. Subsidence around the shores of southern Britain had certainly been active since about 3000 B.C. and continued until about 1600 B.C.,[1] but it was during, or just previous to, the Neolithic phase that a critical stage was reached when Britain ceased to be merely a peninsula of Europe. Possibly at about the same time the climate also was greatly modified. The warm and moist oceanic régime, known as the "Atlantic period",[2] had lasted from about 5500 B.C. until about 2500 B.C., when conditions slowly became less cyclonic and gave way to a warm and dry sub-Boreal period. Except for many detailed changes in coastal topography, the major orographical features of Britain showed little difference from those of the present day. There was, in particular, the same fundamental division between the Highland Zone in the north and west and the Lowland Zone in the south and east, separated roughly by a line joining the mouth of the Tees to the mouth of the Exe. Natural vegetation was, however, more pronounced. Dense oakwood covered both the damp clay vales of the lowland and the coastal flats.[3] More porous soils carried less timber and presented a more open landscape, while the slopes of the Lower Palaeozoic uplands of the Highland Zone were rich in thickets of the small oak (*Quercus sessiliflora*).

Thus from the Neolithic period onward there existed two sets of geographical conditions which influenced the distribution of immigrant cultures into Britain: (1) the position of the shorelands in relation to the Continent, and to the regions from which invasion emanated; (2) bold contrasts in the internal geographical conditions of the island. Until the second half of the first millenium B.C., and, more particularly, until the period immediately previous to the Roman Conquest, it is obviously impossible to know of any groupings of peoples on a tribal or political basis in Britain. When, however, prehistoric monuments and objects of

---

[1] For resumption of subsidence in post-Roman times, see pp. 61 and 94 below.

[2] The terminology here adopted is that of the Blytt-Sernander climatic periods. The sequence is as follows: from *c.* 5500 B.C. Atlantic time (warm and moist); from *c.* 2500 B.C. sub-Boreal time (warm and dry); from 700 B.C. sub-Atlantic time (moist and cold); from *c.* A.D. 1000 present conditions. For recent researches see (1) C. E. P. Brooks, *The Evolution of Climate* (1925); (2) "Post-Glacial Climates and the Forests of Europe", *Quart. Journ. Roy. Met. Soc.* LX, 377 (1934); and (3) T. W. Woodhead, "Post-Glacial Succession of Forests in Europe", *Science Progress*, XXVI, 250 (1931–2).

[3] The remains of these forests are found at low tide around our shores at the present time.

distinctive types and ages are mapped, their distribution patterns reveal definite cultural divisions related to position and physiography.

During the Neolithic period the chief cultural contacts of Britain were by way of the sea routes of the western fringe of the Continent. It was only natural, therefore, that access to Britain at this time was *via* the peninsulas of the north and west. From such contact there developed the "western cultural pattern" in prehistoric distributions. The spread of the Megalithic culture during this time is an excellent example of such a western distribution.[1] This culture, associated with the building of the great stone monuments, was sea-borne. It is established that the distribution of these monuments indicates the line of an early spread of the fundamental arts and crafts of civilisation in a northerly direction from Iberia and the western Mediterranean *via* Brittany to western Britain.[2] The dolmens, and chambered tombs, which form the chief element in the megalithic evidence are situated, for the most part, on the peninsulas and headlands that jut out into western seas, but they are also found upon the lower slopes of the coastal plateaux. They seldom occur at elevations above 600 feet, and they are not part of an upland culture in the strict sense of the word. Furthermore, the details of the distribution, shown on Fig. 1, indicate that there is very little correspondence with the Highland Zone as physiographically defined. The presence in southern Britain of long barrows with megalithic chambers suggests a fairly deep penetration into the Lowlands; in south-western England, especially in the area which was later known as Wessex, the penetration of long barrows is very marked.[3] On the other hand, the fact that the Lowland Zone was soon populated, by Beaker Folk from across the North Sea, suggests that the penetration, except in Wessex, was of a temporary character. It may be possible that Beaker Folk, influenced by megalithic contacts in the Netherlands, built the Medway megaliths.[4] This contact between these important early cultures is held to have been

[1] It must be remembered that the megalithic culture continued to flourish after the Neolithic Phase in Britain. Many of the characteristic monuments, such as the stone circles, undoubtedly belong to the Bronze Age, and an Iron Age date has even been claimed for some of the Scottish circles; see V. G. Childe, "Final Excavations at the Old Keig Stone Circle", *Proc. Soc. Antiq. Scot.* LXVIII, 372 (1933–4).

[2] C. Daryll Forde, "Early Cultures of Atlantic Europe", *American Anthropologist*, XXXII, 28 (1930).

[3] Sir Cyril Fox has suggested that this penetration is due to the proximity of Wessex to the south-western part of the Highland Zone on which the Megalithic culture first impinged (*Personality of Britain*, p. 41).

[4] See, however, Stuart Piggott, "A Note on the Relative Chronology of the English Long Barrows", *Proc. of the Prehist. Soc.* I, 122 (1935).

2

very prominent in the Wessex country, and here, on Salisbury Plain, is the most famous of all megalithic monuments—the great stone circle of

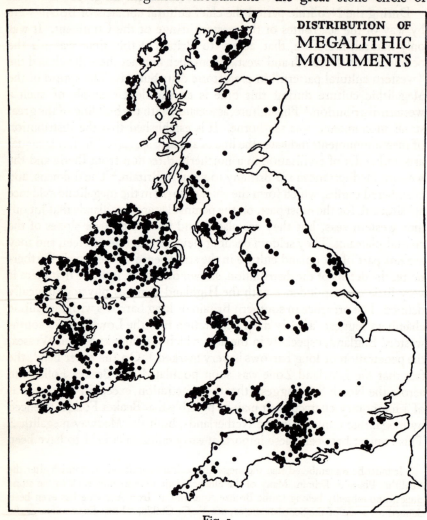

Fig. 1

After L. F. Chitty and C. Fox in *The Personality of Britain* (1932), p. 10, where the detailed sources are given. The map shows long barrows, chambered cairns and dolmens, but it does not include stone circles, alignments and menhirs.

Stonehenge.[1] It is probably, like all other stone circles, later in date than the dolmens, long barrows, and chambered cairns shown on the map (Fig. 1).

[1] H. J. E. Peake and H. J. Fleure, *The Corridors of Time*, VI, 119 (1929)

South-eastern Britain forms the apex of a triangle pivoted on Cape Griz Nez, presenting two sides to a long strip of continental coast from western France to the Danish Peninsula. Invasion emanating from the French coast influenced southern Britain, that from the Flemish and north German coasts impinged on eastern Britain. Cultures arriving on either

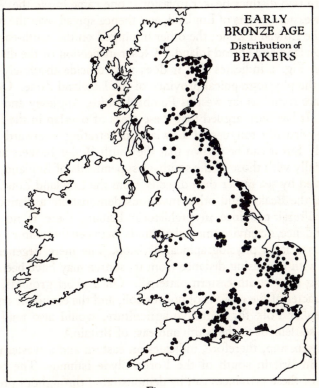

EARLY
BRONZE AGE
Distribution of
BEAKERS

Fig. 2

After L. F. Chitty and C. Fox in *The Personality of Britain* (1932), p. 11, where detailed sources are given. The dots indicate (1) finds of single beakers of all types; (2) sites where two or more beakers have been found together; (3) sites with definite evidence of settlement by beaker peoples.

shore tended to spread generally over the Lowland Zone giving to their associated objects a characteristic "south-eastern pattern". An excellent example of this distribution is that resulting from the Beaker invasions[1] of

[1] The term "Beaker" is restricted to a class of beaker-shaped vases decorated in patterns arranged in horizontal bands; they are found in graves generally with contracted skeletons (see Miles Burkitt and V. G. Childe, *op. cit.* p. 200).

the Early Bronze Age (see Fig. 2). The fact that the eastern shores of Britain are so thickly dotted with Beaker burials suggests that their origin was the region of the Rhine mouth; and British beakers certainly resemble more closely in type the "northern beaker" of the Low Countries than the "atlantic beaker" of Spain and Brittany. It is now recognised that there were two distinct Beaker invasions; one came in *via* the Wash and the north-eastern coasts of England, and thence spread over the north and west of the Lowland Zone; the other impinged on the south-eastern part of the Lowland and spread inland.[1] Closer inspection of the distribution shown on Fig. 2 indicates that it does not coincide absolutely with the limits of the physiographical province of the Lowland Zone. Occasional beakers are found as far west as Pembrokeshire, Anglesey and the Vale of Eden. It has been argued that these areas of overlap in the Highland Zone are virtually tongues of Lowland penetrating westward into the Highland. But it can be shown that the south Wales beakers are linked typologically with those of the south-west peninsula of England, and that they arrived by sea rather than directly from the Lowland Zone.[2] While, therefore, the Beaker and other invasions demonstrate a distinct south-eastern cultural province in prehistoric Britain, there is no absolute correspondence, on environmental grounds, between this cultural province and the associated physiographical division. The mere degeneration of invasion with increasing distance from its source may have been a factor influencing the "south-eastern pattern". On *a priori* grounds, too, the open character of the regions of porous soil, and the economic equipment of the Beaker Folk for primitive agriculture, would also tend to limit intensive settlement to the south and east of Britain.[3]

This discussion, therefore, indicates an eastern and a western cultural province in Britain south of the Forth-Clyde isthmus. The continued existence of these provinces can be demonstrated throughout prehistoric, and even later, times. The encrusted urns of the Late Bronze Age and (much later) the Ogham-inscribed stones of the Dark Ages illustrate clearly the

---

[1] This distinction has been based on typological differences among the British beakers themselves—see V. G. Childe, *The Bronze Age* (1930), p. 156, and J. G. D. Clark "The Dual Character of the Beaker Invasion", *Antiquity*, v, 415 (1931).

[2] C. Fox, "On two Beakers of the Early Bronze Age recently discovered in South Wales", *Arch. Camb.* (1925), i.

[3] The distribution patterns of early cultures in Britain illustrate yet another cultural region. This is the portion of the Highlands that lies north of the Forth-Clyde isthmus. Its distinctiveness is associated with its remoteness, but it will be omitted from further consideration as it lies outside the region dealt with in the scope of this chapter. See C. Fox, *Personality*, p. 34.

"western pattern", while the globular and finger-tipped cinerary urns of the Late Bronze Age and the domestic pottery of the early phases of the Iron Age show the "south-eastern pattern" to advantage.[1]

It is possible to generalise further on the essential differences between these provinces. In the south-eastern area, invasion was frequent, and large groups of men and materials could be ferried over the narrow seas. Hence, successive cultures succeeded one another in rapid succession, each invasion tending to destroy the influences of its predecessor. Fox has termed this region one of cultural replacement. On the other hand, the western region received its new contacts in smaller doses—perhaps more by trading and occasional raiding than by massed invasion. Most frequently, also, the contacts on its eastern margin were with an enfeebled outer wave of an already spent invasion of the south-eastern zone. Here, therefore, already existing cultures tended to absorb the new infiltrations; old customs and old ideas lived long. The west has always been an area marked by a continuity of tradition. This is an important feature of the life of early Britain, and has an important bearing on the chronological table given above. The conditions revealed by Fig. 5, showing the massed finds of the Early Iron Age, illustrate the point. Throughout the latter half of the last millenium B.C., Iron Age cultures had dominated the south-eastern province, and their massed effect in Fig. 5 provides an excellent demonstration of the "south-eastern cultural pattern".[2] But, even after these five hundred years, the new ideas and customs had barely influenced the western province which (as shown on the map) still retained in the main a Late Bronze Age civilisation continuing alongside the newer cultures of the south and east.[3] There was a time-lag. Thus the dating given in the chronological table refers only to the changes in the south-eastern province; considerable modifications are necessary for dating parallel cultures in the western zone, another aspect of the dual character of early Britain.

[1] This is the basis of the argument in C. Fox, *The Personality of Britain* (1932); see p. 28 for a general appreciation.

[2] The Belgic districts of Fig. 6 (p. 21 below) also show a "south-eastern pattern". See G. F. C. Hawkes and G. C. Dunning, "The Belgae of Gaul and Britain", *Arch. Journ.* LXXXVII (1930), and R. G. Collingwood in *Camb. Anc. Hist.* vol. x, chap. XXIII. For the western shores of the Severn see V. E. Nash Williams, *Arch. Camb.* (1933), 237.

[3] It should be noted, however, that on the Continent the La Tène II culture had become associated in Brittany with the western sea routes and Iron Age cultures had arrived by sea in the south-west of England and so to south Wales and the Severn Valley, from the second century B.C. See p. 28 below.

DISTRIBUTION OF SETTLEMENT

It is essential first to outline on *a priori* grounds what most probably constituted the chief limiting factors in prehistoric settlement. Prominent among these must have been the wide existence of natural "damp oakwood" forest with a particularly dense undergrowth. To primitive peoples, with poor material equipment in the way of sharp-edged tools, the penetration and clearing of such a woodland presented considerable difficulties. And it has been argued that early man avoided these regions of dense wood with thick brambly undergrowth, on a water-holding soil, and chose the more open country for settlement. This may well have been very true of the early stages of settlement, though, with improvement in the technique of implement manufacture, a more successful attack on the forested lands probably took place.[1]

Among the varied factors controlling the distribution and composition of prehistoric vegetation, climate and soil must have been all important. The changes of climate which took place were not sufficiently revolutionary in character to modify seriously the soil character.[2] Their influence was primarily felt in modifying directly the vegetation, and hence in influencing the areas suited to early settlement. Conditions in the Highland were quite distinct from those in the Lowland. During "Atlantic" and "sub-Atlantic" times, when oceanic influences were dominant, the north and west obtained a higher orographical rainfall than the south and east; during the more continental period (i.e. the sub-Boreal) the comparative aridity must have made itself felt more markedly in the south and east than in the north and west. The rainfall even in these times was never so low as to prevent tree growth reaching its present altitudinal limits anywhere in Britain but the drier conditions made the forest-free highlands of the north and west more suitable for human occupation than they had previously been. On the other hand, the sub-Boreal climate, especially towards the close of the Bronze Age, may have become too dry to make the chalk ridges and uplands of the south-east really successful pasture lands; the population, accordingly, appears to have been concentrated on the upper valley slopes.

About 700 B.C., with the onset of the sub-Atlantic phase, there was definite change; the climate became more oceanic and, as the rainfall

---

[1] Other factors, such as proximity to sources of raw material (e.g. flint), were naturally important secondary considerations; but their importance decreased as long-distance trade in precious commodities, such as gold and metal, developed.

[2] Apart from peat formation, etc.

increased, conditions apparently became colder than they had been for the preceding two millennia.[1] The highlands of the north and west became waterlogged, rain-swept, peaty, and less habitable. There is evidence that the main centres of settlement moved valleywards; the uplands were used possibly as seasonal pasture lands in summer time. But, on the other hand, in southern and eastern England, the increased rainfall made the open lands of the chalk ridges more fertile than before, and the valley bottoms became more densely clogged with damp-loving forest. Thus in south-eastern England in the early Iron Age the uplands gained a renewed importance, although the advent of iron weapons and tools meant that material culture was gradually giving men all over the country a mastery over the original limitations of nature. The significant point, however, is that the past changes of climate in Britain appear to have been associated mainly with changes in the altitude at which human settlement occurred rather than with any great changes in its surface distribution. In order to discuss these surface distributions greater attention must be given to soil conditions.

Marked differences between north-west and south-east are at once apparent in the nature of the rock surface. The north-west, for the most part, is a region of Palaeozoic rocks, of impervious shales and mud-stones with some partially porous Old Red Sandstone. The south and east is a region of newer rocks. Extensive outcrops of porous limestone form ridges and alternate with clay vales. Light porous soils, such as those derived from chalk, and from sands and gravels, maintain a light forest with a thin undergrowth. Here, the beech and ash flourish, while heavy and impervious soils, derived from clays and mudstones, support a heavy forest growth in which the oak (*Quercus robur*) is associated with the dense damp undergrowth. In the north and west, there are differences. Altitude and exposure to strong sea winds exercise a modify-ing influence, clothing the upland valleys and the lower hill slopes with the small oak forest, and making the narrow coastal strips relatively free from dense woodland.[2] The high winds on the moorlands keep the tree limit rather low here as well. The upper limit of trees was probably every-where below 1500 feet during the Bronze Age. Thus in Wales, in Cumbria, and in Cornwall, the population gathered on the forest-free coastal patches or on valley slopes in the high moorlands; while upon the English Plain the concentration was along the outcrops of pervious strata bearing lighter

[1] See footnote 2 on p. 4 above.
[2] H. J. Fleure and W. E. Whitehouse, "The Early Distribution and Valleyward Movement of Population in Southern Britain", *Arch. Camb.* 1916, p. 101.

forest or open parkland. The location of the areas of settlement, as shown on the maps of massed finds for the three great prehistoric phases, illustrates the general truth of these remarks. Complications occur owing to the presence of glacial drift over large tracts of country.[1] Fig. 14 shows[2] those areas of southern Britain that presented to the early inhabitants the most favourable opportunities for settlement. In many cases the drift forms a light soil-cover, with correspondingly light vegetation, where otherwise would have been damp clays and dense vegetation. But it must be remembered that there are numerous local variations in the drift itself, and detailed study in the field has yet to clear up many of these points.

On Fig. 3, showing massed settlement and burial sites[3] for the Neolithic phase, two concentrations of finds are at once apparent—one on Salisbury Plain and the Cotswolds, and the other on the Yorkshire Wolds. Both are on clearly marked areas of porous soil; and so also are the areas of less concentration—on the Lincoln Wolds and in the Peak District. Other concentrations are upon the northern and southern peninsulas of Wales, and, to a smaller extent, in the Isle of Man, and upon the extreme northern tip of the Cornwall peninsula. The absence of settlements on the uplands of Wales, of Dartmoor, of Cumbria, and of the Pennines, at this period is remarkable. It may be explained in part by the fact that the settlements associated with the sea-borne cultures were confined to the coastal fringes.[4] The full development of the western sea routes in the short Neolithic phase (associated with the megalithic civilisation) explains too the outstanding importance of the Wessex concentration. But Salisbury Plain itself continued to be important throughout prehistoric times,

[1] The study, and even the primary official mapping, of drift soils in Britain has been much neglected and really detailed information is available only for a few small and isolated districts.

[2] Prepared by Dr S. W. Wooldridge on p. 95; see also footnote 1, p. 19.

[3] In the preparation of this map, stone (including flint) axes and arrowheads have been omitted. This is deliberate, as it is thought that the selection of megalithic monuments, flint mines, long barrows, pottery, etc., may be considered as demarcating definite settlement and burial areas more satisfactorily than if the Neolithic axes and arrowheads were included. Use has been made *inter alia* of L. F. Chitty and C. Fox, *The Personality of Britain*, Fig. 1, E. C. Curwen, *Antiquity*, IV, 22 (1930), V. G. Childe and Stuart Piggott, *Arch. Journ.* LXXXVIII, 37 and 67 (respectively), 1931. It is readily acknowledged that the list of Neolithic finds shown on the map is incomplete; but, in the absence of a catalogue (such as is being compiled for the Bronze Age by the British Association), it is impossible to indicate the distribution with great accuracy.

[4] C. Fox and E. G. Bowen, in *A History of Carmarthenshire* (edited by Sir John E. Lloyd, 1935), chapter I, p. 44.

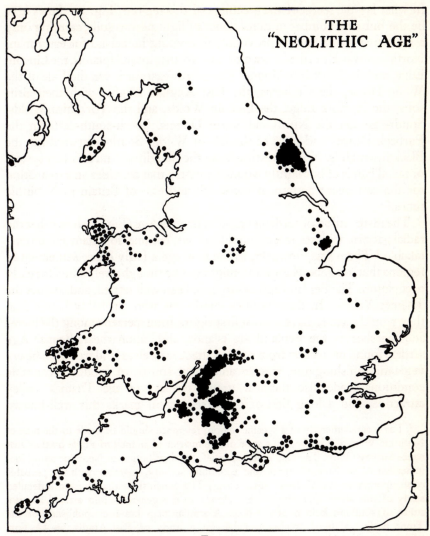

Fig. 3

This map is an attempt to group the main antiquities of the "Neolithic Age" by using the same symbol. It shows the distribution of dolmens, chambered cairns, long barrows, Neolithic camps, flint mines and Neolithic pottery. For further details see footnote 3, p. 12.

and long after the importance of the western sea routes began to decline. In what lay the secret of the greatness of this region? Fig. 14 shows it to be the hub of a number of radial zones of light porous soils, carrying only a light tree cover and so providing an inviting homeland for early man. North-eastwards ran the Cotswolds, the Northampton Uplands, the Lincoln Edge, and the Yorkshire Moors. East-north-eastwards ran the Vale of the White Horse, the Chilterns, the East Anglian heights with their drift soils, the Norfolk Edge, the Lincoln Wolds, and the Yorkshire Wolds. South-east ran the North and South Downs. South-south-east ran the Purbeck Downs and parts of the Isle of Wight. South-westward ran the Blackdown Hills. North-westward ran the Mendips. And it is this system of radial bands of but lightly wooded country that provides an explanation for the settlement pattern of the south and east of Britain in Neolithic times.[1]

The distribution of settlement in the Bronze Age (Fig. 4) shows that the radial pattern, with some local modifications, became dominant and unmistakable.[2] There was, however, in the new age a fairly dense settlement in regions that, on general grounds, might not be thought well suited for early occupation. The Fenland appears to have been well settled, and so does the Thames Valley. In these regions conditions were probably better than they would seem to have been at first sight. Interspersed among the meres and marshes of the Fenland are islands, and, upon these, Bronze Age settlers found protection from sudden attack, as well as land free from heavy vegetation; fishing too was an additional source of food. Favourable conditions made also the low gravel terraces of the Thames Valley attractive, and a long line of dense settlement marks this well-known

[1] For a detailed study of a specific region, reference should be made to the map of Neolithic Wessex published on a scale of one quarter of an inch to a mile by the Ordnance Survey Office. The map indicates relief, the distribution of Neolithic antiquities indicative of settlement, burial, and economic activity, together with the probable contemporary extent of dense forest cover. The conclusions drawn from a detailed study of this sheet approximate very closely to the generalisations arrived at by a study of southern Britain as a whole. A similar map has been published of the Trent Basin. Material is also available for a study of East Anglia, see C. Fox, "Distribution of Man in East Anglia, c. 2300 B.C.–50 A.D.", *Proc. Prehist. Soc. East Anglia*, VII, Pt. II, 149 (1933). The Neolithic map here (by L. F. Chitty and C. Fox) indicates Neolithic stone and flint axes.

[2] The Bronze Age map is that of Chitty and Fox—whose pioneer work remains most outstanding. The corresponding maps for the Neolithic Age and Iron Age have not previously been published, and although the present attempts are necessarily incomplete in the present state of our knowledge they have been prepared to complete the survey.

THE
BRONZE AGE

Fig. 4

After L. F. Chitty and C. Fox in *The Personality of Britain* (1932), Map. C, and p. 55, where details are given. The map shows the massed finds of the Bronze Age; a single dot represents either a single find, a group of associated objects or a barrow of the Bronze Age with its contents.

valley floor in Bronze Age times.[1] The dangers of any dogmatic statement is shown by the enigma of the Cotswolds. A fairly dense occupation of these would be expected on *a priori* grounds; but, very curiously, the area seems to have suffered relative depopulation after Neolithic times. It has been suggested, however, that the Long Barrow people of the Cotswolds remained here through part, at least, of the Bronze Age.[2]

Apart from these exceptions, the denser settlement of the lighter soils of the south and east during the Bronze Age, as compared with the Neolithic Phase, throws into greater contrast the significance of the sparsely populated "midland triangle". Its apex rests near Charnwood Forest and its base reaches from the estuary of the Severn to that of the Dee. This is a lowland region of heavy clay soils, only partially relieved by islands of sand; and the drift soils of its northern parts are uninviting to a light forest cover. The small number of Bronze Age finds mark it as a negative region.[3] Both Fleure and Fox have stressed the importance of this triangle in adding to the cultural isolation of Wales in pre-Roman times.

Despite the decline in the relative importance of the western sea routes, the highlands to the north and west of the "midland triangle" doubtless carried, in common with lowland England, a denser population in the Bronze Age than they did during Neolithic times. The settlement reached higher elevations, too—in Cumbria, in Wales, and in Cornwall. Cultural and economic factors help to account for this; climate may also have been important. As the Bronze Age advanced, the full influence of the drier conditions became apparent, and high moorlands may have become less windswept and sodden, and so more amenable to settlement. So important may these climatic modifications have been that it has been argued that the drier conditions must have made the less watered and more porous soils of the downlands of south-eastern England too dry for permanent settlement;[4] and it has been assumed that the dense cluster of finds shown on these lands masks the fact that *permanent settlement* was on the slope near the spring line and that the uplands themselves were the pasture grounds.

More detailed examination of settlement in the Highland Zone indicates the importance of the limestone uplands of the Peak district, of the drift

[1] E. T. Leeds, "Early Settlement in the Upper Thames Basin", *Geographical Teacher*, XIV, 527 (1927–28).   [2] C. Fox, *Personality*, p. 60.
[3] First demonstrated by O. G. S. Crawford, "The Distribution of Early Bronze Age Settlements in Britain", *Geographical Journal*, XL, 188 (1912).
[4] C. Fox, *Personality*, p. 69.

lands of the Eden Valley, and of the periphery of the Cumbrian highland. These were all well-marked foci of settlement. In Wales, the denser con-

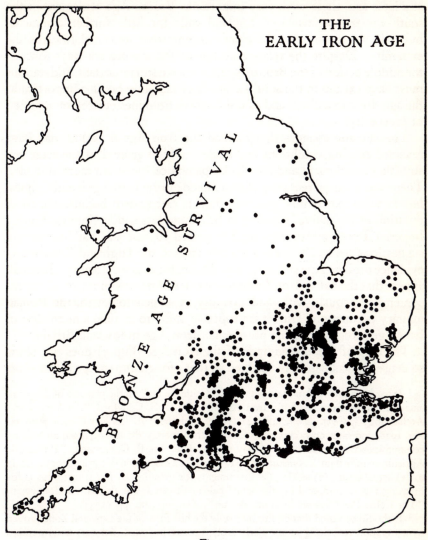

Fig. 5

This map is an attempt to indicate by the same symbol the main antiquities of the pre-Roman Iron Age in England and Wales. For further details see footnote 2, p. 18.

centration in the north may have been due, in part, to its close association with northern Ireland, and the gold-producing area (near Dublin) which was a centre of much trading activity in the Bronze Age. South Wales, on the other hand, had more natural links with southern Ireland; and, at a later age, these were very important. Bronze Age settlement in the south-western peninsula of England calls for little further comment; but it is significant that the chief concentration remained towards the extremity. Despite the relative decline of the western sea-ways towards the middle period of the Bronze Age, they were still important, and stations must have existed in the south-west. They certainly became important as the age drew to a close, and as the Cornish tin mines entered on a period of prosperity.

The climatic changes that marked the Iron Age A period may have reversed existing conditions and made the high ground of the west less suitable to settlement and the chalk uplands of the south more suitable.[1] There was a partial recovery of the climate during Iron Age B and C times, but to what extent this was important is not known because the short duration of the native Iron Age in Britain and its sudden interruption by Imperial Rome afford too brief a period to test the validity of any suggestions here. This is particularly true of the Highland Zone where evidence for Iron Age culture in pre-Roman times is very scanty. It would appear that the native Iron Age culture reached its maximum development contemporaneously with, and possibly in association with, the Roman military occupation. Thus it is profitable, here, to confine a discussion of Iron Age settlement to the pre-Roman Iron Age in lowland Britain.

The map of the massed finds of the Iron Age (Fig. 5) does not show so dense an occupation of the land as that of the Bronze Age.[2] The short

---

[1] It has been shown that the existence of "Celtic Fields" on the plateaux of the south and east datable from the Early Iron Age (see O. G. S. Crawford, "Air Survey and Archaeology", *Geog. Journ.* LIX, 342 (1922), and (with A. Keiller) *Wessex from the Air*, 1928) indicates an occupation of, rather than merely the use of, those uplands by communities with an arable economy at this time. See E. C. Curwen, "Prehistoric Agriculture in Britain", *Antiquity*, I, 261 (1927) for a discussion of the evidence relating to (1) actual grain, (2) sickles, (3) instruments for grinding corn, (4) instruments for breaking the ground, and (5) the actual fields cultivated; see also G. A. Holleyman, "The Celtic Field-System in South Britain", *Antiquity*, IX, 443 (1935).

[2] It should be noted that on this map only the hill-forts of the Lowland Zone proved by excavation to belong to the Early Iron Age are included. An enormous number remain unexcavated and are omitted, as also are the numerous hill-forts, in Wales and on the eastern fringes of the Highlands, which probably belong to the end of this period or to early Roman times. As they remain unexcavated, it is impossible to be precise as to their age. There are also shown domestic and funerary pottery, inscribed coins and currency bars, iron swords, sickles, implements and personal ornaments. Use has been made *inter alia* of L. F. Chitty and C. Fox, *The Personality of Britain*,

time during which the Iron Age cultures flourished in south-east Britain accounts for this apparent low density. It may be due also to the failure of land workers and others to collect rusty iron objects as readily as those of bronze. Distribution alone, therefore, must be considered. The region of Salisbury Plain still maintained its importance, by reason of its relatively forest-free surface, and owing to its proximity to the great ports of the south coast. Generally speaking, the earlier pattern of settlement in lowland England, resting on the lighter porous soils, and focussing naturally on the Salisbury region, continued with little modification. Even the Cotswold area began, apparently, to regain its Neolithic prosperity. But away in East Anglia, the map shows that changes had taken place during the brief period of the native Iron Age. And these changes were destined to be of greater moment for the future development of Britain than anything that was happening on the Cotswolds or even upon Salisbury Plain. The Iron Age A invaders focussed their settlement on the limestone and chalk uplands, but the Iron Age C invaders, who came in immediate pre-Roman times, showed a preference for the lower Tertiaries of Kent, and particularly for the drift soils of Essex, Hertfordshire and Bedfordshire.[1] The concentration in this latter region is one of the most significant changes brought out by Fig. 5. The Breckland and Fenland concentrations of the Bronze Age gave way to these new developments to the south. The Iron Age C invaders, responsible for this great change in the distribution of early population in south-eastern England, were the Belgae. They first arrived in Kent round about 75 B.C. and had, by the time of Caesar's expeditions (55–54 B.C.), penetrated beyond the Thames. They avoided the difficult Wealden lands which appear to have been but

---

Figs. 4, 5, 35, and Plate VII; C. F. C. Hawkes, *Antiquity*, V, 60 (1931); G. C. Brooke, *Antiquity*, VII, 268 (1933); S. Harris, *Arch. Camb.* 1925, pp. 190–1; C. F. C. Hawkes and G. C. Dunning, *Arch. Journ.* LXXXVII, 150 (1930); H. O'N. Hencken, *Cornwall and Scilly* (1932); R. E. M. Wheeler, *Prehistoric and Roman Wales* (1925). It should be noted that the distribution of massed finds as shown on this map *does not give any indication of the distribution of settlement outside south-eastern England in Iron Age times,* for in the Highlands a Bronze Age civilisation survived contemporaneously with the Iron Age cultures of the Lowlands until the Roman Conquest (see p. 9).

[1] It has been suggested that this change in settlement distribution represents an occupation, for the first time, of the clay-covered lowlands, and a mastery over the difficulties of dense forest undergrowth (see C. Fox, *Personality*, p. 79). But Dr Wooldridge's soil map (p. 95) makes it clear that the centre of the settlement lay essentially upon drift land and not upon clay. See also S. W. Wooldridge and D. L. Linton, "The Loam-Terrains of S.E. England and their Relation to its Early History", *Antiquity*, VII, 297 (1933), and C. Fox, *Proc. Prehist. Soc. East Anglia*, VII, Pt. II, 149 (1933).

sparsely settled even by native peoples at this time,[1] and made the centre of their kingdom on the drift soils north of the Thames. Near Wheathampstead, overlooking the valley of the Lea, are the earthen banks of a former Belgic stronghold which may very plausibly be identified with the unnamed *oppidum* of Cassivellaunus (the chief of the Catuvellauni), who organised the resistance to Caesar, and whose stronghold was stormed by the Roman legions.[2] By the end of the first century B.C. the Catuvellauni had conquered the Trinovantes who were Pre-Belgic folk living in Essex, and the capital of Catuvellaunia was transferred to Camulodunum (i.e. Colchester). Meanwhile, about 50 B.C., there had been a second invasion of Belgic folk led by Commius the Atrebate, a fugitive from Caesar's pacification of Gaul. These spread over the chalk plateau of Wessex and constituted a second nucleus of Belgic settlement, this time in south-western England, distinct from the earlier Belgic kingdom in the south-east (Fig. 6).

One of the first general effects of these Belgic invasions was the substitution of larger and more powerful kingdoms for the old, small tribal groups. Each of the older groups may have had a hill-fort as the capital point of its upland territory. Within the new and larger kingdoms, cities such as Camulodunum (Colchester) in the south-east and Calleva (Silchester) in the south-west were new features in the geography of Britain, reflecting the urban life of Gaul.[3] These towns were linked together by tracks which, apparently, may almost be termed roadways. Furthermore, while the Belgic immigration left the already existing upland areas of agriculture intact, it probably increased the valleyward movement of the population, and enlarged the acreage under the plough in the lower lands. This growing importance of agriculture under the Belgae was probably associated with the introduction of the wheeled plough, coulters of which have been found in the native (Romanised) town of Great Chesterford, Essex.[4] These changes must have been well advanced when Caesar landed, as he was impressed by the dense population of Kent and the ease with which he was able (when left free to do so) to reap enough native grown corn to supply his troops.[5] The Catuvellaunian dynasty, in the person of Cunobelinus, held sway in the

[1] C. F. C. Hawkes and G. C. Dunning, "The Belgae of Gaul and Britain", *Arch. Journ.* LXXXVII, 255 (1930).

[2] R. E. M. Wheeler, "The Belgic Cities of Britain", *Antiquity*, VII, 21 (1933). This contains a useful distribution map of these settlements.

[3] C. F. C. Hawkes and G. C. Dunning, *art. cit.* p. 299.

[4] Quoted by C. Fox, *Proc. Prehist. Soc. East Anglia*, VII, Pt. II, 159 footnote (1933).

[5] Caesar, *De Bello Gallico*, Liber V, Cap. 12, 3; also IV, 31, 2 and V, 17, 2.

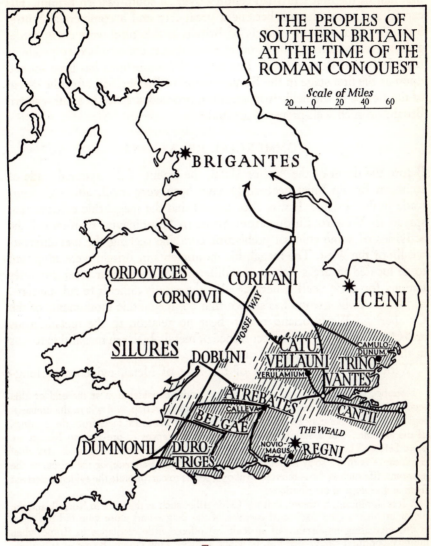

Fig. 6

Redrawn from C. F. C. Hawkes and G. C. Dunning, "The Belgae of Gaul and Britain",
Arch. Journ. LXXXVII, 317 (1931), and on information kindly communicated by Mr
Hawkes. The Belgic districts, known and presumed, are marked with heavy and light
shading respectively. The tribal names of the active opponents of Rome are underlined,
whereas an asterisk marks those whose rulers became allies or vassals of the Imperial
power. The black lines and arrows mark Roman roads and lines of advance.

3

first half of the first Christian century over all south-eastern Britain. His capital at Camulodunum became a great city and a commercial centre. Strabo[1] noted that the exports of Britain at this time were corn, cattle, gold, silver, and iron; the imports were almost entirely luxury products from Italy and Roman Gaul. The reign of Cunobelinus has been looked upon as the first stage in the Romanisation of Britain, for, after the defeat of the Belgae in the field, Romanisation proceeded apace in south-eastern Britain on such well-prepared ground.[2]

COMMERCIAL RELATIONS

Before the dawn of the Age of Metal, the extent of the external trade of southern Britain must necessarily have been very small, although some trade in flint existed. It was once argued that the megalithic culture, that spread in Western Europe from Spain to Scandinavia, represented the activities of some colossal prehistoric combine seeking the metalliferous wealth of the West. The megalithic monuments in Britain were supposed to be located in areas rich in metalliferous veins, or near river estuaries famous for their pearl fisheries.[3] But the case is difficult to substantiate, and R. G. Collingwood has shown that the megalithic monuments of the Cumberland-Westmorland region bear no relation to the metalliferous areas,[4] and instances of the occurrence of metal objects in megalithic tombs in this country are extremely rare.

With the full development of the Age of Metal, evidence of long-

[1] Strabo, *Geography*, IV, v, 2. The period under consideration at the end of this section is one where prehistory and written history overlap, and where the archaeological distributions may be supplemented by a few historical facts from the writings of the ancients. For a summary of the literary references to pre-Roman Britain see p. 30 below. The sum total of the information is however extremely small. Its chief significance is that a stray sentence here and there about climate, or the crops, or the economic life such as those referred to above, seems to corroborate the evidence derived from archaeological distributions.

[2] It is significant, however, that the Celtic tribes such as the Iceni in Norfolk and the Regni in Sussex (the traditional enemies of the Belgae and those who did not share in this increasing importance of agriculture) offered little resistance to Rome when military conquest came. See Fig. 6, which shows "how closely Belgic distributions correspond to the main areas of resistance to Rome, while the non-Belgic tribes [except those among the Welsh hills] appear for the most part pro-Roman or neutral"— C. F. C. Hawkes and G. C. Dunning, *art. cit.* p. 317.

[3] W. J. Perry, "Megalithic Monuments and Ancient Mines", *Manchester Lit. and Phil. Soc. Mem.* (1915), No. 1; see also *ibid.* (1921), No. 13.

[4] R. G. Collingwood, *Trans. Cumberland and Westmorland Antiq. and Arch. Soc.* XXXIII, 176–7 (1933).

distance trade becomes clearer. Considerable attention was paid to the alluvial gold of the Wicklow Hills in south-eastern Ireland, as well as to the copper deposits of Kerry. Flat axes of copper abound in Ireland, and implements of Irish copper and bronze spread into western Britain in the Early Bronze Age. These scattered metal objects in southern Britain must be regarded as evidence of trade in finished materials rather than of Irish immigration at this time.

A study of the Irish gold trade, however, provides more detail as to the nature of this early commerce. It proceeded from small beginnings in the early phase of the Bronze Age to reach vast proportions at the end of the era. Chitty and Fox have provided the statistical evidence, assigning 47 gold objects found in Britain to the Early Phase, 91 to the Middle Phase, and 310 to the Late Phase of the Bronze Age.[1] To what extent Britain can be said to have "purchased", by exchange or otherwise, the gold objects found within its area is, of course, not known; a very large number were undoubtedly lost in transit towards the Continent, where objects of Irish gold of the Bronze Age are frequently found.[2] Thus Britain may be said to have participated at this time in trade relations with both Ireland and the Continent. The characteristic gold objects of the Early Bronze Age are the lunulae. Two were found at Harlyn Bay, Cornwall, associated with a flat axe of that period.[3] Their distribution in Britain indicates that they were handled in the Highland Zone. Scotland has three, whence those of Denmark are, presumably, derived. Wales has one, found in the Lleyn peninsula. Cornwall has three, whence contact with Brittany was maintained.[4] This and other evidence appears to indicate that the commercial relations of Ireland in the Early Bronze Age were with northern rather than southern Europe. Spain and the south of France at this time were becoming particularly arid and warm, and civilisation there was decadent.[5]

At the end of the first thousand years of the Age of Metal a profound change came over the commercial relations of Britain. The local trade in

[1] C. Fox, *Personality*, p. 38.

[2] W. Fitzgerald, *The Historical Geography of Early Ireland* (1925), pp. 69–70.

[3] O. G. S. Crawford, "The Ancient Settlements at Harlyn Bay", *Antiq. Journ.* I, 294 (1921).

[4] Fig. 7A illustrating this distribution is a well-known map originally produced by George Coffey nearly a quarter of a century ago. Slight modifications and additions of detail may be noted from maps showing the same distribution by (1) O. G. S. Crawford, *art. cit. Geog. Journ.* XL, 195 (1912), and (2) E. C. R. Armstrong, *Guide to the Collection of Irish Antiquities* (1920).

[5] H. J. E. Peake and H. J. Fleure, *The Corridors of Time*, VII, 72 (1931).

Irish copper and bronze implements faded away, and Britain, presumably, relied on its own resources of copper and tin, or else imported them from continental sources. Most significant of all, the flourishing Irish gold trade reached the Continent by overland routes, that is, by routes running across the width of the English plain rather than across the tips of the western peninsulas as before. Gold objects (now best represented by the Irish torcs) became scarce in the Highland Zone during the Middle Bronze Period, but their number greatly increased in lowland Britain. Chitty and Fox have calculated that lowland Britain had roughly 72 per cent. of the total number of these objects found in Britain and assigned to this period, as compared with only 36 per cent. of the total of such objects assigned to the Early Bronze Age.[1] To understand some of the causes of this great shift in the commercial orientation of Britain it is necessary to consider the contemporary changes that were taking place on the Continent. The development of the great amber routes gave Europe transcontinental links,[2] and the older links by way of the western seas suffered accordingly, though it must be remembered that their importance was not completely obscured.[3] The ports of the mid-Channel, along the Hampshire coast, were still active. They were admirably situated geographically to share in oceanic as well as continental contacts, and there is evidence of their activity from an early period. Their prosperity increased with the growing trade in the Channel, and, in any case, it was bound up with that of the hinterland around Salisbury Plain. In the Middle Bronze Age there can be discerned the beginnings of the up-Channel shift of the important trade routes, which were destined to be of such importance later on. But this movement of trade up-Channel was far from being complete as yet, and the last thousand years of the pre-Roman age saw a clear revival in the importance of the western sea routes. In the Late, as in the Early, Bronze Age, Ireland again exported finished metal objects, while the gold trade reached its greatest development. The Gold Ribbon Torc best represents Irish trading activities at this time. The Highland Zone in Britain was once again the

[1] C. Fox, *Personality*, p. 38.

[2] J. M. de Navarro, "Prehistoric Routes between Northern Europe and Italy defined by the Amber Trade", *Geog. Journ.* LXVI, 481 (1925).
See also the map showing European trade in the second millenium B.C. by R. E. M. Wheeler in *European Civilisation: its Origin and Development*, edited by Edward Eyre, II, 210 (1935). This map shows in an admirable manner the relationship of Britain to the main European trade routes of the time.

[3] See C. Fox, *Personality*, p. 23, for a suggestion about the late formation of the Straits of Dover which might have affected the importance of the western routes to the Continent. The evidence can, however, be disputed.

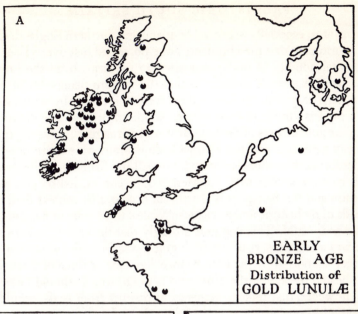

Fig. 7

A. After G. Coffey *The Bronze Age in Ireland* (1913), p. 55. For a discussion see footnote 4, p. 23.

B. After L. F. Chitty and C. Fox in *The Personality of Britain* (1932), p. 38, where details are given.

C. After L. F. Chitty and C. Fox in *The Personality of Britain* (1932), p. 39, where details are given.

In maps B and C, ● represents a single object; ■ two or more associated objects; symbols in outline show that the locality is only approximately known.

receiving area, especially in central Scotland and northern England. Chitty and Fox estimate that the Highland Zone possessed just over 72 per cent. of the total number of finds, as against 27 per cent. from the Lowland Zone.[1] It is reasonable to suppose that the western peninsulas of Wales (Lleyn, Anglesey and Pembrokeshire) were the routes by which a great deal of the Irish traffic in gold entered southern Britain. The distribution map of the gold objects of the Late Bronze Age suggests, however, that the Galloway peninsula and the Forth-Clyde isthmus were more important. The Pembrokeshire route led on to Cornwall and Brittany, while the north Welsh route, together with the Mersey-Dee entries, led to Derbyshire, Yorkshire and the Baltic. It is worthy of note in this context that, of the five finds of prehistoric amber beads in Wales, three were in Anglesey, and another near Mold. This suggests strongly that the amber trade from the Baltic was a counterpart of the Irish gold trade from the west. This view is further supported by the Late Bronze Age date of three of these Welsh amber beads; the age of the other two is unknown. It should be remembered, however, that folk movements, as distinct from trade, would tend to move gold one way and amber the other. Jet from Yorkshire is recorded also from four sites in Anglesey, two of them dating from the very beginning of the Bronze Age and two from the Late Bronze Age.[2] This implies that the contacts of Yorkshire and Anglesey in the Bronze Age were based not only on the transit trade in Irish gold, but also upon internal intercourse. Behind this return to prosperity of the western sea-ways is to be noted the series of Late Bronze Age invasions of the Lowland of Britain. They must have caused, among other things, a disruption of the economic life of the south-east and thus created conditions under which the western routes could regain their importance. It would appear that whenever the Lowland Zone was overrun by invaders, civilisation and trade shifted westwards.

Our knowledge of the internal trade of early Britain is naturally fragmentary. Evidence exists for an interesting output and sale of stone axes from the Graig Lwyd "factory" in Carnarvonshire in the Neolithic Age,[3] but in the Late Bronze Age a new sidelight is thrown on internal economic conditions by the large number of founder's hoards belonging to this period. These probably represent the stock-in-trade of travelling

---

[1] C. Fox, op. cit. p. 38.

[2] R. E. M. Wheeler, Prehistoric and Roman Wales (1925), p. 288.

[3] S. Hazzeldine Warren, "A Stone-Axe Factory at Graig Lwyd, Penmaenmawr", Journ. Roy. Anth. Inst. XLIX, 342 (1919); also T. A. Glenn, "Distribution of the Graig Lwyd Axe and its associated cultures", Arch. Camb. 1935, p. 189.

tinkers and smiths and may have been lost or buried by the wayside. Spearheads and axes, often damaged, and broken pieces of swords and other odds and ends characterise their contents.

By 500 B.C., as Western Europe was turning from the habitual use of bronze to that of iron, the Irish gold trade was already on the wane; but it would appear that its place was taken in western commerce by the tin exported from the tin-bearing alluvium of the Cornish rivers. As the bronze in use in Western Europe contained at least 10 per cent. tin, there had always been a great demand for the metal, and new supplies were continually being sought after. There is evidence that Cornish tin was already important in the Late Bronze Age; when the Carthaginians closed the Straits of Gibraltar to the Greek merchants in the sixth century B.C., the tin supply of north-west Spain was cut off, and the attention of the Greeks was directed to the Gaulish trade routes and their Atlantic connections.[1] There followed a period of great prosperity for the Cornish tin trade which seems to have moved from Cornish ports, especially that at the foot of St Michael's Mount (*Ictis*) to Corbilo, at the mouth of the Loire, and thence by land or sea to the Garonne and so *via* the Carcassonne gap to Massilia. The Cornish tin trade had Irish connections as well. The classical author Diodorus, in giving an account of the tin working of the early Iron Age in western Cornwall, mentioned that the metal was carried to Narbonne and Massilia. He also stated that the natives beat the metal into masses shaped like 'astragali' and transported it in wagons to the port (*Ictis*). A tin ingot has been found at St Mawes near Falmouth, but it is unlikely that it is one of the blocks shaped like 'astragali'.[2] The mention of wagons suggests the existence of well-defined trackways. Two transpeninsular routes crossed Cornwall, and are well authenticated by finds and forts of the Iron Age, and, at the same time, pass directly across the important centres of prehistoric tin mining. These were the routes from the Camel to the Fowey estuary, and from St Ives Bay to St Michael's Mount.[3]

The importance of this region in the last millennium B.C. is seen by the way in which its metalliferous wealth attracted newcomers from over the seas. At the end of the Bronze Age, people belonging to the western branch of the Deverel-Rimbury folk whose culture had associations with Brittany and, possibly, with the Pyrenean lands, settled in the extreme west of Cornwall. Judging by the distribution of their pottery, they

[1] H. O'N. Hencken, *Cornwall and Scilly* (1932), p. 185.   [2] *Ibid.* p. 171.

[3] It is characteristic of the western lands that both route-ways were important in earlier and later periods. In this context they suggest the Irish links of the Cornish tin trade.

avoided eastern Cornwall and Devon, and concentrated on the metal-liferous regions of the west.[1] Then in the Iron Age, during the second century before Christ, iron-using Celtic peoples, presumably from Atlantic France, moved into the peninsula where they built their hill fortresses.[2]

Another significant change at this time was Caesar's conquest of the Veneti, and the probable complete destruction of the port at Corbilo in the middle of the last century B.C., which was a severe blow to the prosperity of the western sea routes and to the international character of the Cornish tin trade.[3]

By this time, the Belgic peoples had made great progress towards a complex civilisation in northern Gaul and on the borders of the Rhineland. Their successful invasions of south-eastern England, as we have seen, resulted in the establishment of a kingdom in Essex and Hertfordshire. These developments led to an increase in the importance of the Straits of Dover and of the Rhine-Thames crossing, especially after the disruption of the Cornwall-Corbilo-Massilia route. It is clear that now most of the external trade of Britain was by way of the ports of the south-eastern quadrant of the Lowland. This reflected the tendency for prehistoric trade with the Continent to move up-Channel, and was connected with the growing economic importance of the south-east of Britain as well as with the progressive pacification of Gaul by Rome. The stage was set for the coming of the Romans.

BIBLIOGRAPHICAL NOTE

It is obvious that a chapter of this kind can be little more than a brief summary of the works of others. My indebtedness is, therefore, very large, but more particularly to the works of Sir Cyril Fox, Prof. H. J. Fleure, and Mr O. G. S. Crawford, who have always stressed the importance of the geographical factor in their archaeological studies of early Britain. I have endeavoured to make full acknowledgment to their work and that of others in the footnotes, but it is essential to mention *The Personality of Britain* (1932) wherein the prehistory of the island has already been excellently portrayed from the geographical viewpoint. From this work I have borrowed ideas as well as many maps which were originally the work of Miss Lily F. Chitty. These maps were up-to-date, as far as published evidence was available, in July 1932. Since

[1] H. O'N. Hencken, *op. cit.* p. 164.
[2] It has been further conjectured that the demand for iron ores led these Celtic folk across the Severn sea to the South Wales plain and the lower Severn valley, whither they were attracted by the iron ore deposits of the Forest of Dean. See C. Fox, *Personality*, p. 39.
[3] H. O'N. Hencken, *op. cit.* p. 174.

that date there has been a considerable accession of material, but this does not fundamentally alter any of the distribution patterns on which the above arguments are
based. T. D. Kendrick and C. F. C. Hawkes, *Archaeology in England and Wales,
1914–1931* (1932), contains a useful account of the progress of investigations up to
date, while H. O'N. Hencken's discussion of the tin trade in the *Cornwall and Scilly*
(1932) volume of the Methuen County Archaeologies has a general interest. Special
mention should also be made of the various publications of the Ordnance Survey
Office relating to prehistoric distributions, such as the Neolithic maps of Wessex and
of the Trent Basin, and the map of Celtic Earthworks on Salisbury Plain, as well as
to the archaeological material in the various Professional Papers. The 1936 edition
of *Field Archaeology* by O. G. S. Crawford (published for the Ordnance Survey Office
by H.M. Stationery Office) contains an up-to-date list of useful books and articles
classified under headings.

Chapter II

THE HUMAN GEOGRAPHY OF ROMAN BRITAIN

E. W. GILBERT, B.Litt., M.A.[1]

The Roman conquest of Britain took place only a few years after the death of Christ, and for the succeeding three and a half centuries the island was one of the provinces of the Roman Empire. During this period a remote land, situated at the western extremity of the known world, came at last into the light of written history. It is true that there are literary references to Britain before Claudius invaded the island in A.D. 43. The first definite and scientific account of Britain was brought back to the Mediterranean by Pytheas, after his famous voyage undertaken in or about the year 330 B.C.[2] Julius Caesar's descriptions of Britain, which were based on personal observations made during his military expeditions of the years 55 and 54 B.C., the accounts written by Diodorus Siculus in about 45 B.C. and by Strabo in about 30 B.C. or later, all contributed to the geographical knowledge of Britain possessed by the ancient world. But the sum total of the information concerning Britain which can be obtained from these sources is very slight, and any description of the geography of pre-Roman Britain must be compiled almost entirely from archaeological material. With the Roman conquest, Britain's history, if not recorded in detail, is at any rate a part of the history of the whole Empire. A reconstruction of the geography of Roman Britain is aided by literary as well as by archaeological material; and a good deal of the latter, consisting as it does of inscriptions,

---

[1] This chapter was written when I was Lecturer in Historical Geography in the University of Reading, and I am indebted to the following of my former colleagues, who kindly read through the manuscript and made many useful suggestions: Professor F. M. Stenton, Professor of Modern History; Professor H. L. Hawkins, Professor of Geology; Miss Norah C. Jolliffe, Lecturer in Classics, now of the University of London. I am grateful to many others for assistance and especially to the following: Mr F. T. Howard of University College, Exeter; Miss S. Westaway of Chichester; Lieut.-Col. C. D. Drew of Dorchester; Dr A. Raistrick of Armstrong College, Newcastle; and Mr W. F. Grimes of the National Museum of Wales.

[2] Clements R. Markham, "Pytheas, the discoverer of Britain", *Geog. Journ.* I, 504–24 (1893).

may be classed as literary.[1] The purpose of this chapter is to describe the human geography of Roman Britain, but it must be made clear that there was not one geography, but many geographies. During the four centuries of the Occupation, the face of the country was constantly altering, largely owing to the activities of man. Unfortunately, the sources, both archaeological and literary, are of such a character that it is only possible to write one account of the human geography, and to suggest a few of the ways in which changes probably occurred.

There are no exact meteorological records of the weather in Roman Britain, and without such figures it is impossible to describe the climate of the period with any certainty. Such fragments of evidence which exist lead to the assumption that the climate differed but little from that of to-day. During the past thirty years many scientific investigators have endeavoured, with the help of geological, botanical and zoological evidence, to compile a history of the climatic changes which are said to have occurred since the recession of the last ice-sheet. A theory has been formulated that the climate of north-west Europe has undergone a series of fluctuations from warm and dry at one period to cold and wet at another. Historical climatologists assert that a period between roughly 850 B.C. and A.D. 300, which is called sub-Atlantic time, was one of moist and cold climate, and that it was succeeded by a period of improved climatic conditions which have persisted up to the present day.[2] If this hypothesis is accepted, it would seem that Roman Britain endured a climate somewhat wetter and colder than that of the present. This supposition should be compared with the brief descriptions of British climate written by contemporary Roman authors. Caesar recorded that the climate was more temperate than that of Gaul, the cold being less severe,[3] and a panegyrist who wrote in the fourth century A.D. included the absence of extreme heat or cold as one of the blessings with which nature had

[1] The literary evidence about Roman Britain is scattered but quite considerable in the mass. At least forty-nine writers in Greek and seventy-nine in Latin make some reference to the British Isles. Of these, the account written by Tacitus is perhaps the most valuable historically, but the author had little interest in topography and the geographical information contained in his works is tantalisingly meagre. The *Agricola* of Tacitus contains only eleven British place-names, four of which are tribal names. See C. G. Stevens, "Ancient Writers on Britain", *Antiquity*, I, 189–96 (1927).

[2] See the following works by C. E. P. Brooks: "The Climate of Prehistoric Britain", *Antiquity*, I, 412–18 (1927); "The Evolution of Climate in North-West Europe", *Q. J. R. Met. Soc.* XLVII, 173–90 (1921); *The Evolution of Climate* (1925); *Climate through the Ages* (1926); "Post-Glacial Climates and the Forests of Europe", *Q. J. R. Met. Soc.* LX, 377–95 (1934).   [3] Caesar, *De Bello Gallico*, V, 12.

endowed Britain.[1] Strabo remarked that the atmosphere was more subject to rain than to snow, and that even on clear days the mists continued for a considerable time, so that the sun was only visible for three or four hours.[2] Tacitus noted that Britain's climate was gloomy with frequent showers of rain and murky fogs, but free from excessive cold, and also added that olives and grapes would not grow as the climate was not warm enough.[3] He observed that the days were longer than in Italy, and that British crops ripened slowly. A non-scientific observer coming to visit England from his home in the Mediterranean might easily describe the climate of to-day in language similar to that used by the classical authors. While this evidence does not seem to imply any great change from present conditions, the emphasis laid on the prevalence of rain can be used to support the view that the climate of Roman Britain was slightly wetter than the climate of to-day. It should be remembered, however, that whether the climate of Roman Britain was wetter or not, the country must have been much boggier and damper before the days of modern drainage, and before the woodland was cleared. Tacitus remarked on the great humidity of the soil as well as that of the climate.[4] It is true that the mere appearance of much damp ground might have led a Roman visitor to exaggerate the wetness of British climate. It has also been urged that the intensive arable cultivation of the downlands of southern England in the Early Iron Age and the Roman period is proof that a wetter climate existed.[5] Without very damp conditions it is stated that the downlands would not have been suitable for cultivation on a large scale. On the other hand, it seems probable, in spite of Tacitus' statement to the contrary, that the vine was actually cultivated in Britain, and such a fact would not be consistent with a wetter climate than that enjoyed at present. In the excavations at Woodyates, Dorset, General Pitt-Rivers found that the bottom of a well, undoubtedly used in Roman times, was no less than sixty feet above the level of modern wells in the district.[6] The lowering of the level of the water-table in the chalk since the Roman occupation is shown by evidence of this character, and has been used as a further argument to prove that the climate was wetter in Roman times. This assumption is not conclusive without considerable corroborative evidence, as the lowering of the level of the water in the chalk may be due to entirely artificial causes such as the establishment

---

[1] *Panegyrici Latini*, VII, 9.   [2] Strabo, *Geography*, IV, v, 2.

[3] Tacitus, *Agricola*, 12.

[4] *Multus umor terrarum caelique*, Tacitus, *Agricola*, 12.   [5] See pp. 11 and 18.

[6] See note by J. P. Williams Freeman on "Meteorology and Archaeology", *Antiquity*, II, 208–10 (1928).

of modern pumping stations, even if they are situated at a very considerable distance away from the ancient wells. The whole evidence concerning
the climate of Roman Britain is so fragmentary that it is impossible to
make any conclusive statement about it, but it is reasonably safe to assume
that conditions approximated to those of the present day.

Some of the physical elements of English scenery have changed but
little since the Roman occupation, but it is a mistake to imagine that
important changes have not taken place. Relief, drainage, coastline and
the distribution of woodland all undergo change, but it is extraordinarily
difficult to describe the state of any one of these at periods before the days
of accurate mapping and exact observation. Nature and man together
have altered the relief, the drainage and the coastline. The drainage of
marshes has had considerable effect on the appearance of the land,
but the modification of the natural vegetation has been man's chief contribution to the alteration of the scenery. The Roman period was a time
during which woodland was being rapidly cleared,[1] so that the country's
appearance must have differed considerably at the end of the occupation
from that which it presented to the original invaders. The work of natural
forces on the other hand is most evident in the changes which have
occurred in the outline of the English coast since Roman times. Marine
erosion, silting, and the changes of the relative level of land and sea have
considerably altered the appearance of the coastline during the past two
thousand years. The limitations of space do not permit a separate
treatment of the physical geography of Britain during the Roman period.
In the following analysis of the human geography of Roman Britain,
it has only been possible to indicate some of the more important ways in
which the physical geography of the period differed from the geography
of to-day.

THE EXTENT OF ROMAN BRITAIN
AND ITS DEFENCES

Before describing the human geography of Roman Britain, the actual
limits of the area must be considered (Fig. 8). The boundaries of the
Province did not remain constant during the long period of occupation,
and it is necessary to discuss these changes of boundary in order to appreciate the fact that Britain did not mean exactly the same region at

[1] Cyril Fox, *The Archaeology of the Cambridge Region* (1923), p. 224. It is only
on detailed maps of small areas, such as those constructed by Sir Cyril Fox, that it is
possible to suggest the approximate amount of clearing carried out by the Romans.
See maps on pp. 46 and 47 of Cyril Fox's *The Personality of Britain* (1932).

different dates.[1] The Romans never occupied the whole of Britain and it was an exceedingly difficult task to find a satisfactory northern boundary. Britain is a long and narrow island. It extends for over 600 miles from north to south, but it has two slender waists, one from Tyne to Solway Firth and the other from Forth to Clyde, the latter being less than 40 miles in width. These two lines were both used at different times as limits of the Roman occupation, and the zone between them was a frontier region in which much fighting took place.

The greater part of the plain of south-eastern England was quickly overrun by the Romans in the first four years of their conquest and, in A.D. 47, they were able to establish a temporary frontier south of the rivers Trent and Severn, probably along the line of the Fosse Way from Lincoln to Cirencester[2] (Fig. 8). It took 40 years to subdue the highlands of the west and north. Agricola, after completing the conquest of Wales in A.D. 78, subdued the north of England in the following year, and placed the Roman boundary along the line between Firth and Clyde in A.D. 80. The advance was followed by a withdrawal until, in A.D. 126, Hadrian's wall between Bowness-on-Solway and Wallsend, a length of 73 miles, was constructed as a more southern boundary. In A.D. 143 the narrower northern waist, from Forth to Clyde, was once more used as the boundary by the building of what is generally known as the Antonine wall, rather more than 37 miles in length, the most northerly limit of the whole Roman Empire. This northern extension was shortlived, as the Antonine wall was finally abandoned during the troublous times, between A.D. 180 and 185, when both walls were captured. When peace was restored, Hadrian's wall became the northern limit of Britain, and remained as such until it was finally abandoned in about the year A.D. 383.[3]

Within these varying northern limits, Roman Britain can be divided into two regions, a military and a civil zone[4] (Fig. 8). On the north and west, mountainous or hilly country was occupied almost entirely for military purposes. This zone can be subdivided into three separate regions, Wales, northern England, and the country between the two walls, which may be

[1] See R. G. Collingwood, "The Roman Frontier in Britain", *Antiquity*, I, 15–30 (1927).

[2] R. G. Collingwood, "The Fosse", *Journal of Roman Studies*, XIV, 252–6 (1924).

[3] For Hadrian's wall, see J. C. Bruce, *The Roman Wall* (3rd ed. 1867); R. G. Collingwood, "Hadrian's Wall: A History of the Problem", *J.R.S.* XI, 37–66 (1921), and "Hadrian's Wall: 1921–1930", *J.R.S.* XXI, 36–64 (1931). For the Antonine wall, see George Macdonald, *The Roman Wall in Scotland* (2nd ed. 1934).

[4] F. Haverfield and G. Macdonald, *The Roman Occupation of Britain* (1924), pp. 149–52.

THE FRONTIERS OF ROMAN BRITAIN

● – TOWN
⊡ – LEGIONARY FORTRESS
■ – FORT
⊕ – SIGNAL STATION

10 0 10 20 30 40 50 *Miles*

THE ANTONINE WALL

MILITARY
ZONE
BETWEEN THE
WALLS

HADRIANS WALL

MILITARY

ZONE

OF

NORTHERN

ENGLAND

MALTON■

■HUNTCLIFF
⊕GOLDSBOUGH
⊕RAVENSCAR
⊕SCARBOROUGH
⊕FILEY

●YORK
COLONIA

■CAISTOR

MILITARY

ZONE

OF

WALES

CHESTER⊡

LINCOLN●
COLONIA

R.TRENT (TRISANTONA)

■HORNCASTLE

■BRANCASTER

R.SEVERN (SABRINA)

THE

FOSSE

R.AVON

AD TRISANTONAM ET SABRINAM FLUVIOS

WAY

CIVIL

ZONE

BURGH
CASTLE

WALTON
CASTLE

COLCHESTER●
COLONIA

●GLOUCESTER
COLONIA
CIRENCESTER●

St.ALBANS ●
MVNICIPIVM

BRADWELL

CAERLEON⊡

CARDIFF ■

LONDON●

RECULVER

RICHBOROUGH■

DOVER

THE SOUTH-WEST PENINSULA

EXETER●

PORTCHESTER■

■CARISBROOKE
CASTLE

PEVENSEY■

STUTFALL
CASTLE

Fig. 8

The forts and signal stations which were built to defend the east and south
coasts, are marked on the map. The existing coastline is indicated.

called Roman Scotland. The highland region of Wales contained about twenty-five forts while in northern England (which may be defined as the Pennine country south of, and including, Hadrian's wall), about seventy forts were built. In the southern uplands of Scotland, if the Antonine wall is included, more than thirty forts were constructed during the Roman occupation. Thus, well over a hundred forts were established at strategic points in Wales, the Pennines and the southern uplands. It must not be imagined that all these forts were built or occupied at one and the same time, but when considered as a whole, they show that Britain contained a large military zone, occupied by no less than 40,000 troops, who formed perhaps as much as one-tenth of the whole army of the Empire. This military zone acted as a screen to protect the lowlands of the south-east from incursions by tribes who inhabited the mountains, much in the same way as the North-West Frontier Province of India shields the plains of the Indus and the Ganges. Behind the elaborate chains of forts were the three great legionary fortresses of York, Chester, and Caerleon, which acted as the bases for the defence of the frontier.

To the south-east of these fortresses lay the English plain, essentially an outlying portion of the great plain of Northern Europe, and it was this part of Britain, so easily overrun by a continental invader, which received the civilising influences of the Roman conquest. This region was the zone of civil occupation and contained the towns, the villas, and the other marks of the peaceful occupation of Britain. After the early years of the conquest, the whole of the civil zone needed no forts, and archaeological remains of a military nature are rarely found in this area. The one exception to this statement is the establishment of a system of military defences along the shores of the civil zone. An island, even if it is part of a powerful empire, must always regard its coasts as part of its frontier. In the third and fourth centuries the southern and eastern coasts of Britain began to receive the attacks of Saxon marauders, and at the same time, Irish invaders began to land on the shores of Wales. To meet this danger to the civil zone, a scheme of coastal defence had to be devised. A number of forts were built along the coast between the Wash on the north and the Solent on the south. The sites of ten of these forts have been discovered at Brancaster (BRANODVNVM), Burgh Castle (GARIANNONVM), Walton Castle, Bradwell (OTHONA), Reculver (REGVLBIVM), Richborough (RVTVPIÆ), Dover (DVBRIS), Stutfall Castle (LEMANIS), Pevensey (ANDERIDA), and Portchester (Fig. 8). An official called the Count of the Saxon Shore was in charge of the defences as a whole, and a list of the names of nine of the forts which were under his command is preserved in the *Notitia Dignitatum*.[1] Remains of a fort have been discovered at Carisbrooke Castle, only one mile from the tidal part of the Medina estuary, and it has been suggested that another unit of the Saxon Shore defences was established there.[2] A fort of a similar nature has been found at Cardiff, and was probably built to protect the Severn entrance to the rich region of the Cotswolds at the head of the estuary; other sites found in Wales, notably at Holyhead, may have been designed as a western system of coastal defence against the Irish. The actual date of the construction of the forts of the Saxon Shore is very uncertain, but it seems that they were built at different dates probably in the last 25 years of the third century. Each fort was large enough to contain between 500 and 1000 men[3] and another feature common to the whole group is the fact that they are all built on the edge of sheltered harbours which command important entrances into the country. Brancaster guarded the Wash. Burgh Castle was constructed with the object of commanding the immense shallow estuary which then extended to the north of it and included what are now called the Broads. Walton Castle[4] and Bradwell together protected the several estuaries that might lead an invader to Colchester and St Albans. Bradwell and Reculver stand on either side of the Thames estuary.

For the Romans the coast of eastern Kent was the most important part of the shore of Britain as it commanded the narrowest section of the Channel and therefore the chief routes to and from the Continent. A reconstruction of this coast is therefore of especial interest in the study of the communications and harbours of Roman Britain.[5] While an accurate

[1] For the forts of the Saxon Shore, see article by F. Haverfield under "Saxonicum Litus" in Pauly-Wissowa's *Real-Encyclopädie*, which gives full references to various archaeological periodicals for descriptions of each fort, and G. Macdonald, *Roman Britain, 1914–1928* (1931), pp. 53–73, for the most recent discoveries and theories about this system of defence. Jessie Mothersole's *The Saxon Shore* (1924) is a popular account of the forts. It is possible that forts at Caistor and at Horncastle, in Lincolnshire, were also units of the Saxon Shore defences (*Arch. Journ.* XCI, 129–33 (1934)). [2] *Antiquity*, I, 476 (1927).

[3] Each fort enclosed an area of between 6 and 10 acres. Pevensey and Stutfall Castle, the largest forts, were about 10 acres, while Burgh, Bradwell and Richborough were all about 6 acres in size.

[4] Walton Castle, which was situated south of the river Deben, near modern Felixstowe, has been completely destroyed by marine erosion, the last remaining wall falling into the sea between 1732 and 1766.

[5] See T. Rice Holmes, *Ancient Britain and the Invasions of Julius Caesar* (1907), which contains on pp. 518–52 an essay on "The configuration of the coast of Kent in the time of Caesar", with numerous references and maps.

4

map of the Roman coast of Kent cannot be drawn, it should be remembered that the alluvial areas of the modern geological map provide a rough indication of the approximate extent of the arms of the sea in the Roman period (Fig. 9). Although silting had already begun to fill up some of the inlets which had been formed as a result of the Neolithic submergence, the indentation of the coast must have been still quite considerable. The tides ebbed and flowed up these inlets and up the estuaries of small rivers. The most noticeable feature of the coast was the existence of a strait between what was then an actual island (Thanet) and the mainland. The greater and lesser Stour with their tributaries emptied their waters into this channel which was later called the Wantsum.[1] It was probably a narrow and shallow strait and may not have been practicable for navigation at all conditions of the tide, but it provided good sheltered harbours.[2] At each entrance to this channel the Romans established a fort, at Reculver (REGVLBIVM) in the north, and at Richborough (RVTVPIÆ) in the south. At Reculver the cliffs, on which stood the Roman fort, have been eroded so that a considerable portion of the Roman remains have disappeared beneath the sea.[3] The Roman harbour was probably on the south side of the fort whose commanding position made it impossible for any vessel to pass through the strait unobserved. At Richborough silting has had the effect of building up two and a half miles of land between the old Roman harbour and the present coast. To-day, the coast between Beachy Head and Hythe is very even, but in Roman times it contained two great indentations, the Pevensey levels and the greater part of Romney Marsh, while the modern Dungeness did not exist.[4] These two bays were covered by water, and tides found their way up the estuaries of the rivers Brede, Tillingham and Rother. On the northern side of Romney Marsh, the Romans established the fort of Lympne[5] while the fort of Pevensey

---

[1] See G. P. Walker, "The Lost Wantsum Channel: Its importance to Richborough Castle", *Arch. Cantiana*, XXXIX, 91–111 (1927).

[2] The channel was still open in the eleventh century, when Sandwich was the chief base for English naval concentrations.

[3] As late as 1540 Leland observed that Reculver was within a quarter of a mile of the sea. This proves the rapidity of the erosion. *The Itinerary of John Leland*, ed. L. Toulmin Smith (1909), Pt. VIII, p. 59.

[4] W. V. Lewis, "The Formation of Dungeness Foreland", *Geog. Journ.* LXXX, 309–24 (1932).

[5] LEMANIS must have been situated at Stutfall Castle which is below modern Lympne. This site was at the edge of the water in Roman times, but is now one and a half miles from the sea. Dr Rice Holmes concluded that the port of LEMANIS was not at Stutfall (*Ancient Britain and the Invasions of Julius Caesar*, pp. 544 ff.), but Professor F. Haverfield was not convinced by Rice Holmes' arguments (*J.R.S.* II, 204 (1912)).

guarded the entrance to Pevensey levels, where the sea penetrated at one time for about five miles inland. This fort, like Richborough, was built on a small island. The group of four forts at Richborough, Reculver, Dover and Stutfall Castle (all built on good harbours), watched the narrow straits

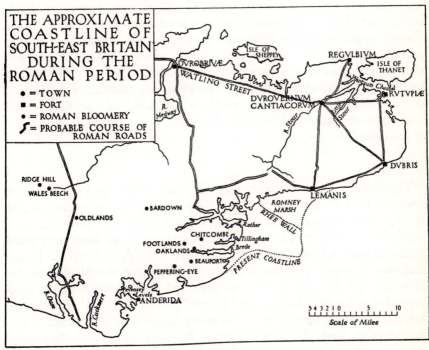

Fig. 9

The inland boundary of the alluvium is here marked as the coastline; it is not suggested that this is an accurate representation of the outline of the coast in Roman times. The eastern part of Romney Marsh had probably been reclaimed by the end of the Roman period. It has been asserted, without much evidence, that the Rhee Wall was built by the Romans. Sites of bloomeries after E. Straker, *Wealden Iron*; course of Roman road from London across the Weald to the river Ouse after *Antiquity*, VI, 350–6 (1932).

of the English Channel, and maintained communications with the Continent. The intricate network of inlets in the neighbourhood of Portsmouth was controlled by Portchester. Each of the ten forts stood on comparatively low ground, but good use was made of any minor elevations in choosing the actual site so that the harbour could be more effectively guarded.[1] The voyages of the Saxons must have become much bolder and longer during the fourth century, so that even the Yorkshire coast was not free from danger. To meet these incursions, a number of small military posts were built on the high headlands of Yorkshire, probably as late as A.D. 360 and even later. One at Huntcliff overlooked the entrance to the Tees estuary; others were erected at Goldsborough, Ravenscar, Scarborough and Filey, while some may have been built on the coast south of Flamborough, which has been heavily eroded since Roman times.[2] It is possible that the fort of Malton was the strategic base for what was obviously a chain of signal stations, constructed to give the alarm when a large fleet of invaders was sighted.[3] The whole organisation of the defence of the coast of Britain shows, as Haverfield said, "the tenacity of the imperial government and its struggle to do its duty even on the shores of the ocean".[4]

The boundaries of the administrative units of Roman Britain are not known. In A.D. 197 Britain was divided into two provinces called BRITANNIA INFERIOR and BRITANNIA SVPERIOR. York, AESICA on Hadrian's Wall, and Lincoln are all known to have been in 'Lower Britain', while Chester and Caerleon were in 'Upper Britain'. It was formerly believed that the dividing line ran from the Mersey to the Humber, but the discovery of an inscribed altar at Bordeaux in 1921 proved that Lincoln was in the 'Lower' part of Britain.[5] Dr Wheeler has therefore suggested that the Watling Street may have acted as the boundary between the two Provinces, but there is no definite proof.[6] About one hundred years later, Britain was sub-divided, in Diocletian's general reorganisation, into four provinces, namely, PRIMA, which certainly contained Cirencester, SECVNDA, FLAVIA CAESARIENSIS, and MAXIMA. It appears that a fifth division called VALENTIA was

[1] The construction of the Martello towers in the early nineteenth century as a line of defence against the threatened invasion by Napoleon is a later instance of the same coast being fortified in a manner very similar to that used by the Romans; the most northerly tower was built at Aldeburgh.

[2] For the Yorkshire signal stations, see F. Haverfield, "Notes on the Roman coast defences of Britain, especially in Yorkshire", *J.R.S.* II, 201–14 (1912), and G. Macdonald, *Roman Britain, 1914–1928*, pp. 67–9 (1931).

[3] P. Corder and J. L. Kirk, "Roman Malton", *Antiquity*, II, 69–82 (1928).

[4] F. Haverfield, *op. cit. J.R.S.* II, 214 (1912).

[5] *J.R.S.* XI, 101–7 (1921).

[6] *Royal Comm. Hist. Mon., London*, III, 60–1 (1928).

made in A.D. 369. Nothing is known about the boundaries of these provinces, but Giraldus Cambrensis, writing in about the year A.D. 1205, stated that York was in MAXIMA, London in FLAVIA, and that Kent formed part of SECVNDA.

The exact areas of the smaller administrative units are equally unknown, but they must have consisted of three kinds. Each of the towns with official status probably ruled a region that may have been about the same size as an English county. The Imperial Domain included the mines, but the actual extent of the Imperial properties is unknown. Outside these areas the country was divided into tribal territories, which already existed when the Romans entered Britain, and which probably enjoyed some form of local self-government.[1]

ROMAN TOWNS

The Roman Empire was the creation of a city. It was therefore quite natural that the establishment of new cities and towns in the conquered provinces was one of the chief tasks of empire-building. The ancient civilisations of the Mediterranean were based on urban life. The Agora or the Forum was the centre of the economic and social activities of a large district. Peasants whose daily work called them out to the fields some miles from their town, returned at night to the protection of its walls and the society of its market-place. "Among our numerous debts to Rome," said Haverfield, "not the least is that connected with its legacy of town life",[2] and a study of the towns of Roman Britain is therefore one of the main tasks that lie before the historical geographer. It must be clearly understood that all Roman towns were small when compared with the leviathan-like cities that have arisen in the last two centuries. The population of ancient Rome itself has been estimated at about half a million, and there were few towns with more than 100,000 citizens in the Roman Empire. Nevertheless, several towns of Roman Britain compared very

[1] Professor F. Haverfield believed that there were few existing traces of the minor administrative boundaries of Roman Britain. He summarised the vague and fragmentary evidence which had been brought forward to show the possible limits of the *territorium* of the *Colonia* of Colchester. "Centuriation in Roman Britain", *Eng. Hist. Rev.* XXXIII, 289–96 (1918). Further efforts to discover proofs of centuriation in Britain have been made in recent years. See p. 115 below, and Gordon Ward, "A Roman Colony near Brancaster", *Norfolk and Norwich Arch. Soc.* XXV, 373–85 (1935), but none of these attempts is very conclusive.

[2] F. Haverfield and G. Macdonald, *The Roman Occupation of Britain*, p. 186.

favourably in size with their contemporaries in other parts of the Empire, and if they are judged by medieval and not by modern standards they will not be considered negligible. When the Romans conquered Britain they obtained possession of an island with a predominantly rural population. The racial elements of that population were many and varied, but it must be emphasised that the last invaders of Britain before the Romans were a civilised people, who established several large urban settlements in the south-eastern part of the country. The Belgic tribes arrived in Kent in the early part of the first century B.C., spread over Essex, Hertfordshire, and parts of Oxfordshire, while other groups of the same stock entered England later by the Southampton gate and spread over Hampshire, Berkshire, Wiltshire, Dorset and part of Sussex. These tribes found that the existing inhabitants of Britain had built great hill-forts in commanding positions as their capitals. The Belgae superseded the old system of hill-forts by establishing their *oppida* in lower woodland sites,[1] and in some cases built their new towns in the valley actually below the old upland fort.[2] It is certain that the eastern Belgic peoples constructed large towns at VERVLAMIVM and at CAMVLODVNVM, while the southern group founded CALLEVA and probably VENTA BELGARVM. The origin of these four towns and possibly of others must be attributed to the Belgae rather than to the Romans. The Roman system of urban life was in fact superimposed on an already existing set of towns, and the Romans also continued the process of moving settlements from the hilltops to the valleys. In Gaul the Romans are known to have compelled the Gallic tribes to leave the hill-forts and establish new towns in the valleys as in the well-known example of Bibracte.[3] In some cases the Romans certainly planted towns on virgin sites, but in several instances they enlarged or modified the existing settlement or merely moved its site a short distance from an old town.

During the second half of the first century and the early part of the second century the foundation of towns, with rectangular street plans, town halls, public baths, and other urban amenities, was part of the general policy of Romanisation adopted by the conquerors of Britain.[4] The towns

---

[1] For Caesar's description of these towns, see *De Bello Gallico*, v, 21.

[2] For the settlements made by the Belgic peoples, see C. F. C. Hawkes, "Hill-Forts", *Antiquity*, v, 60–97 (1931); C. F. C. Hawkes and G. C. Dunning, "The Belgae of Gaul and Britain", *Arch. Journ.* LXXXVII, 150–335 (1930). See pp. 20–21 above for map and discussion.

[3] F. Haverfield and G. Macdonald, *op. cit.* pp. 196–7.

[4] *Hortari privatim, adiuvare publice, ut templa fora domos extruerent*, Tacitus, *Agricola*, XXI.

of Roman Britain may be grouped into classes according to their status. Only five British towns received the title of either *municipium* or *colonia* with the full rights of Roman citizenship. VERVLAMIVM was the one place with the rank of a *municipium*, which was usually given to "pre-existing native towns which had reached by natural progress some size and some civilization of a Roman (or Italian) kind, and which seemed to merit from the central government the grant of a definite charter and of an urban constitution".[1] Colchester, Lincoln, Gloucester, and York were the only *coloniae* or settlements of discharged Roman soldiers. The second class of towns is a group of twelve, which were the capitals of old tribal districts. Some of these must have been replanned by the Romans upon the existing sites.[2] The majority of these places are the natural capitals of large blocks of scarpland in south-eastern England and are generally situated in a valley below their upland belt of territory. As their names indicate they were the centres of tribes; for instance, Cirencester was the capital of the Dobuni and the natural focus of the Cotswold region. In addition to these places, Britain contained about fifty other towns most of which were very small market centres or mere posting stations, but this class also included the great commercial centre of London, the spa of Bath, and Corbridge and Carlisle which can be regarded as permanent military bases. It should be noticed that, with the exception of the last two places, all the towns were situated in the lowland or civil zone. The military zone contained no real towns. The following description of the towns is based on a division of Britain into six regions, and the towns of each region are discussed as a group. Eastern England included the towns that were earliest established and the settlements of this region will be first considered.[3]

The estuaries of the Essex rivers, although they do not penetrate Britain so deeply as the Thames, formed a useful gateway into the country.

---

[1] F. Haverfield and G. Macdonald, *op. cit.* p. 188.

[2] Caistor-by-Norwich (VENTA ICENORVM), Canterbury (DVROVERNVM CANTIACORVM), Silchester (CALLEVA ATREBATVM), Winchester (VENTA BELGARVM), Chichester (REGNVM), Dorchester (DVRNOVARIA), Exeter (ISCA DVMNONIORVM), Cirencester (CORINIVM DOBVNORVM), Caerwent (VENTA SILVRVM), Wroxeter (VROCONIVM CORNOVIORVM), Leicester (RATÆ CORITANORVM) and Aldborough (ISURIVM BRIGANTVM). Most of these tribal names are given in the lists of the Ravenna geographer. (See Bibliographical Note, p. 83).

[3] A number of references for each town together with a list of the best available plans is given in the Bibliographical Note at the end of this chapter. No map of the towns and the roads has been provided because it is assumed that the reader will consult the O.S. *Map of Roman Britain* (2nd ed. 1928).

At the head of one of these rivers, the Colne, a promontory of rising ground provided a naturally protected area on which a town might grow up. Between the winding stream of the Colne on the north, and its tributary the Roman river on the south, a plateau of about 100 feet in altitude stretches in an east to west direction. The sides of this plateau are quite steep, and natural marshes on the east and south and woodland on the north and west made it into a defensible position. Earthworks about $3\frac{1}{2}$ miles in length had been built across the western neck of the promontory in pre-Roman times and the area enclosed by the rivers and artificial defences was about 12 square miles. Within this protected island, on the banks of the Colne, was built CAMVLODVNVM, the most important of the native British towns. It was a mercantile settlement and imported goods from the Roman Empire in the fifty years before the conquest. This *oppidum* was the seat of Cunobelinus who exercised a kind of overlordship over all the Belgic tribes of south-east Britain. The centre of administration had been moved from VERVLAMIVM to CAMVLODVNVM at some time during the fifty years prior to the Roman invasion. As CAMVLODVNVM was the capital and leading town of Britain, it was naturally the first objective of the Roman legions; here the expeditionary force set up their camp,[1] and later, in about A.D. 50, there was established the first *colonia* of discharged soldiers. The site chosen for the new Roman town of COLONIA VICTRICENSIS was situated upon the plateau southeast of British CAMVLODVNVM. The place was utterly destroyed in the disastrous Boudiccan revolt of A.D. 61, but later it was rebuilt and given a circuit of walls 3100 yards in length enclosing an area of 108 acres and having the usual town plan with streets running at right angles to each other. The colony was soon connected by roads to London and St Albans, and it has been suggested by Dr Wheeler that the purpose of the Balkerne Gate at Colchester may have been to commemorate the completion of this triangle of communications.[2] Colchester contained the great temple of Claudius, and may therefore have been, for the early part of the Occupation at least, the religious centre of the province. After the first century the town did not grow and, as the area contained within its walls was always sufficient to hold it, there is no doubt that it was inferior in size to at least five other towns of Roman Britain. London obviously soon captured Colchester's position as the leading mercantile centre of the country, and although Colchester was the focus of two main roads, for the greater part of the

[1] This camp was discovered in 1931. See *J.R.S.* XXII, 211 (1932).
[2] Dr Wheeler described the Balkerne Gate as a *porte monumentale*. *R.C.H.M. Essex*, III, 21.

Occupation it was merely a small country town carrying on a trade of modest dimensions with the Continent.[1]

A casual examination of the map of Roman Britain might lead to the conclusion that VERVLAMIVM possessed a strangely unfavourable geographical position for a town which was politically unique as the only one on which the Romans conferred the dignity of a *municipium*. The place was situated in wooded country at some distance away from good water communications, and apparently enjoyed no obvious natural advantages. It must be remembered, however, that the Roman settlement was the heir of British towns which had grown up in the same locality, and the position must be considered in relation to pre-Roman conditions if its strategic value is to be realised. Dr Wheeler has suggested that the great earthworks at Wheathampstead, 5 miles north-east of VERVLAMIVM, which enclose an area of about 100 acres and command a crossing of the Lea, mark the site of the first Belgic town in the neighbourhood.[2] He considers that the Belgic tribe, the Catuvellauni, during the course of their invasion of Britain, made their way up the river Lea, the first major tributary of the Thames, and established their *oppidum* at Wheathampstead. It is believed that this place was probably Cassivellaunus' capital which was stormed by Julius Caesar's troops. Later in the first century B.C., daughter towns from Wheathampstead were established near fords across the rivers Ver and Macan. Welwyn commanded the crossing of the Macan, and VERVLAMIVM, which controlled the ford over the Ver, was also a good base for extending Belgic influence north-westward into Northamptonshire and therefore grew rapidly. The Belgic town of VERVLAMIVM, which was situated at Prae Wood, at an altitude of about 130 feet above the western banks of the Ver, has recently been excavated. It was a large town and was the capital of King Tasciovanus, who issued coins from the place. With the increase of trade between Belgic Britain and the Continent, the centre of administration was moved to Colchester, but VERVLAMIVM maintained some of its importance, and the road between the two towns must have been a busy highway. The Romans moved the site of the town further down the hill towards the river Ver, and, after Boudicca's rebels completely destroyed the settlement in A.D. 61, it was rebuilt and an area of about 150 acres was surrounded with walls. In the first half of the second century, VERVLAMIVM was enlarged by a southward extension and spread right down to the river valley. The size of the new Roman town was about 200 acres and the total length of the circumference of its walls was

---

[1] Modern Colchester, which stands on the site of the *Colonia*, is still a small port.
[2] R. E. M. Wheeler, "Belgic cities of Britain", *Antiquity*, VII, 21–35 (1933).

about 2 miles. It was not until Roman times that VERVLAMIVM became an important focal point in the communications of the country, by the construction of the Watling Street and Akeman Street. The town plan which is now being revealed by excavation was rectangular and dependent on the line of the Watling Street which ran through the town. Belgic VERVLAMIVM had begun to lose its position as the leading town to Colchester, and in the same way its Roman successor was eventually overshadowed by London. The town was the regional capital of a rich agricultural district and in the second century enjoyed considerable prosperity, but the third century was a period of decay.

The Roman regional centre of the area now called Norfolk was VENTA ICENORVM (Caistor-by-Norwich), which is only 3 miles away from its modern counterpart, Norwich. This town was the cantonal capital and administrative centre of the important tribe of the Iceni, and its walls enclosed an area of about 35 acres, but it was probably larger than this figure suggests, as a number of buildings were erected outside the circuit of the walls. It appears that no pre-Roman town existed in this place and that the town was planned in the years after the Boudiccan revolt. The town stood to the east of a stream called the Tas which is a tributary of the Yare. It was therefore quite possible for small vessels to come up the river almost as far as the town and it should be remembered that Norwich is still a port. Aerial photography has revealed the complete plan of the streets which were arranged in the conventional rectangular pattern.[1] The discovery of pottery kilns and an annealing-furnace for glass are the only evidences of industry which have been found at this place. Caistor was in fact a small country town, the focus of the northern half of East Anglia, a region whose economic interests must have been mainly concerned with agriculture and pasturage. Two smaller and less important towns of eastern England require only brief mention; both were situated at road junctions. At Great Chesterford, a town, whose Roman name is unknown, occupied a rectangular area of about 35 acres and was surrounded by a strong wall. There is abundant evidence that it was inhabited during the whole of the Occupation and it seems to have controlled several valley routes which led northwards to the Icknield Way. At Cambridge, where two Roman roads crossed one another, was situated an enclosure of about 28 acres, containing a town which commanded a convenient crossing of the river Cam.

Two large Roman towns and one of no great importance were placed on the western edge of the Fens at points where the great road to the

---

[1] See *Antiquity*, III, 183 (1929), for an air photograph of Caistor-by-Norwich.

north, Ermine Street, was forced to cross the Ouse, the Nene and the Witham. Godmanchester was situated on the south bank of the Ouse, not far from modern Huntingdon, which is on the north bank. It was a small posting and market town, and little is known about its exact extent or history. At the crossing of the Nene the Romans planted towns on either side of the river. The southern town was surrounded by a rampart which contained an area of about 44 acres now known as "the Castles" near Chesterton. It seems to have been established during the first century, the great period of Roman town-planning, but it was not so important as the settlement on the north bank of the river at Castor (probably DvROBRIVÆ), 4 miles west of Peterborough. The exact area of this place is not known but it contained large buildings with evidence of wealth, and mileage seems to have been reckoned from it rather than from "the Castles". The two places together were described by Haverfield as "one extensive straggling settlement",[1] and they were certainly the centres of the potteries which existed on both sides of the Nene valley and produced the remarkable form of ware known as "Castor".

The limestone scarp of Lincoln Edge, which runs in a north-to-south direction beyond the western limits of the Fens, is breached by the river Witham. The gap made by the river is of obvious strategic importance and the Fosse Way which extended diagonally across England was directed towards this point. The Ermine Street followed the Lincoln Edge northwards from Ancaster to the Witham and crossed that river at Lincoln, which therefore commanded the river crossing used by two great roads. The river itself must have been navigable for small vessels as far as the Roman town. Lincoln was probably founded by the Romans as a legionary fortress, at the northern end of the frontier of the Fosse, in the early days of the Occupation, and for about a quarter of a century it was the home of the Ninth Legion. After this legion had been moved forward to York, a *colonia* or settlement of discharged soldiers was established in about A.D. 75 probably on the site of the old legionary fortress. The settlement lay on the northern side of the river upon the top of the scarp, 200 feet above sea-level. It was a rectangle of about 41 acres and included the ground now occupied by the castle and the greater part of the cathedral, but the street plan is unknown. It is probable that a suburb of the Roman town later extended down the steep slopes towards the river, but little is known of its exact area.

---

[1] *V.C.H. Northants*, 1, 166.

The second group of towns to be considered were all situated in Kent and they deserve separate treatment, because together they commanded the main approach to Britain (Fig. 9). Channel ports were established at Reculver, Dover, Lympne, and Richborough, but only the last of these places was a town. Reculver can be disregarded as it was purely a fort, built during the fourth century, and in no sense a town. The harbour of Lympne, though it may have been used for trade, was primarily a naval station. In the fourth century a fort was built at Lympne, but any civil settlement that may have existed there, can never have been large. Dover must also be regarded as a base for the Channel fleet rather than a packet-station for goods and passengers. The Roman harbour of Dover was a lagoon now silted up, at the mouth of the river Dour. The port must have been a very small one, but it was well protected by high chalk cliffs on either side of the entrance. These two hills were both crowned by Roman lighthouses, and, as the lights were placed at an altitude of 380 feet above sea-level, they must have proved very valuable to navigators, and were probably visible on a clear night from the French coast.[1] The Roman fort of the Saxon Shore occupied an area of about 5½ acres on the eastern side of the Dour. Traces of civil occupation in the centuries before the construction of the fort certainly do occur, but they do not justify the inclusion of Dover among the towns of Britain; throughout the Occupation Dover was evidently a naval station. Richborough was the principal Channel port to which the crossing from the Continent was usually made from GESORIACVM (Boulogne). There must also have been a considerable amount of traffic between Richborough and the Rhine mouth, and it is obvious that Richborough was better placed for this trade than Dover. The Roman port of RVTVPIÆ was built on a small island at the eastern entrance to Wantsum channel, and possessed a splendid harbour, probably protected by a natural breakwater. There are many literary references to RVTVPIÆ and the adjective *Rutupinus* was coined by poets as an alternative to Britain in much the same way as the phrase "the cliffs of Dover" is now used to imply England.[2] Richborough was one of the landing-places of the expeditionary force of A.D. 43, but it soon became important in a civil as well as a military sense. A town grew up, part of which was probably situated on the eastern side of the existing remains. This port must have shared with London the bulk of the trade with the Continent. There is some evidence that troops were shipped direct

[1] R. E. M. Wheeler, "The Roman Lighthouses at Dover", *Arch. Journ.* LXXXVI, 29–46 (1929); *Antiquity*, IV, 352–6 (1930).
[2] F. Haverfield, "A note on Rutupinus", *Class. Rev.* XXI, 105 (1907).

from the Rhine ports to the Tyne,[1] and York may have been used as a port, but the bulk of the commercial traffic must have belonged to Richborough and London. In the third century Richborough became once more a place of military importance as the site of one of the forts of the Saxon Shore, and later as the station of the Second Legion which was transferred there from Caerleon.

Canterbury (DVROVERNVM CANTIACORVM) was placed at a convenient crossing of the Stour and at a point which may be regarded as the navigable limit of that river in Roman times (Fig. 9). The name of the place is Celtic, but there is little direct evidence of pre-Roman origin. Canterbury was the capital of the Cantii, a Belgic tribe which inhabited Kent, and they or the Romans may have moved the centre of local life from Bigberry Camp down to the river valley site, only 2 miles away.[2] The Romans built a road from each of the four Kentish ports to Canterbury, which thus became the focus of all Channel traffic. It was, however, a small country town of only 40 to 50 acres in size, and, although little is known of its plan, it was apparently less wealthy than several of the other tribal capitals. The Watling Street from London to Canterbury is compelled to cross the river Medway. The Romans established the town of Rochester (DVROBRIVÆ) at a point where the Medway can be conveniently bridged, and even there the river is 150 yards wide. A town of about 23½ acres in size grew up round the eastern end of the bridge on a good gravel and chalk site, but not much is known of its street plan. The western side of the Medway was marshy and the Romans had to build a causeway across the narrowest part of the marsh.

The third group of towns were those of southern England. Five of these, Silchester, Winchester, Chichester, Dorchester and Exeter, were tribal capitals, and it will be noticed that the last four are still important regional centres to-day, being in fact the county capitals of Hampshire, West Sussex, Dorset and Devon. Silchester (CALLEVA ATREBATVM), on the other hand, is now a deserted place and it is the only one of the five towns that was not situated on a river site. The present regional capital of the area once partly dominated by Silchester, is the valley-settlement of Reading, 10 miles north-east of the Roman town. Silchester, unlike the four other southern tribal capitals, was not covered by the buildings

[1] R. G. Collingwood, *Arch. Aeliana*, ser. iii, XX, 61–2 (1923).
[2] R. F. Jessup, "Bigberry Camp, Harbledown, Kent", *Arch. Journ.* LXXIX, 87–115 (1932).

of medieval and modern times, and the excavations carried out between 1890 and 1913 were therefore able to reveal "a fuller knowledge of the town than of any other Roman site in the western provinces".[1] It has been proved that Silchester possessed a native mint and used fine Arretine ware before the Roman invasion, and it was probably the most important *oppidum* of the southern Belgic tribes, when it was the seat of the Atrebatic dynasty of Commius. The Romans took over this pre-existing town and allowed it to retain its function as the administrative centre of the Atrebates, but in addition they made it into a great road centre. The general east-to-west direction of the lower Thames is continued westward by the Kennet, whose valley forms the obvious route across England from the Thames estuary to the Severn. Several towns have grown up in the western part of the London basin, at places where routes diverge from the main westward course of communications. In the Roman period CALLEVA became the point where the west road from London divided into four branches.[2] From it roads ran north to Alchester, and on, presumably to a junction with Watling Street, westwards to Bath and Cirencester, and south-west to Winchester and to Salisbury. The site of Silchester is on a spur of the Bagshot beds, which are here covered with plateau gravel, and is well above the wooded valley of the Loddon. The position was a relatively strong one; most of the town was above the 300-foot contour which almost encircled it except on the west. Distant views can be obtained from Silchester towards the east and the south. The population depended on wells for its water supply, and may have experienced difficulties in a dry summer. The town is in country that is devoid of ordinary building stone and the material used to construct the walls consisted of flints and flattish slabs of oolitic limestone, probably of Corallian age,[3] which must have been transported from the Oxford heights, possibly from Headington nearly 30 miles away. Silchester was a small place and the 104 acres within its walls contained only eighty houses.[4] The Romans town-planned the site, possibly at two periods, first beginning with a central block of 40 acres, and later extending the plan to cover the whole ground. The town contained a forum, four temples, a Christian church, and public baths; the street-plan was rectangular. Like other places of the same type it possessed an amphitheatre outside its

---

[1] F. Haverfield, *Ancient Town-Planning* (1913), pp. 127–8.
[2] CALLEVA may be said to have combined the present-day functions of both Basingstoke and Reading in the network of British railways.
[3] According to information received from Professor H. L. Hawkins.
[4] The area of Silchester is usually given as 100 acres, but 104 is more accurate.

walls, but it also contained a large building that is believed to have served as an inn, a necessity for such a considerable route centre. The houses do not appear to have been planned in relation to street frontages, and Silchester has been described as a collection of country houses grouped together into a "garden-city". Haverfield thought that the plan of the town revealed an "attempt to insert urban features into a country-side",[1] and it seems that the Britons were not as fully municipalised by the town-planner as the Romans had intended. Silchester was a market town with a large dyeing industry.

Three tribal capitals, Chichester, Winchester and Dorchester, were all placed very close to the boundary between the chalk and the Eocene rocks of the Hampshire basin, and at points on rivers which could be reached by very small boats. Unfortunately very little is known about these three towns. Dr E. C. Curwen believes that Chichester (REGNVM) arose during the pre-Roman period as the Belgic successor of the old upland tribal settlement on the Trundle near Goodwood and became the capital of the Romanised king Cogidubnus in the early days of the Occupation.[2] In any case, the Romans town-planned the site in the seventies of the first century and the area within the walls (103 acres) is almost identical with that of Silchester, which town it also closely resembles in outline; but nothing is known of its street plan. A creek runs inland from the sea about as far as Chichester, which therefore could be used as a port. REGNVM was the capital of the tribe called Regni whose territory probably included the South Downs and the narrow coastal plain at the foot of the hills. Winchester (VENTA BELGARVM) was another country market town and tribal centre. The Roman town was built on gravel above the flood-plain of the western side of the Itchen, and it commanded a good ford across that river. Direct access was possible to the coast at Southampton or to the Portsmouth Harbour region, and many important roads led to Winchester. If the Roman walls coincided with the medieval defences, the area of the VENTA BELGARVM was 138 acres; but it is quite possible that it was smaller.[3] A hypothesis that the Belgae destroyed a hill settlement on St Catherine's Hill (328 feet), outside Winchester, and then established a town in the valley on the site of the modern city, has not been proved. There is, however, considerable evidence that a pre-Roman Belgic settlement existed at Winchester, which may have been a mere village unlike the large *oppidum* at Silchester. Near the mouth of the Itchen a small

---

[1] F. Haverfield, *Ancient Town-Planning* (1913), p. 129.
[2] E. C. Curwen, "Excavations in the Trundle, Goodwood, 1928", *Sussex Arch. Coll.* LXX, 76–7 (1929).   [3] *V.C.H. Hants*, I, 286.

settlement grew up on the right bank of the river. Little is known of this station of CLAVSENTVM which is surrounded on three sides by a loop of the tidal river and is now covered by the buildings of Bitterne Manor. It may have been the port of Winchester, and two Roman pigs of lead, which were dug up here in 1918, "on the line of the wall which marked the limit of high tide in Roman times",[1] seem to indicate that Mendip lead was exported. Dorchester is placed on the south bank of the Frome, 16 miles from the entrance of that river to Poole Harbour. The Roman town of DVRNOVARIA was definitely a valley site and the north-eastern part of it was little above the level of the river. The Belgic tribes seem to have entered this region comparatively late, but there is some evidence that they occupied this site before the Roman occupation. The great upland "camp" of Maiden Castle[2] is only 2 miles south-west of Dorchester and the lowland settlement may have been built to supersede the older highland town. The outline of the Roman walls at Dorchester is known and the area which they contained is 86 acres, but nothing is known of the street plan. The "walks" which surround the town represent the line of the walls and somewhat resemble the boulevards of continental towns. More than a dozen mosaics and the other remains discovered led Haverfield to believe that Dorchester "counted for far more" than Winchester,[3] in spite of its smaller size. An aqueduct which is attributed to the Romans leaves the river Frome at Notton, 7 miles upstream from Dorchester, and finds its way by a winding course to the town and roughly follows the 300-foot contour.[4] The town of Exeter (ISCA DVMNONIORVM) was the western terminus of the Roman road system. The highlands of Exmoor and Dartmoor did not attract the Romans, and finds of their occupation in Devonshire are exceedingly rare on the western side of the Exe. A few discoveries have been made on the Devonshire coast west of Exeter, but there is more evidence of Roman influence in distant Cornwall than in Devonshire.[5] Exeter is placed at the head of the estuary of the Exe, about

[1] *Proc. Soc. Ant.* 2nd ser. XXXI, 37 (1918).
[2] "The most stupendous fortification of any age in England", *Camb. Anc. Hist.* (1934), X, 800.   [3] F. Haverfield and G. Macdonald, *op. cit.* p. 214.
[4] Major Coates, "The Water Supply of Ancient Dorchester", *Proc. Dorset Nat. Hist. and Arch. Soc.* XXII, 80–9 (1901).
[5] The signs of Roman occupation in Cornwall are most evident on the sea coast in the neighbourhood of harbours or navigable rivers. Communications with the north and south coasts of Cornwall must have been maintained by sea, as the area occupied by Exmoor, Dartmoor and Bodmin Moor was not crossed by any Roman road. The recent discovery of a Roman villa near Camborne shows that Cornwall was occupied in the civil sense. *Antiquity*, V, 494–5 (1931).

10 miles from its mouth and high up above the eastern bank of the river. The position was one of considerable natural strength and was, presumably, at the approximate limit of navigation and at the lowest convenient crossing of the river. It is probable that Exeter was a tribal centre before the Roman conquest, and it is certain that the Romans soon occupied the site and made it the administrative capital of the Dumnonii. The Roman walls were probably along the line of the medieval fortifications and are well preserved on the south and near the castle.[1] The area thus enclosed was about 91 acres, and the mosaics, a bath, and other remains, show that Exeter was a civilised settlement, even though it marked the limit of town life in south-western Britain.

A fourth group of towns were placed on both sides of the mouth of the river Severn. The land that stretches inland from this estuary, especially its south-eastern side, was prosperous and wealthy, and it is not surprising that it contained several important towns. Cirencester (CORINIVM DOBVNORVM) was the second largest town in Britain and was exactly the same size as Cologne (COLONIA AGRIPPENSIS). It became the capital of BRITANNIA PRIMA, one of the subdivisions created by Diocletian's administrative reorganisation at the end of the third century, and it may be regarded as the chief town of western Britain. Its geographical significance in Roman times is easy to understand. The town is placed at the eastern end of a good route across the Cotswolds to Gloucester and the Severn. The Roman road ran for 10 miles north-west from Cirencester until it reached an altitude of 700 feet and then dropped over the edge of the scarp at Birdlip. Cirencester became a great road centre. The Fosse Way ran through the town, while Akeman Street from St Albans and Alchester, and the west road from London and Silchester both converge on this point in order to cross the Cotswolds. Apart from its importance in the road system, Cirencester was the capital of one of the richest regions in Britain. Perhaps it is not inapt to call the Cotswolds "the Dukeries of Roman Britain" because they certainly contained an unusually large number of magnificent country houses. These great villas probably flourished on the wealth gained from agriculture and the woollen industry, whose products were marketed at Cirencester. A road known as the White Way ran from the town to the large villa of Chedworth. Cirencester was also the administrative capital of the tribe

---

[1] According to information received from Mr F. T. Howard.

called Dobuni, whose territory seems to have extended from Hereford-shire to Oxfordshire. The site of the town was on gravel on the left bank of the Churn, an upper tributary of the Thames, and the area within its two miles of walls was about 240 acres. The remains which have been found are substantial and bear witness to the high state of civilisation attained by this town. A great town hall and numerous mosaics of an elaborate character show that it was a place of far greater wealth and importance than places like Silchester. There is no evidence for a pre-Roman town on the site and Cirencester, which seems to have begun life as a military post, certainly flourished as a town from about the year A.D. 70;[1] it was the greatest and the most civilised of the dozen tribal capitals of Britain.

Gloucester is situated on the eastern side of the Severn at a point where the river is divided into two branches and is joined by its tributary the Leadon. A westward road from Gloucester can therefore cross the Severn more easily by means of the three river crossings than by way of a longer crossing over the united river lower down. The Severn bore, so dangerous for small ships, is less powerful here than it is further downstream. The Roman town therefore commanded what was the lowest convenient bridging point of the river and was also favourably placed for navigation. Very little is known of Gloucester's history; it is possible that a legionary fortress was established here and later moved to Caerleon but there is no direct evidence. A *colonia*, or settlement of time-expired soldiers (possibly from Caerleon), was established at Gloucester (GLEVVM) at some time between the years A.D. 96 and 98. Although the town enjoyed municipal rank, it seems to have been a much smaller and less prosperous place than its neighbour at Cirencester. If the alignment of the medieval walls followed the line of the Roman defences (and this has recently been proved on the eastern side), the colony enclosed an area of about 46 acres and was therefore rather larger than the colony of Lincoln.

A small town with a character that was unique in Britain was situated in the deep valley of the Avon and was almost encircled by a great loop of that river. Bath (AQUÆ SVLIS) was a remote place on the banks of an unnavigable stream and possessed no strategic importance; its only function was to serve as a spa for invalids who suffered from gout, rheumatism, and other similar ailments, which afflicted the ancients even more heavily than the moderns of to-day. A Roman town grew up round the abundant hot mineral springs at some time before A.D. 77 and had

[1] Cirencester was apparently founded during Nero's reign (A.D. 54–68). F. Haver-field, *Archaeologia*, LXIX, 194 (1917–18).

become quite a prosperous place by the end of the first century. The surrounding walls were three-quarters of a mile in length and the area which they contained only amounted to 23 acres. The principal buildings were a suite of baths, whose original area may have been as much as 1½ acres, and a temple to the goddess Sul, who gave her name to the place. The pediment of this temple was adorned by a gorgon's head, whose originality and astonishing power are remarkable, and which indicates that a Celtic spirit survived in Romano-British art. It has been proved that visitors came to Bath not only from different parts of Britain—York, Chester, Caerleon, Gloucester and Cirencester—but also from northern Gaul, as far away as the modern towns of Trier, Metz and Chartres.

The small town of Alchester is a long distance from the Severn, but like Cirencester, its larger neighbour 40 miles away, it is placed at the foot of the gentle rise up to the limestone scarp, and for that reason may be included in this group rather than with the midland towns. Alchester was situated at a point where Akeman Street intersected a road from Silchester to Watling Street and covered an area of 26 acres. The place was occupied throughout the whole of the Roman occupation.

The task of conquering the Silures of southern Wales was one that lasted for more than 30 years. At some date soon after the conclusion of this military achievement, probably between A.D. 70 and 80, the Romans endeavoured to pacify this warlike tribe on the western side of the Severn by establishing a town amongst them as a civilising influence. The shape and size of the place have given rise to the conjecture that it may have originated as a military fortress but there is no real evidence. Caerwent (Venta Silvrvm) is situated below a hill-top camp in Llanmelin Wood and, like many other towns of Roman Britain, it may have been the lowland successor of an upland town.[1] Caerwent is 8 miles east of the fortress of Caerleon and is placed in the middle of the fertile coastal plain, which extends between the Wye and the Usk and is bounded by wooded hills on the north and by the Severn on the south. The place occupied a good geographical position as an administrative centre for the civilised part of southern Wales, and belonged to the class of small country towns that were established by the Romans as tribal capitals. The complete plan of Caerwent has been revealed by excavation which shows that its size was about 44 acres and that it possessed a rectangular street pattern. The interior of the town was divided into twenty blocks or *insulae*, and it contained a forum, public baths, shops, and other features of civilised life. The houses were quite closely packed in the limited space within the walls.

[1] R. E. M. Wheeler, *Prehistoric and Roman Wales* (1925), p. 291.

and the town does not seem to have extended beyond its fortifications. Caerwent was a remote place, one of the extreme western outposts of Roman city life in Britain and far from the main channels of trade.

Two other small towns grew up on the western side of the Severn. A Roman road from Wroxeter to Caerleon crossed the Wye south of Kenchester, and the town of MAGNA[1] commanded the crossing. Here a rampart enclosed an area of about 22 acres. The earthworks of Credenhill, near Kenchester, are 720 feet above sea-level and may have contained the upland ancestor of the lowland town, but this is pure conjecture. The first definite record of a town on this site can be dated between A.D. 70 and 80. MAGNA was not a tribal capital but merely a small market town, the regional centre of a large part of Herefordshire. The Anglo-Saxon settlement of Hereford is only 4½ miles away and it now serves a similar function to that formerly performed by MAGNA, but it is more conveniently situated for communications up the valley of the Lugg. An east-to-west road crossed the Wroxeter-Caerleon road at MAGNA, which was therefore a road junction. This Roman town was a small place, but the streets, the drainage, and the mosaics form a remarkable witness of the penetration of Roman urban civilisation into an outlying corner of Britain. A Roman town which has generally been equated with ARICONIVM was placed near modern Weston-under-Penyard, 3 miles from Ross-on-Wye. Excavation has not been very extensive and it is impossible to state the area occupied by the town, but there is sufficient evidence to show that its character was somewhat unusual. There is abundant proof that iron smelting was carried on here, presumably using the raw material provided by the mines of the Forest of Dean, and the place was certainly occupied from the first to the fourth century. In 1821 an archaeologist described ARICONIVM as a "Roman Birmingham", and a recent writer has stated that "no other Roman town in the whole province, perhaps, had quite so specialised an industrial character".[2]

The great legionary fortress which protected the rich area of the Severn and the Cotswolds was situated at Caerleon (ISCA) and although the place must not be regarded as a town, it may be useful to consider its geographical significance. Caerleon is about 3 miles from modern Newport, which enjoys the strategic advantages of position formerly held by the Roman fortress. In recent years the Roman site was actually being overgrown by the modern expansion of Newport, and excavation has

---

[1] The name may be either MAGNA or MAGNI or MAGNAE, as it is only known in its ablative form MAGNIS.

[2] R.C.H.M. Hereford, III, Introduction, liii.

been taking place from 1925 onwards in order to make every possible discovery before it becomes too late. The Roman fortress was on the north bank of the tidal river Usk, just below the junction of the tributary Afon Lwyd with the main stream, and was splendidly situated for the military control of Wales. Troops could be taken down the Usk and landed on any part of south Wales or they could march westward along the Glamorgan coastal road; while the valley of the Usk itself provided an excellent route into the interior of the Welsh highlands. The fortress, which enclosed an area of about 50 acres, was probably founded in A.D. 75 and its amphitheatre was certainly built in A.D. 80. A small civilian element appears to have existed on its south-western side between the fortress and the river. While the fortress was well placed as a base to control the hill-tribes of Wales, it was in a bad position to check the Irish coastal raiders who arrived in the third and fourth centuries. In order to protect the civilian zone which had grown up on the coastal plain of south Wales from the attacks of maritime marauders, a fort of the Saxon Shore type was built at Cardiff, 14 miles south-west of Caerleon. It seems likely that the Second Legion was moved from Caerleon to Richborough at the same time as the construction of the fort at Cardiff.[1] The two events together ended the days of Caerleon's importance.

There were few real towns in the fifth region, the midlands. Leicester (RATÆ CORITANORVM) was in fact the only considerable settlement that was situated within the region proper. Most of the others were placed on the outer edge of the midland woodland zone. Leicester was near the limit of navigation on the Soar, a tributary of the Trent, and was built on a gravel terrace on the east bank of the river. It was therefore between the Northamptonshire uplands and the Charnwood Forest. The area within the medieval walls was 105 acres and the Roman town is believed to have covered the same ground. Leicester was one of the tribal capitals and served as a market town for the agricultural produce of the region and as a centre for the administration of the Coritani. It is not known whether a pre-Roman town occupied the site; and the first Roman settlement was of a military nature, but only lasted for a few years. The town appears to have developed rapidly, like so many others, in the seventies, although it was probably founded earlier. It became, as the mosaics show, a prosperous place, but it remained a mere country town with an uneventful

[1] R. E. M. Wheeler, *Prehistoric and Roman Wales* (1925), pp. 235–6.

history. Leicester was on the Fosse Way 12 miles north-east of VENONÆ, where that road crossed the Watling Street. A road from Leicester to Mancetter enabled travellers to reach western districts by a short cut without going round by VENONÆ. Two small towns on Watling Street deserve a brief mention. Towcester and Wall (LETOCETVM), probably 35 and 30 acres respectively in area, were both stations on this great road. Wall, which was at the junction of the Rycknield Way and the Watling Street contained a large suite of baths. Another small town was situated at Irchester (20 acres) on the south bank of the Nene, about 2 miles from modern Wellingborough, and this settlement must have been surrounded by densely wooded country. The Roman town of Wroxeter (VROCONIVM CORNOVIORVM)[1] was placed on the western edge of the midland region, and is 5 miles south-east of Shrewsbury. Like its medieval and modern successor, Wroxeter was a position of considerable strategic importance and a natural route centre. The Severn valley provided a western route into Wales, while the Tern and Roden tributaries opened a way to the north. Southwards it was easy to cross the Severn and go to MAGNA by way of the Church Stretton pass, the route followed by the modern railway. Wroxeter commanded a good ford over the Severn, and it is interesting to notice that there is no recognised Roman crossing of that river between Wroxeter and Gloucester. Five miles east of the town is the Wrekin, whose summit (1335 feet) may have contained the original settlement that was transferred by the Romans to Wroxeter, but there is no definite evidence. Wroxeter was originally chosen as the site of a legionary fortress, but its use for military purposes cannot have lasted long. The actual fortress has not yet been discovered, and the troops were probably soon moved forward to Chester. The town was built on the gravel flats above the Severn and covered an area of 170 acres. Between the years A.D. 80 and 90, public baths of extraordinary magnificence and size for such a town were in the process of erection but were not completed. Later, a forum was built but was destroyed by fire in about A.D. 160. An inscription, the largest ever found in Britain, was discovered in 1924 and proves that the Cornovii enjoyed a form of local government, and that VROCONIVM was their administrative capital,[2] and it must also have served as a large market town. The Fourteenth Legion was moved forward from Wroxeter to Chester, which was made into a fortress. The

[1] The spelling usually adopted (as on the O.S. *Map of Roman Britain*) is VIRO-CONIVM but Professor F. M. Stenton, on philological grounds, considers that VROCONIVM is the correct form.

[2] D. Atkinson, "Civitas Cornoviorum", *Class. Rev.* XXXVIII, 146–8 (1924).

date of Chester's foundation is not known, but eventually it became the stonebuilt fortress of the Twentieth Legion and enclosed an area of 56 acres. It was not a town, but its fine strategic position must be mentioned. Chester (DEVA) controlled the lowest safe crossing of the Dee and was at the same time a suitable landing place. The three estuaries of the Dee, the Mersey, and the Ribble, must have altered considerably in outline since the Roman occupation. Ptolemy only included two estuaries in his tables for this part of the coast, and the southern of these appears to be the Dee, while the other may be the Ribble on which the Romans established the fort of Ribchester. On this evidence Sir J. A. Picton argued that the Mersey did not exist in Roman times, and concluded that "either the Romans displayed in reference to the Mersey an apathy or ignorance which attached to them in no other instance or that the estuary in its present form did not exist".[1] While this theory about the Mersey is not supported by geological evidence, it is certainly true that the Dee estuary was by far the most important of these three inlets in the Roman period. The estuary of the Dee must have then been a wide open arm of the sea at least as far as Chester.[2] Silting has been amazingly rapid here, and much land has been reclaimed, as is proved by the maps of the last three hundred years.[3] The harbour of Chester had been affected by silting before 1449,[4] and subsequent ports have been built at different periods further down the northern side of the river, as the channel filled up, until finally the Dee's economic importance was eclipsed by the rise of Liverpool on the Mersey.[5] It can be safely assumed that the largest vessels of the Roman period were able to sail up to the walls of the Roman fortress. The site of Roman Chester is on the north bank of the river, upon good foundations of red sandstone, 100 feet above sea-level. The outline of the fortress corresponded in a remarkable way to the exposure of the sandstone and avoided the glacial clays. This fortress commanded important routes both to the north by way of Lancashire and along the north coast of Wales to the lead mines of Flint and the copper mines of Anglesey. The legionary tile works at Holt, 8 miles south of Chester, made it un-

[1] J. A. Picton, "The changes of sea levels on the west coast of England during the historic period", *Proc. Lit. and Phil. Soc. Liverpool*, v, 113 (1849).

[2] It is still marked as such on the map of Cheshire made by Saxton in 1577.

[3] See A. Strahan, *The Geology of Flint, Mold and Ruthin*, Mem. Geol. Survey, 1890, opposite p. 158, for a set of maps of the Dee in 1684, 1839, and 1887, showing the alterations and reclamations of the estuary.

[4] A. Strahan, *ibid.* p. 156 for references. See p. 296 below.

[5] It was not until the foundation of Liverpool by King John that the Mersey's possibilities were recognised.

necessary for the army to import pottery and tiles from the south-eastern civil zone.

Town life extended even as far as the north of England, the sixth region. The vale of York formed a natural avenue to the north and a Roman road ran along its western border from Tadcaster to Catterick, but another northward route crossed the Humber to Brough and followed the western edge of the Wolds to Stamford Bridge and York. The marshy vale is crossed by two terminal moraines which rise about 50 feet above the general level. One of these ridges runs south-west from Sand Hutton, about 2 miles north of Stamford Bridge, to York and Tadcaster and has always provided a good means of communication between the rich districts of eastern Yorkshire and the Pennines. York is situated at a point where the Ouse breaks through the morainic ridge, and is therefore at a crossing of the east to west route by the moraine and a north to south route by the river. In Roman times the tidal Ouse probably carried ships up to the town; to-day the tide reaches Naburn lock, 5 miles below York. In addition to the general advantages of position, York also enjoyed a well-protected site. The fortress occupied by the Ninth Legion was built on the eastern side of the Ouse in the angle formed by the junction of the Foss with the main river, and it was therefore surrounded on three sides by water. York (EBVRACVM) as a fortress was a successor of Lincoln. The northward extension of Roman power made it convenient to establish a great military capital at York, which was better placed than Chester as a base for the troops on the Wall. The fortress was established between A.D. 71 and 74; its exact area varied somewhat during the occupation but ultimately it was about 50 acres in size. A town was eventually built on the south-western side of the Ouse, on the site of the modern railway station, and was separated from the fortress by the river. This settlement was a *colonia* of time-expired soldiers; the year of its foundation is uncertain but the first datable record is not until the year 237. Very little is known of the town, but it may have covered an area of 50 acres or more. The most northerly of the tribal capitals was at Aldborough (ISVRIVM BRIGAN-TVM), a short distance south of the river Ure, and 16 miles north-west of York. A pre-Roman centre of tribal life existed at Stanwick near Darlington, 30 miles away, and it is possible that Aldborough was founded to succeed this older settlement,[1] probably during the seventies of the first

[1] R. G. Collingwood, "The Roman Town at Aldborough", *Excursion Handbook Q of the British Association for the Advancement of Science*, Leeds Meeting, 1927, p. 4.

century. The Roman town controlled the important crossing of the Ure by a bridge. The town walls enclosed an area of 60 acres and the numerous mosaic pavements, baths, and other finds show that this capital of the Brigantes enjoyed a certain amount of civilisation and prosperity and existed until the end of the fourth century. In the far north, just south of the Wall, were two places which cannot be regarded as real civil towns, but which were described by Haverfield as "playgrounds during a happy holiday, for soldiers garrisoning Hadrian's Wall".[1] Corbridge (COR- STOPITVM) was about 40 acres in size and was placed on the north bank of the Tyne, less than 3 miles south of the Wall, while Carlisle (LVGVVAL- LIVM), over 50 acres in size, lay on the south bank of the river Eden and was very close to the Wall. These two places therefore commanded important river crossings made by roads leading from Chester or York to the Wall, and were permanent military storehouses rather than towns.

The most important port and town has been left to the last. London (LONDINIVM) was the greatest town in Britain and was actually the fifth largest in the northern provinces of the Empire.[2] It seems probable that, although the name London is Celtic in origin, no settlement of any significance had been established there before the time of the Roman invasion; London is definitely a Roman foundation. At first it seems strange that a site which possessed so many natural advantages should not have been used in pre-Roman times. The Thames estuary is the greatest inlet into Britain, and forms a natural avenue of entry from the Continent. It faces the many channels by which the Rhine, the Scheldt, and their tributaries, discharge their waters into the North Sea.

A series of interesting discoveries in the Thames estuary has proved that the level of the land has sunk about 15 feet relatively to the level of the sea since the Roman period. The level of the high tide must therefore have been about 15 feet below the present high-water mark. From the evidence as a whole, Captain T. E. Longfield has estimated the rate of submergence at about 9 inches per 100 years.[3] Part of Bradwell Fort (OTHONA), one of the forts of the Saxon Shore, has sunk below the sea

[1] F. Haverfield and G. Macdonald, op. cit. p. 182.

[2] London (330 acres) was only exceeded in size by Nîmes (790 acres), Trier (704 acres), Autun (494 acres) and Avenches (roughly 350 acres) in Gaul, Germany and Britain.

[3] T. E. Longfield, The Subsidence of London, Ordnance Survey, Professional Paper, New Series, No. 14 (1933).

and proves that the land at this point was higher in Roman times than now. At Tilbury the remains of a small Romano-British settlement of circular huts of the first and second centuries of our era have been found at a depth of 13 feet below the high-water mark.[1] In 1928 a pile dwelling of the Roman period was discovered at Brentford below the present low-tide level of the Thames.[2] In the marshes of the Medway other discoveries have been made which support the view that "the level of the land has sunk considerably since Roman times, and that what is now drowned marsh was then habitable land".[3] It must be remembered that submergence of this character affects the position of the tidal limit, which has progressively advanced westward up the river in historic times. It seems probable that the tides did not go far beyond London Bridge in Roman times, and that "Roman London was established at, or only a short distance below, the tidal limit at the time".[4] London was also the lowest point on the river where a bridge could be built without very serious difficulty. This is also the lowest spot where gravel terraces on both sides of the river approach one another at all closely.[5] The patch of gravel on the northern bank of the river forms a large well-drained area, easily cleared of its vegetation and suitable for human settlement. The gravel on the south bank formed a useful means of approach to the higher and larger northern area of gravel, and also provided a good foundation for the southern bridgehead. In spite of these facts, the position of London had certain disadvantages which probably account for its comparatively late development. To the north of the gravel on which Roman London was built, lay a broad belt of London clay which was undoubtedly densely wooded and difficult to clear. A strong centralised government and a long period of peace were necessary if roads were to be made through the woodlands to the lowest bridging point of the Thames at London. The actual site of the town lay on the north bank of the river between the river Fleet on the west, which still joins the Thames at Blackfriars bridge, and the Tower of London on the east. This patch of gravel was divided into two nearly equal portions by a stream called the Wallbrook. On each side of this stream were small hills rising about 50 feet above the level of the Thames.[6] The two hills were about three-

[1] R.C.H.M. Essex (South-East), IV, 38–9 (1923).
[2] R. E. M. Wheeler, "Old England", Antiquity, III, 20–32 (1929).
[3] V.C.H. Kent, III, 132.   [4] R.C.H.M. London, III, 14 (1928).
[5] See Fig. 85 on p. 532 below.
[6] See H. Ormsby, London on the Thames (1924), for very clear contoured maps of the city of London in relation to the Roman wall. See especially Figs. xiii, xiv and xv. The contours are drawn at a vertical interval of 5 feet.

quarters of a mile distant from each other; the western hill is now crowned by St Paul's, while Leadenhall market stands on the eastern summit.

It appears that London must have grown with amazing rapidity in the first seventeen years after the Roman invasion, but in A.D. 61 the town was completely destroyed and its citizens were "massacred, hanged, burned and crucified" by the rebels led by Boudicca. This early town of London was built on the eastern of the two hills round what is actually the highest point in the whole site. The eastern hill is directly north of London Bridge, the factor which really determined the growth of the city. After the Boudiccan revolt was crushed, London was rapidly rebuilt within the circuit of a protecting wall, over 3 miles in length, which surrounded an area of 330 acres and included the hills on either side of the Wallbrook. On the southern side of the river, close to the bridge, there was a small unwalled settlement of 15 acres, which was probably a suburb, and not, as some writers have suggested, the original nucleus of London. The town was planned with the regular streets and chess-board pattern of the buildings in the normal Roman fashion; it contained on the eastern hill the greatest British Basilica or town hall,[1] which was in fact one of the largest buildings in the whole empire. Dr Wheeler describes the Roman town in the following words:

It was a civilised city, a comfortable one, with an efficient drainage system and an adequate water-supply. There were probably more buildings of stone and brick than at any subsequent period until after the Great Fire of 1666. There were more adequate and attractive facilities for bathing than ever until the latter part of Queen Victoria's reign. The Roman city-surveyor, standing in the midst of his simple street-system would have laughed at our curiously deformed inheritance from the Middle Ages.[2]

London became the great route centre of Britain, and six main roads eventually converged on its bridge, the hub of British communications. Both the Watling Streets, from Chester, and from the Channel ports, the Stane Street from Chichester, the road from Silchester, the Ermine Street from York, and the Colchester road from Colchester, all led to London.

The site gave the town a long frontage on the river and undoubtedly London's primary function was that of a trading port and distributing centre of Roman Europe. Its interests were continental rather than

[1] The total length of the building was certainly 350 feet and possibly 420 feet. The lengths of basilicae at other British towns were as follows: Cirencester, 333 feet; Silchester, 276 feet; Wroxeter, 250 feet; Caerwent, 180 feet.
[2] R. E. M. Wheeler, *London in Roman Times* (1930), pp. 156–7.

insular. As early as A.D. 60 London was described by Tacitus as *copia negotiatorum et commeatuum maxime celebre*.[1] Archaeological evidence has proved that quantities of oil and wine, those great staples of Mediterranean commerce, were imported into London, as well as pottery from Gaul and works of art from Italy. The exports included corn, slaves from the western hills, and possibly some minerals. Corn was probably the chief export, and it should be noticed that London was the natural focus of the rich agricultural regions both of Essex and Hertfordshire in the north and of Kent in the south.[2] Like all large modern ports, London itself was the home of some important industries; there is plenty of evidence of shoe-making, and of glass working, and Roman pottery kilns have been found on the site of St Paul's.

The question has often been asked whether London was the capital of Roman Britain. Colchester was certainly the capital during the early part of the Occupation, and London never appears to have enjoyed the status of a *municipium* or of a *colonia*. As might be expected, this great port was the seat of the chief customs officers of Britain. In the fourth century it contained the offices of the chief of the treasury,[3] and was obviously the centre of the financial administration; it had contained a mint from the third century. In the fourth century London received the title of "Augusta", and in the same period it is known to have been the seat of a bishopric. None of these facts, however, are sufficient to prove that London was the capital of Roman Britain, but they do show that it was the most important commercial town in the province. Although it is not proved, it is quite possible that London may have been the administrative centre of the country between A.D. 60 and 197, but the division of Britain into two provinces in the latter year, and its later subdivision into four or five parts, make it unlikely that London enjoyed a complete metropolitan control over the whole region in the last two centuries of the Occupation. In some respects modern London is badly situated as a capital for Great Britain or even for England. In Roman times it certainly shared some of the functions of a capital with York, the great military base in the north. Nevertheless, London was already in Roman times an international city rather than a national capital. Nature made it a focus of European traffic, and modern

---

[1] Tacitus, *Annals*, XIV, 33.

[2] See S. W. Wooldridge and D. J. Smetham, "The Glacial Drifts of Essex and Hertfordshire, and their bearing upon the Agricultural and Historical Geography of the Region", *Geog. Journ.* LXXVIII, 259 (1931), for an interesting discussion of this point of view.

[3] The *Notitia Dignitatum* refers to a London official called *Praepositus thesaurorum*.

industrialism and Atlantic trade have merely widened that function so that it is now a metropolis of the world. The Romans appreciated the importance of London's European position, and their establishment of the city is one of the greatest achievements of a race of practical geographers. In Dr Wheeler's words, "twenty miles of London shipping and an enlarged Watling Street still bear witness to the unerring imagination of the Roman founders of modern Britain".[1]

It is now possible to summarise the urban acreage of Britain. Of the thirty-three purely civil towns described, the area of five cannot be given and they were probably small places, Richborough and Castor being the largest. The total area within the walls of the remaining twenty-eight towns is 2287 acres, so that it is unlikely that the sum total of urban Britain much exceeded 4 square miles (2560 acres). The twelve tribal capitals occupied 1226 acres, an average of about 102 each. The four *coloniae*, if York is estimated as 50 acres, make up 245 acres. London and St Albans together amount to 530 acres, while a group of ten smaller places makes a total of 286 and therefore average nearly 30 acres apiece.

VILLAGES, VILLAS, AND AGRICULTURE

The main facts about the rural settlement of the country must now be briefly summarised. Outside the towns, the Romans established numerous villas or farms. At present there is no evidence to prove the existence of a pre-Roman prototype of the "villa". It is believed that a villa was the estate of a landowner who lived in a large house, cultivated the adjacent fields, and possibly let off the remainder to semi-serf tenants.[2] It has been suggested that the town life imposed by the Romans reached its highest development during the first and second centuries and then decayed. The life of the villas, on the other hand, was more prosperous in the third century or even later, although it must be remembered that a few are known to have existed in the first century. For this reason it appears likely that the Roman civilisation was slowly communicated from the towns to the countryside. The economic life of the villas must have depended almost entirely on agriculture, and the geographical distribution of these estates, as shown on the Ordnance Survey *Map of Roman Britain*, indicates the areas that were agriculturally productive in the Roman period.

[1] R. E. M. Wheeler, *London in Roman Times* (1930), p. 159.
[2] R. G. Collingwood, *The Archaeology of Roman Britain* (1930), pp. 113–36, gives a description of the different types of villas. See also T. D. Kendrick and C. F. C. Hawkes, *Archaeology in England and Wales, 1914–1931* (1932), pp. 260 ff.

Villas were frequently grouped around towns, where they must have marketed their produce. The number of villas is especially large on the Hampshire uplands, on the dip-slopes of the North and South Downs, in the Isle of Wight, on the Cotswolds, in the region surrounding Bath, and finally in the neighbourhood of Yeovil and Ilchester. Smaller groups of villas are also to be found on the Lincoln wolds, on the rim of the vale of Pickering, and on the lowlands of southern Wales, but very few existed in the midlands, west of Leicester.

The other elements in the rural life of Britain were the native villages,[1] and the isolated farmsteads,[2] both of which had existed in pre-Roman times. The influence of Roman civilisation can be traced in many of these villages and farms, but the general mode of life of such settlements was very little affected by the Roman Occupation. The villages were thickly distributed over the whole of the areas now called the Marlborough Downs, Salisbury Plain, and Cranborne Chase, a region which contained few villas. Native villages appear to have continued their existence in all parts of the western and northern highland regions of Britain during the period of the Roman military occupation, while the inhabitants of Devonshire and Cornwall went on living in round stone huts. The hill-forts of Wales continued to serve as native settlements, for the Romans did not compel their inhabitants to move down into the lowland.

The main occupation of the people was agriculture, and the Celtic inhabitants of Britain were exporting corn in Strabo's time before the Roman Occupation.[3] There is no doubt that the Celtic system of agriculture persisted throughout the Roman period, and that the Romans stimulated agricultural production.[4] There is plenty of evidence that British corn was exported to the Rhineland in the fourth century. Julian reorganised the inland transport of corn in the Rhine valley and increased the fleet which brought the corn over from Britain. The British farmer also possessed a market for his produce in the army of occupation, which was concentrated in the infertile regions of the north and west. The areas of production were those in which villas or villages were thickly distributed. Air photography

---

[1] General Pitt-Rivers, *Excavations in Cranborne Chase* (1887–98), 4 vols., is still the best account of life in Romano-British native villages; see also O. G. S. Crawford, "Our Debt to Rome?", *Antiquity*, II, 173–88 (1928).

[2] Lowbury Hill on the Berkshire Downs is a good example of a pre-Roman farmstead which continued its existence throughout the Roman occupation, possibly as the upland cattle farm of an estate in the Thames valley below. See D. Atkinson, *The Romano-British Site on Lowbury Hill* (1916).

[3] Strabo, IV, v, 2.

[4] E. C. Curwen, "Prehistoric Agriculture in Britain", *Antiquity*, I, 261–89 (1927).

has revealed a network of small rectangular fields in many parts of the chalk lands of southern Britain.[1] "In Romano-British times", says O. G. S. Crawford, "practically the whole of Salisbury Plain, Cranborne Chase, and the Dorset uplands were under plough."[2] Recent research in the Fenland of eastern England makes it appear probable that this region also was an area of intensive agricultural activity during the Roman period.[3] Air photographs show that native villages were thickly scattered over parts of the Fenland. Little is known of British agricultural practice, but Pliny states that chalk was used to fertilise fields.[4]

INDUSTRIES

Britain is a region well suited by nature for pastoral industries and especially for the rearing of sheep. In medieval times England was one of the leading areas of sheep production in Europe, and wool was the keystone of her foreign trade. At the present day Great Britain contains no less than 25 million sheep, which is approximately the same as the number found in New Zealand and is actually one-quarter of the number found in Australia. It is known that the Celtic inhabitants of Britain were pastoralists and it can be assumed that the Roman peace promoted the increase of this industry. Bones of sheep have been found on many sites, and there are literary references to British flocks. There is sufficient evidence to show that Roman Britain was probably a producer of wool, but it is scarcely sufficient to prove that the trade in the commodity was considerable. Nevertheless, it is reasonable to assume that raw wool was a staple product of Roman Britain. It would be expected that the supply of wool would be made up into cloth locally, and evidence that a woollen manufacturing industry existed is not wanting. In three places, archaeologists have discovered remains which are believed to be *fullonicae* or fulling mills.[5] Fulling is the process of treating and washing textiles and steeping and

[1] See O. G. S. Crawford, "Air Survey and Archaeology", *Geog. Journ.* LXI, 342–66 (1923); O. G. S. Crawford, *Air Survey and Archaeology*, Ordnance Survey Professional Paper No. 7 (2nd ed. 1928); O. G. S. Crawford and A. Keiller, *Wessex from the Air* (1928); see also p. 18 above, footnote 1.

[2] O. G. S. Crawford and A. Keiller, *op. cit.* p. 9.

[3] Gordon Fowler, "The Extinct Waterways of the Fens", *Geog. Journ.* LXXXIII, 30–9 (1934); Catalogue of *Fenland Survey Exhibition of Early Maps and Air Photographs* (1934), pp. 27–30.

[4] Pliny, *Nat. Hist.* XVII.

[5] G. E. Fox, "Notes on some probable Traces of Roman Fulling in Britain", *Archaeologia*, LIX, 207–32 (1905), contains an account of the villas at Chedworth, Darenth and Titsey.

rinsing them in water. The large villa of Chedworth, 7 miles from Cirencester, contains buildings that seem to have been used for fulling and also
possibly for wool scouring. At Darenth in Kent, a part of a very large
villa, nearly 3 acres in extent, has been recognised as a fulling establishment.[1] The hypocaust of this villa is said to be similar to those used in tile
kilns, where great heat was necessary. At Titsey in Surrey, a small villa,
originally a dwelling-house, seems to have undergone considerable alterations, which appear to have been made with a view to converting its
buildings into a fulling mill.[2] Fulled goods had to be dried in the open air
or by means of artificial heat. The British climate made artificial heat
essential, and rooms for this purpose may have existed at Titsey. At
Silchester a large portion of the site was occupied by dyers and contained
furnaces which may have formed part of a dyer's establishment.[3] The
Notitia includes the name of an official entitled *procurator gynaecii in
Britannis Ventensis*, which has been translated as "administrator of the
imperial weaving works at VENTA".[4] Thus it can be assumed that an
imperial weaving factory existed in or near Winchester in the fourth
century. In other places loom weights and bobbins are further indications
of a woollen industry.

The supply of raw wool for these undertakings undoubtedly came from
the locality in which they were placed. It is natural to assume that the
Cotswolds (Chedworth), the Hampshire Downs (Winchester), the North
Downs of Surrey and Kent (Titsey and Darenth), and the Berkshire
Downs (Silchester), were already at that time four of the principal areas
of sheep production in Britain.[5]

The great size of the villas at Chedworth and Darenth leads to the
assumption that more cloth was manufactured than was needed locally.
Wool was apparently exported from Britain, and, as it is mentioned in the
Eastern Edict of Diocletian of A.D. 301, it must have travelled as far as
Eastern Europe. British products were certainly used in Gaul. There are
sufficient scraps of information to satisfy us that Britain exported manufactured wool, but we must not rashly assume that the great medieval
woollen trade of England had already been born in Roman times.

[1] *V.C.H. Kent*, III, 111–13.
[2] *V.C.H. Surrey*, IV, 367–9.
[3] *Archaeologia*, LIV, 460 (1895).
[4] *V.C.H. Hants*, I, 292.
[5] A glance at a map of the distribution of sheep in England to-day shows that the
chalk and limestone pastures of the scarplands of south-eastern England are still
leading producing areas.

The minerals of Britain (Fig. 10) were described by Tacitus as "the reward of victory".[1] Although he singled out gold for special mention, the production of this mineral could never have been great in Roman Britain. A Roman gold mine is believed to have existed near Dolaucothy House, 10 miles north-west of Llandovery, but even in this case the Roman origin of the mine is not absolutely proved.[2] British tin was

THE MINING REGIONS OF ROMAN BRITAIN

● GOLD MINE
● COPPER MINE
▨ LEAD MINING AREA
▨ IRON MINING AREA

Scale of Miles

HADRIAN'S WALL BENWELL

YORKSHIRE LEAD MINES

YORKSHIRE LEAD MINES

GT ORMES HEAD COPPER MINE
PARYS MOUNTAIN COPPER MINES

WILDERSPOOL DERBYSHIRE LEAD MINES

FLINTSHIRE LEAD MINES

LLANYMYNECH HILL COPPER MINE

COLSTERWORTH

SHROPSHIRE LEAD MINES

DOLAUCOTHY GOLD MINE

• TIDDINGTON

PETERSTOW ARICONIVM
WHITCHURCH FOREST OF DEAN
SPEECH HOUSE IRON MINES
LYDNEY

MENDIP LEAD MINES

THE IRON MINES OF THE WEALD

CARNANTON
TIN MINES

Fig. 10

Fert Britannia aurum et argentum et alia metalla, pretium victoriae.
Tacitus, *Agricola*, XII.

mentioned by Caesar[3] and by many ancient writers. Nevertheless, the archaeological evidence of tin mining during the Roman period is not considerable. An ingot of tin with a Roman stamp was found at Carnanton, not far from Newquay. It appears to be certain that the tin trade of

[1] Tacitus, *Agricola*, XII.
[2] R. E. M. Wheeler, *Prehistoric and Roman Wales* (1925), pp. 273–4.
[3] Caesar, *D.B.G.* V, 12.

6

Cornwall was negligible during the early part of the Occupation, but that it revived in the third and fourth centuries.[1] The numerous cakes of copper found in the island of Anglesey indicate that the ores of the Parys Mountain near Amlwch were worked, and similar finds on the adjacent mainland have led to the suggestion that Roman mines existed in Carnarvonshire.[2] There was a Roman copper mine on the Great Orme's Head. Copper mines of supposed Roman origin are found at Llanymynech Hill in Shropshire, on the borders of Montgomeryshire.[3] Spain was undoubtedly of greater importance than Britain as a producer of tin and copper. The abundant British woodlands supplied the Romans with wood and charcoal as fuel, but coal was also used for smelting and heating. The Romans worked the outcrops of coal in several regions and traces of the use of this mineral have been discovered in places at some distance from a coalfield. Coal was probably worked at Benwell near Newcastle, and the numerous remains of coal found on Roman sites in Wiltshire have led to the supposition that it was mined in the Somerset coalfield or in the Forest of Dean. A third-century literary reference to the use of coal ("stony balls") as fuel in the temple of Minerva at Bath confirms this view.[4] British iron was mentioned by both Caesar[5] and Strabo,[6] and it was certainly mined in many parts of the country. The smelting of iron was also carried on in widely scattered districts. Furnaces for iron smelting have been discovered at Tiddington near Stratford-on-Avon,[7] at Wilderspool on the south bank of the Mersey,[8] at Colsterworth in Lincolnshire,[9] and at numerous other places. There is some evidence that ironstone was worked in Northamptonshire, and Northumberland, but there is no doubt that the chief areas of production were, as in medieval England, the Weald, and the Forest of Dean.

The Weald contains few traces of the Roman Occupation, but it was a region of some economic importance, because of the presence of iron ore

[1] *V.C.H. Cornwall*, II, 3 ff.

[2] R. E. M. Wheeler, *Prehistoric and Roman Wales*, pp. 270–2, for a list of the copper cakes.

[3] *V.C.H. Shropshire*, I, 266–7.

[4] M. E. Cunnington, "Mineral Coal in Roman Britain", *Antiquity*, VII, 89–90 (1933), and by the same author in *Wilts Arch. Mag.* XLV (1930), for a list of Roman sites in Wiltshire where coal has been found. W. J. Fieldhouse, T. May and F. C. Wellstood, *A Romano-British Settlement near Tiddington* (1931), p. 15, contains a list of references to the use of mineral coal by the Romans in Britain.

[5] Caesar, *D.B.G.* V, 12.   [6] Strabo, IV, 5, 199.

[7] Fieldhouse, May and Wellstood, *op. cit.*

[8] T. May, *Warrington's Roman Remains* (1904).

[9] *Ant. Journ.* XII, 262–8 (1932).

found in the Hastings beds. It appears certain that the Romans were merely continuing an industry which had already existed in the Weald for many years before the conquest. The workings which have been proved to be Roman were used soon after the invasion. The method of production was the direct or bloomery process, by which a small lump of wrought iron was made in one operation. Wrought iron contains very little carbon and can be made into a tool or weapon immediately after production. The method requires a small furnace, some form of bellows and abundant fuel. The iron produced is impure and the cinder which it contains has to be extracted. As the cinder always contains a high percentage of iron, the method is very wasteful. Cinders were frequently used by the Romans as foundation for roads. Heaps of cinders mark the sites of bloomeries of both Roman and medieval dates, and, of the hundred which have been discovered in the Weald, only nine can be proved to be Roman by the coins and pottery found associated with the remains. It should be remembered that the method of producing cast iron by blast furnaces did not begin until Elizabethan times, but Roman cast iron has been found at Beauport in Sussex, at Warrington, and at Tiddington. The ores used were a clay ironstone which occurs towards the bottom of the Wadhurst Clay and a bed of shelly calcareous ironstone found a few feet above the Ashdown Sand. The necessary fuel was readily available in the Weald, and charcoal was made from the trees of the forest. It has been proved that birch, oak, hazel, maple, ash, and plum, were all used by the Romans in this region for the production of charcoal. Birch grows well on the sands of Ashdown Forest and appears to have been much used in both Roman and medieval times, and is used in Sweden for the same purpose at the present day. Of the nine bloomeries,[1] which coins and pottery prove to be Roman in origin, a group of five has been found reasonably close to one another at Footlands, Chitcombe, Oaklands, Beauport and Peppering-Eye (Fig. 9). They are all situated within a few miles from the coast and are sufficiently large to have produced more iron than was required locally. The sites are all not far from the great tidal estuaries which existed on the south coast in Roman times, and it is quite possible that the ore was sent to Lympne and exported from that harbour. A Roman road has been proved to run westwards from Lympne to within a few miles of this group of bloomeries. Three of the other bloomeries are at Ridge Hill, Walesbeech, and Oldlands, situated further west near the Ashdown Forest, between East Grinstead and Crowborough. Ridge Hill, the most northerly

[1] Three additional sites are listed in *V.C.H. Sussex*, III, 30–2 (1935), as Roman iron mines.

site, is only 28 miles from London Bridge and the iron could have been taken to the port of London by a road which ran across the Weald (Fig. 9). The pottery from this site has been dated between A.D. 100 and 300, while the isolated bloomery at Bardown near Wadhurst contains pottery fragments of the second half of the second century A.D. It is curious that no Roman buildings have been found at the workings, and Mr Ernest Straker has suggested that the industry "was a seasonal one, carried on in the drier parts of the year only, with temporary shelters".[1]

A considerable quantity of iron was produced by the Romans in the Forest of Dean and the surrounding region. The exact position of the Roman mines is uncertain, as it is impossible to distinguish them from medieval and later workings. A mine that has been described as "the first British iron-mine which can, on conclusive evidence, be assigned to the Roman period"[2] was recently discovered at Lydney, which appears to have been a place of industrial activity in the middle of the third century. Large quantities of scoriae have been found associated with Roman remains at Whitchurch, at Peterstow,[3] at Speech House[4] and in other parts of the Forest. ARICONIVM, 3 miles east of Ross-on-Wye, was an important centre of iron-smelting and the industry seems to have flourished at this place in the third and fourth centuries.[5] Charcoal made of oak, birch, elder, willow, and hazel, and mineral coal probably from the Forest of Dean coalfield were used for smelting iron at ARICONIVM.[6]

Lead was the most important of the mineral products of Roman Britain. The working of the Mendip mines began very shortly after the invasion; two pigs of Roman lead have been dated to the year 49 by their inscriptions. The export of lead must have soon formed a considerable item in British trade, as Pliny, writing before A.D. 77, refers to a law which limited the production of lead in Britain.[7] This quota system may have been imposed at the request of the proprietors of the lead mines in Spain, after they had begun to feel the effects of British competition. Sixty

[1] Ernest Straker, *Wealden Iron* (1931), p. 10.

[2] R. E. M. Wheeler and T. V. Wheeler, *Report on Excavations in Lydney Park, Gloucestershire* (1932), p. 21.

[3] H. G. Nicholls, *Iron Making in the Olden Times: as instanced in the Forest of Dean* (1866); George Wyrall, "Observations on the Iron Cinders found in the Forest of Dean", *Trans. Bristol and Glos. Arch. Soc.* II, 216–34 (1877–8).

[4] Tacitus, *Agricola* (ed. Furneaux and Anderson, 1922), p. 181, in note by F. Haverfield.

[5] G. H. Jack, *Excavations on the site of Ariconium* (1923).

[6] G. H. Jack, *op. cit.* p. 31.

[7] Pliny, *Nat. Hist.* XXXIV, 17, 164.

pigs of British lead, each of which can be dated, and at least four others have been found in different parts of the country.[1] These pigs had evidently been lost, or stolen and afterwards forgotten. Several have been found within a short distance of a main Roman road and were probably lost when a packhorse had strayed. The number of these pigs suggests that the trade was a large one. The Romans were certainly prodigal in their use of lead; the floors of the baths at Bath are covered with lead, 40 lb. to the foot, while a Roman coffin made of lead weighs 400 lb. The distribution of the pigs on a map (Fig. 11) shows not only the position of the five chief mining areas, but also the routes by which the lead was presumably conveyed to the ports.

1. The Mendips.[2] The Mendip area was the region where lead was first worked and where the industry apparently continued longest. The centre of production was near a farm called Charterhouse, south of Blackdown. Eleven inscribed pigs have been found in or near the Mendips, if one at Bath and two at Bristol are included. A pig found near Bossington (Hants), not far from the Roman road between Salisbury and Winchester, had come from the Mendips.[3] Two pigs found at CLAVSENTVM, near Southampton,[4] suggest that Mendip lead was exported from this harbour, and the discovery of another pig at St Valéry-sur-Somme shows the probable line of transport across the Channel.[5]

2. Flintshire.[6] The Halkyn Mountains of Flintshire are rich in lead, and the ores were worked by the Romans in A.D. 74, if not before. Half a dozen lead-smelting furnaces were discovered in 1923 at Pentre near Flint,[7] which was probably the exporting centre. No pigs have been found in the producing region, but three have been found at or near Chester. Twenty pigs were reported by Camden as having been discovered on the Cheshire side of the Mersey near Runcorn.[8] These pigs may have been lost in transport to Warrington. It is quite possible that they were discovered in

---

[1] In addition to these sixty-four pigs, several others, that were uninscribed, have been found and described as Roman. Three such pigs were found at Saham Toney in Norfolk in about 1819 and "were sold to the village plumber", but there is no satisfactory evidence that they were Roman and they are not plotted on the map. *Archaeologia*, XXIII, 369 (1831); *Arch. Journ.* XVI, 37 (1859).

[2] *V.C.H. Somerset*, I, 334–44; J. W. Gough, *The Mines of Mendip* (1930).

[3] *V.C.H. Hants*, I, 323–4.

[4] *Proc. Soc. Ant. Lond.*, 2nd Ser. XXXI, 36–9 (1919).

[5] F. Haverfield, *Arch. Journ.* XLVII, 258 (1890).

[6] M. V. Taylor, "Roman Flintshire", *Flint Hist. Soc. Journ.* IX, 1–39 (1922).

[7] *J.R.S.* XIV, 251 (1924).

[8] Camden, *Britannia* (1590), p. 488.

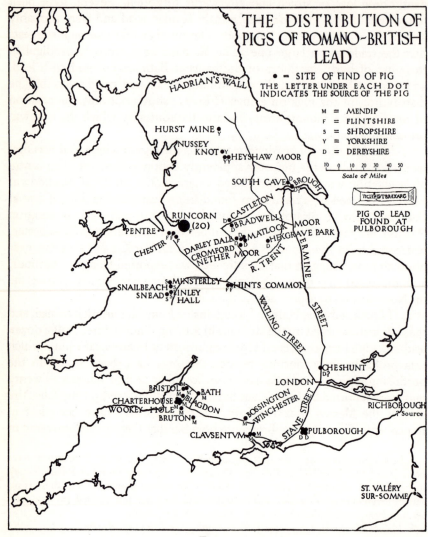

Fig. 11

The roads marked are those mentioned in the text.

medieval times in Flintshire and shipwrecked in a vessel of that period. Two pigs found on Hints Common in Staffordshire were obviously being transported to London along the Watling Street.[1]

3. Shropshire.[2] The lead-mining region of Shropshire was situated in the neighbourhood of the villages of Shelve and Minsterley. Four pigs have been found near these places and all bear the inscription of the Emperor Hadrian (A.D. 117–38). It appears from other archaeological evidence that activity in this area did not last very long.

4. Yorkshire.[3] Two mining areas were situated in Yorkshire, one near Greenhow Hill, between Pateley Bridge and Grassington, and the other at Hurst mines in upper Swaledale, about 10 miles west of Richmond. One pig has been found in each of the mining regions, while two pigs found on Hayshaw Moor above Nidderdale and dated as early as A.D. 81 were being transported to Aldborough.

5. Derbyshire.[4] Fourteen pigs, including four which cannot be dated, have been attributed to the mines of Derbyshire. Five pigs have been found near Matlock, and two near Castleton. The whole area between Wirksworth on the south and Castleton on the north shows much evidence of ancient mining. One Derbyshire pig found in Hexgrave Park, 8 miles east of Mansfield, was probably being taken to the Trent for transport by river. One pig found at South Cave and an inscribed fragment found at Brough[5] make it possible to conjecture that Derbyshire lead was sent by sea from the Humber estuary,[6] while four Derbyshire pigs found at Pulborough in Sussex indicate another route of transport to the Continent. There remain two pigs which contain nothing to prove from which mining region they came. A pig possibly from Derbyshire, found in Theobald's Park, near Cheshunt, shows that lead was sent along the Ermine Street to London.[7] A partially used pig, with Nerva's inscription, was found in 1922 at Richborough. It has been suggested that it had been "buried, possibly after having been stolen, and the hider was never in a position to recover it".[8]

---

[1] V.C.H. Staffs, I, 190.　　　　　　[2] V.C.H. Shropshire, I, 263–5.
[3] A. Raistrick, "Notes on Lead Mining and Smelting in West Yorks.", Trans. Newcomen Soc. VII, 81–96 (1927); and "A Pig of Lead in Craven Museum", Yorks. Arch. Journ. XXX, 181–2 (1930).
[4] V.C.H. Derby, I, 227–33.
[5] It is possible that this fragment came from the Yorkshire mines. O. Davies, Roman Mines in Europe (1935), p. 164.
[6] F. Haverfield, Arch. Journ. XLVII, 257–8 (1890).
[7] V.C.H. Herts, IV, 153.
[8] J. P. Bushe-Fox, Report on Roman Fort at Richborough, I (1926), p. 42.

The Romans extracted silver from lead and the export of this mineral was probably considerable. Twelve of the lead pigs bear the letters EX ARG, or EX ARGENT which probably means EX ARGENTARIIS, "from the silver works". These pigs all come from either Mendip or Derbyshire, but metallurgical analysis of pigs from other areas shows a low silver-content in them also.[1] It is obvious that the Romans successfully de-silverised their lead, and silver refineries have been discovered at Silchester, Tiddington, and elsewhere.

A map of the distribution of Roman potteries and tile kilns (Fig. 12) shows that a pottery industry existed in over a hundred widely scattered places, chiefly in the civil zone. Important potteries were situated in the Nene valley and the New Forest. The Nene valley potteries extended over an area of 20 sq. miles and used boulder-clay to produce the famous Castor ware.[2] An unusual type of pottery was made in the New Forest kilns, which were most active from about A.D. 250 to 350.[3] Only one kiln where Samian ware was produced in Britain has so far been found, namely at Colchester.[4] An interesting series of nine potteries and tile kilns has been found at Holt in Denbighshire, not far from Chester. The site was overlain with boulder-clay, but the potters used the alluvial deposits of the Dee. Holt has been described as "the works-depôt of the twentieth Legion".[5] The Romano-British pottery industry as a whole must have employed a large amount of labour.

POPULATION

The geographical distribution of the population as shown by the Ordnance Survey *Map of Roman Britain* has now been discussed. The map also provides the means by which an estimate of the total number of the population can be made. In an illuminating article Mr R. G. Collingwood explains how he reached his conclusion that the population of Roman

[1] *Ant. Journ.* III, 265 (1923), note by Professor Henry Louis; G. C. Whittick, "Roman Mining in Britain", *Trans. Newcomen Soc.* XII, 73–4 (1931–2) and table on p. 79.

[2] *V.C.H. Northants*, I, 206–13; *V.C.H. Hunts*, I, 225–48.

[3] Heywood Sumner, *Excavations in New Forest Roman Pottery Sites* (1927).

[4] *J.R.S.* XXIV, 210–11 (1934).

[5] W. F. Grimes, Report on "Holt, Denbighshire", *Y Cymmrodor*, XLI (1930), contains a list of Romano-British pottery sites and map. Fig. 12 is based on Mr Grimes' map, and on further information which he has kindly supplied.

Britain was approximately half a million.[1] He arrived at this result by several lines of argument, but the most convincing was based on a study of the map and is briefly summarised in the table on p. 78. This

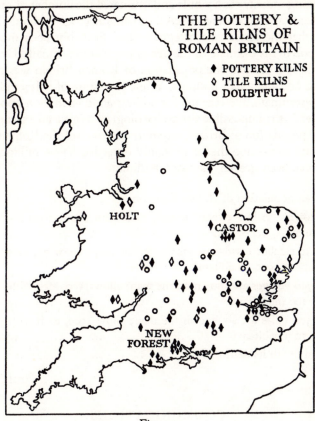

THE POTTERY &
TILE KILNS OF
ROMAN BRITAIN

♦ POTTERY KILNS
◊ TILE KILNS
o DOUBTFUL

HOLT

CASTOR

NEW
FOREST

Fig. 12

After W. F. Grimes, "Holt, Denbighshire", *Y Cymmrodor*, XLI (1930), with additional information received from Mr Grimes.

paper was followed by an interesting controversy. Mr H. J. Randall considered that this estimate was too small, on the ground that Collingwood had underestimated the numbers of an agricultural population which was able to produce surplus corn for export.[2] This argument was supported

[1] R. G. Collingwood, "Town and Country in Roman Britain", *Antiquity*, III, 261–76 (1929).
[2] H. J. Randall, "Population and Agriculture in Roman Britain", *Antiquity*, IV, 80–90 (1930).

by Dr R. E. M. Wheeler, who also stated that the population which lived on the fringes of the province in the military zone of Wales and northern England had been neglected by Collingwood.[1] He added a further 30,000 for the mining industry as a whole and estimated the number of middlemen and merchants at 100,000 and the civil servants at 5000. It should be noted, however, that the latter two categories had probably been included by Collingwood in his estimate of the urban population. Finally Dr Wheeler categorically asserts that the population of Roman Britain was not half a million but one and a half millions. While the figure of half a million may be an underestimate, the reasons for adding another million appear to be very slender.[2] It is impossible for archaeologists to give an exact enumeration of the population of Britain, a remote province of the Roman Empire, but it is safe to assume that it probably exceeded half a million and can scarcely have been greater than one million.

1. Estimated population of Roman London 25,000
2. Large towns. Allow twenty towns, each with about 5000 inhabitants 100,000
3. About 50 smaller towns are shown on the map; allow 75, each with about 1000 inhabitants 75,000
4. About 700 villages are shown on the map; allow 1500, each with about 100 inhabitants 150,000
5. About 500 villas are shown on the map; allow 1000, each with about 50 inhabitants 50,000
6. The army with its dependants 100,000

Total 500,000

COMMUNICATIONS

The roads of pre-Roman Britain consisted of trackways, most of which followed the lines of the south-eastern scarps. Thus, the Icknield Way ran from the Wash, just below the chalk escarpment of the Chilterns and the

[1] R. E. M. Wheeler, "Mr Collingwood and Mr Randall", *Antiquity*, IV, 91–5 (1930).
[2] It is interesting to notice that the South Island of New Zealand (58,092 sq. miles) is about the same size as Roman Britain (England and Wales, 58,340 sq. miles), and its estimated population in 1933, including Maoris, was 544,030. South Island, New Zealand, like Roman Britain in the past, is an exporter of agricultural and mineral produce. Such a parallel cannot be exact, but if the farms and towns of South Island, New Zealand, were now destroyed, it would be easy for an archaeologist to overestimate the former population of this remote Dominion of the British Empire.

Berkshire Downs, to Salisbury Plain. There were many similar tracks which led to the same region. The Romans entered Britain from the south-east and pushed their way towards the north and north-west across the natural grain of the country and across the British trackways. The existing ways were no doubt used by the Romans and in some cases they may have been straightened, but it was essential to construct an entirely new system of communications. The influence of woodlands on movement within Britain has been exaggerated by historians, and it is clear that the vegetation did not create any difficulty for the Roman road builder. The Romans constructed roads across the clay vales within the early years of the Occupation, and eventually at least four roads ran across the Weald. It is impossible to give a detailed description of the Roman roads and a few general observations must suffice. As Haverfield often pointed out, the Roman road system as a whole should be compared with the modern railways rather than with the roads of medieval England. This parallel is a just one, provided that it is not pressed too closely. The Watling Street ran from Canterbury to London, from which many roads radiated and three of these were of especial importance. The Watling Street[1] proper followed a route which is approximately the same as that now taken by the L.M.S. Railway (formerly L.N.W.R.); it went by way of St Albans, Towcester, High Cross to Wroxeter. From this place a road branched off to Chester and so to the military districts of north Wales and northern England. A Great West road ran from London to Silchester, Cirencester, Gloucester and south Wales, while branches left Silchester for Winchester and for Dorchester and Exeter. The Ermine Street corresponds to the L.N.E. Railway (G.N.R.) system and ran by Braughing, Castor, Lincoln and Doncaster to York, and so to the military region of the north. This road was compelled to make a wide circuit by way of Littleborough, Doncaster, Castleford and Tadcaster in order to avoid the marshes of Thorne and Hatfield.[2] The course of the road is a striking example of the fact that marshes provided a much greater physical obstacle to movement than woodland. From London, another road ran north-eastward to Chelmsford, Colchester and Caistor-by-Norwich along more or less the same route as that taken by the L.N.E. Railway (G.E.R.); while Stane

[1] The names of the roads, Watling Street, Ermine Street, and the rest, are post-Roman. The Roman names of the roads are unknown. See vols. of English Place-Name Society (ed. A. Mawer and F. M. Stenton), III, *Beds and Hunts* (1926), pp. 1–7; IV, *Worcs* (1927), pp. 2–4; X, *Northants* (1933), pp. 4–6.

[2] Colonel E. Kitson Clark suggested that the road was built to cross the Trent as high up as Littleborough, to avoid "the ravages" of the aeger. *Arch. Journ.* LXXVIII, 395 (1921).

Street provided a direct route to Chichester and the Hampshire ports as an alternative to the more circuitous route by way of Silchester and Winchester.

The Romans established a number of ferries. The mouth of the Wash was probably not so wide in the Roman period as it is to-day. It is believed that a Roman ferry was carried across the Wash, then only 10 miles wide instead of 14 miles as at present.[1] The terminals of the Roman roads, in Norfolk at Holme-next-the-Sea and in Lincolnshire near Skegness, face one another across the Wash and seem to indicate the existence of a ferry. Other ferries were maintained by the Romans across the Severn from Sea Mills, and across the Humber from Winteringham to Brough. The latter provided an alternative route from Lincoln to York, instead of that by way of Doncaster and Castleford, while the ferry over the Severn provided a line to southern Wales from Bath shorter than the other route by way of Gloucester. Bridges were constructed of wood or stone piers, and paved fords were made to assist in the passage of streams.[2] The centrality of London in the network of Roman communications must not blind the reader to the fact that several important cross-country routes existed in Roman Britain, and these cannot be equated with modern railways. The Fosse Way is one great Roman road which does not lead to London. Its course of 222 miles is from Exeter to Lincoln by way of Bath, Cirencester, and Leicester, and it was probably designed in the first instance to serve as a frontier.[3] This road follows the natural grain of the country and for the most part runs along the edge of a belt of woodland and coincides with a well-drained Jurassic sandy soil.[4] Another cross-country route was the Rycknield Way which left the Fosse at Bourton-on-the-Water and took a northward course over the site of modern Birmingham, as far as Wall, and then north-eastward to Littlechester and so to the Ermine Street near Doncaster. An east-to-west road joined Colchester, Braughing, and St Albans; while Akeman Street continued the general direction of this road as far as Alchester and Cirencester.

The road system as a whole was planned primarily for strategic purposes. Roads were made to take the shortest route across the lowlands to the military and mineral districts of the northern and western highlands. The military regions themselves were penetrated by roads. A network of

[1] C. W. Phillips, "The Roman Ferry across the Wash", *Antiquity*, VI, 342–8 (1932).
[2] An illustration of a paved ford at Benenden is given in *V.C.H. Kent*, III, 139.
[3] See above, pp. 34–5. The direct distance between Exeter and Lincoln is 205 miles.
[4] Dr Wheeler has suggested that the Fosse Way was "like the Icknield Way, a route of pre-Roman origin", *Antiquity*, III, 127 (1929).

this character existed in Wales and one great road skirted the eastern edge of the Welsh hills and linked the two fortresses of Chester and Caerleon. A Roman road ran on either side of the Pennines, one from Littlechester to Carlisle by way of Manchester and Ribchester, and the other from Lincoln to Tadcaster, Aldborough, Catterick, and Corbridge. The Pennines themselves were crossed by a number of east to west roads, which followed the natural passes. One road ran through the Tyne Gap from Newcastle to Carlisle, while other routes joined the valleys of the Tees and Eden, Aire and Ribble, Calder and Mersey. The total length of the roads shown on the Ordnance Survey *Map of Roman Britain* exceeds 5000 miles; this figure can be compared with the 20,408 miles of railways in Great Britain of 1931, or more aptly with the 1782 miles of railways in South Island, New Zealand.

The Romans must have used the rivers for purposes of transport, and the Car Dyke, which skirted the western edge of the Fens, and the Foss Dyke may have been used as canals. These two waterways combined with the rivers would have made direct water communication possible from York to Cambridge. It has even been suggested that they were the means of transporting fenland corn to the military districts of the north, but this is pure conjecture, and they were both possibly built to serve as catchwater drains rather than as canals.[1]

CONCLUSION

Historical geographers often contrast the remote and isolated position of Britain at the verge of the habitable world of classical and medieval times, with the central situation in the whole land hemisphere that was given to it by the Great Age of Discovery.[2] This contrast is apt to mask the important fact that south-eastern England is very definitely a part of Europe and has never been entirely isolated from the main current of European affairs. The period of the Roman occupation is the most striking historical example of Britain's essential connection with the mainland. The great Empire of the Mediterranean may have extended beyond its proper limits by including the distant island within its territory. Britain was not as highly developed as Gaul and Spain, but nevertheless Rome left many marks on the island province, and there are some which time has

[1] Cyril Fox, *The Archaeology of the Cambridge Region* (1923), contains a map of the course of the Car Dyke on p. 163. See also *Arch. Journ.* xc, 117–223 (1934).

[2] See H. J. Mackinder, *Britain and the British Seas* (2nd ed. 1907), chap. I, "The Position of Britain".

not yet erased. The permanent legacy of the Romans to the geography of Britain can be seen on "that most wonderful of all palimpsests, the map of England".[1] In the first place the system of communications which was established by the Romans has lasted up to the present day. The most important Roman routes are now busy with motor traffic. The predominance of that part of England which lies south and east of the rivers Trent and Severn in the economy of the island lasted from the Roman period until the industrial revolution and is once more re-establishing itself in the present century. The Romans discovered many of the sites which were most suited geographically as nuclei for population, and the majority of their towns have modern successors. Three great towns, Manchester, Cardiff and Newcastle, are built on the sites of Roman forts, and, although these cities owe their size to modern industrialism, the Romans were the first to realise the strategic value of their situations. Perhaps the most significant achievement of the Romans was the establishment of London, whose overwhelming pre-eminence among English towns has never been eclipsed. Although the Roman towns may have been destroyed or abandoned, and the roads may have fallen into disuse, it was inevitable that they should once more come into their own. It has been fortunate for England that the foundations of so much of her geography were laid by the practical genius of the Romans.

BIBLIOGRAPHICAL NOTE

The material available for a study of the Geography of Roman Britain is overwhelming in amount and widely scattered. The object of the following bibliography is to provide a select list of books and papers that may be useful to a geographer. Some of the works already quoted in the footnotes are not included.

MAP

The *Map of Roman Britain* (2nd ed. 1928), published by the Ordnance Survey, is indispensable; it is by far the most accurate and complete map of Roman Britain. It is virtually impossible to plot the exact extent of the woodlands of Roman Britain, but the area that they probably covered is shown on the second edition of this map. The distribution of dense and open woodland has been reconstructed on a geological basis. The reader is warned against the tract *De Situ Britanniae*, which is supposed to have been written by one Richard of Cirencester. This book is a forgery, and was in fact compiled by Charles Bertram of Copenhagen in the middle of the eighteenth century; it contains many fictitious Roman place-names, whose only authority is Bertram's imagination, but which still, unfortunately, find their way into local guide-books. (For Bertram see H. J. Randall, "Splendide Mendax", *Antiquity*, VII (1933), 49–60.)

[1] *The Collected Papers of F. W. Maitland* (1911), ed. H. A. L. Fisher, II, 87.

Bertram's work tainted many of the earlier maps of Roman Britain, but has, of course, been entirely eliminated from the Ordnance Survey production.

The main sources from which Romano-British place-names have been obtained are the following: (1) Ptolemy's *Geography*, which was compiled in about A.D. 150 and consists of a gazetteer of names of capes, rivers, towns, and tribes. The latitudes and longitudes attached by Ptolemy make it possible to draw a map of Britain, which shows the geographical conceptions of the period. See H. Bradley, "Remarks on Ptolemy's Geography of the British Isles", *Archaeologia*, XLVIII (1885), 379–96, which contains an excellent reconstruction of Ptolemy's map of Britain. (2) The *Itinerarium Antonini* is a Roman road-book, which was compiled at the beginning of the third century and contains fifteen routes in the section on Britain and no less than 111 names of British places. (3) The Peutinger Table is a thirteenth-century map, which is believed to be a copy of a map made in the middle of the fourth century. The western section of the map is mutilated, but a portion containing south-eastern England survives and includes several roads and names. See O. G. S. Crawford, "A note on the Peutinger Table and the fifth and ninth Iters", *J.R.S.* XIV (1924), 137–41, which contains an illustration of the British section of the map. (4) A document called the *Notitia Dignitatum*, which was prepared in Rome and Constantinople in the fifth century, resembles a modern Army or Civil Service list, and contains names of officials and some topographical information. The date of the British section of the *Notitia* is a matter of dispute, but it is probably a fourth-century text inserted in the fifth-century handbook, represented by the *Notitia* as a whole. (5) An anonymous Geographer who worked at Ravenna in the seventh century, compiled, probably from some earlier authority, a list of names of places and rivers in the Roman Empire. The Ravenna Geographer includes in his lists the names of the rivers of Britain, the names of the smaller surrounding islands, and also a catalogue of the forts of both the northern Walls. In the case of ten place-names, he adds the name of the tribe to the name of the British town, and thus indicates what was probably the chief settlement of each tribe. (6) It should be remembered that the early germs of place-names show that the names of many Romano-British sites survived the decay of the Province. For example, Bede's LYCCIDFELTH (Lichfield) contains the Romano-British name LETOCETVM (Wall).

GENERAL

The study of historical geography owes an immense debt to that great authority on Roman Britain, the late Professor F. J. Haverfield, who fully realised the importance of geography in archaeological research. He was keenly interested in the revival of geography at Oxford under Mr (now Sir) H. J. Mackinder and was a member of the committee of the Oxford School of Geography. Haverfield's opinion that "the character and history of Roman Britain, as of many other Roman provinces, were predominantly determined by the facts of its geography" (*Camb. Med. Hist.* I, 367) permeates his writings and those of his pupils. The following are among his more important works: (1) *The Romanization of Roman Britain* (4th ed. 1923). (2) F. Haverfield and Sir George Macdonald, *The Roman Occupation of Britain* (1924). See chap. II on "the Geography of Britain". (3) *Roman Britain in 1913* (1914). (4) *Roman Britain in 1914* (1915). (5) *Ancient Town-Planning* (1913). The results of more recent research will be found in (1) Sir George Macdonald, *Roman Britain, 1914–1928* (1931).

(2) T. D. Kendrick and C. F. C. Hawkes, *Archaeology in England and Wales, 1914–1931* (1932), pp. 209–302 on "Roman Britain". (3) Annual summaries compiled by R. G. Collingwood and M. V. Taylor in the *Journal of Roman Studies* from 1921 onwards. The following will be found useful: (1) R. G. Collingwood, *Roman Britain* (1932), which is the best short introduction to the subject. (2) M. Rostovtzeff, *The Social and Economic History of the Roman Empire* (1926), pp. 212–16. (3) R. G. Collingwood, *The Archaeology of Roman Britain* (1930). (4) A. H. Lyell, *A Bibliographical List of Romano-British Architectural Remains in Great Britain* (1912).

TOWNS

The following list includes references to the best available map for each of the Roman towns, which are arranged in the same grouping as in the text.

EASTERN BRITAIN

(1) COLCHESTER (108 acres). R. E. M. Wheeler and P. G. Laver, "Roman Colchester", *.JR.S.* IX (1919), 139–69 with map of district, street plan and plan of finds; *R.C.H.M. Essex*, III (1922), xxv–xxvii and 20–32, map of earthworks, p. 72; *Antiquity*, II (1928), p. 354 for street plan; *Antiquity*, IV (1930), p. 363 for plan in relation to Celtic town.

(2) ST ALBANS (200 acres). *V.C.H. Herts*, IV, 125–39; *R.C.H.M. Herts* (1910), pp. 190–1 with plan of outline of wall. These are now superseded by Dr Wheeler's recent excavations; *J.R.S.* XXI (1931), 227–32 with plan; *J.R.S.* XXII (1932), 207–10 with plan of earlier and later Roman town; *J.R.S.* XXIII (1933), 198–202; *J.R.S.* XXIV (1934), 207–9 with plate XV showing street plan; R. E. M. Wheeler, "A Prehistoric Metropolis: the first Verulamium", *Antiquity*, VI (1932), 133–47, and "Belgic Cities of Britain", *Antiquity*, VII (1933), 21–35 with map of town in relation to former woodland.

(3) CAISTOR-BY-NORWICH (35 acres). *V.C.H. Norfolk*, I, 288–93; *Antiquity*, III (1929), p. 182 for air photographs of site and plan; D. Atkinson, *Norfolk and Norwich Arch. Soc.* XXIV (1929), 93–139.

(4) CAMBRIDGE (28 acres). Cyril Fox, *Archaeology of the Cambridge Region* (1923), pp. 174–5 and plan on p. 246.

(5) GREAT CHESTERFORD (35 acres). *R.C.H.M. Essex*, I, xxiii and 113; Cyril Fox, *Archaeology of the Cambridge Region*, pp, 173–4.

(6) GODMANCHESTER. *V.C.H. Hunts*, I, 252–4 with plan.

(7) LINCOLN (41 acres). G. E. Fox, "Recent Discoveries of Roman Remains in Lincoln", *Archaeologia*, LIII (1892), 233–8 with plan; F. Haverfield, *Ancient Town-Planning* (1913), p. 117 for plan; A. Smith, *Catalogue of Roman inscribed stones found in the City of Lincoln* (1929); C. W. Phillips, "Present State of Archaeology in Lincolnshire", *Arch. Journal*, XCI (1934), 124–6 with plan.

(8) CASTOR. E. Trollope, "Durobrivae", *Arch. Journal*, XXX (1873), 127–40; *V.C.H. Northants*, I, 166–78 with plan.

(9) "THE CASTLES" (44 acres). *V.C.H. Hunts*, I, 228–48.

KENT

(1) CANTERBURY (40–50 acres). *V.C.H. Kent*, III, 61–80 with plan.
(2) ROCHESTER (23 acres). *V.C.H. Kent*, III, 80–8 with plan.

(3) RICHBOROUGH. *V.C.H. Kent*, III, 24–41 with plan; J. P. Bushe-Fox, *Report on Excavation at Richborough* (Repts of Research Com. of Soc. of Ant. of London), VI (1926), VII (1928) and X (1932). Most recent plan at end of vol. x.

(4) RECULVER, DOVER, LYMPNE. *V.C.H. Kent*, III, 19–24, 42–59 with plans; E. G. Amos and R. E. M. Wheeler, "The Saxon-Shore Fortress at Dover", *Arch. Journal,* LXXXVI (1929), 47–58 with plan.

SOUTHERN BRITAIN

(1) SILCHESTER (104 acres). *V.C.H. Hants*, I, 350–72 with plan; the reports of the excavations at Silchester appeared in *Archaeologia*, LII–LXII (1890–1910), LXI (1909) containing the best plan of completed excavations, opposite p. 486; *The Short Guide to the Silchester Collection at Reading Museum* (7th ed. 1927) is useful.

(2) WINCHESTER (138 acres). *V.C.H. Hants*, I, 285–93 with plan; C. F. C. Hawkes, J. N. L. Myres and C. G. Stevens, *St Catharine's Hill, Winchester* (1930), pp. 169–88 for "St Catharine's Hill and the origin of Winchester"; O. G. S. Crawford, *Air Survey and Archaeology* (1928) for map of Winchester at end.

(3) CHICHESTER (103 acres). E. Curwen, *Sussex Arch. Coll.* LXX (1929), 76–7 and LXXII (1931), 198–205; Ian Hannah, "The Walls of Chichester", *Sussex Arch. Coll.* LXXV (1934), 107–29 with plan; T. H. Hughes and E. A. Lamborn, *Towns and Town-planning* (1923), for plan of REGNVM on p. 37; *V.C.H. Sussex*, III, 9–19 with plan.

(4) DORCHESTER (86 acres). H. J. Moule, "Notes on the Walls and Gates of Durnovaria", *Proc. Dorset Nat. Hist. and Ant. Field Club*, XIV (1893), 44–54; for map see O.S. 6 in. Dorset, sheet XL. S.E.

(5) EXETER (91 acres). Thomas Kerslake, "The Celt and the Teuton in Exeter", *Arch. Journal*, XXX (1873), contains a plan on p. 22.

(6) CLAUSENTVM. *V.C.H. Hants*, I, 330–9 with plan.

SEVERN AND COTSWOLD REGION

(1) GLOUCESTER (46 acres). St Clair Baddeley, "Some Evidences of the Defences of Roman Gloucester", *Arch. Journal*, LXXVIII (1921), 264–70 with plans; F. T. Howard, "Gloucester", *Geog. Teacher*, XII (1923), 110–23; *J.R.S.* XXII (1932), p. 214 for plan.

(2) CIRENCESTER (240 acres). F. Haverfield, "Roman Cirencester", *Archaeologia*, LXIX (1917–18), 161–209 with plans.

(3) BATH (23 acres). *V.C.H. Somerset*, I, 219–60 with plans; A. J. Taylor, *The Roman Baths of Bath* (10th ed. 1933), with introd. pp. 8–12 by F. Haverfield.

(4) ALCHESTER (26 acres). *V.C.H. Oxon*, II, 320 with map; *Ant. Journal*, VII (1927), 155–84 with plan; *ibid.* IX (1929), 105–36; *ibid.* XII (1932), 35–67.

(5) CAERLEON (Legionary Fortress, 50 acres). R. E. M. Wheeler and T. V. Wheeler, "The Roman Amphitheatre at Caerleon, Monmouthshire", *Archaeologia*, LXXVIII (1928), 111–218 with plan of Caerleon and surrounding district; Excavation Reports by V. E. Nash Williams and C. Hawkes in *Arch. Camb.* 1929–32 with plans; *J.R.S.* XXIV (1934), plate 6 is a plan of fortress as revealed by excavations of 1850, 1908 and 1926–33.

(6) CAERWENT (144 acres). Reports of the excavations of 1899–1913 in *Archaeologia*, LVII (1901), 295–316; LVIII, 119–52; LIX, 87–124; LX, 111–30, 451–64; LXI, 562–82; LXII, 1–20; LXIV (1913), 437–52. The results of the excavations of 1923–5 and the most complete plan are to be found in *Archaeologia*, LXXX (1930), 229–88.

(7) ARICONIVM. G. H. Jack, *Excavations on the site of Ariconium* (1923); *R.C.H.M. Hereford*, II, 208–10.

(8) KENCHESTER (22 acres). *V.C.H. Hereford*, I, 175–83; *R.C.H.M. Hereford*, II, 93–5 with plan.

THE MIDLANDS

(1) LEICESTER (105 acres). F. Haverfield, "Roman Leicester", *Arch. Journal*, LXXV (1918), 1–46 with plans.

(2) WROXETER (170 acres). *V.C.H. Shropshire*, I, 220–56 with plan; J. P. Bushe-Fox, *Excavations on site of Wroxeter* (Repts of Research Com. of Soc. of Ant. of London), I (1913), II (1914), IV (1916). Plan in Report IV (plate XXXIII); a summary of Professor Atkinson's excavations of 1924–7 is in G. Macdonald, *Roman Britain, 1914–1928* (1931), pp. 89–97.

(3) CHESTER (Legionary Fortress, 56 acres). F. T. Howard, "The Geographical Position of Chester", *Geog. Teacher*, X (1919), 94–100 with maps; for best plan, P. H. Lawson, "Schedule of the Roman Remains at Chester with Maps and Plans", *Chester Arch. Journal*, XXVII (1928), 162–89.

(4) TOWCESTER (35 acres). *V.C.H. Northants*, I, 184–6 with plan.

(5) WALL (30 acres). *V.C.H. Staffs*, I, 193–6 with map; *J.R.S.* XI (1921), 207; W. F. Blay, *Letocetum* (1925).

(6) IRCHESTER (20 acres). *V.C.H. Northants*, 178–84 with plan.

NORTHERN BRITAIN

(1) YORK (Legionary Fortress, 50–53 acres and *Colonia* area unknown). See *J.R.S.* XV (1925), 176–94; *ibid.* XVIII (1928), 61–99; G. Home, *Roman York* (1924) contains map; F. and H. W. Elgee, *Arch. of Yorkshire* (1933) contains plan on p. 161; The British Association published in 1932 *A Scientific Survey of York*, which contains A. V. Williamson, "York in its Regional Setting", pp. 3–8, and A. Raine, "Roman Excavations at York", pp. 52–6.

(2) ALDBOROUGH (60 acres). H. Eckeroyd Smith, *Reliquiae Isurianae* (1852); F. and H. W. Elgee, *Arch. of Yorks* (1933) contains plan on p. 164; R. G. Collingwood, *Excursion Handbook Q* for British Association Meeting at Leeds, 1927, contains "The Roman Town at Aldborough" with plan.

(3) CARLISLE (50–56 acres). R. C. Shaw, "Romano-British Carlisle", *Cumb. and West. Ant. and Arch. Soc. Trans.* XXIV (1924), New Series, 95–109 with plan; good plan in same periodical, Old Series, XII (1893), 334.

(4) CORBRIDGE (40 acres). In *A History of Northumberland*, vol. X, see F. Haverfield on CORSTOPITVM, pp. 474–522 with plan.

LONDON

LONDON (330 acres, with Southwark (15 acres)). F. Haverfield. "Roman London", *J.R.S.* I (1911), 141–72; W. Page, *London: its Origin and Early Development* (1923), both now superseded by the following: *R.C.H.M. London*, vol. III, *Roman London* (1928). See especially Section 3 by R. E. M. Wheeler on "The Geol. and Geog. Setting" with plans C and D of London in relation to relief, geology and natural vegetation, street plan on p. 68 and plan of structural remains at end; *London in Roman Times*, London Museum Cat. No. 3 (1930) with geographical introduction and maps.

TRADE AND INDUSTRY

(1) M. P. Charlesworth, *Trade-Routes and Commerce of the Roman Empire* (1924), chap. XII, "Britain". (2) L. C. West, *Roman Britain: The Objects of Trade* (1931) with very full references. For mining see (1) F. Haverfield, "Minerals in Early Roman Britain", pp. 173–82 in Tacitus, *Agricola*, ed. by H. Furneaux and J. G. C. Anderson (2nd ed. 1922). (2) W. Gowland, "The Early Metallurgy of Silver and Lead", *Archaeologia*, LVII (1901), 359–422. (3) G. C. Whittick, "Roman Mining in Britain", *Trans. Newcomen Soc.* XII (1931–2), 57–84. (4) O. Davies, *Roman Mines in Europe* (1935), especially chap. V, pp. 140–64 on "British Isles".

ROADS

(1) T. Codrington, *Roman Roads in Britain* (3rd ed. 1918). This book should be used with discrimination. See a review of it by F. Haverfield in *Eng. Hist. Rev.* XXXIV (1919), 245–7. (2) Fresh stretches of Roman road are often discovered and reports of such finds are given in *Antiquity* or *J.R.S.* (3) R. G. Collingwood, *Arch. of Roman Britain* (1930), chap. I.

REGIONAL LITERATURE

The greater part of this is to be found classified by counties. The county divisions are often very unsatisfactory for the purpose of describing Roman Britain. (1) The following volumes of the *Victoria County History* contain sections on Roman Britain, many of which were written by F. Haverfield: *Bedford*, II, 1–15; *Berkshire*, I, 197–227; *Buckingham*, II, 1–19; *Cornwall*, published as a separate part of vol. II; *Derby*, I, 191–263; *Hampshire*, I, 265–372; *Hereford*, I, 167–97; *Hertford*, IV, 119–72; *Huntingdon*, I, 219–69; *Kent*, III, 1–176; *Leicester*, I, 179–219; *London*, I, 1–146; *Norfolk*, I, 279–323; *Northampton*, I, 157–222; *Nottingham*, II, 1–36; *Rutland*, I, 85–93; *Shropshire*, I, 205–78; *Somerset*, I, 207–371; *Stafford*, I, 183–98; *Suffolk*, I, 279–323; *Surrey*, IV, 343–78; *Sussex*, III, 1–70; *Warwick*, I, 223–49; *Worcester*, I, 199–221. (2) The volumes of the *Royal Commission on Historical Monuments* on *Buckingham, Essex, Hereford, Hertford, Huntingdon* and *London* contain useful material. (3) The volumes of *The County Archaeologies* contain chapters on Roman Britain. *Middlesex and London, Kent, Berkshire, Somerset, Surrey, Cornwall* and *Yorkshire* have already appeared in this series. (4) For Wales see R. E. M. Wheeler, *Prehistoric and Roman Wales* (1925), chap. VII, "The Roman Occupation of Wales". (5) Cyril Fox, *The Archaeology of the Cambridge Region* (1923), deals with a large part of eastern Britain. (6) The following regional articles will also be found useful: M. V. Taylor, "Roman Flintshire", *Flint. Hist. Soc. Journal*, IX (1922), 1–39; P. Manning and E. T. Leeds, "An Arch. Survey of Oxfordshire", *Archaeologia*, LXXI (1921), 227–65 with detailed map of Roman Oxfordshire; M. E. Cunnington, "Romano-British Wiltshire", *Wilts. Arch. and Nat. Hist. Soc. Mag.* XLV (1930), 166–216; C. W. Phillips, "The Present State of Archaeology in Lincolnshire", section on "The Romano-British Period", *Arch. Journal*, XCI (1934), 110–35.

Chapter III

THE ANGLO-SAXON SETTLEMENT

S. W. WOOLDRIDGE, D.Sc., F.R.S.[1]

Since J. M. Kemble wrote his *Saxons in England*, the character and circumstances of the Anglo-Saxon settlement have been actively, though intermittently, debated by historians for nearly a century. During the last thirty years, archaeologists and students of place-names have joined the debate, and have contributed a large body of new facts which supplement the scanty literary sources. Even at the risk of being regarded as an intruder in the field, the geographer cannot forgo the task of bringing another technique, and still other facts, into the realm of the discussion.

The influence of the physique of Britain on the settlement of the Saxon peoples has been very largely ignored by historians, save in its broader and more evident aspects. A notable exception was J. R. Green, who supplemented the limited literary sources on which he depended by a deliberate attempt to obtain guidance from the facts of physical geography as known to him. The standpoint and mode of treatment adopted here could hardly be more aptly expressed than in his own words:

The ground itself, where we can read the information it affords, is, whether in the account of the Conquest or in that of the Settlement of Britain, the fullest and most certain of documents. Physical geography has still its part to play in the written record of that human history to which it gives so much of its shape and form.[2]

From the viewpoint of agricultural settlement it is not only the familiar and relatively static outlines of hill and valley that are important, but also the original vegetation, the soils, and the water-supply. For the Saxon settlement was a pioneering venture by an agricultural people, and we must seek to see the country in its former state through the eyes of a practically-minded immigrant farmer.

[1] I am indebted to the following colleagues for their kindly assistance in reading the manuscript or in general discussion: Mr H. A. Cronne, Professor F. J. C. Hearnshaw, Professor A. P. Newton and Professor Ll. Rodwell Jones. My special thanks are due also to Mr E. T. Leeds for his suggestions, and to Mr D. L. Linton for permitting me to draw upon our joint work which is acknowledged below.

[2] *The Making of England* (1882), p. vii.

THE PHYSIQUE OF SAXON ENGLAND

The customary major physical divisions of southern Britain are the high-lands of the oceanic border, and the English plain, separated roughly by the Exe-Tees line. But within the plain itself there is a fundamental contrast in geographical features between the midland zone and the south-eastern zone. The latter is at its broadest—more than 100 miles—between the coast of Kent and the Chiltern Hills. Quite apart from the early emergence of London as some sort of regional capital, this area possessed unique geographical qualities, largely due to the dominance of the Chalk formation and of certain types of drift deposit. It was within the south-east that most of the earlier episodes of the Saxon settlement were enacted. Beyond the south-east, penetration and settlement continued into later times, and there arose the three states, unconfined within rigid bounds of area, whose struggles for supremacy filled two troubled centuries of history, and led, under the further stimulus of the Scandinavian inroad, to the emergence of some measure of national unity.

The south-eastern region was not uniform in character. It had sharper contrasts in physique, considered from the standpoint of pioneer settle-ment, than any other part of the English plain. Water from springs, and shallow wells or water-holes, was widely available, save over the higher portions of the Chalk outcrop and in the areas of heavy clay land; but the soils showed many local differences significant in the history of settlement. The basis of these variations is geological, but we must nevertheless abandon where necessary the geologists' stratigraphical terms and seek to group like with like, on a basis of character, not of age. It has been said that the Saxon pioneer was not an archaeologist; still less was he a geologist, and if we are to see matters through his eyes, it is lithology, not strati-graphy, which must be our guide (Figs. 13 and 14).

Among the less inviting areas, the heavy clay lands and the tracts of sandy upland stood prominent. Large areas of unrelieved heavy clay are comparatively few in the London Basin, occurring only in south-east Essex, in north-west Middlesex, and in parts of Surrey and Berkshire on the margins of the "Bagshot country". They are more extensive in the Weald, comprising the Weald Clay lowland, and the large areas of Wadhurst Clay in and around the valley of the East Sussex Rother. All these clay regions were probably heavily wooded in Saxon days; even in the vicinity of London it is doubtful whether Belgic and Roman clearings had made much progress in the woodlands.

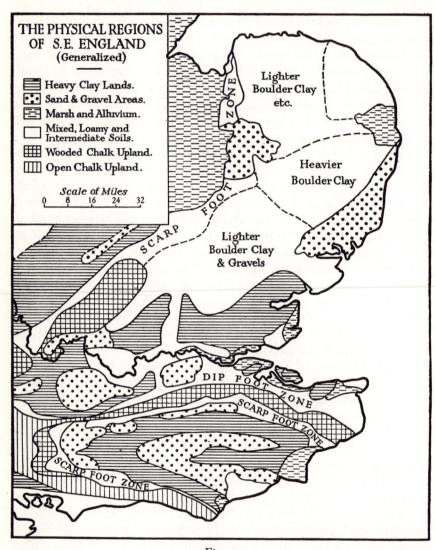

THE PHYSICAL REGIONS
OF S.E. ENGLAND
(Generalized)
———

⊟ Heavy Clay Lands.
⊡ Sand & Gravel Areas.
⊟ Marsh and Alluvium.
☐ Mixed, Loamy and
   Intermediate Soils.
⊞ Wooded Chalk Upland.
⊠ Open Chalk Upland.

Scale of Miles
0   8   16   24   32

ZONE

Lighter
Boulder Clay
etc.

SCARP FOOT

Heavier
Boulder Clay

SCARP

Lighter
Boulder Clay
& Gravels

DIP FOOT ZONE

SCARP FOOT ZONE

SCARP FOOT ZONE

Fig. 13

For the soil distinctions in the Fenland, see Fig. 73, p. 445 below, where also
the fen islands are marked.

The sandy and gravelly areas standing at high or moderate elevations are to-day the home of heath and pinewood. This was not their earlier condition. The pinewood is "sub-spontaneous" and of comparatively recent date, while the heath often represents the degeneration, under pasturing and burning, of dry oak or birch-wood. But whatever their former state, they successfully resisted early Saxon settlement, not only because their soils were sterile and acid, but because the water-table was generally far below the surface. These were the "villainous heaths" of Cobbett in later times; and among them we may number the higher parts of the Central Weald, the "Bagshot country", the western parts of the Wealden Lower Greensand (excluding the outcrop of the Sandgate Beds), the New Forest, the "Blackheath plateau" of south-east London,[1] the more elevated tracts of glacial gravels near Beaconsfield and Chelmsford, the Suffolk sandlings[2] and the Breckland.[3] The last-named region lacks the plateau character of its analogues, but is otherwise comparable with them.

Very different are certain sandy and gravelly soils at lower elevations where the water-table is much nearer the surface. This is the condition of much of the valley and terrace gravel areas of the Thames and its tributaries. Extreme dryness of surface is not the rule, and there is commonly an admixture of earthy or loamy elements in the soil. These lowland gravels often afforded ledges and "islands" of dry ground favouring cultivation amid alluvial marsh or clay forest. For similar reasons, large areas of Tunbridge Wells Sand in the Central Weald are not to be ranked as of heathland type; the soils contain a large element of clay and the finer sandy grades, thus retaining water and rendering normal cultivation possible. An instance of this is the tract between the Battle ridge and the coast near Hastings, a low dissected plateau with clay "bottoms" and cultivable sandy ridge-tops. Water is readily available at shallow depths, and from all points of view the area invited early settlement.

Even more important among the soils inviting pioneer cultivation were those of the "intermediate" or loamy class. These provide an essential clue in interpreting much of the primary Saxon settlement.[4] The soils of

---

[1] S. W. Wooldridge, "The Physiographic Evolution of the London Basin", *Geography*, XVII, 13 (1932).

[2] P. M. Roxby, article "East Anglia" in *Great Britain* (ed. A. G. Ogilvie, 1928).

[3] See Cyril Fox, *Archaeology of the Cambridge Region* (1933), pp. 5–6, 64, 230, and 308, for Breckland distributions.

[4] See S. W. Wooldridge and D. L. Linton, "The Loam-Terrains of South-east England in their Relation to its Early History", *Antiquity*, VII, 297 (1933), where their influence upon even earlier settlement is discussed. See below, p. 109, footnote 2.

this readily arable type are derived from a number of geological formations of widely different ages, a fact which obscures the natural unity and historical significance of the group. They include not only the valley and plateau brickearths—the true representatives of "les sols de cultures faciles" (loess and limon) of the Continent,[1] but large areas of essentially similar soils derived from the Lower Greensand, the Lower Eocene deposits and the glacial drifts. Of the true brickearth areas, there are three major representatives, the Sussex coastal plain, the Norwich loam region, and the plain of south-west Middlesex. The first two are coastal in situation and, together with the smaller loam plateaux of Southend and the Tendring Hundred, functioned in some sense as landing stages for immigrant peoples. Less considerable, though still important, loam areas lie in the lower Lea valley and in the Stour and Medway valleys within the Weald.

The Thanet Sands of East Kent give rise to essentially similar soils and are indeed found associated with much actual valley brickearth. The Hythe Beds of east Kent also give rise to a large area of loamy soils, as do parts of the Sandgate Beds of Sussex and Hants, and the Bargate and associated beds of Surrey. Even more important are the large areas of boulder-clay in Herts, Essex, and Norfolk; these form the basis of loamy soils, favouring tillage. It is here particularly that the geological map is likely to be misunderstood for not all the boulder-clay is of this type. Its character reflects that of the rock on which it rests, as well as the provenance of its far-travelled constituents. Thus it inevitably varies in character, and no general statements can be made about boulder-clay as such. But it is true to say that the chalky boulder-clay of Norfolk, east Herts, and east Essex gives relatively light and loamy soils, while that of Suffolk and west Essex gives heavier soils, probably best grouped with those of the Mesozoic and Tertiary clays.[2]

Next, the very varied characteristics of the "Chalk country" must be noted. The true downlands of the present landscape were almost certainly open country in Saxon times. Celtic cultivation flourished at considerable elevations on the downlands of Wessex and Sussex,[3] but the Saxon settlers found little attraction or profit in the thin red and black downland soils,

---

[1] Vidal de la Blache, *Tableau de la Géographie de la France, passim* (vol. 1 of E. Lavisse, *Histoire de France*, 1911).

[2] Wooldridge and Linton, *Antiquity*, VII, 302 (1932); and C. Fox, "Loam-Terrains", *Antiquity*, VII, 473 (1933).

[3] See pp. 18 and 66 above.

preferring valley sites, as O. G. S. Crawford has so conclusively demonstrated.[1]

Those extensive Chalk areas in which the plateau surface is covered with clay-with-flints stand in an entirely different category. Clay-with-flints yields cold and intractable soils which still bear woodland in profusion, and they certainly must have been heavily wooded at the time of the Saxon settlement. The greater part of the Chiltern plateau and the North Downs must have been wooded. In both areas, scarp-crest trackways are absent or discontinuous, and it is probable that in early times the only naturally open ground was on the scarp face.[2] Though later settlement penetrated the clay-with-flints plateaux to some extent, early Saxon settlement made little progress within them.

At the foot of both the scarp- and the dip-slopes of the Chalk escarpments, conditions of soil and water-supply were much more favourable to settlement. The lower portion of the dip-slopes is often closely coincident with the sub-Tertiary surface, from which the Eocene beds have been stripped, leaving, however, considerable relics to enrich the soil, which is quite different from that of the downlands and the upland clay plateaux. Moreover, the Eocene junction is often a spring-line and in any case water is obtainable from the Chalk at a shallow depth. These conditions are best marked in Kent, where the lower portions of the Chalk dip-slope may be appropriately grouped, from the standpoint of settlement geography, with the areas of intermediate Tertiary soils and the brickearths, in a "dip-foot zone" (Fig. 13).[3]

Conditions at the foot of the Chalk escarpments were equally favourable to settlement. The marly Lower Chalk provided good arable soils where slopes were moderate; even where steep chalk slopes rise above clay land, the clay below the scarp face is lightened by chalky wash, which assists in flocculating the clay aggregates and rendering the soil crumby or granular. Where present, the Upper Greensand also provided a bench of easily worked soils, and water was generally available from one or more lines of springs.

In the Weald, the "intermediate" soils of parts of the Lower Greensand in Kent and Sussex can be included in the "scarp-foot zone", for the

[1] *Air Survey and Archaeology*, H.M. Stationery Office, and Ordnance Survey, 1924.

[2] There is no reason to doubt that the clay-capped Chalk hills of eastern Hampshire were also heavily wooded; as indeed is suggested by the extent of the Wealden woodland as recorded in the Anglo-Saxon Chronicle under the year A.D. 893.

[3] Such a zone extends westward into Surrey, where it is narrower; but does not occur in Essex, Herts, or Bucks, where glacial deposits cover, or abut against, the Tertiary edge. It is not typically developed in Berkshire, but it reappears in Hants and Dorset.

intervening Gault outcrop is generally very narrow and often covered by "scarp drift". Thus locally extended, the "scarp-foot zone" marks a peripheral belt of early settlement around the Weald. Even more striking is the "scarp-foot zone" of the Chiltern escarpment (in the wide sense) which retains its character as a well-watered and more or less dissected chalk-marl platform from Hunstanton to Devizes—a distance of more than 150 miles. Throughout the greater part of its course it is followed by the Icknield Way. In the region north of Goring Gap, the zone can be extended, as in the similar Wealden cases, to include the various light-soil areas around Oxford.

Finally come the numerous tracts of fluviatile and marine alluvium which are conspicuous elements in the present agricultural landscape. As is well known, their present aspect is the sequel to a long history of reclamation. In the Saxon phase there is reason to believe that many of them were actually water-covered.[1] In southern England at least, we are probably justified in treating all present areas of coastal alluvium as water-covered at high tide, during the period of the Saxon entry.

From this survey, it is apparent that the cardinal feature in the physique of the south-east during the Saxon immigration was the isolation of north Kent and the lower Thames valley by a composite barrier ring, comprising the clayland of south Essex, the Chiltern plateau, the Bagshot country, and the Weald. Within the ring lay the nuclear area of the Kentish kingdom, and areas of early settlement near London. Outside the ring lay the regional nuclei of the South Saxon, the West Saxon and the East Saxon kingdoms. The ring was in no sense unbroken, being penetrated by the Hogs Back ridgeway in Surrey, and by the valleys of the middle Thames and lower Lea, as well as by some half a dozen Roman roads. These avenues of penetration limited, but did not destroy, its isolating function.[2]

[1] The main submergence of the British coast is dated as lasting from 3000–1600 B.C. Thereafter, the silting up of the coasts and estuaries proceeded. But there has been an appreciable resumption of subsidence since Roman times. It is generally supposed that this movement has been slow and steady but there are indications that part of the submergence took place rapidly during or just after the Roman occupation. See A. Major, "Surrey, London and the Saxon Conquest", *Proc. Croydon Nat. Hist. and Sci. Soc.* IX, 1 (1920). At Prittlewell in Essex, second-century relics are covered by marine silts; A. G. Francis, "On Subsidence of the Thames Estuary since the Roman Period, at Southchurch, Essex", *Essex Nat.* XXIII. 151 (1932). A similar post-Roman submergence is evidenced on the Belgian coast. See also pp. 3–4 and 61–2 above.

[2] The barrier broke down earliest towards Essex and East Anglia, for London in due course became the focus of the East Saxon kingdom, of which the original heart lay in the driftlands beyond the London Clay woodland.

The Midland Plain, with its south-eastern scarpland border and its extensions east and west of the Pennines, provided less favourable con-

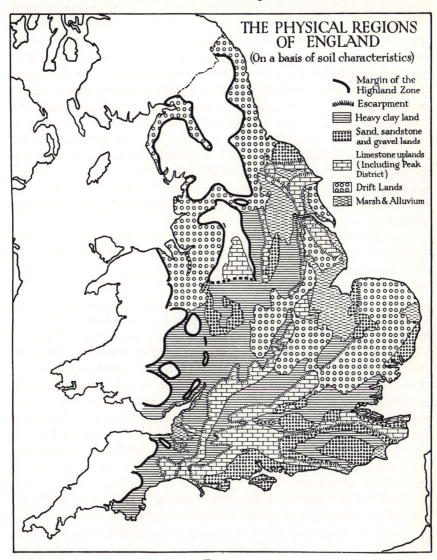

THE PHYSICAL REGIONS
OF ENGLAND
(On a basis of soil characteristics)

Margin of the Highland Zone
Escarpment
Heavy clay land
Sand, sandstone and gravel lands
Limestone uplands (Including Peak District)
Drift Lands
Marsh & Alluvium

Fig. 14

ditions for early settlement than south-east England. Many of the same elements of soil and general physique are present, but they are combined in different proportions, with heavy clay land and alluvial marsh in a dominant rôle. It is particularly notable that large areas of "intermediate" soils are absent. Here, if anywhere, the English plain must have appeared to the Saxons as to their predecessors "a damp green woodland with meres and marshes".[1] There is general agreement that the region was thinly peopled in Roman times, and many centuries later it still showed large areas of unaccounted waste, particularly in the west. In seeking to assess its condition in Saxon times, it is better to attempt deductions from the substantially invariant physical controls than to argue backwards from its later medieval condition. To follow the latter course is to encounter a host of difficulties, notably the fact that the eastern portion of the region suffered much of the brunt of the Scandinavian immigration. Fig. 14 shows the approximate limits of both the eastern and the western drifts in the Midlands.[2] These two great drift areas rarely figure in geological maps of Britain and their existence tends to be forgotten. They are, however, factors of prime importance in the human geography of the area; not least in some of its earlier stages.

Starting in the south-east, the great Upper Jurassic Clay vale, though varying in width, is a continuous feature from the Vale of Blackmore to the Thame valley. It presents soil conditions identical with those of the heavy clays of the south-east; it was densely wooded and, until the eighth century, constituted the western boundary of Wessex. It is subdivided both in Somerset and Berkshire by the narrow belt of the Corallian limestones; but only in the upper Thames region, south and east of Oxford, is it appreciably diversified by lighter soil elements. Here are tracts of Lower Greensand and Portlandian Limestone, as well as the widely spread limestone gravels of the Thames itself. Conditions for farming settlement were favourable both from the viewpoints of soil and of water supply. Farther north-east, in the Ouse valley, there is an extensive covering of chalky boulder-clay and its associates. The ecological and hydrological influence of such drifts within the midland province was relative to the character of the surrounding rocks. Within a clay terrain they offered on the whole lighter and better soils, and enclosed considerable bodies of underground water. In a limestone area they were, relatively, a heavy soil element, and tended to remain wooded.

The Oolitic edge is a well-marked and continuous feature only between the Mendips and the Vale of Moreton. Its higher parts, exposed and subject to severe soil drought, repelled the settlement of valley-seeking peoples,

[1] G. M. Trevelyan, *History of England*, p. 3 (1929).
[2] The interpretation of the physique here is gravely hindered by the incomplete nature of the existing geological maps of parts of the area.

but the "brashy" soils of the lower dip-slope, and particularly those of the little Cornbrash bench, were inviting and readily worked. Here indeed we find a dip-foot zone comparable with those of the Chalk areas. True Cotswold characteristics recur at moderate elevation in Dorset and Somerset (the Sherborne district), an area which became in due course an important appendage to the West Saxon kingdom. But it was in the Northampton region that some of the most favourable sites for early settlement occurred, within the Oolitic belt. In this region of low dip and pronounced dissection, the underlying clays are laid bare in the valleys; springs break out at the base of the Northampton Sands, while above the spring-line are drier cultivable slopes. The lower parts of the dip-slope here are covered with boulder-clay, and the ancient forests of Whittle-wood, Salcey, etc., formed the western edge of the "Bruneswald" on the clays of Bedfordshire and Huntingdonshire.

Beyond the Oolitic edge, the chalky boulder-clay and glacial sands and gravels are widely spread in Leicestershire and east Warwickshire. Inspection of the geological maps indicates how close is the relation between village sites (not all of them early) and the patches of water-bearing sand and gravel. It is perhaps more than a coincidence that the tracts of drift-land on either side of the Northamptonshire Edge together constituted the larger part of the Middle Anglian region as it emerged in the sixth century. Farther south-west, the great plain of Liassic and Keuper claylands is unrelieved, save by the river drifts of the Severn and Avon, of which the extent and distribution have yet to be accurately ascertained. Here was Arden, the core of perhaps the largest tract of wooded clayland in the country.

In the north-west Midlands there is an interesting series of physiographic relations, of which the historical significance has not been fully appreciated. Fig. 14 shows certain considerable areas of sandstone and sandy or gravelly glacial drift as being grouped together. While the higher parts of this tract (Cannock Chase, etc.), like the sandy uplands of the south country, repelled settlement, the lower slopes and the sandy drift areas were the first expanses of reasonably light soils encountered by immigrants entering by the Trent valley. The Trent flows through a broad flood-plain, locally margined, it is true, by habitable terraces, but it leads into a wide tract of unrelieved Triassic marls, where the woodland cover must long have remained dense and impenetrable. It was beyond this wooded belt that the nucleus of the Mercian kingdom arose, within the area of lighter soils noted above. Westwards again lay difficult country on the marls and sandstones of the Staffordshire-Shropshire border. Here was the pre-

Norman Forest of Blore; while beyond was the western drift-plain, across the southern portion of which there runs a belt of hummocky morainic hills, interspersed with lowland peat-mosses, formerly lakes. Against this considerable and diversified tract of negative country the "march" of the early Mercians lay; the Celtic kingdom of Theyrnllwg, ancestor of the later Powys, was recorded as extending from Clwyd to Cannock Chase.

Another element in the midland terrain included the great tracts of alluvial plain in the Fens and the Somerset levels. The condition of these tracts in the Saxon phase is difficult to reconstruct, but both areas fulfilled in their time the function of regions of refuge.[1] In the Fens the great rivers served as avenues of entry in an early phase, but later the marshland itself sustained the rôle of common frontier barrier between Lindsey, East Anglia and Middle Anglia. In Somerset and over most of the Fenland, the Saxon settlements occupied marginal or insular sites, showing that the greater part of the area, if not actually water-covered, supported lowland peat or other types of marsh.

In north-east England the coastal ledge is drift-covered as far south as Teesmouth, but the boulder-clay is coarse and stony, and the country in no sense compares with the southern drift-lands. Nevertheless, it was settled in the northern part; and there is no obvious physical reason for the fact, attested alike by the eighth-century record of Simeon of Durham and the archaeological distributions, that Durham long remained waste. Farther south, in the coastal zone, there are three upland areas—the North York Moors, and the Wolds of Yorkshire and Lincolnshire. It should be noted that the first-named area is not, as too often stated, a limestone upland. Its greater part is an elevated sandstone plateau, re-calling in almost every detail the Millstone-Grit moorlands to the west. It is, in fact, an outlier of the Highland Zone, and though its margins were later colonised by Scandinavian peoples, it was hardly entered upon during the early English settlement. Only along its southern margin, near Pickering, is there a belt of limestone soils; in the lower portion of this there are "dip-foot zone" conditions of the Cotswold type.

The Chalk areas of east Yorkshire and Lincolnshire were almost islanded in marsh; but there was early settlement upon their lower slopes, as well as in the eastern portion of the Holderness drift-plain, less studded

[1] H. C. Darby has recently discussed in some detail the Saxon history of the Fen-land. See "The Fenland Frontier in Anglo-Saxon England", *Antiquity*, VIII, 185 (1934).

with marsh than the lands of the Hull valley. Central Lincolnshire, west of the Wolds, is a drift-filled vale, with axial tracts of marsh; and much drift, too, lightens the clayland west of Lincoln Edge. These drift-lands are, of course, comparable with those of East Anglia, not with the northern type.

In the great interior vale, the real continuation of the Midland Zone, the central feature was the Humberhead Marsh, which lay across the southern boundary of the old Deiran kingdom, and which narrowly constricted the approach from the south, west of Hatfield Chase. Farther north, towards York, the alluvial lands are discontinuous; inliers of sand and gravel rise from beneath the marsh. North of York the vale is a hummocky drift-plain with rather heavy land formerly occupied by the Forest of Galtres. To the west is an elevated margin of sandy drift, while southwards the Magnesian Limestone ridge forms an elevated causeway between the marsh and the low-lying, wet and formerly wooded lands of the Middle Coal Measures.

Since the Saxon wave ultimately reached the edge of the Highland Zone, and locally passed beyond it, some mention must be made of the edge as presented to the plain.

The Welsh massif shows marked contrasts along its eastern boundary. From the Vale of Powys northwards, the plateau descends sharply to the plain and is breached only by the Vale of Llangollen. South of the Powys opening, the hills of south Shropshire project east as a distinctive Welsh "prong", which, together with the great forest to the south of it, for long divided the Welsh border into two distinct sections. South of the prong, the highland boundary does not in reality follow the line of the Malvern Hills and the Forest of Dean, but swings back to Radnor Forest and the Black Mountains. The Hereford plain is an adjunct of the English plain —a vestibule region, on the route to Wales; and, though heavily wooded, it presented no other obstacle to English penetration from the east.

A somewhat similar contrast exists between the northern and southern parts of the Devon boundary. Here it is the Exe valley which functions as the vestibule region. There is no northern coastal plain, and the natural lines of entry to the interior are the southern coastal ledge and the depression which leads westward from Crediton.

Further, the Pennine margin, on the east and south, also shows features of note in this connection. There are easy entrances from the south, *via* the Derwent and Dove valleys, to the limestone plateau of the Derbyshire dome and its lowland enclaves. On all other sides, this region

is isolated by a great inward-facing grit-stone scarp; and the physique of
the region clearly foreshadowed its history as a definite province or com-
partment during the Saxon settlement. On the eastern side of the south
Pennine range, a belt of sandstone scarplands on the Lower Coal Measures
intervenes between the high moors of the Millstone Grit and the heavy
low-lying land of the Middle Coal Measures. It was this belt, particularly
at its northern end, which afforded a home for the Celtic kingdom of
Elmet. Its eastern defences of marsh and woodland were formidable, and
must have played no small part in assisting its unexpectedly late survival.
But, unlike the others, the Pennine upland as a whole presented no
ultimate barrier to the penetration of the English from the east, or later,
to that of the Vikings from the west.

ON THE NATURE OF THE CONQUEST

The older and more familiar interpretation of the Saxon conquest, based
as it was on the record of the Anglo-Saxon Chronicle, treats of the history
of a number of separate invading bands, responsible for establishing small
kingdoms which grew into effective contact in the later phase of struggle
for supremacy. To Professor Chadwick, however, it was "incredible that
such a project as the invasion of Britain could have been carried out except
by large and organised forces".[1] He saw in the later Danish invasions a
possible analogy for a wide concerted operation, and he pressed into the
service of his argument the vaguely constituted "imperium", or "bret-
waldaship", held in turn by certain of the English kings. Thus was
initiated a group of theories which, as they developed, abandoned even
more completely reliance on the Chronicle as a true account of the invasions. Sir Charles Oman, dismissing the early Chronicle entries, some of
them evidently garbled, in favour of the more nearly contemporary evidence of Gildas and others, deduced an early widespread overrunning of
the country, checked by a British rally under Aurelius Ambrosianus at the
Battle of Mount Badon,[2] which he believed to have taken place near the end
of the fifth or near the beginning of the sixth century. Then, and only
then, in his view, did the advancing wave recoil and "crystallise" into
separate kingdoms in the eastern part of Britain. "The moment of settlement was a moment of political disruption."[3]

[1] *The Origin of the English Nation* (1907), p. 12.
[2] There is no clear evidence as to the site of "Mons Badonicus", but presumably it
was in the west of the country, possibly at or near Bath.
[3] *England before the Norman Conquest* (1929), chapters XI, XII.

There is, however, a considerable difference of emphasis between the theories of Chadwick and Oman. They agree in abandoning the early Chronicle story; but an early overrunning does not necessarily imply a large concerted operation.[1] The attribution of Napoleonic guile and wide strategic design to the Saxons is quite unconvincing. An early phase of widespread fighting seems to be implicit in the account given by Gildas, and we cannot lightly dismiss the possibility thus indicated, but there is a very evident reaction against these more modern historic theories in recent years. R. V. Lennard[2] has given cogent expression to the objections which may be made to the views of Chadwick and, to some extent, to those of Oman. Other writers have sought to vindicate the essential credibility of the early Chronicle entries. There is, for instance, O. G. S. Crawford's demonstration that the details of Cerdic's landing in Wessex, and of the ensuing battles as recorded in the Chronicle, can be reasonably reconstructed on the ground along the line of the "Cloven Way" and continuing trackways.[3]

There is probably a measure of truth in both the older and the newer theories. The evidence favours two distinct phases of Saxon immigration separated by the defeat of Mount Badon. But even if so much can be distilled from the verbiage of Gildas, and from the "facts of distribution" noted below, matters still remain very vague. The site of the battle is unknown, and its date is uncertain. The episodes of the struggle which preceded it involved relatively small numbers, and took place in a country still generally wooded. They have thus left hardly more sign on the ground or the map than in contemporary literature. Mount Badon may, in a real sense, have punctuated the historical story, but it is beyond our present means to show, save locally and doubtfully, that it interrupted the process of settlement. In a word, separate maps of the phases before and after Mount Badon cannot be produced, and it is inevitable that the stereotyped argot of political history misses much of the truth when it speaks of kingdoms being "crystallised" or "carved out". Such terms may serve in the history of kingdoms as institutions, but not in the geography of kingdoms as areas or regions. It is the successive stages of settlement that form the subject of the present study. Since archaeology and place-name

[1] The latter conception has been considerably elaborated by such writers as (1) P. T. Godsal, *The Storming of London and the Thames Valley Campaign* (1908); see also *The Conquests of Ceawlin, the Second Bretwalda* (1926), and the review in *History*, XI, 82 (1926–7); (2) G. P. Baker, *The Fighting Kings of Wessex* (1931), p. 63.

[2] "The Character of the Anglo-Saxon Conquests: A Disputed Point", *History*. XVIII, 204 (1933–4).

[3] "Cerdic and the Cloven Way", *Antiquity*, V, 441 (1931).

study, which provide the facts of distribution, cannot afford a detailed and accurate dating, a relative rather than an absolute chronology must be adopted. There was an *Entrance Phase*, the skeletal infiltration of the immigrant people to the areas of early settlement. Then followed an *Expansion Phase*, or secondary colonisation around and within the early settled tracts as well as territorial expansion by conquest. Finally there came a *Terminal Phase* in which the settlement plan filled out and assumed, for the time being, a relatively static condition. A given phase was not necessarily contemporary throughout the country; but the time-lag between one region and another was not large, and the facts of changing geography can be correlated with the written history.

THE ENTRANCE PHASE: ARCHAEOLOGICAL EVIDENCE

Actual occupation areas of the early Saxon period have come to light in this country only during recent years. A notable example is that of Sutton Courtenay (Berks), where a small village of "hut-floors" has been investigated.[1] The original walls of the huts were composed of non-durable materials, probably layers of mud and straw, so that in this, as in other cases, nearly all overt signs of the settlement have perished. Other hut-sites have been described at Bourton-on-the-Water (Gloucs), at Peterborough and St Neots (Hunts), and at Waterbeach (Cambs),[2] while recently an interesting group of sites has been found exposed in the cliffs near Selsey,[3] eloquently attesting the large effects of marine erosion on this coast. These scattered records provide an interesting glimpse of the small and unpretentious houses of some of the early settlers, but they are too small in number to afford any picture of early settlement as a whole. For this, we are largely dependent on the distribution of burial grounds of the pagan period (A.D. 450–650) and of certain place-names of early type. In both cases it may reasonably be assumed that inhabited sites existed in the vicinity and, though we cannot exactly locate the actual settlements, we may treat the regions in which these evidences occur as those of Saxon occupancy.

Fig. 15 denotes the areas in which burial grounds are of frequent occurrence, showing also the location of the more important isolated finds in

[1] E. T. Leeds, "A Saxon Village near Sutton Courtenay, Berkshire", *Archaeologia*, LXXIII, 147 (1922–3), and LXXVI, 59 (1926–7).

[2] T. D. Kendrick and C. F. C. Hawkes, *Archaeology in England and Wales, 1914–31*, pp. 320–5 (1932), and references there quoted.

[3] G. M. White, "A Settlement of the South Saxons", *Ant. Journ.* XIV, 393 (1934).

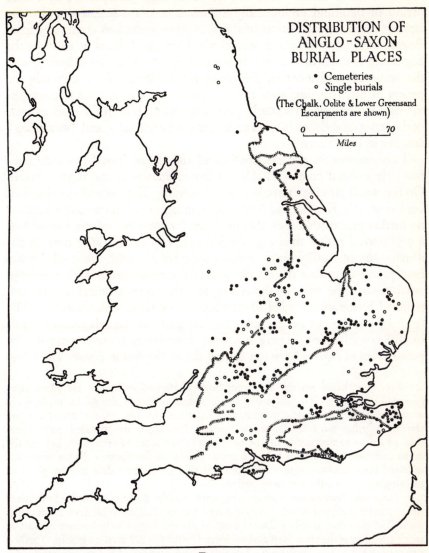

Fig. 15

The map is based on the Ordnance Survey *Map of Britain in the Dark Ages* (*South Sheet*), 1935. Single burials are shown only where they afford some suggestion of the extension of the areas of occupancy beyond the limits indicated by the cemeteries. See footnote 1, p. 104.

the intervening tracts.[1] This map gives an aggregate picture of a long period of time and undoubtedly embraces several successive and over-lapping distributions. Some of the sites show unbroken continuity from Romano-British times; and several yield jewellery dated as early as the beginning of the fifth century. In these cases we are no doubt dealing with the scattered settlement of the phase before Mount Badon. While the period of the map thus straddles an episode of vital historic importance, yet its distributions, as a whole, give more than a hint of the nuclear areas of the states established in embryo before the British rally, and "swarming" steadily in later years.[2]

Two features in the distribution of the pagan burial grounds have formerly evoked comment. One is the avoidance of the Roman roads. On the whole this is strikingly true between the Thames and the Humber, but less so in the Kentish and Northumbrian regions. This probably requires no further explanation than that the "eye for country" of the Roman was a geometrical rather than a geological eye; and it paid little heed to the distribution of readily arable soils or supplies of water. Secondly, it has been pointed out that the distribution of the burial grounds is in evident relation with the plan of the rivers, and that both cultural divides and political boundaries tended to develop on, or near, water-partings. The truth of this cannot be gainsaid, but the basis of the relationship needs careful consideration. There has been a tendency to over-estimate the possibilities of *ingress* by water during the earlier Saxon phase. The lower

[1] Fig. 15 is based on the Ordnance Survey *Map of Britain in the Dark Ages (South Sheet)*, 1935. To this authoritative source reference should be made for a more complete survey of the archaeological data. In Fig. 15 the cemeteries are shown, together with certain single burials. The latter are included only where they afford some suggestion of the extension of the areas of occupancy beyond the limits indicated by the cemeteries. Barrow burials are not shown. Pagan interment survived the introduction of Christianity so that no definite date can be given as marking the end of the cemetery period.

[2] These distributions are necessarily incomplete in detail and it is possible to mistrust the argument *a silentio* which tacitly governs the conclusions to be drawn from such maps. Yet they do provide a limited basis of assured fact, which deserves at least as much attention as the records of garbled legend or third-hand partisan gossip. Further, one cannot fail to be struck by the fact that a large number of the more recently dis-covered sites fall within the tracts of high frequency, as outlined on earlier maps, and thus confirm rather than extend the boundaries of the nuclear tracts. Moreover, there is one eminently reasonable test that can be applied. Were the high frequency areas geographically favourable to early settlement? Were they separated by physically negative regions? To this question the answer, as we shall see, is an emphatic affirma-tive. But see R. H. Hodgkin, *A History of the Anglo-Saxons* (1935), I, 372.

parts of the larger streams were certainly navigable, and the estuarine entrances were wider and deeper than at present, owing to the post-Roman submergence. But there is grave reason to doubt whether the upper parts, and the tributaries, of the main streams were navigable, even by the ancestors of the shallow draft Viking boats.[1] It has often been stated without reason or authority that the rivers were larger then, but in this connection it must be remembered that the still wide spreading woodlands must have diminished the run-off, and helped to choke the streams with their debris. Even with the larger streams it is probably safe to assume, in default of good evidence to the contrary, that they were not normally navigable above the first pronounced break in their longitudinal profile, sometimes the head of the former estuary, and often marked subsequently by an important medieval town site. Apart altogether from water transport, the valley bottoms possessed notable advantages, where the flood-plain was margined by gravel terraces, affording spring water, and giving dry and level ground for trackways and habitations. The associated valley loams often afforded, too, the best of arable soils.[2]

The distribution of the nuclear areas as shown in Fig. 15 is definite and significant. They are eight in number, five situated upon the coast, and three inland. In the south, there was an occupation of the Sussex coastal plain, and of the lower part of the Ouse valley. In Kent, burial grounds are numerous in Thanet and southwards on the lower parts of the Chalk country. They extend also along the coast south of the downs, and along the Stour valley, and, avoiding the area of the Blean Forest, occur at intervals along the dip-foot zone as far as the Medway. Along the valley of this river they also extend south of the downs. In west Kent and in Surrey lies a group of sites in a very similar physical setting on Thames-side and in the Darent and Wandle valleys. These latter burial grounds show Saxon rather than Jutish characters, and must be regarded as forming a separate group. The Essex and south Suffolk area was clearly penetrated

[1] There is little difference in style between the Nydam boat dating from the third or fourth century, and the Gokstad ship dating from the eighth or ninth. See G. Baldwin Brown, *The Arts in Early England* (1915), IV, 590. A. F. Major, on the authority of Dr H. Schetelig of Bergen, states that the latter ship, loaded, would require about 3 feet of water to float it, the Nydam example possibly less. *Trans. Croydon Nat. Hist. and Sci. Soc.* IX, 1 (1920).

[2] The force of these considerations will appear when we recall the fact that burial grounds occur close to the Warwick Avon and its tributaries; while it has never been credibly demonstrated that the invaders reached these sites *via* the Bristol Channel. The existence of these sites seems to imply, on the contrary, a considerable element of land movement.

by way of the Blackwater, Colne, Stour, and Gipping-Orwell valleys which lead to the drift-land margin. The London Clay area in south Essex is devoid of evidence, though we must note the Prittlewell site on the loam plateau in the south-east. Central Suffolk, another area of heavy soils, again yields practically no evidence, but the lighter soils of Norfolk and the Lark valley area show signs of widely spread early settlement. Finally, there is the concentration on the Chalk and boulder-clay lands stretching west from Bridlington. Here burials occur also in pre-existing barrows. All these coastal tracts are areas of light or intermediate soils, and although they are not coextensive with the kingdoms as they emerged after 597, they can evidently be regarded as the nuclei of Sussex, Kent, Essex, East Anglia and Deira respectively.

The inland areas present a number of problems. The upper Thames region, and its structural continuation in the vales of Thame and Aylesbury, stand out prominently as an area of high frequency, and it is here that E. T. Leeds would seek the original nucleus of the kingdom of Wessex. As noted above, this region is in no sense an unmodified clay vale, but gave ample encouragement to settlement. Beyond the headwaters of the Lea, however, this zone of settlement is broken; Bedfordshire yields little, except the important burial ground at Luton, and a group on the Ouse near Bedford. Beyond, in the headwater region of the Cam, there is another group, perhaps an appendix to East Anglia, but better regarded as a separate area. The two remaining areas are complex. A group of burial grounds lies in the dissected plateau region of Northants, sited in evident relation to the spring-line below the Northampton Sands. Beyond, in the Avon valley, is another group, which though in part early in date, was no doubt affected by the sixth century expansion of Wessex under Ceawlin. In any case, these represent the beginnings of settlement in a true clay land, though most of the actual sites occur in association with river-drift. Farther north there is clear evidence of penetration by the valleys of the Trent, Soar, and Derwent, the latter giving access to the heart of the Derbyshire dome, where burials in earlier barrows are numerous. At a later date, before the emergence of the greater Mercia, the midlands fell into three distinct provinces, the Hwiccian, the Middle Anglian, and the original Mercian areas; but the relations of these three could hardly be deduced from Fig. 15.[1]

[1] Other features of the map include the evidence of an early Jutish settlement of the Isle of Wight.—E. T. Leeds, *Archaeology of the Anglo-Saxon Settlements* (1913), p. 116, but the opposite mainland affords, as Leeds has insisted, only the sparsest evidence of early settlement, nor is it at all likely that future discoveries will materially affect the

This is the evidence of the pagan burial grounds as a whole. We must now pass to the more particular, though controversial, conclusions which have arisen from a study of the earlier sites within the groups.

The small but definite differences in material culture between Angle, Saxon, and Jute, have a bearing of some importance on the argument now to be presented. More than twenty years ago in a study of saucer- and applied-brooches, E. T. Leeds recognised that distinctively Saxon material figured among the grave goods of the mid-Anglian area and, further, that the upper Thames region stood out as an important area of early settlement.[1] He postulated an entry to the latter region *via* the lower Thames, and supposed that there was a separate entry of Saxon elements by way of the rivers draining to the Wash. In 1925 Mr Leeds developed these views in a well-known paper[2] in which he stated his conclusion that the story of a West Saxon entry across the southern chalk lands from Southampton Water must be entirely abandoned. He presented cogent arguments in favour of penetration from the Wash region by the Icknield Way which links the Mid-Anglian and upper Thames regions. Recently, he has restated the archaeological evidence for this hypothesis in full.[3] He notes in the first instance the indications of the distribution of cremation in the burial grounds. This practice steadily decreased as time went on, and it is reasonable to argue that cemeteries in which cremation is preponderant or common were amongst the earliest in use. Within the Anglian region, there are many large cemeteries showing cremation only, not only near the coast, but well inland in the Trent and Avon valleys, and at York where the site was continuously in use from Romano-British times. Farther south, cremation is prominent in a group of cemeteries near Cambridge, and it also occurs to some extent in the lower Thames region. Even more important is the fact that it is conspicuous in the upper Thames region, as at Long Wittenham, Frilford, Brighthampton and Fairford, and in the Avon valley, as at Bidford and Baginton, near Coventry. We are thus led to infer a deep and early penetration of the south-east midlands, balancing that of the Trent valley.

position here. On the other hand, new evidence has recently been obtained in Lincolnshire. See C. W. Phillips, "The Present State of Archaeology in Lincolnshire", *Archaeol. Journ.* XCI, 97 (1935). Lastly, there is a large blank area north of the Tees in Durham, suggesting the gap between the original Deiran and Bernician nuclei.

[1] "The Distribution of the Anglo-Saxon Saucer Brooch in relation to the Battle of Bedford, A.D. 571", *Archaeologia*, LXIII, 159 (1911–12).

[2] "The West Saxon Invasion and the Icknield Way", *History*, X, 97 (1925).

[3] "The Early Saxon Penetration of the Upper Thames Area", *Ant. Journ.* XIII, 229 (1933).

This conclusion is enforced by the distribution of distinctively early pottery and jewellery. Early decorated pottery occurs as far west as Sutton Courtenay. The earliest type of cruciform brooches occurs both in the upper Thames and mid-Anglian regions, and the same is true of certain equal-armed brooches regarded as dating from the latter part of the fifth century. Certain other early objects, such as the window-vases studied by F. Roeder[1] occur in the mid-Anglian, but not in the upper Thames region. For the former region indeed there is a large mass of irrefutable evidence of early settlement extending as far west as Bedfordshire; and since some of the most characteristic of the early objects reappear in the upper Thames region, only 40 miles to the south-west along the same readily penetrable zone, the hypothesis of direct cultural contact between the regions is at least eminently reasonable. At the slightly later date represented by the saucer- and applied-brooches, the chain of linking records between the regions becomes even stronger. On the other hand, connection with the lower Thames region only becomes apparent in the borrowing of certain Kentish decoration motives from brooches dated by N. Aberg[2] as not earlier than the middle of the sixth century.

Such, in brief outline, are the facts which, taken with the absence of evidence of early settlement in the traditional Wessex, recommend the hypothesis of penetration by the Icknield Way. This hypothesis is evidently not lightly to be turned aside. As noted above, the physique of the Icknield zone marks it as an outstandingly favourable route of entry, and, while this does little to prove the theory, it contributes greatly to its feasibility. A number of objections have been put forward, however, and some of these must be noted. T. C. Lethbridge[3] points out that the cemeteries to the north of Cambridge imply the existence of a body of Anglian settlers who must have impeded or prevented access to the Icknield Way *via* the Cam. In reply, Leeds has suggested that the Saxons circumvented this Anglian block by passing up the Ouse to the neighbourhood of Huntingdon, whence they could have passed by way of Ermine Street to the area south of Cambridge or north-westwards into Northants by the road to Leicester. M. W. Hughes has also joined issue in spirited fashion on the question of the settlement of the Chiltern-foot area.[4] Leeds is prepared to abandon the Chronicle account of Cuthwulf's

[1] *XVIII. Bericht der Röm.-German. Kommission* (1928), p. 149.

[2] *The Anglo-Saxons in England* (Upsala, 1926).

[3] "Recent Excavations in Anglo-Saxon Cemeteries in Cambridgeshire and Suffolk", *Camb. Ant. Soc. Quarto Publications*, n.s. III, 35 (1931).

[4] "Grimsditch and Cuthwulf's Expedition to the Chilterns in A.D. 571", *Antiquity*, V, 291 (1932).

Chiltern campaign of 570, but Hughes seeks to establish the credibility of this entry. He certainly places his finger on a weak spot in the archaeological evidence in pointing out that the Buckinghamshire cemeteries do not reveal evidence of a date earlier than that assigned to Cuthwulf's campaign, and that there is thus a pronounced break in the chain of early evidences between Bedfordshire and the upper Thames. This fact is not in question, so that we can perhaps fairly conclude that there were at least two phases of settlement in the vale at the Chiltern foot. Even if it be claimed that Leeds is not justified in rejecting the Chronicle account of the Chiltern campaign, it is none the less true that Saxon elements had entered the upper Thames region long before, and were also established in the region to the north-east. They may thus have passed by the Icknield routes without prejudice to Hughes' theory of Cuthwulf's campaign. There is time and space for both happenings. Moreover, it is to be noted that even if we reject the theory of the Icknield entry in favour of any other, say that of the lower Thames, it still remains true that a considerable body of early settlers inhabited the Thames valley above Goring Gap, and that this region has evident claims to be regarded as the nucleus, or one of the nuclei, of Wessex. Crawford's skilfully handled brief for Cerdic and the Cloven Way[1] contributes little to the early peopling of the Chalk lands of southern Wessex; it suggests rather the mode and route of entry of one element in the later conquering host of Wessex. We must defer any attempt at a verdict on the difficult questions here involved till we have considered the evidence of place-names, but sufficient has perhaps been said to indicate the considerable force of the archaeological arguments.

THE ENTRANCE PHASE: EVIDENCE OF PLACE-NAMES[2]

As long ago as 1849, Kemble[3] demonstrated the importance of place-names containing the suffix -ing. He listed over twelve hundred such names, which he regarded as representing communal settlements or "marks", dating from the earliest phase of the Saxon settlement. In later years his theory of the mark became discredited and J. H. Round[4] and others

[1] Op. cit. p. 441.
[2] The material in this section is an extension of work undertaken jointly with Mr D. L. Linton. See "Some Aspects of the Saxon Settlement in South-East England considered in relation to the Geographical Background", Geography, xx, 161 (1935).
[3] The Saxons in England (1849).
[4] "The Settlement of the South and East Saxons", in The Commune of London and Other Studies (1899).

showed that he had gravely misinterpreted some of the names. Neverthe-less, the more careful work of modern place-name students has gone far to establish the essential truth of Kemble's central idea.

Many of the *-ing* names listed by Kemble prove to be those which in their O.E. form terminate in *-ingas*. It is now generally agreed that these are among the oldest of the English names, referring in the first instance to groups of people rather than to places. With the growth of a natural topographic nomenclature, such a system of names would naturally go out of use. The great age of the names in *-ingas* is indicated not only by the character of the suffix, but also by the fact that it is frequently combined with personal names never used in this country but represented only in continental literature. Professor Stenton concludes that such stems fell out of use by the seventh century at latest.[1] Further, the geographical distribution of the names in *-ingas* strongly enforces their claim to be regarded as early.

Other names which have been regarded as "early" in the general sense are those ending in *-ham* and *-ton*. These two suffixes, however, show strikingly different distributions. Equivalents of *-ham* names are common on the Continent, but in Britain they are confined to the south-eastern area. On the other hand, place-names in *-ton* are virtually absent on the Continent, but extend in Britain far beyond the limits of earliest settlement. While, therefore, some of the names in *-ton* belong no doubt to the earlier group, it is clear that the suffix remained in use until a much later time. Such names cannot safely be taken as marking primary settlement; as a whole they seem rather to belong to the phase of expansion. We are probably justified in taking many of the *-ham* names as early, but there is risk of confusion with the non-habitative terminal *-hamm*, signifying an enclosure, usually of pasture and often of water-meadow. The two forms are liable to confusion, even in O.E. documents, though there is little doubt that they can be to some extent geographically differentiated if regard is paid to the characters of the country in which they occur.

Still other names of probable high antiquity are those in which the suffix *-ham*, or in some cases *-ton*, follows a compound in *-inga*. In this case again, we are probably dealing with the settlements of groups of people, though the names may be somewhat later than those in *-ingas*. Stenton thinks it improbable that any names of this kind arose after the end of the eighth century, and that most of them are probably far earlier.[2]

[1] A. Mawer and F. M. Stenton, *Introduction to the Survey of English Place-Names* (1924), I, Part I, p. 51.
[2] A. Mawer and F. M. Stenton, *ibid.* p. 54.

Finally, there is the small group of place-names involving the names of Germanic gods such as Tiw, Thunor, or Woden, or giving other indications of association with places of heathen worship. Many such names cannot have survived the conversion to Christianity but the few that remain give definite evidence of early settlement. Isolated instances help us little; but in one case at least (p. 116) a group of such names can be significantly related to the geography of the country.

Pending the completion of the monumental work of the Place-Name Society, a really thorough review of the geography of English place-names is impossible. This survey therefore has to rely upon Professor Ekwall's study of the names in -ing and -ingham.[1] His list of such names covers the whole country and, though not quite complete, is representative. Fig. 16 shows the distribution of names in -ingas and -ingaham as noted by Ekwall. In the southern area, where evidence of this type is perhaps most valuable, the distribution presents the same major features as were indicated by Kemble's necessarily imperfect pioneer work. There is a marked concentration in the south-east; four-fifths of the names lie south of a line joining Southampton Water and the Wash, and only a very few west of a meridian passing close to Southampton, Oxford and Nottingham. The names occur in distinct groups, isolated in large measure by physical barriers—a further indication, if such be needed, that the actual occupation of the country involved a process of piecemeal penetration. In particular, the influence of the "barrier ring" which encircles London is clearly seen on the map. These groups of early names must be examined separately with regard to the physical setting of each group.

Along the south coast, both the personal names and the local topographic nomenclature of Sussex are so distinctive as to be entirely consistent with the historical traditions of a separate early conquest and settlement. Physiographic isolation no doubt contributed to the survival of archaic elements. The local vocabulary shares certain ancient words with that of Kent, but there are few links with Hampshire, a region of which the nomenclature is of distinctly later type, containing *inter alia* an

---

[1] *English Place-Names in -ing* (Lund, 1923). It should be noted that Fig. 16, based upon names listed in this monograph, does not include the names ending in -ing (singular). These, however, are shown on a small map published with the Ordnance Survey *Map of Britain in the Dark Ages (South Sheet)*, 1935; and this should be compared with Fig. 16.

DISTRIBUTION OF
CERTAIN EARLY
PLACE-NAME ELEMENTS
(-ing and -ingham)

The Chalk, Oolite and Lower
Greensand Escarpments are shown

0   70
Miles

Fig. 16

This map is based upon the names listed in E. Ekwall, *English
Place-Names in -ing* (Lund, 1923). See footnote, p. 111.

appreciable Celtic element virtually absent in Sussex.[1] A large number of names in -*ingas* occur, implying a relatively dense population on the loam terrain of the coastal plain, but extending also through the gaps to the scarp-foot zone. On the whole, therefore, the early names confirm the indications of the pagan burial places as to the areas first settled, but it is significant that the latter are found up to considerable elevations on the Downs, while the former occupy lower elevations on plain and valley slope.

If the former embayed outline of the coast[2] is restored by taking the inner edge of the mapped alluvium as the strand-line, it is at once evident that about half the names in -*ing* are on, or quite close to, the former water-edge. West of the Arun, names in -*ing* are largely replaced by those terminating in -*ham*, but the latter show a similar relation to the water-edge. Between the Adur and the Ouse there are no early coastal settlements; the probability is that they existed but have succumbed to marine attack: some of the "lost" Sussex manors of Domesday were no doubt situated here. At the western end of the littoral, the early names, with one possible exception in the Avon valley, all fall within the region which archaeology shows to be Jutish. The place-names thus support the burial grounds in an eloquent silence concerning any early West Saxon settlement here. Indicating landward penetration farther east, there are one or more settlements with names of early type in each of the gaps. They are generally situated on low spurs above the alluvium, which here, no less than nearer the coast, represents former tide-water. Names in -*ing* also extend into the scarp-foot zone and reappear in the region of loamy soils sited upon the Sandgate Beds of the Rother valley.

Perhaps the most interesting group of early names in Sussex is that which occurs in the Rape of Hastings between the Brede valley and the coast. Names in -*ing* are accompanied by a considerable number in -*ham*, and though some of these may be cases of O.E. -*hamm*, this hardly invalidates the claim of the group as a whole to be deemed early. The physique of this region favoured early settlement.[3] There can be little

[1] A. Mawer, F. M. Stenton and J. E. B. Gover, *The Place-Names of Sussex*, Part I (1929), Introduction.

[2] In order to appreciate in more detail the location of the early settlements, we must remind ourselves that frequently where we see "alluvium" to-day we can probably read "water" for the early Saxon phase (see p. 39). Further, there has been pronounced clipping back of the edge of coastal plain by marine erosion, leading to a great loss of land, save at the western end of the coastal plain, where in the lee of the Isle of Wight, the coast still retains much of its original "submergent" aspect.

[3] See p. 92 above.

doubt that here is the region of the "gens Hestingorum"[1] of Simeon of Durham, the "Haestingas" of the Chronicle.

In Kent, conditions on the north coast afford both a comparison and a contrast with those of Sussex. Here again creeks, formerly deeper and less encumbered with mud, penetrated the coastal zone, but the promontories and islands on this coast, with the single exception of Thanet, are masses of London Clay, affording little attraction to settlement. There are no names in -ing at the creek heads, but Faversham, Teynham and Ludden-ham, are no doubt quite early settlements. Beyond, in the dip-foot zone, names in -ing duly appear; and they extend eastwards into the similarly constituted tract near Canterbury, which was approached, no doubt, by way of the Wantsum Channel and the estuaries of the Great and Little Stour. Names in -ing occur south of the Downs in a region where there is a group of burial grounds.[2] Scarp-foot settlement, marked by names in -ingham, extends west to the Stour gap, while beyond it and still in the scarp-foot tract, names in -ing occur at intervals as far as the Medway valley. Here, in a region favoured by the juxtaposition of drift loams and the "intermediate" soils of the Hythe beds, is a considerable group of names in -ing. Westwards there is a break in the chain of early settlements, but they recur in west Kent and east Surrey, in the Thames, Darent and upper Wandle valleys, and in the scarp-foot region, as at Chevening. It is open to doubt whether these settlements were made by Jutish or Saxon elements (cf. p. 105).

Kent and Sussex are two of the most geographically distinct of the early settlement areas, and in both the evidence of burial grounds and place-names is, in general, mutually confirmatory, and consonant with the surviving scraps of written history. In each case there is a first nuclear area near the coast, and a second in the scarp-foot terrain within the Weald. On the other hand, in Essex and the Thames valley region, the problems are more difficult, and the place-name evidence is accordingly more critical.

In Essex, names of early type are conspicuously rare in the London Clay area of the south (Fig. 16). The avoidance of this tract may just be a simple consequence of its physique, and an example of what is practically a general rule; but it should be noted that R. C. Coles claims some

---

[1] J. E. A. Jolliffe, in *Pre-Feudal England* (1933), would regard this settlement area as falling originally within the Jutish province; but at present there is no archaeological evidence to support this hypothesis.

[2] See p. 105 above. It is reasonable to urge that here we must look for the territory of the Limenwara, who gave their name to the lathe.

evidence of Roman centuriation in the London Clay area, and thinks that a surviving Romano-British population may have assisted in diverting the Saxons northward.[1] However this may be, the only groups of early Saxon names near the coast occur in the small patches of "good ground" provided by the loams of the Tendring Hundred, and of the Southend plateau with its extension along the coast to the Blackwater estuary. With these may be included also, Mucking, Fobbing and Corringham, similarly sited on the selvage of river-drift on the northern side of the Thames estuary. A further group of early names lies in the neighbourhood of the Roman road from Colchester, the southern edge of the drift-lands, where gravel, loam and boulder-clay are intimately juxtaposed. This region was evidently approached from the Colne and Blackwater estuaries. Beyond, on the main mass of the boulder-clay, the early names thin out, but they recur in force further west, in the region of "the Rodings". This is a relatively undissected plateau underlain by boulder-clay of heavy type, and the early settlements were plateau, not valley, villages. This region was evidently not entered from the south-east, nor is there any suggestion of penetration from the Thames up the Roding valley. In all probability access to it was gained from the east *via* Stane Street leading directly from Colchester.

It is evident from Fig. 16 that the early East Saxon penetration stopped short of the higher boulder-clay tracts behind the Chalk escarpment, and also failed to cross the Lea. The area north-west of London is remarkably free from names of early type, and there is no evidence of early penetration up the valleys of the Colne and Lea. However, within the area of the Thames river-drifts near London, and in the well-marked gravel-loam plain of south-west Middlesex, there are Barking, Ealing, and other names in -*ing* and -*ham*. The southern tributaries of the Thames show some evidence of early penetration and names in -*ham* are very common south of London. These facts evidently have a bearing upon the interesting and difficult problem of the state of London itself during the Dark Ages. As is well known, medieval London enjoyed privileges in respect of widely spread hunting grounds in the Chilterns, Middlesex and Surrey, which, in the opinion of L. Gomme,[2] represented the effective survival of the "territorium" of Roman London. Both F. M. Stenton[3] and R. E. M.

[1] "The Past History of the Forest of Essex", *Essex Naturalist*, XXIV, 121 (1933).
[2] *The Governance of London* (1907), p. 106; *The Making of London* (1912), p. 70.
[3] F. M. Stenton and others, *Norman London* (1934), p. 6.

Wheeler,[1] moreover, regard these hunting rights as deriving from a state of affairs older than the not uncommon organisation of Saxon towns as civic centres for country areas. Wheeler has adduced evidences of the survival of "sub-Roman" London itself, and he has also brought into the service of his argument the interesting and novel hypothesis that the "Grims Ditches" of Harrow Weald and the Chilterns, together with the "Faestendic" near Bexley, are boundary marks between the surviving sphere of London's influence and the surrounding lowland farming lands of immigrant Saxons, particularly of those "thrusting southwards unchecked until they reached the reverse slopes of the hills fringing the London Basin". That this was the state of affairs at least north of London seems very probable. There is evidence from more than one source which suggests the survival of a Romano-British population in the Chiltern area. On the other hand, there were, as we have seen, early Saxon settlements in the valleys of the Thames and its southern tributaries near London, evidenced both by burial grounds and place-names. The surviving "territorium", if such it was, must have been Anglo-British, rather than a Romano-British from a very early date.

A further question arises: did the presence of London, surviving intact and unreduced, bar progress up the lower Thames, as has sometimes been supposed? Too little attention has been given to A. Major's suggestion, perfectly feasible on geological or physiographic grounds, that parts of the alluvial stretch south of the river between Limehouse and Lambeth reaches would have been water-covered at high tide, and that, as a result, London could have been effectively bye-passed by vessels passing upstream.[2] Whether by this route or others, upstream river-side areas were certainly occupied early; as witness the settlements of the Middlesex loam plain and those in the Mole and Wey valleys (Dorking and Woking). Farther afield there is a significant group of names in -ing upon the favoured Bargate loams of west Surrey; while upon the outcrop of the Bargate or Sandgate beds occurs a notable group of heathen names (Thursley, Tuesley and Peperharrow). Farther westward, penetration is indicated as far as the neighbourhood of Farnham (near which a *Bintungas* is recorded), and so on to Basing and Worting along the margin of the Chalk outcrop. These settlements were isolated from the Thames valley by the great barren tract of the Bagshot country. In the middle Thames region,

---

[1] "London and the Grim's Ditches", *Ant. Journ.* XIV, 254 (1934). For a discussion, see also *Antiquity*, VIII, 290, 437, 443 (1934).

[2] *Proc. Croydon Nat. Hist. and Sci. Soc.* IX, 1 (1920). See also T. Codrington, *Roman Roads in Britain* (1905), p. 56. See above p. 62.

names in -*ing* are relatively few (Sonning, Reading, Wokingham, Goring, Ealing on the Pang, Wasing on the Kennet). All these westerly settlements may be interpreted as marking penetration from Thames mouth, but it is fair to add that a hypothesis of penetration from the west is in many respects equally reasonable. In south-west Middlesex and western Surrey particularly, it is possible reasonably to regard the settlement as preceding, or immediately following, the West Saxon conflict with Kent in 568, as recorded in the Chronicle; downstream penetration from the upper Thames region would have led naturally to many of the occupied areas. But this theory raises certain difficulties which must be discussed now.

It is a striking fact that beyond Goring Gap, names in -*ing* are exceedingly few and far between. There is no continuous string of such names in the Chiltern scarp-foot vale and, curiously enough, the isolated Buckinghamshire group of Tring,[1] Halling, Wing and Oving, occurs precisely in the region where evidence of the earliest burial grounds fails us. Here then is a case in which the place-name evidence seems definitely at variance with that of the burial grounds, and as a result, the enigmatical origins of Wessex recede still farther into obscurity. A. Mawer and F. M. Stenton state that "place-names which suggest early settlement seem to decrease in passing from Oxfordshire through Buckinghamshire into Bedfordshire, and to increase in passing from Bedfordshire through Cambridgeshire into Suffolk".[2] Thus it may be claimed that the place-name evidence enforces certain doubts which arose in our discussion of the archaeological evidence. Both lines of enquiry indicate the existence of early settled areas at either end of the vale, and no doubt, certain contacts were maintained between them. But there is no evidence that the intervening tract was extensively settled prior to the date assigned to Cuthwulf's much debated campaign. The upper Thames region has been regarded latterly as the true original nucleus of the West Saxon kingdom; we shall not find cause to dissent from this view, but such a conclusion must find substantiation upon grounds other than that of place-names alone.

On Fig. 16 the separate settlement of the Norfolk and Suffolk areas is plainly shown; the belt of unsettled country between them extends from the coast at Southwold, *via* Harleston and Attleborough to the Fenland. In the west, this belt includes the Breckland, while in the east it traverses

---

[1] Prof. Ekwall informs me that further material now available suggests that this is not a name in -*ing*.

[2] A. Mawer and F. M. Stenton, *The Place-Names of Bedfordshire and Huntingdonshire* (1926), p. xv.

9

the heavy land on the Suffolk boulder-clay. Within the Suffolk region there is a clear distinction between an eastern and a western group of settlements. In east Suffolk and south-east Norfolk, the settlement lies, not upon the coast, but on the landward side of the barren heaths of the "sandlings". The estuarine entrances sunk appreciably below the level of the heathland plateau, lead through it to the boulder-clay margin beyond; the morphology of the coastal zone is in complete contrast to that of Kent or Sussex. The early names are most thickly clustered between Ipswich and Halesworth, but thin out rapidly westward. In the extreme west, and extending into Cambridgeshire, there is another distinct group of names. We can hardly doubt that this region was entered from the west *via* the Fenland rivers. The burial grounds of the Lark valley would in any case indicate an early entrance from this direction, but the names in *-ing* do not occur with the burial places on the low ground, but upon the wooded boulder-clay plateau. Sir Cyril Fox would therefore regard these settlements as dating from the later Christian period,[1] and this cannot be ruled out as impossible. But similar sites with early names are found elsewhere in East Anglia and in Essex, and in view of the consistent "geographical behaviour" of the names in *-ing* as a whole, it may well be that these names signify early settlement on the plateau.

The main features of the Suffolk settlement are reproduced in Norfolk, though with some significant differences. Names in *-ing* are common on the eastern margin of the boulder-clay plateau, but disappear towards the forested interior. In the Norwich loam region, however, the conditions are different and recall those of Sussex. Here are roughly a score of names in *-ing* or *-ingham*, many lying near the marshland edge, the former strand-line. North of the Cromer moraine is another small group of names in *-ing*, on a boulder-clay ledge between the gravel ridge and the coastal marshland. Another small group occurs in the Lower Greensand region and this is in some sense the counterpart of the west Suffolk settlement area from which it is separated by the Breckland.

In the Mid-Anglian region the outstanding fact is the sparseness of early settlement judged by this criterion. Appearances may be misleading, however, in some degree, since it is difficult to assess the effect of later Danish influence. A small group of early names occurs on the Fenland silt between Spalding and Boston, and a rather larger group in Kesteven, west of the Car Dyke. Several of the latter are on or near the Cornbrash out-

---

[1] *The Archaeology of the Cambridge Region* (1923), p. 275.

crop, the zone of dip-foot springs. Further groups occur in the Welland valley above Stamford and in the Ouse valley of Huntingdonshire. But it is in the valleys of the Nene and its tributaries about Wellingborough that the most considerable evidences of early settlement are found. It is here that several of the ancient folk-names mentioned in the Tribal Hidage are probably to be identified. Moreover, in addition to names in -*ing* (such as Billing and Kettering), there are other names of ancient type (such as Oundle and Naseby). Harrowden and Weedon contain elements relating to heathen worship, and there is reason to regard Wellingborough Orlingbury and Kislingbury as probably early.[1] Moreover, it will be recalled that the region affords abundant evidence of pagan burials.

To the north, the interpretation of the evidence becomes even more dubious. We do not know the extent to which early Anglian names were replaced by those of Danish or Norwegian type.[2] Thus the suffix "worth", a generally early Anglian type, is fairly common in Durham and the West Riding of Yorkshire, where Scandinavian influence was slight, but it is almost absent in the North Riding which suffered a more thorough Scandinavian settlement. The obvious inference is that it was replaced in the latter region and the same may be true of other early names. Moreover, it is far from certain that all the surviving early Anglian names represent sixth-century settlement as they probably do in the south. Certain names in -*ing* and -*ingham* in West Yorkshire can hardly have been given before the fall of the Celtic kingdom of Elmet in the seventh century, a fact which suggests the later survival of this type in the north. While, however, it is clearly necessary to bear these doubts in mind, it cannot be gainsaid that the distribution of names in -*ing* and -*ingham* presents a coherent and reasonable picture, remarkably accordant with that based on the pagan burial grounds, and in the coastal regions at least, we may take these names as indicating primary settlement with a high degree of probability, without insisting on literal contemporaneity with the south.[3]

Fig. 16 shows that the names in -*ing* and -*ingham* cluster at the marsh-edge at the eastern foot of the Lincolnshire Chalk, and occur also in the broader southern end of the drift-filled vale west of the Wolds. They occupy scarp-foot and dip-foot sites along the Oolitic belt, and extend

[1] J. E. B. Gover, A. Mawer and F. M. Stenton, *The Place-Names of Northamptonshire* (1933), p. xvii.   [2] See Eilert Ekwall, pp. 137 ff. below.

[3] The secondary or expansion settlement tends to occur around the primary nuclear areas and a suggestion is conveyed, that the Scandinavian settlement in some measure filled the gaps between the Anglian groups.

along the Trent valley as far as Nottingham. In this direction, as in the upper Thames area, place-name evidence fails to indicate so deep a penetration as is suggested by the distribution of burial grounds.

Farther north there are settlements on the sheltered landward edge of the Holderness peninsula and on the drift-lands around Great Driffield. They also appear in the scarp-foot zone of the Wolds, below the western edge of the Howardian Hills, and upon the margins of the great Pickering depression. Lastingham, north of the latter, is neatly sited at the foot of the great Corallian escarpment, south of the sandstone moorlands. The latter area is destitute of evidence, save for the name Fyling south of Whitby, and the same is true of the whole northern portion of the Vale of York, and southern Durham, where Billingham is the only name plotted on the map. Still farther north, however, between the Cheviots and the sea, there is a distinct group of names, marking no doubt the chief southern nucleus of the Bernician kingdom.

West of the Vale of York and the coastal lowland in Durham, there is a string of settlements showing a certain consistency of siting. It extends from near Pontefract to the Tyne, following roughly the line of the Magnesian Limestone (drift-covered in the north) and the Millstone Grit hills of west Durham. Though changing its character in detail, this zone marks in reality the raised western border of the great Triassic depression and its northern coastal continuation. It is, in some sort, the foothill zone of the Pennines, and in north Yorkshire it is followed by the Roman road (Leeming Lane), close to which many of these settlements lie. Penetration farther west may be indicated in the north by Bellingham, and by Beltingham which lies in the Tyne gap between Hexham and Haltwhistle. Similar westward penetration *via* the Craven gaps may be marked by the names Manningham, Bowling, and possibly by Cowling, in Airedale and its connections, though it is very doubtful whether the last two names are genuine derivatives from compounds in -*ingas*, and in any case there are good reasons for regarding the names as relatively late. Westwards again, names in -*ing* and -*ingham* are thinly scattered, in Cheshire and Lancashire; and some occupy coastal situations in the latter county and in Cumberland. There seems no safe warrant for regarding these names as indicative of genuinely early settlement,[1] and certainly not as evidence of settlement from the west.

[1] These names are not included on the map (Fig. 16).

THE EMERGENCE OF THE KINGDOMS

Before pursuing the story of the Saxon settlement farther in space and time, some attention must be paid to the record of written history in so far as it bears on the events of the Early Period. The two lines of argument based upon burial places and place-names generally coincide, indicating nuclear areas of settlement. But these were far from being coextensive with the kingdoms of the Heptarchy as they ultimately emerged in the light of history. Calculations from the surviving royal genealogies indicate dates of foundation for many of the kingdoms consistent with the archaeological and other evidence; dates, moreover, which place them, as current theory has demanded, in the phase following Mount Badon. But that definite boundaries arose so early, as supposed by Oman, is most improbable; in any case, they were certainly not the later boundaries of the kingdoms, and we cannot claim knowledge of where they lay. The whole tenor of the archaeological and place-name evidence favours the conception of growth from nuclei. That a number of small kingdoms arose in place of one or more large states is a fact reflecting the comparatively small numbers of the immigrant peoples, coupled with the scattered distribution and well-marked physiographic separation of the tracts favouring early settlement.

The coastal kingdoms were all of relatively early foundation. The Kentish and South Saxon kingdoms can be assigned, following the Chronicle, to the latter part of the fifth century. According to the evidence outlined above, it is likely that both occupied relatively restricted areas at an early stage, though the coastal belt between their respective heart regions seems to have been quickly settled. This conclusion admittedly runs counter in part to the findings of J. E. A. Jolliffe,[1] who brings new evidence to bear on the problem. We cannot here do justice even in briefest summary to his closely reasoned argument. It must suffice to state that in respect of settlement type, manorial structure and custom, he finds many points of affinity between Kent, Sussex, south Hants and east Surrey, and concludes that "the first great phase of the English invasion was a general settlement of England south of the Thames, as far west as the Hampshire Avon, by a homogeneous people", whom he identifies as the Jutes. He supports archaeological conclusions in emphasising the essential unity in culture and organisation between the Jutes and the Franks, and accounts for the admitted Saxon element of the Sussex coastal plain by supposing that it may represent a Saxon *enclave* swept into a

[1] See above p. 114, footnote 1.

Frankish enterprise, and derived from the region of supposed Saxon settlements on the coast of Picardy.

This new and challenging theory demands very careful attention for its general suggestions are not incompatible with the conclusions of history, archaeology or place-name study. Moreover, it falls harmoniously into line with Collingwood's conclusions that certain of Arthur's battles were fought within the Weald.[1] Looking at the matter from a geographical point of view, it is not enough to conclude that the wide distribution of hamlets of Kentish type is simply a reflex of water-supply or other physical conditions. In other places these factors have no doubt been paramount in controlling village morphology, but in the area studied by Jolliffe, hamlet settlement extends over terrains of very diverse physique, and a fair weighing of the evidence must surely enforce the conclusion he himself reaches that the phenomenon is essentially "racial" in some sense or other. Nevertheless, "a general settlement of England south of the Thames" seems to be too large a claim. In the light of other evidence, it would probably be much truer to say "a general settlement of the habitable fringes of the Weald"—these being taken as including the dip-foot and scarp-foot zones of the Chalk together with the coastal lands between Hastings and Folkestone. Early settlement within the central Weald has yet to be clearly attested; the extensive and systematic exploitation of the appurtenant woodlands on the sandstone core behind the encircling girdle of clay forest must surely have waited upon the phase of secondary settlement or expansion.

Essex appeared first in history as a sub-kingdom under Aethelbert of Kent (560–616), but genealogical arguments indicate a possible first consolidation about 520–30. There is no evidence that it extended westwards to include London or eastern Hertfordshire until a century later. Genealogical arguments similarly indicate a date in the early sixth century for the foundation of the East Anglian kingdom, or of the two early sub-kingdoms from which it was compounded. There is every reason to regard the foundation of Lindsey as early, but the definite emergence of Bernicia and Deira as states is deferred, so far as present evidence goes, till the years 547 and 560 respectively. Royal genealogies admittedly carry us back to the beginning of the fifth century, and the coastal settlements are evidently early, but their scattered distribution supports the idea that a number of small communities existed long before the foundations of the major states.

[1] "Arthur's Battles", *Antiquity*, III, 292 (1929). See also O. G. S. Crawford, "Arthur and his Battles", *Antiquity*, IX, 277 (1935).

While the coastal kingdoms emerged early, circumstances were far otherwise in the interior of the country. In the Midland Plain, and in the isolated region of the lower Thames basin, the emergence of definite kingdoms was deferred until a later date—a fact of the highest significance in both the history and geography of Saxon England. It is at this point that we must at last confront the enigma presented by early Wessex. Archaeological evidence very strongly indicates the upper Thames region as the essential nucleus of Wessex,[1] and force is added to this contention by the founding, in later years, of the West Saxon bishopric at Dorchester. Nevertheless, the place-name evidence which is so consistent in the coastal regions, here fails almost completely to indicate early settlement. Similarly, in the north-west Midlands, the early place-names do not extend so far inland as the burial grounds. This seeming conflict of evidence may afford an important clue to the nature and sequence of events in the fifth and early sixth centuries. In both the inland regions in question, there existed, if we may trust the archaeological dating, an appreciable English element, at the very time when names in -ing and -ingham were spreading nearer the coast. This seems to imply a difference in the social organisation of the settlers in the inland and coastal tracts respectively. It seems, indeed, reasonable to suggest that in the original Wessex no less than in the original Mercia, the settlers were remnants of invading war-bands, living on terms, if not in amity, with surviving British elements, and probably mixing with the latter to a considerable extent. Nearer the coast, the English element was reinforced by many emigrant groups from the homeland, and some form of community or kindred settlement followed as a normal consequence.

There is a certain amount of positive archaeological evidence in favour of this picture of the upper Thames and Midland areas. Continuous occupancy from Romano-British times onward, on or near the same sites is attested by burials or associated antiquities at Long Wittenham, Reading, and Theale.[2] Hanging bowls with ornamental motives carried out in enamel in the fashion of Celtic and Romano-British production are especially numerous in the Midlands, and are interpreted by Baldwin Brown[3] as evidence of surviving Celtic influence. A scattered English population may thus have mingled with surviving British elements.[4]

[1] See p. 107 above.
[2] R. A. Smith in *V.C.H. Berkshire*, I, 229 (1906).
[3] *The Arts in Early England* (1915), IV, 475.
[4] This conclusion arises, as we have seen, from a comparison of the distribution of early place-names and pagan burial grounds. Several writers have been led to similar

The emergence of Wessex as an important state was deferred until the conquests of Ceawlin, carried out with the help of a mixed host of fighting men—the Gewissae. One element of this host no doubt accompanied the leader and real founder of the West Saxon dynasty from across the southern Chalk hills which Cerdic had reached by his advance along the Cloven Way. These Chalk lands, however, were not effectively fused with the upper Thames nucleus till later times. Similarly, and at a still later date, Mercia emerged under Penda, who united with the original Mercian nucleus the scattered settlers of the midland plain, whose original partition into small groups is attested by the Tribal Hidage.[1] It is not unlikely that the territory around London passed through a similar phase.[2]

These facts and arguments may be summarised in the form of a map (Fig. 17). Here are shown the Celtic areas of the west and the areas of early English settlement in the east, while between them two transitional zones are distinguished (2 and 3). The boundary between the latter is based upon Zachrisson's survey of the Celtic place-name elements *pen*, *bre*, *cruc*, etc. Three or more such names occur in the following counties:

conclusions by other arguments. See especially G. Sheldon, *The Transition from Roman Britain to Christian England* (1932). T. D. Kendrick suggests that the early settlements in the upper Thames region were headquarters of river pirates, comparable with those established much later by the Vikings on the banks of the Seine and Loire in the heart of the Frankish kingdom (*Archaeology in England and Wales, 1914–31*, pp. 304–5). See also R. E. Zachrisson, *Romans, Celts and Saxons in Ancient Britain* (Upsala, 1927), on the general question of fusion between the Romano-British and Saxon peoples.

[1] The Tribal Hidage enumerates the hides of the several districts tributary to the Mercian king in the 7th century. It dismisses the southern kingdoms briefly under single headings but gives details for a number of small tribes or regions in the Midland area: the Peak dwellers, the Wrekin dwellers, Gyrwa, Wixna, Hicca, Sweordora, etc. Though the hidation figures are puzzling, the form of the record plainly indicates the small-scale partition of settlement in the Midlands. See J. Brownbill, "The Tribal Hidage", *Eng. Hist. Rev.* XL, 497 (1925).

[2] We have noted the arguments for the essential continuity of London as an occupied centre retaining something of its Roman tradition, and there is also more than one indication of the survival of British elements in the Chiltern area. English settlements occur, however, close to London, and we can hardly escape the conclusion that an amicable *modus vivendi* between the two peoples must have been reached. It seems probable that Middlesex and Surrey represent the original northern and southern provinces of the London "city state", the two halves later falling apart and coming to own allegiance to the neighbouring expanding powers. See H. J. Mackinder, *Britain and British Seas* (1925), pp. 182, 193; and A. Mawer and F. M. Stenton, *The Place-Names of Buckinghamshire* (1925), p. xiv. See also the Introduction to the Surrey volume of the Place-name Society (1934).

Yorks, Derby, Staffs, Northants, Worcs, Gloucs, Oxon, and Wilts; they
are notably numerous in Lancs and Dorset, but rare or absent east of the
regions indicated. The line as drawn on the map must be regarded as at
best a generalised boundary, separating regions in which an appreciable
Celtic population lingered from those in which purely English outposts of

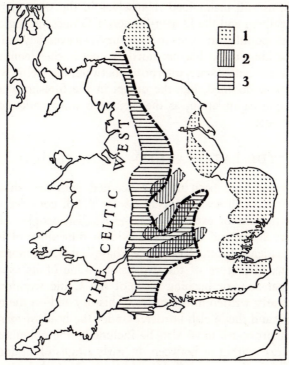

Fig. 17

1 indicates the coastal areas of Early English place-names; 2 indicates the inland areas of frequent burial grounds; 3 indicates the
distribution of certain Celtic place-name elements.

the coastal settlements were widely, though sparsely, scattered. The map
shows, further, in a general fashion, the position of the inland areas of
numerous early burial grounds. These lie within the "sub-Celtic" region,
or athwart its eastern boundary. Their situation thus goes far to explain
their rôle as regions of Anglo-Celtic fusion. So far as they go, the facts
shown in Fig. 17 may be regarded as geographical evidence in favour of
the "Mount Badon theory",[1] though it would be rash to claim that they

[1] See p. 101 above.

proved it. It is certain, however, that our picture of the political organisation and racial geography of southern Britain during the early part of the sixth century must take account of a "Celtic West" more extensive than it soon afterwards became. We know that in addition to the kingdoms of the "inner fastnesses", Damnonia, Demetia, Gwynedd and Reged, two Celtic kingdoms still maintained themselves in the west of the English plain—Theyrnllwg, and the Dorset-Somerset-Gloucester region. A considerable Celtic population seems to have survived even beyond the eastern limits of these kingdoms. Farther north, Elmet in the Pennine foothills, was another small buffer state, interposed between the Celtic west and the English east, and doomed, like the others in the foreshortened view of history, to rapid capitulation, as the Teutonic wave gathered westward impetus once again.

THE LATER ANGLO-SAXON PERIOD

In this study we have been more concerned with first than with last things, and space will scarcely avail us for a full consideration of the secondary settlement or expansion of the English peoples. Paradoxical though it may seem, and despite the evident difficulties of interpreting some of the evidence for the "Entrance Phase", it is still more difficult to present a detailed picture of the expansion. One of its aspects, actual territorial expansion, or pushing back of the Celtic frontier, is indeed sufficiently clearly written in the pages of history and on the maps of the West Country and the Welsh border. Under the head of expansion two other distinct processes must also be included—the colonisation of the "negative" areas within or between the early English kingdoms, and the secondary colonisation or subdivision of the early settled tracts.

Territorial expansion was resumed before A.D. 650 both in the south and the north of the country, so that some of its results may be included on the maps of the "Entrance Phase". There was continual warfare between the petty states of the Heptarchy. In the south, Wessex under Ceawlin turned first against Kent in A.D. 568. This campaign has some bearing on the extent of Kentish influence or territory at that date, and it was perhaps accountable for a certain amount of West Saxon settlement in west Surrey and south-west Middlesex, though it must be noted that *Wibbandun* cannot be identified with Wimbledon, as the site of the Kentish defeat.[1] There followed the much debated Chiltern campaign in 571; and it seems

[1] J. E. B. Gover, A. Mawer, F. M. Stenton and A. Bonner, *The Place-Names of Surrey* (1934), p. 38.

reasonable to accept the outline of the Chronicle story, and to suppose that this campaign, for the first time, welded the Chiltern scarp-foot region (the territory of the Chilternsaetna) into the structure of growing Wessex, although scattered English elements had entered the region long before. Of even greater importance was the western expansion signalised by the English victory at Dyrham, in A.D. 577, and the fall of Bath, Cirencester and Gloucester. This extended the West Saxon kingdom westwards along the scarp-foot vale and the Cotswold dip-slope, and also for a time, in the Severn plain beyond the Cotswolds. The latter region, however, soon asserted a provincial independence as the Hwiccian sub-kingdom, which in its later form was probably roughly coextensive with the diocese of Worcester. The region, however, is a clay-land, and there is no suggestion that it was completely settled, even in outline, as a result of Ceawlin's conquest. The place-names of Worcestershire are generally late in type; there are a few, such as Arrowfield and Weoley, of heathen origin, but many are formed from feminine personal names—a distinctively late feature.[1] It is notable that the heathen burial grounds are confined to the south of the county. In later phases it is clear that Anglian elements from the northeast entered the region. The names Phepson and Whitsun are derived respectively from the folk-names *Feppingas* and *Wixan* of the Mid-Anglian region, and imply the movement of Anglian peasantry into the area, probably after 628, when it became tributary to Penda of Mercia.

In the north, Aethelfrith overran Reged, and made much of the Celtic country subservient to his realm, but it is doubtful whether this involved the pushing west of a definite English frontier. There are reasons for regarding both Bernicia and Deira as Anglo-British rather than purely Anglian states at this stage, and the recognition of the Bernician overlordship by Celtic states may not necessarily have involved the movement into them of large bodies of English settlers.[2] Of more significance was the victory of Chester in 613, a geographical analogue of the Dyrham campaign. The Northumbrian power laid its hand on the Cheshire plain from across the Pennine barrier, presumably by way of the Craven gaps, and thus descended on it from the north. Mercia had not as yet emerged. Nevertheless, the region ultimately passed to Mercia, for the Northumbrian hold was short-lived, and relaxed to the spirited British rally under Cadwallon. In any case, the victory of Chester only gave temporary emphasis to the existence of distinct northern and southern provinces in

[1] A. Mawer, F. M. Stenton and F. T. S. Houghton, *The Place-Names of Worcestershire* (1927), Introduction.
[2] G. Sheldon, *op. cit.*, pp. 101 *et seq.*

the Celtic west, provinces of which the basis was geographical in the simple sense. South Wales was within sight and easy reach of Damnonia, but north Wales was separated from the northern uplands by a broad expanse of forested and marshy plain. But the Northumbrian conquests were characteristically incoherent and ephemeral owing to the strategic shortcomings of the narrow and lengthy coastal plain upon which the state was based, and, owing to the marked separation of the Deiran and Bernician nuclei, only sporadically at one with each other. It needed the emergence of Mercia to complete the conquest of the English plain.

The creation of Penda's Mercia involved the co-ordination of the scattered Anglo-British areas of the forested Midlands, and there is little need to speak here of its significance in early English history. It held apart for a time the northern and southern compartments of English Christendom. Backed by the Welsh upland and in alliance with its people, it may have arrested for the time being the completion of the conquest of the plain as a whole, only perhaps to render it more secure and final in due course. The western expansion of Wessex was perhaps resumed, following the battles at Burford and Penselwood (A.D. 658). It is clear that the wooded belt below the western brink of Salisbury Plain formed the march of Wessex until this date, and it is probable that little progress was made beyond it until the wars of Ine. Ultimately, however, the inviting "Oolitic country" beyond it was entered upon; the creation of the bishopric of Sherborne marks the effective settlement and consolidation of this tract. The English had probably settled in the Exe valley vestibule by A.D. 700 or shortly afterwards; Exeter was in their hands by A.D. 690. The settlement of west Devon then slowly proceeded until the eleventh century and later. To a considerable extent, the place nomenclature of north and south Devon differs, suggesting separate settlement from Somerset and Dorset respectively, by way of the two distinct lowland routes lying north and south of Dartmoor. The most remarkable feature of the region is the prevailingly English character of the nomenclature as a whole; the Celtic element is less marked than in Dorset or Somerset.[1] It seems probable that the English expansion involved no considerable displacement of Celtic peoples. The region was thinly peopled in Romano-British times and was probably heavily depopulated by the sixth-century exodus of the Britons to Brittany. With the fall of the last Celtic power in Dorset and Somerset, there was little to oppose the English expansion.

Shortly before Wessex was at war on its Penselwood frontier, the

[1] J. E. B. Gover, A. Mawer and F. M. Stenton, *The Place-Names of Devon*, Part I (1932), Introduction. See also the review by J. E. A. Jolliffe, *History*, XVIII, 38 (1934).

Mercians obtained a lodgement on the Hereford plain and settled an area which became the diocese of Hereford. More than a century later, Offa of Mercia devastated parts of Wales and established his boundary mark in the form of Offa's Dyke. In the north, this placed the whole of the English lowlands in his hands and included parts of present-day Wales, but in the south it failed to embrace the whole of the Hereford plain, which, as Fox has shown, was still in large part heavily wooded.[1]

The problem of the secondary Saxon settlement, in and between the early settled tracts of the plain, presents many difficulties. It is perhaps best studied in south-east England, where the earlier settlement affords a basis for comparison, while the complication of Scandinavian settlement is slight or absent. It is easy to beg the question in dealing with this matter. Clearly, if the place-name maps for the Entrance Phase could be claimed as complete, all other names occurring within or between the early settled tracts would represent subsequent expansion. Perhaps in a broad sense this is true, but the matter cannot remain there, for, apart from the fact that these maps certainly omit certain names whose early date cannot be proved, we wish to exclude also any late names. These difficulties make it impossible to do more than offer, as a contribution to the problem, certain maps definitely limited in basis, but from which nevertheless several tentative conclusions can be drawn.

The suffix -ton continued in use long after the early stages of the settlement,[2] as witnessed in its numerous occurrences in the west Midlands in combination with personal names. Moreover, the related pairs of names Winteringham, Winterton (Lincs), Nottingham, Sneinton (Snotingatun)[3], convey a clear suggestion that the -tun was sometimes at least a secondary settlement. It is not impossible that some of the names in -ton are early, for a longer life as a suffix does not necessarily imply a later birth. Nevertheless, it is worth while to adopt the working hypothesis that the names in -ton, as a whole, represent the expansion phase. In Fig. 18 names in -ing, -ingham, -ham, and -ton are plotted for south-east England; comparison with Fig. 16 brings to light some interesting facts. The chief expansions of settlement judged by this criterion are in the following areas: the Wessex Chalk area and the western portion of the South Downs, the Sussex levels and the Rother valley (west Sussex), the Thames and Lea valleys and the vales of Oxford and Aylesbury. Beginnings of settlement in the central

---

[1] *The Personality of Britain* (1932), p. 52.   [2] See p. 110 above.
[3] A. Mawer and F. M. Stenton, *Introd. Surv. Eng. Place-Names*, Part I, pp. 45–6.

Weald, the Bagshot country, the Chiltern plateau and the Middlesex and
Hertfordshire uplands, are also apparent. While the comparison formally
proves nothing, it suggests a great deal and the picture it presents is
essentially credible geographically. In detail, the distribution of the names

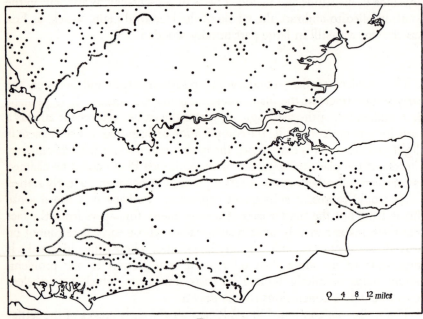

Fig. 18

The distribution of place-names ending in *-ing*, *-ingham*, *-ham*, and *-ton* in south-
east England. The lines of the Chalk and Greensand escarpments are indicated,
together with the River Thames.

in *-ton* is both interstitial and peripheral with respect to groups of
supposedly earlier names. The areas of early settlement have grown
normally, and many of the significant blanks of the early map have been
filled up. But the main negative regions surrounding London in the
earlier phase still remain clearly marked.

The suffixes *-ley*, *-den*, *-field*, etc., which are the names of topographic
features, have some significance in this connection. Presumably they are
of a relatively late date. In Essex, for example, there is a markedly
differential distribution between names of this type and the earlier group.[1]

[1] See S. W. Wooldridge and D. J. Smetham, "The Glacial Drifts of Essex and
Hertfordshire, and their Bearing upon the Agricultural and Historical Geography of
the Region", *Geog. Journ.* LXXVIII, 260 (1931).

Not only do the "topographic" names occur amongst the earlier names in the north of the country, but they are notably concentrated in the London Clay area of the south, marking, with little doubt, the spread of secondary settlement in the heavy forest.[1] Very similar results are obtained

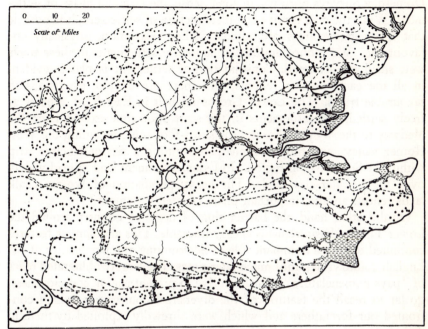

Fig. 19

The distribution of Domesday vills in south-east England. The physiographic boundaries shown may be recognised by reference to Fig. 13, or to the Ordnance Survey ¼-inch Geological Maps. Alluvium is shaded.

by plotting the distribution of the names in -*field* and -*ley* for a portion of the Wealden area. It appears that certain elements in the place-nomenclature of the central Weald are of high antiquity. That the partial occupancy of the area was early, there is indeed no reason to doubt. It is possible that there has been a tendency to over-estimate its repellent

[1] R. C. Coles, *Essex Naturalist*, XXIV, 115 (1934). Dr Coles has shown that the distribution of forest in Essex during Saxon times as deduced from place-names giving evidence of forest, open place, or woodland clearing, yields results consistent with those deduced from a study of the Perambulation of the King's Forest in A.D. 1225. The Place-Name Society's volume on Essex (by P. H. Reaney, 1935) gives some interesting distribution-maps of place-name elements important in this connection.

qualities as a "trackless waste", but the facts shown in Fig. 18, if they are to bear any analogy with those of neighbouring regions, indicate the central Weald as an area of secondary settlement.

The map (Fig. 19) showing the distribution of vills mentioned in the Domesday Surveys portrays, in some sense, the terminal phase of settlement. While the map is not a complete picture of settlement, it is notable that the negative areas are, almost without exception, physically unfavourable. We cannot doubt that any early settlements in these tracts were small and scattered. Consolidation of settlement is plainly evident in all the early nuclear regions, and there are definite extensions in favourable tracts. The Lower Greensand belt in East Kent appears as freely settled, and settlements in the scarp-foot zone extend from the Medway to the Wey. In Sussex, the scarp-foot zone, including the West Rother valley, was fully occupied, constituting a second nuclear strip within the South Saxon region. In Essex, settlement had spread widely over the boulder-clay lands. The heavy population of Wessex chalk lands reflects another important phase of later settlement.[1]

On the other hand, the barrier-ring of negative country around the Lower Thames basin remains clearly traceable. This and similar features continued to play a dominant rôle in determining the personality of the English countryside, until much later times. The largely obliterated relics of "pays nomenclature", which survive in present use or consciousness, go far to recall the features of the diversified countryside which confronted our forefathers and which were shrewdly exploited by them in their first settlement of a foreign land.

BIBLIOGRAPHICAL NOTE

The footnote references given in this chapter constitute a skeleton bibliography for the subject. They may be supplemented with advantage by (1) the invaluable publications of the English Place-Name Society to which reference has been made; (2) the current relevant archaeological journals. The following two studies should be consulted: R. H. Hodgkin, *A History of the Anglo-Saxons*, 2 vols. (1935), and R. G. Collingwood and J. N. L. Myres, *Roman Britain and the English Settlements* (1936).

[1] See, also, Professor Stenton's discussion of "-field" terminations in south-east Berkshire in A. Mawer and F. M. Stenton, *Introd. Surv. Eng. Place-Names*, p. 37. The whole chapter on "The English Element" is very relevant. See p. 179 below.

Chapter IV

THE SCANDINAVIAN SETTLEMENT

EILERT EKWALL, Ph.D.

It was early in the ninth century that the Scandinavian attacks on the coast of the British Isles commenced in earnest.[1] Shortly after the middle of the century, the Danes began to winter in England, now in Kent (851), now in Mercia (868), now in Northumbria (869); and these winterings formed the prelude to systematic settlement. The chief source of information is the Anglo-Saxon Chronicle which, year by year, records the movements of the Danish army in England. Unfortunately, after the first settlements had been made, the chronicler apparently ceased to take an interest in the districts occupied by the invaders, except to mention occasional raids and to note the recapture of the Danelaw. In any case, his statements are to some extent misleading, unless interpreted in the light of other evidence.

In 875 a portion of the Danish army under Healfdene marched into Yorkshire (which seems to have been under Danish sway since 867), and finally occupied the land south of the Tyne. In the next year Healfdene "dealt out the lands of Northumbria; and they began to plough and till theirs (i.e. their land)": thus briefly the Chronicle records the Scandinavian settlement in Yorkshire. In 877 another division of the army seized part of Mercia, leaving the remainder to the Mercian king, Ceolwulf. In 880 a third part of the army settled in East Anglia and "divided the land". No further settlement by the host in England is recorded in the Chronicle.[2]

[1] Under the year 787, the Anglo-Saxon Chronicle records how: "this year king Beorhtric took to wife Eadburg, Offa's daughter; and in his days first came three ships [of Northmen out of Haeretha-land]. And then the reeve rode to the place, and would have driven them to the king's town, because he knew not who they were: and they there slew him. These were the first ships of the Danishmen which sought the land of the English nation". The bracketed words are not in the Parker Chronicle.

[2] But the anonymous *Historia de Sancto Cuthberto* (c. 1050) apparently records a fourth systematic settlement in Durham between the years 912–15. There, Ragnall, later king of Waterford in Ireland, divided the villages belonging to St Cuthbert beyond the Tees, and gave one part from Eden to Billingham (on Tees) to one Scula, the other from Eden to the Wear to one Onlaf-ball. See *Symeonis Monachi Opera Omnia* (ed. T. Arnold, 2 vols. 1882–5, Rolls Series), i, 209. But the statement refers only to the coastal district, though it is just possible that the interior parts were included. Scula's name may enter into School Aycliffe north of Darlington.

The boundary of the Danish territory (i.e. the Danelaw) is not mentioned in the Chronicle, but the treaty concluded between King Alfred and Guthrum, king of East Anglia,[1] shows that the Scandinavians had seized not only East Anglia proper, but also much land to the south in Essex, Herts, Beds, and perhaps Bucks; the boundary between Danish East Anglia and the English kingdom was to follow the Thames, the Lea to its source, a straight line from that point to Bedford, and then the Ouse as far as Watling Street.[2] On the boundary between Danish and English Mercia there is no contemporary evidence. But a passage in the so-called Laws of Edward the Confessor (c. 1135) refers to the boundary as Watling Street and 8 miles beyond it;[3] the counties included under Danish law were those of York, Lincoln, Nottingham, Leicester and Northampton; Derbyshire must have been omitted by mistake. The boundary to the west is unknown. Some scholars include in the Danelaw parts of Staffordshire, Cheshire and Lancashire; but it may be doubted if a definite line was ever fixed in the west where mountains provided a natural frontier zone. In the north, the Danish territory extended to the Tyne, over Yorkshire and Durham; while Northumberland is stated, in a thirteenth-century source, to have been outside the Danelaw.[4] Whether the Danelaw included also Westmorland and Cumberland or not, is not known. There are some indications that at least some portions of these counties did belong to it.

The reconquest of the Danelaw really began with the Peace of 886. A large part of Essex was retaken in 913. East Anglia followed three or four years later. The exact dates are, of course, uncertain; but by 919, the whole of Mercia was in English possession. After that year, it is unlikely that any fresh settlements were made south of the Humber. The period of primary Scandinavian settlement then, in the southern area, fell between 877 and 913–19. But the northern kingdom of York, including Yorkshire and the adjoining districts, remained independent much longer; and the period of settlement was correspondingly prolonged. Here the Danes were sometimes independent, at other times under English suzerainty. Some of their

[1] The exact date of the treaty is not known. F. Liebermann, *Die Gesetze der Angelsachsen* (Halle, 1898–1912), III, 84, assigns it to the year 880, or at any rate to the time before 884. English scholars generally give the date as 886.

[2] On later statements concerning Bucks, see below, p. 154.

[3] In the earlier version, the word "miles" (*milliaria*) is omitted, but it is found in most MSS. of the later version. F. Liebermann thinks eight counties beyond Watling Street are meant. *Die Gesetze der Angelsachsen*, II, 347.

[4] See H. Lindkvist, *Middle English Place-Names of Scandinavian Origin* (Upsala, 1912), p. xxx, for a discussion of this point.

independent kings were scions of the Scandinavian royal line at Dublin. It was not until 954 that they finally submitted to Eadred, when they were left apparently in undisturbed possession of their land, to retain their own customs and laws. And it was not until after Sweyn (1013–14) and Canute (d. 1035) had made themselves kings over the entire country that an independent Danish political power ceased to exist in England.

While the Danes sailed southward along the east coast of Britain, the route taken by the emigrants from Norway was westward and north-westward. And so, along the western shores of the British Isles, there was another and a rather different Scandinavian immigration. Settlements of Norsemen were early established in the Orkneys and in the Shetlands, as well as in the Scottish isles. Eventually, the Norse power became so strong that earldoms were created in these outlying fringes; and in the Isle of Man there was a line of Norse kings. Powerful Norse kingdoms were also founded in the coastal towns of Ireland. The kingdom of Dublin was particularly strong, and its kings ruled both in Dublin and in York during the first half of the tenth century. It was therefore impossible for the western shores of England and Wales to escape from Scandinavian influence in the ninth and tenth centuries. It is evident in Cheshire, in Lancashire, in Cumberland, in Westmorland, and in Wales. The informa-tion is scanty enough, but these immigrants were probably Norwegians, who came over from older colonies in Ireland, the Isle of Man, and the Hebrides. The intimate connection between the kingdom of York and the Norse kingdom of Dublin may have been a factor in this influence. An Irish source provides an important statement on the settlement in Cheshire.[1] According to this, one Ingemund with his men was expelled from Ireland about 901 or 902, and he was given land near Chester (*Castra*) by Æthel-fled, Lady of the Mercians. For the rest of the western region, however, there is no contemporary evidence; but it may be supposed that the Norwegian settlement in general took place about the same time as that of Cheshire. Further, it is generally held that much of the Norwegian influence was peaceful in character; the settlers either bought land, or exploited hitherto unoccupied country.

This short summary, based on the meagre documentary evidence, leaves many questions untouched, for a conquest by arms does not necessarily imply actual settlement. One would like to know not only what districts were under Scandinavian sway but also where Scandinavians actually settled; and one would like to know, further, not only what were

[1] J. O'Donovan, *Annals of Ireland: three fragments copied from the ancient sources....* (Irish Archaeol. and Celtic Soc. 1890.)

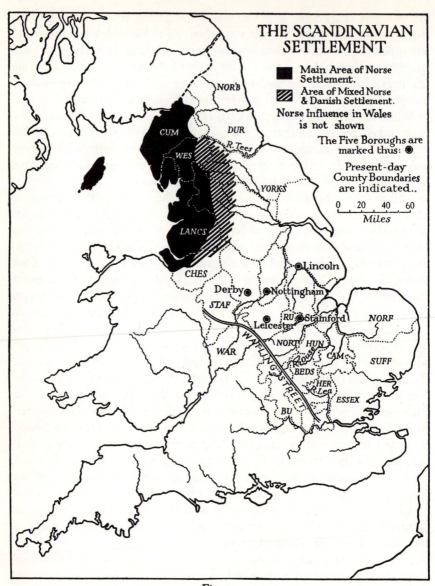

THE SCANDINAVIAN
SETTLEMENT

■ Main Area of Norse
Settlement.

▨ Area of Mixed Norse
& Danish Settlement.

Norse Influence in Wales
is not shown

The Five Boroughs are
marked thus: ◉

Present-day
County Boundaries
are indicated..

0   20   40   60
Miles

NOR'B

DUR

R.Tees

CUM

WES

YORKS

LANCS

CHES

Lincoln

Derby ◉ ◉ Nottingham

STAF

RU ◉ Stamford

◉ Leicester

NORF

NORT HUN

WATLING STREET

R.Ouse

WAR

CAM

SUFF

BEDS

HER R.Lea

ESSEX

BU

Fig. 20

It is impossible to assign a definite boundary to the area of mixed settlement. There
were also scattered Norse colonies in the Danish areas, but their extent cannot be
safely determined, and they are not marked on this map. The Ridings of Yorkshire
and Lincolnshire are marked but not named. Note that Norse influence in Wales is
not shown.

the geographical variations in the intensity of their settlement but also what were the relations between the new settlers and the earlier population. Answers to these and to other questions are provided by the evidence of place-names (together with personal names and field-names) and by the conclusions of economic and social history. It is the former source that provides most material for a geographical survey of the settlement. But the latter source, so important from other points of view, also throws much light upon the territorial problems.

Some general notes on place-names as historical and geographical evidence are very necessary by way of introduction.[1] The interpretation of place-name material is a very intricate task. It is not enough simply to count Scandinavian and English names on the map, and to draw conclusions from numerical relations. Indeed, it is not always easy to distinguish definitely between the two sets of names. Earlier scholars often ascribed a Scandinavian origin to place-names or place-name elements that are undoubtedly English. Finally, a distinction must be made between various kinds of Scandinavian influence. Three types of names may be recognised.

(1) A great many place-names are Scandinavian names in the strictest sense, that is, names formed by people who spoke a Scandinavian language. Absolutely certain examples are names containing a Scandinavian inflexional form, such as genitives in -ar, as Amounderness, Litherland, Scorbrough (OScn *Agmundar nes, Hlíðarland, Skógarbúð*), or -a from -ar, as Hawerby (*Hawardabi*, twelfth century, from ODan *Hāvarthabȳr*), or -s, as Laceby, Rauceby (OScn *Leifs bȳr, Rauðs bȳr*). Here, generally, belong names containing two Scandinavian elements. Sometimes a strictly Scandinavian name was to some extent Anglicised later, as when *Askabȳr* (ME *Askeby*) became Ashby. With this group also may be included names formed by Scandinavians but containing an Old English place- or personal name, which is often inflected in the Scandinavian way. Thus Aller-, Enner-, Miter-, Nidderdale contain Scandinavian genitive forms in -ar of OE (originally British) river-names (Ellen, Ehen, Mite, Nidd). Similarly, Atterby, Audleby and Barnetby contain Scandinavian genitives in -a of the OE personal names *Ēadrēd, Aldwulf, Beornnōþ*.

(2) A second and very important group is formed by Scandinavianised names. Many English names contained sounds or sound-groups unfamiliar to Scandinavians, who were apt to substitute sounds familiar to themselves. This was a very common phenomenon, and many English place-names have come

---

[1] Note the following abbreviations used in this chapter: OScn = Old Scandinavian; ODan = Old Danish; ON = Old Norse; OE = Old English; OIr = Old Irish; Swed = Swedish; ME = Middle English; DB = the Domesday Book; H = Hundred.

down to us in a Scandinavianised form. Examples are numerous names in *Sk-*, as Skelton, Skipton, and names such as Keswick, Kildwick, Louth, or Mythop, for what should have been Shelton, Shipton, Cheswick, Childwick, Loud, and Middop. The English sounds *sh*, *ch*, and *d* between vowels were unknown to Scandinavian of the time. In Skipwith and Tockwith, *ch* (OE *c*) was replaced by *th*; in Sessay, English *dg* by *ds*, later *ss* (Sessay contains the word *sedge*). A still more common phenomenon was the substitution of a Scandinavian word for an English synonym, whether the word offered phonetic difficulties or no. Thus OE *brād* "broad" was often replaced by OScn *breiðr*, as in Braithwell; OE *circe* "church" by OScn *kirkia*, as in Kirton; OE *cyning* "king" by OScn *konungr*, as in Coniscliffe, Coniston; OE *middel* by OScn *meðal*, as in Melton, Methwold; OE *rēad* "red" by OScn *rauðr*, as in Rawcliffe, Rockcliff; OE *stān* "stone" by OScn *steinn*, as in Stainley, Stainton. Eagle (*Aycle* DB) is a Scandinavianised form of OE *Āclēah*; Beckwith one of OE *Bēcwudu* "beech wood"; Howden one of OE *Hēafoddenu* "chief valley"; Watton one of OE *Wētadūn* "wet place", OScn *eik*, *viðr* "wood", *hǫfuð* "head", *vātr* "wet" having replaced OE *āc*, *wudu*, *hēafod*, *wǣt*. Carl(e)ton is probably as a rule a Scandinavianised OE *Ceorlatūn*. Sometimes a Scandinavian word was even substituted for an unrelated OE word. Thus -*by* has frequently replaced OE -*burg*, as in Badby, Naseby. Very likely -*by* has sometimes been substituted for OE -*tūn*, as in the common name Willoughby. *Willow* is unknown in Scandinavian. Willoughby may be a Scandinavianised Willoughton.

(3) A third group is formed by English names containing a Scandinavian element. The most important among these are names containing an OScn personal name and an English word such as *tūn*, *feld*, Grimston, Kettlestone, Thurstonfield, etc. But *tūn* was a common place-name element in Norway and Iceland, and a name such as Stainton found in Cumberland or Westmorland, etc., may well be strictly Scandinavian. Other hybrids are of less interest. Many Scandinavian words with a topographical meaning were introduced into English, and freely used in forming place-names, being often combined with English words, as in Altofts "old tofts", Beanthwaite, Goldshaw Booth. Such names mostly denote minor places of comparatively late origin. They are rarely names of villages. Many Scandinavian words came to be widely used in English dialects, especially in the North, and they may have spread to districts where Scandinavians never settled. But though names of this kind are of minor importance on the whole, a considerable percentage of them in the place-nomenclature of a district may be worthy of consideration.

Of the Scandinavian elements found in place-names, the most important is *by* (OScn *bȳr*), which in Danish means "village" or "town", in Norwegian chiefly "homestead". It is most common in the Danelaw. Names in -*by* presuppose a population that spoke a Scandinavian language. *Thorp* is also important, but OE *þorp* (*þrop*) was by no means rare, and names consisting of *þorp* and an English element may be English. The word was rarely used in Norway and may be looked upon as a Danish testword. In Danish, *thorp* means "a hamlet or a daughter settlement dependent on an older village". English villages frequently had their *thorp*, as Barkby with Barkby Thorpe in Leicestershire. Clearly *thorp* is here used in its Danish sense and, in general, places with names in *-thorp* are probably to be looked upon as relatively later settlements than those with names in *-by*. Among other elements are *toft* "homestead"; *lathe* "barn" (OScn *hlaða*); *thwaite* "clearing" (OScn *þveit*); *garth* "enclosure" (OScn *garðr*); *fell*, *how*, "hill, mound" (OScn *haugr*); *meol* "sandhill" (OScn *melr*); *holm*; *wray* "remote place" (OScn *vrá*); *wath* "ford" (OScn *vað*); *scough*, *with* "wood" (OScn *skógr*, *viðr*); *lund* "grove"; *beck*, *tarn* (OScn *bekkr*, *tiǫrn*); *crook* (OScn *krókr*); *gate* "road, street" (OScn *gata*); OScn *leir* "clay"; *rá* "landmark"; *trani* "crane". Some elements may be used as tests of Danish or Norwegian provenance. Danish testwords, besides *thorp*, are *bōth* "booth", *hulm* "holm". Norwegian testwords are *breck* "hill, hillside" (ON *brekka*), *būth* "booth" (ON *búð*), *gill*, *slack* "valley" (ON *gil*, *slakki*), *scale* "hut" (ON *skáli*).

In drawing conclusions from this kind of material, it should be remembered that at this period names arose spontaneously and were not given deliberately. On the whole, they would be given by neighbours rather than by the inhabitants of the places themselves. Place-names thus indicate the predominant element in a district, but such predominance need not imply numerical superiority. Inferiority in numbers may have been counterbalanced by political or social superiority. A few Scandinavian settlers in an English district would hardly leave other traces on its place-nomenclature than names such as Grimston and Thurgarton. Names of this kind need not point to a considerable percentage of Scandinavians; indeed, they rather suggest that the English element predominated. On the other hand, strictly Scandinavian and Scandinavianised names indicate a strong Scandinavian element in the district, which may have been limited in extent. Even a single place-name of this kind is worthy of notice. It would not have arisen, unless there was an appreciable Scandinavian element in the population. But even a considerable Scandinavian settlement would not necessarily leave strong traces on the place-nomenclature in a district. When Scandinavians settled in old villages, they would frequently take over the old name, which might, if it offered no phonetic difficulties, be preserved unchanged.

THE NORTHERN DANELAW

Yorkshire. The Scandinavian element in the place-names of Yorkshire is very strong, though Scandinavian names do not generally cluster so thickly and so obviously outnumber the English as they do in parts of Lincolnshire and Leicestershire. The district is characterised by an unusual number of Scandinavianised names, while hybrids of the type Grimston are numerous only in the East Riding. There are a considerable number of names in -*by*, altogether some 250, of which about 150 belong to the North Riding. Some of these, however, like Halnaby and Jolby, are of late origin. The place-names in general give the impression that the Danes dealt out the lands of Yorkshire, and settled practically everywhere upon the lowlands and along the larger valleys, taking over many names from an already existing numerous English population. The nucleus of Danish Yorkshire was the capital city of York, whose name is itself Scandinavianised. The city must have been strongly garrisoned and, no doubt, numerous Danes settled in the city as tradesmen and as merchants. The strong Danish element may still be seen in old street-names ending in -*gate* (OScn *gata*), as Coppergate, Fishergate, Fossgate, Goodramgate, Skeldergate; Coney Street, "King Street", may be a Scandinavianised form of OE *Cyninges-stræt*.[1] The immediate neighbourhood of York, however, is perhaps less Scandinavian than might be supposed.[2]

The North Riding. The most considerable settlements in Yorkshire, to judge by place-names, were made in the broad valleys of the Ouse and its tributaries, the Swale and the Ure—that is, the central part of the North Riding. Names in -*by* are very numerous here, and, in some districts they are found in clusters. A good example is provided by the old Hallikeld wapentake between the Swale and the Ure, through which runs a Roman road, Leeming Lane.[3] In Domesday times, Scandinavian names outnumbered English ones, but it is only fair to say that the latter were not few.[4] High up along these rivers, Scandinavian names do occur, but in

[1] See H. Lindkvist, "A Study on Early Medieval York", *Anglia*, XXXVIII, 345, (1926).

[2] But, even so, close to York we find Rawcliffe, Skelton, Haxby, Holtby, Scoreby; and a little to the north on the edge of a wold, there is a cluster of names in -*by*: Huby, Moxby, Brandsby, Dalby, Skewsby, Stearsby, Whenby.

[3] Here we find Milby, Kirby Hill, Leckby, Asenby, Baldersby, Ainderby Quernhow, Sinderby, Roxby, Swainby, Gatenby, Ounesby (lost) east of the road; Melmerby, Exelby, Normanby (lost), west of the road; together with Holme, Howe (OScn *haugr* "mound"), Upsland, Wath and a few names in -*thorpe*.

[4] Examples are Burton (now Humburton), Brampton, Hutton, Middleton, Dishforth, Cundall, Pickhill.

decreasing numbers; thus Askrigg and Aysgarth are high up the Ure. To the north, many are found along the Tees, and they are generally common in the Stokesley district.[1] But, again, many English names are interspersed with the Scandinavian ones. Old Scandinavian names are found along the sea, sometimes forming small clusters (as that around Whitby); they are quite frequent north of the Derwent and in the Rye valley, though here they are more scattered.

The East Riding offers rather difficult problems. Strictly Scandinavian names are on the whole scattered, though small clusters can be noted here and there, and hybrids of the type Grimston are relatively common. But there are numerous Scandinavianised names. There are groups of Scandinavian names: (1) near the coast in the north;[2] (2) east of the Derwent near Malton, itself a *Middeltūn* in disguise;[3] (3) around the lower Derwent;[4] and (4) near the Humber west of Hull.[5] The old Holderness hundred or wapentake has relatively few strictly Scandinavian names, but Holderness itself is Scandinavian, as are Ellerby, Thirtleby, Eske, Wassand and Ravenser Odd, the old name of a promontory. It also contains numerous Scandinavianised names, as Carlton, Coniston, Skeckling, Skeffling, Skirlaugh and Skirlington.

The West Riding is large in extent, but much of it is hilly country, sparsely peopled in Anglo-Saxon days; old names of villages are restricted to the lowlands in the east, and to the river valleys. Here also we find old Scandinavian names indicating that the Danes must have settled practically all over the areas under cultivation. They have left strong traces in the part between the lower Aire and the Ouse, where, however, we do not find many strictly Scandinavian names, but several Scandinavianised name-forms;[6] and also along, or near, the Don.[7] Scandinavian names are also

---

[1] E.g. Rudby, Faceby, Busby, Newby, Swainby, Dromonby, Thoraldby, Kirby, Ingleby Greenhow, Easby, Battersby, Coulby, Thornaby, Ingleby Barwick, Maltby, Stainsby, Tollesby, Carlton, and Skutterskelfe.

[2] E.g. Bessingby, Carnaby, Sewerby, Hunmanby, Flotmanby, Willerby, Flamborough, and others.

[3] Firby, Kirby Underdale, Thoralby, Uncleby, Skirpenbeck, Scrayingham, numerous thorpes, and others.

[4] Howden, Skelton, Belby, Asselby, Barmby on the Marsh, Gunby, Loftsome, Bubwith, Barlby, Osgodby, Thorganby, Lund, Cottingwith, Skipwith, and others.

[5] Anlaby, Ferriby, Tranby, Willerby, Wauldby, Hessle (OScn *hesli*), Melton, and others.

[6] Selby, Lumby, Brayton, Carlton, Gateforth, Scarthingwell.

[7] Barnby on Don, Balby, Cadeby, Denaby, Hellaby, Maltby, Braithwell, Bramwith, Conisbrough, Stainforth, Stainton.

frequent along the Aire and the Calder.[1] The district on both sides of the Nidd is less strongly Scandinavian than might be expected, but scattered names of the strictly Scandinavian type, together with some Scandinavian-ised names, show that there were Scandinavian colonies here also. The districts on the upper Aire and the Ribble are markedly Scandinavian; names such as Skipton, Coniston, Flasby, Earby, Kirkby Malham, may be mentioned. But in this region the Danes met with Norwegians from the west, to whom the Scandinavian settlements are partly due.[2]

The names already mentioned belonged to the original settlement and, roughly speaking, they are those found in the Domesday Book. But the Scandinavians eventually pushed up also into the more difficult hilly country. The Cleveland district in the North Riding, for instance, is very rich in Scandinavian names of hills and minor places. So are the hilly districts in the western part of the North Riding and the West Riding. Very likely the first settlements in these areas were for the most part effected after the Scandinavians came, and by them. Many names of small streams and lakes in these parts are Scandinavian;[3] and names in -garth, -thwaite and the like are frequent. But, it should be added that in the southern part of the West Riding, on the borders of Derbyshire, traces of Scandinavian settlement are slight.

Durham seems to have had Scandinavian settlements on a large scale only in the south along the Tees, where there was one wapentake, Sadberge.[4] To the north, Scandinavian names are very scattered. Durham (*Dunholm*) has at least a partly Scandinavian name, and in it there are many old street-names in -gate (OScn gata). Not far from it, too, are to be found Ornsby, Raceby, Rumby, Copeland, and the river Gaunless. One name in -by, Follingsby, is found on the Tyne near Jarrow; other isolated names include Ouston, Swainston and Throston. Scandinavianised names are absent except along the Tees; and, to sum up, the distribution of place-names suggests that the chief Scandinavian colonisation in Durham was connected with that in the adjoining North Riding of Yorkshire, rather than with Ragnall's activities.[5]

[1] Thus the Huddersfield district has Quarmby, Linthwaite, Slaithwaite, Crosland, Thurstonland, and others; and on, or near, the upper Calder are Sowerby, Stainland, and others.

[2] See p. 158 below.

[3] Kex Beck, Skell, Skirfare, in the West Riding; Greta and Blean (tarn), in the North Riding.

[4] And names such as Aislaby, Killerby, Raby, Selaby, Ulnaby, further Carlton, Coniscliffe, the river name Skerne—all west of Stockton and Billingham.

[5] See footnote 2, p. 133 above.

Northumberland has only slight traces of Scandinavian influence, except possibly in the Newcastle district, where Byker and Walker (*-ker* from OScn *kiarr* "marsh") may be noted. The places stand opposite to Follingsby in Durham. Lucker, near the coast in the north (? OScn *ló-kiarr* "sandpiper marsh"), is noteworthy. Otherwise there are only Coupland and a few hybrids,[1] together with a few names containing widely distributed words such as *crook*, *trane* "crane", and some others.[2] Scrainwood looks like a Scandinavianised name, but is probably to be explained differently.[3]

Westmorland consists of two parts, separated by fells: the old barony of Appleby (or of Westmorland) in the north, and the old barony of Kendal in the south. In the northern part, at least in the upper Eden valley, the Scandinavian element is distinctly Danish in character: there are no village-names of a Norwegian type, and the Irish influence typical of Cumberland place-names is absent. The three names in *-thorpe* (Cracken-, Hack-, Melkinthorpe) point to Danish colonisation. There are many names in *-by*, clustering especially round Kirkby Stephen, near Brough under Stainmore.[4] From this we must conclude that there was a strong Danish colony on the upper Eden, and that it was founded by settlers coming from the North Riding by way of the Roman road that runs from Rokeby, past Bowes, to Brough and the upper Eden district. The district was evidently occupied by a considerable English population before, for, although Scandinavian names preponderate, it has many Old English place-names.[5] There can be little doubt that this district belonged to the Danelaw, and the Barony of Appleby may well have its origin in this Danish colony. In the areas more distant from the Eden (those on the Lowther, Ullswater, and Hawes Water) Norwegian features are prominent in the place-nomenclature. But it is very likely that the earliest Scandinavian settlers were Danes here too, and names such as Bomby, Colby, and the Scandinavianised Yanwath, may be due to them. The Norwegian elements may represent later immigration from the west. It is likewise doubtful whether the Danes in the Appleby district spread further down the Eden. In the lower Eden valley a Norwegian place-nomenclature begins; Glassonby, Kirkoswald are found near the West-

[1] Such as Gunnerton, Nafferton, Ouston (two) and Dotland.
[2] E.g. Crookdean, Crookham, Crookhouse and Tranwell.
[3] See p. 154 below.
[4] E.g. Brough Sowerby, Crosby Garrett, Nateby, Rookby, Soulby, Waitby, with Asby not far off; Appleby, with Colby, and Kirkby Thore and Temple Sowerby, some way down stream; and Crosby Ravensworth, Newby and Thrimby, a little to the west.
[5] E.g. Brough, Brampton, Hilton, Marton, Murton, Musgrave, Cliburn.

morland border. The southern part of Westmorland seems to be mainly a Norwegian district, but there may quite well have been an earlier though limited Danish colonisation. The names in -*thorpe* (Clawthorpe, Milnthorpe), however, may be English.

The neighbouring district of north Lancashire, around Lancaster itself, has some possible traces of Danish colonisation; Hornby contains the personal name *Horne*, which seems to be exclusively Danish and Swedish, and the lost Thirnby may contain the Danish personal name *Thyrne*. One name in -*thorpe* occurs in the district. But even if no Danish colonisation of importance took place in Cumberland, south Westmorland, and north Lancashire, these areas may well have been reckoned in with the Danelaw.[1]

THE DANISH MIDLANDS

The Danish settlements in Mercia clustered round five boroughs, Lincoln, Stamford, Leicester, Nottingham, Derby, and are therefore often referred to as the district of the Five Boroughs.

Lincolnshire. In Lincolnshire we see the Scandinavian influence at its highest. Of the three divisions of the shire, Holland, a low-lying district belonging to the Fenland was, apparently, not to a considerable extent colonised by Danes at the first settlement.[2] It has not a single name in -*by* or -*thorpe*. But there are a few Scandinavianised names, as Kirton, Skirbeck, Scrane, or names in -*toft*, as Brothertoft, Fishtoft, Wigtoft, also Bicker, which seems to be OScn *bȳkiarr* "village marsh". Elloe wapentake has no Scandinavian village name at all. Very likely the Danish settlements in Holland mostly belong to a somewhat later period.

In the rest of the county, Danes settled extensively. The place-nomenclature is characterised by a great number of names in -*by* (altogether nearly 250), by a fair number of Danish names in -*thorpe*, by several other

---

[1] It is significant that carucation (a marked feature of the Danelaw) was found in Lancashire and in parts of Cumberland and Westmorland.

[2] Isaac Taylor wrote in *Words and Places* (1902), p. 111: "The fens which border the Witham, the Welland, and the Nen effectually guarded the southern frontier of the Danish settlers; and this natural boundary they do not seem to have crossed in any considerable numbers". The older authorities all stress this, e.g. G. S. Streatfeild's *Lincolnshire and the Danes* (1884); and J. Beddoe, in a Memoir of the Anthropological Society, asserted as early as 1869 that in Lincolnshire the Danish element in the physical appearance of the people was particularly strong as far as the borders of the Fens (p. 252).

strictly Scandinavian names,[1] and by several Scandinavianised names.[2] Hybrids of the type Grimston are few.

The key to the distribution of the Scandinavian names is provided by the Roman roads:

it will first of all have to be borne in mind that the first settlements would be made by the army in the central parts, for instance Lincoln, from which Roman roads branched out in all directions, and Stamford. The settlements were not founded, as Streatfeild held, by Scandinavians landing on the coast and penetrating inland. The different frequency of Scandinavian names in the various districts is no doubt due to more circumstances than one. Considerations of a military nature would play an important rôle in the earliest colonisation. It would be desirable that the host could be easily collected and quickly moved from one spot to another. The chief means of communication were the Roman roads, and it is evidently not due to chance that there seems to be a certain connection between Scandinavian place-names and the Roman roads.[3]

Lindsey is divided into three ridings. In the West Riding a Roman road (Ermine Street) runs from Lincoln to the Humber at Winteringham. Danish place-names are common on both sides of the road, but, except in the north, Danish names are few near the Trent. In Axholme, there are only to be found Keadby and some such names as Lound and Beltoft; but Axholme is a Scandinavian name itself, as may be Haxey (= Ax-). The Danish settlements in the West Riding are thus seen to be comparatively scattered, but some groups of names suggest a compact settlement within a limited area. The central point of the South Riding is Horncastle, which was easily reached from Lincoln by Roman roads, though not by a direct one. The country round Horncastle, particularly in its eastern portion, is the most strongly Scandinavian part of Lincolnshire, and indeed of the whole Danelaw. In the wapentakes of Bolingbroke, Candleshoe, Calceworth,

---

[1] E.g. Stainfield (OScn *Steinþveit*), Timberland (*-lund*) in Kesteven; Bratoft, Langworth (*-wath* "ford"), Skegness in Lindsey.

[2] E.g. Brigsley, Carlton, Louth in Lindsey, Casewick, Eagle, Scopwick, Scredington, Skillington in Kesteven.

[3] Eilert Ekwall, "The Scandinavian Element", in *Introduction to the Survey of English Place-Names* (ed. by A. Mawer and F. M. Stenton, 1929), p. 83.

[4] East of the road are Firsby, Saxby, Owmby, Normanby, Caenby (in a group), Atterby, Snitterby (together), Scawby, Appleby; west of it are Carlton, Scampton, Kirton, Brattleby, Manby, Risby, Roxby. Nearer the Trent are Saxilby, Ingleby, Bransby, Normanby, Kexby, Somerby, Aisby, Ashby, Brumby, Crosby, Scunthorpe, Hornsby, Gunness (the last five in a group), Normanby, Conesby, Darby, Thealby, Coleby, Haythby (in a group near the Humber and the Trent).

Hill and Horncastle, which represent about one-half of the riding, there are nearly seventy names in -*by*, almost all of which are in the Domesday Book. And there are also other Danish names, some in -*thorpe*; the river-name Bain, too, is Scandinavian. The names in -*by* cluster in a comparatively limited area in the centre of the district. They outnumber the English names, and evidently indicate a compact Danish colony.[1] The -*thorpes* are mostly found along the sea, and are doubtless later settlements. In the rest of the riding, the Danish element, though not so strong, is still very prominent, especially around Wragby and Louth.

In the North Riding the Danish settlements really continue unbroken from the Louth district northwards between the Wolds and the Humber. Few Danish names are found in the low-lying tracts along the Humber, where, incidentally, Old English names too are rare. Names such as Great and Little Coates are clearly late. Great Grimsby, with Itterby, Weelsby and Thrunscoe, however, are on or near the river. Farther inland, Danish names are numerous, though much interspersed with English ones. In the north of the riding, English names are in the majority; but on the western slopes, and at the foot, of the wolds east of the Ancholme, where a Roman road from Horncastle to South Ferriby may have run, Danish names are again very numerous.[2]

In Kesteven, Danish names cluster round the two Roman roads, Ermine Street from Stamford to Lincoln, and King Street from Castor *via* Sleaford to Lincoln. This cannot be due to chance. The first Danish settlements were made within easy reach of these important roads. Ermine Street runs partly on a wold, and settlements were here often made some way from the road. On choosing the 15 miles stretch of the Ermine Street from the southern boundary to Ancaster, we find within 4 miles of the road very many names in -*by*.[3] Along King Street, between

---

[1] A square six miles round Spilsby provides a good example. Here we find eighteen names in -*by*, Ashby, Aswardby, Candlesby, Claxby, Dalby, Eresby, Firsby, Grebby, Gunby, Hanby, Hundleby, Irby, Raithby, Scremby, Skendleby, Spilsby, Sutterby, Willoughby, also Bratoft, Keal (OScn *kiǫlr* "ridge"), Dexthorpe, Sausthorpe. English are here Halton, Langton, Toynton, Welton, Partney, Steeping.

[2] Examples are Tealby, Risby, Walesby, Otby, Osgodby, Kirkby, Kingerby, Owersby, Normanby le Wold, Claxby, Usselby, Fonaby, Audleby, Clixby, Grasby, Owmby, Searby, Somerby, Bigby, Barnetby, Wrawby, Kettleby, Worlaby, Bonby, Saxby, South Ferriby.

[3] Gunby, Stainby, Harrowby, Barrowby, Great Gonerby, west of the road; Corby, Osgodby, Ingoldsby, Boothby Pagnell, Old Somerby, Welby, Humby, Braceby, Haceby, Dembleby, Aunsby, Aisby, Oasby, Kelby, east of the road; in addition there are some -*thorpes*, as Lobthorpe, Londonthorpe, and Skillington.

Bourne and Sleaford (a distance of some 16 miles), there is another group of Danish names within 4 miles of the road.[1] In Winnibriggs wapentake, on the Leicester border, the -*bys* are all within 4 miles of the road. Farther west, there are only occasional names in -*thorpe* and Stenwith; most of the place-names are English. And in Loveden wapentake, north of Winnibriggs, there are hardly any certainly Danish village-names beyond the 4-mile limit.

Leicestershire had for its centre the stronghold of Leicester. But the chief Danish settlements do not cluster around Leicester itself. They are found to the east and north-east, in the country around the Wreak, which itself has a Danish name. Nor do the Danish names reach the Lincolnshire border. The centre of the district north of the Wreak is Melton Mowbray, which is an Old English *Middeltūn* Scandinavianised. The wapentake is Framland, named from Great Framlands near Melton (*Franelund* is a Danish name). Here, Danish names are found in an unbroken series along and near the Wreak.[2] English names are few except in the northern part. South of the Wreak, a similar Danish district is found, but there English names are more common, especially upon the wolds. Around Leicester, though not in its immediate vicinity, are smaller clusters of Danish names.[3] Few Danish names are found along Watling Street; Bittesby is an exception. In the remaining parts of the county, only small clusters of Danish names are found. But near Market Bosworth are Cadeby, Kirkby Mallory, Carlton. A larger cluster occurs in the district of Ashby-de-la-Zouch: this includes Appleby, Blackfordby, Kilwardby, Boothorpe, Osgathorpe, Thringstone and others. This Danish colony clearly once formed a whole with that in the adjoining Repton district in Derbyshire. A good deal of the northern part consists of Charnwood Forest, where Danes did not settle.

Rutland has several names in -*thorpe*, as Gunthorpe, Ingthorpe, Kilthorpe, Tolethorpe and one Normanton. But the Danish settlement here seems to be comparatively late.

[1] Bulby, Kirkby Underwood, Keisby, Hanby, Aslackby, Osbournby, Scott Willoughby, Aswarby, Swarby, Silk Willoughby, South Rauceby, west of it; Hacconby, Dunsby, Dowsby, Graby, Spanby, Asgarby, Kirkby Laythorpe, east of it. It should be noticed that the fens begin not far east of the road.

[2] Examples are Hoby, Shoby, Saxelby, Asfordby, Old Dalby, Wartnaby, Abkettleby, Welby, Sysonby, Brentingby, Wyfordby, Freeby, Saxby, Goadby, Stonesby, Saltby, Bescaby.

[3] Barkby, Beeby, to the north-east, Scraptoft, Bushby, Thurnby, Oadby, Kilby, Arnesby, Shearsby, Knaptoft, to the east and south-east; Willoughby, Ashby, Cosby, Blaby, Enderby, to the south and south-west; Kirby Muxloe, Ratby, Groby, to the west; and Sileby, Swithland (-*lund*), to the north.

Northamptonshire certainly belonged to the district of the Five Boroughs. Watling Street, generally important as a boundary of the Danelaw, cuts off the south-western portion from the rest. To the north of Watling Street, particularly along the Welland bordering Leicestershire, there are numerous Danish names, and, quite clearly, there was a considerable Danish colony here. There are some Danish names in the country between the Welland and the Nene, but they decrease in number near the latter.[1] South and east of the Nene there are only some isolated names, e.g. Castle Ashby, Wigsthorpe, Knuston, Strixton. The immediate neighbourhood of Northampton is free from Old Danish names, and the street-names of the town are purely English. It seems to have remained an English centre. In the district around Peterborough and adjoining Lincolnshire are few Danish village-names. Maxey, Gunthorpe, Peakirk, may be mentioned. But several smaller places have Danish names, as Cathwaite, Northolme, Lound; and many old street-names in Peterborough contain OScn *gata* "street", as *Bondegate*, whose first element is OScn *bondi* "free house-holder". Peterborough seems to have had a considerable Danish popula-tion. This is in accordance with the fact that a late tenth-century document dealing with the Peterborough area, and recording a great number of its inhabitants, mentions many with Scandinavian names, about a third of the whole number (37 out of some 110).[2] In the region to the south-west of Watling Street are two small clusters of names in *-by*;[3] Kilsby and Barby are near Rugby, and so link up this group of Danish names with the small group in *Warwickshire*. This latter group comprises Rugby (originally an OE name in *-burg*), Monks Kirby, Willoughby (recorded already in 956), Thurlaston with Toft, and Wibtoft—all on or near Watling Street. This Danish district is to be identified with the zone 8 miles beyond Watling Street that was mentioned by the Laws of Edward the Confessor as belonging to the Danelaw.[4]

Nottinghamshire has Nottingham as its nucleus. Danish names are here

---

[1] Along or near Watling Street are Holdenby, Long Buckby, Yelvertoft and others, and along the Leicestershire border Cold Ashby, Thornby, Naseby (both originally names in *-burg*), Sulby, Sibbertoft, Clipston, East Carlton, Corby, Kirby. Nearer the Nene, are Mears Ashby, Wilby, Scaldwell, not far from Castle Ashby.

[2] W. de Gray Birch, *Cartularium Saxonicum*, 3 vols. (1883–93), Charter no. 1130. The date is 972–92.

[3] Viz. Ashby St Ledgers, Barby, Kilsby close together, and Badby, Catesby a little farther south, near Daventry.

[4] See p. 134 above. It is true that none of these places are as much as 8 miles from the road; but some sites are 4 or 5 miles away, and the village territories naturally stretch beyond the line of the sites.

on the whole scattered, but they are fairly evenly distributed, except in Sherwood Forest, which was, doubtless, but little inhabited in the ninth century. Later names of Danish origin, however, are found also in Sherwood. The nomenclature is characterised by numerous hybrids of the type Grimston, and by names in *-thorpe*. Names in *-by* are few, altogether some seventeen. So are Scandinavianised names; Scarrington and Screveton may be mentioned. Smaller clusters of Danish names occur in the north.[1] Danish names are fairly common round about Nottingham and Bingham; and also along the northern bank of the Trent. Some are found west of Sherwood Forest and on the Derbyshire border: Skegby, Kirkby in Ashfield, Eastwood (really *-thwaite*). Altogether, the general impression that emerges is that of a considerable, but scattered, Danish settlement in the districts under cultivation.

Derbyshire. In Derbyshire, as in Nottinghamshire, the Scandinavian place-nomenclature is quite different in character from that of Lincoln or Leicester. Derby was originally *Norþworþig*. The Anglian place seems to have been a village, and Derby as a town is a Danish creation. The site had strategical importance, being situated near where the Trent is crossed by Riknield Street. The Danish settlements in Derbyshire were scattered and were hardly of great importance. Yet a compact colony is indicated by Bretby, Ingleby, Smisby, Foremark (*Fornewerk* "old fort") in Repton hundred, which adjoins the Ashby-de-la-Zouch district of Leicestershire;[2] it may be worth mentioning that in 874 the Danish army wintered at Repton. Apart from this area, names in *-by* are few. Besides Derby itself and Little Derby (near Derby), there are only Denby and Stainsby; but Domesday mentions two *-by's* now lost. However, there are several names in *-thorpe*, and a fair number of hybrids of the type Grimston, and also a few other names showing Danish influence. A fairly Scandinavianised district is Scarsdale wapentake, which has a Danish name itself; near the Notts border, in the district east and south-east of Chesterfield, there are Stainsby, Langwith, two Normantons, Lound, Hardstoft and others. The other names of the county include examples like Normanton, Stenson, Thulston, Kedleston, Thurvaston, near Derby. Remarkable are Hoon (OScn *haugum* "at the hills or mounds"), and the Scandinavianised Scropton near the river Dove in the far west. In north-west Derbyshire, Danish influence is slight, but there are Griffe (OScn *gryfia* "pit") near Wirks-

---

[1] Here are Bilby, Ranby, Barnby Moor, Scrooby, Serlby, Lound, Ranskill, Carlton near the Ryton and the Idle; and Budby, Thoresby, Walesby, Kirton, on or near the Maun.

[2] See p. 147 above.

11

worth, and Rowland (*Rālund* "roe wood") north of Bakewell. The Peak district seems to be quite free from Scandinavian influence.

Staffordshire, Cheshire, South Lancashire. It is doubtful if any part of Staffordshire belonged to the Danelaw, and the place-names do not give much help. A considerable Danish immigration certainly did not take place, but there are some traces of Danish influence. On or near the Derbyshire border, there are Swinscoe near Ashbourne, Croxall near Burton on Trent,[1] and Drointon ("the *tūn* of the drengs"). Isolated hybrids such as Gunston near Brewood, or Croxton near Eccleshall, do not carry much weight. But it is surprising to find Danish names in the district of Newcastle and Stoke on Trent: Knutton, Normacot, Thursfield, and particularly Hulme. This group must be considered in connection with a number of names in east Cheshire[2] and Lancashire. There is a remarkable cluster of Danish names near Manchester, where a small Danish colony must have existed, on the northern bank of the Mersey, as indicated by the names Flixton and Urmston (*Flik*, and *Urm* are Danish, not Norwegian), Hulme, Levenshulme and Davyhulme. The colony also extended into Cheshire where *Hulmes* are located; but in other parts of Cheshire Danish names are very rare. The difficulty is to explain how these Danes got there. The distribution of the names suggests that the Mersey valley was the nucleus of the colony, but it is far from the nearest Danish colonies in Derbyshire. Yet the district is within easy reach of Derby *via* the Roman road from Derby by Buxton to Manchester, and the Roman road south-west from Manchester runs close to the Urmston district, Knutsford and Rostherne. It might also be suggested that the Danish colony was founded by some of the followers of King Sihtric from Dublin, who, according to Symeon of Durham, occupied Davenport in Cheshire in 920,[3] but we should rather expect Sihtric's followers to have been Norwegians, and in any case there can hardly be any doubt that this district belonged to the Danelaw.

[1] This is found in a source of A.D. 942. See Birch, *op. cit.* Charter no. 773. It seems to contain the word *crook*, or a personal name *Krōkr*.

[2] Not far north of Stoke and Newcastle, there are Hulme Walfield and Church Hulme on the Dane; Swanscoe and Kettleshulme north-east of Macclesfield; Knutsford and Rostherne, which seem to have Danish personal names as first element; Cheadle Hulme near the Mersey; and a few others.

[3] Davenport is close to Church Hulme and Hulme Walfield.

DANISH EAST ANGLIA

In the Laws of Edward the Confessor, only Norfolk, Suffolk and Cambridgeshire are included in Danish East Anglia. But, after the Peace of 886, it seems that the Danish kingdom of East Anglia included Norfolk, Suffolk, Cambridgeshire, Essex, Huntingdonshire, parts of Bedfordshire and Hertfordshire.

Norfolk. The chief folk settlement in East Anglia took place in Norfolk, and Danish place-names are widely distributed all over the county. But in most parts, the place-nomenclature is characterised chiefly by names in *-thorpe* and by very frequent hybrids of the type Grimston, while Scandinavianised names are on the whole not common,[1] and names in *-by* hardly occur except in the area around the Broads.[2] In these eastern hundreds, names in *-by* are common. There is a remarkable cluster in East and West Flegg on the coast north of the Bure; and the adjoining hundreds contain several *-by* names also.[3] Another strongly Danish district lies to the north of Norwich,[4] and it is clear that the districts to the east and north were strongly Scandinavianised. Norwich itself may have been a Danish centre: it has street-names in *-gate*, as Cow Gate, Mountergate, Pottergate. Some

[1] We may note some Scandinavianised names, as Scarning (near East Dereham) and Methwold in the south-west; further, Holme Hale near Swaffham, which is pretty certainly the *Holm*, mentioned in the Anglo-Saxon Chronicle under the year 902 and in Birch *op. cit.* 1064, thus a very early name, and the clusters of Danish names in Gallow H round Fakenham (Alethorpe, Bagthorpe, Pensthorpe, Sculthorpe, Croxton, Helhoughton, Kettlestone); in Wayland H west of Wymondham (Caston, Griston, Scoulton, Thompson); or in Mitford H south-east of East Dereham (Flockthorpe, Whinburgh, Garveston, Reymerston, Thuxton). Some hundreds have Danish names, as Brothercross, Guiltcross, Forehoe, Grimshoe, Wayland (*-lund*), Flegg, perhaps Gallow, Greenhoe. In some of these, the Danish element is prominent in place-names; in others it is not.

[2] Wilby near Shropham is an exception.

[3] In the Flegg hundreds are Filby, Herringby, Mautby, Ormesby, Scratby, Stokesby, Thrigby (East Flegg); Ashby, Billockby, Clippesby, Hemsby, Oby, Rollesby (West Flegg). English place-names are comparatively few here. In neighbouring hundreds there are Aldeby, Kirby, Haddiscoe, Toft Monks in Clavering H, south-west of Yarmouth; Ashby, Thwaite, Carleton, Kirstead in Loddon H; Kirby and Rockland in Henstead H, near Norwich.

[4] South Erpingham H, north of Norwich, with Alby, Colby, Corpusty, Thwaite, Scottow, Skeyton, and several names in *-thorpe*. The adjoining North Erpingham H has Felbrigg, Matlask, Gunton, Thurgarton; and Tunstead H, east of these, Crostwight (really *-thwaite*), Holme, Keswick. In Eynsford H, north-west of Norwich, are Tyby, Guestwick (from *-thwaite*), and Themelthorpe.

Scandinavianised names are also found to the south and west of it.[1]
Finally, the medieval records contain many other Scandinavian names
which have now disappeared; this is particularly true in the case of field-
names and personal names[2] which, it should be noted, suggest considerable
Scandinavian admixture through the entire region.

The general implications of these distributions I have summed up
elsewhere:

The obvious inference would seem to be that the Scandinavian colonies were
founded in the first instance in the tracts on the lower Waveney and that from
there settlers found their way up along the rivers. But what we know of the
Scandinavian colonisation tells us that this must be a wrong conclusion. The
victorious army would not march right through Norfolk and settle on the lower
Waveney. More probably the centre of the settlements would be Thetford,
where the army had wintered in 870. The explanation of the curious distribution
of Scandinavian place-names is probably simply this. The Scandinavians settled
about equally thickly all over (or over most of) the district. But in most parts
there was a considerable English population, and the Scandinavians were not
numerically strong enough to affect the place-nomenclature very seriously
except in the very low-lying district on the lower Waveney, which was probably
not much inhabited before the Scandinavian time. In most of the districts the
Scandinavians to a great extent adopted names already in use, but when new
settlements were founded, probably at a somewhat later period, these often got
names with suffixed *Thorpe*. It is possible that the large number of Scandinavian
names in the lower Waveney district may to some extent be due to a later influx
of Scandinavian settlers, who might have been induced to come over after the
conquest had been made by the army.[3]

The rest of East Anglia. In *Suffolk* a strong Danish influence is noticeable in the low-lying district south of Yarmouth, which adjoins the
strongly Danish parts of Norfolk around the mouth of the Waveney.[4]
Another Danish area is the hundred of Thingoe, to the west of Bury St
Edmunds, which has a Danish name and includes Risby. But generally in
Suffolk, Danish names are more scattered and contain very few Scandinavian names in the strictest sense. They are mostly found along, or not

[1] E.g. East Carleton, Keswick, Great and Little Melton.
[2] See Walter Rye, "Pedes finium relating to Norfolk, 3 Richard I to the end of the
reign of John", *Norf. and Norwich Archaeol. Soc.* (1881). See p. 162, footnote 1 below.
[3] *Introd. Surv. Eng. Place-Names* (ed. by Mawer and Stenton), p. 82.
[4] Here are Ashby, Barnby, Lowestoft, Lound, Carlton Colville, Kirkley, and some
hybrids such as Corton, Flixton, Gunton, Somerleyton. But there are several Old
English names in the district.

far from, the Waveney or the Little Ouse in the north,[1] and here and there along the coastal districts.[2] Two hundreds near Ipswich have Danish or partly Danish names—Carlford and Colneis. In the interior only some isolated Danish names occur, as Thorpe Morieux, Flowton, Kettlebaston, Kettleburgh. It is clear that there was a sprinkling of Danish settlers in the greater part of the county, but that a fairly dense occupation was restricted.

Cambridgeshire has only slight traces of Danish colonisation. There is a Toft south-west of Cambridge; Carlton and Conington are Scandinavianised names, and Croxton has a Danish personal name as first element. Stourbridge in Cambridge seems to be "Styr's bridge", *Styrr* being a common Scandinavian name. The places mentioned are scattered over the county, and no definite conclusions can be drawn from this material.

In *Essex* there is a small cluster of Danish names in the district of Walton on the Naze (Kirby and Thorpe le Soken). Otherwise there are no traces of Danish influence in village-names.[3] Even Clacton, which has generally been held to be a Danish name, is certainly English. The Danish occupation of Essex seems to have been purely a military one.

Huntingdonshire offers puzzling problems. It belonged to the East Anglian Danish kingdom, but its Danish place-names are few. There are Conington (*Cunictun* 957, a Scandinavianised Kington) and Holme in the north (Normancross H), Keyston near the Nene, and Coppingford (probably a Scandinavianised OE *Cēapmanna ford*) near Ermine Street in Leightonstone H, Holland (*Haulund*, 1252) and Warboys (*Wardebusc*, 1077) in Hurstingstone H. But two hundreds themselves have Scandinavian names: Normancross in the north, and Toseland (*-lund*) in the south. There must have been a certain amount of Danish settlement in Hunts, but it is difficult to judge of its extent from the material.

In *Bedfordshire* there is a similar state of affairs. Only the part east of a line from the source of the Lea to Bedford and north of the Ouse remained in Danish possession after the Peace of 886. The few Danish names are found mostly in this area (Holme, Toft), while Carlton is on the southern bank of the Ouse. Clipstone in the south-west does not carry weight. Thurleigh is Leigh in early sources and Tingrith is native. From

---

[1] Here occur Flixton, Thrandeston, Coney Weston (an OE *Cyninges-tūn*), Ilketshall, Thwaite, Wickham Skeith.

[2] Westleton, Minsmere (*Min-* from OScn *mynni* "river mouth"), Carlton, Melton, Eyke, Grimston, Kirton (near Felixstowe), Thurlston (near Ipswich).

[3] It is therefore improbable that "Layer" represents Scandinavian *leir*, clay; it is probably OE *leger* in some sense as "camp" or "grave-yard".

place-names, if from nothing else, it is evident that Bedfordshire was on the outskirts of Danish settlement.

Only a small strip of *Hertfordshire* remained in Danish hands after 886, and even there no Danish names are found. It is surprising to find Dacorum hundred (*Danais*, DB) right in the west of the county, chiefly on both sides of Watling Street. The name seems to mean "the Danes' hundred". Possibly some Danes had settled here before 886, but no Danish place-names are found in the hundred. In all these counties it would no doubt be easy to collect some names of minor features with at least partly Scandinavian names, and a Scandinavian immigration is also indicated by the occurrence of Scandinavian personal names in early sources.

Buckinghamshire is sometimes stated in twelfth-century, and later, sources to have belonged to the Danelaw,[1] but it seems doubtful if these statements can be correct, and they may refer to the time before 886. Few place-names indicate Danish influence. Ravenstone may be English. Turweston (north-west of Buckingham) seems to have a Danish personal name as its first element. These hardly constitute evidence, for names of this kind are occasionally met with in the very south and south-west of England. Such are Swainston in the Isle of Wight, and Thurloxton in Somerset. These places may have been named from members of the King's Danish body-guard, who had manors bestowed on them. East Garston, in Berkshire, was named from Danish *Esgar stalre*, who is mentioned in the Domesday Book. Tolpuddle, in Dorset, got its distinctive addition from *Tola*, a lady of Danish descent, who was married to Urc, Edward Confessor's houscarl. But in the south-west of Bucks, near Henley, is a place called Skirmett, which looks like a Scandinavianised form of OE *scīrgemōt* "shire moot", and would seem to indicate a Danish colony in the Chiltern district. It can hardly be regarded as a safe case for Scandinavianisation, because *sk-* for OE *sc-* is occasionally met with in the place-names of counties where Danish influence is quite out of the question, as in Devon (Scarhill, Skillaton, etc.) and Kent (Scadbury, Scray Lathe, Skid Hill, etc.).

NORWEGIAN SETTLEMENTS IN THE WEST

The Scandinavian settlements upon the western shores of Britain never attained the importance of those upon the eastern coast. The influence was also different, for here the place-nomenclature is characterised by

[1] See J. C. H. R. Steenstrup, *Normannerne* (Copenhagen, 1876–82), Part IV, p. 36.

Norwegian test-words.[1] The Norsemen came mainly from Ireland and from the Celtic fringe generally. It is apparent that they had adopted a certain number of Irish words or personal names;[2] and some place-names show the Celtic peculiarity of "inversion-compound" in which the generic element is placed first, and the defining element last.[3]

Cheshire had a compact Norwegian colony in the Wirral peninsula[4] which was clearly the result of the immigration of Ingemund in 901 or 902.[5] But Wirral was by no means uninhabited when the Scandinavians came, for there are many Old English names in the district (e.g. Eastham, Hooton, Willaston and Heswall). The Norse villages are mostly situated in the coastal regions, which may have been sparsely inhabited before; but Raby and Whitby are a good way inland.

Lancashire. South of the Ribble, Old Norse place-names[6] occur chiefly along the coast, as in Cheshire.[7] In the hundreds of West Derby and Leyland there is a particularly strong Norse element,[8] especially upon low-lying regions, which were probably but sparsely inhabited before the Scandinavians came. From the western districts some Norsemen found their way inland; and in the eastern districts there are obvious traces of Norse influence in names of minor places. These contain such elements as *scale, slack*, or Scandinavian personal names. Even two names in *-erg* are found in Salford H (Anglezark and Sholver).

---

[1] See p. 139 above.

[2] A common element in these districts is ON *erg* "a shieling", which is from OIr *airge*.

[3] Examples are Brigsteer, "Styr's bridge", and Aspatria (*Askpatrik*), "Patric's ash-tree".

[4] There are several Norse names, as Frankby, Greasby, Irby, Kirby, Pensby, Raby, Whitby, Meols, Tranmere (*-mel*), Larton, Storeton. In the centre of the district is Thingwall, which must have been the meeting-place of the colony. Arrowe represents the southernmost *erg*. Greasby is really a Scandinavianised name (*Gravesberie*, DB). Just outside the Wirral peninsula proper, near Frodsham, is an isolated *-by*, —Helsby.

[5] See p. 135 above.

[6] See E. Ekwall, *Place Names of Lancashire* (Manchester, 1922).

[7] On Danish names in the eastern hundreds, see p. 150 above.

[8] Near and north of Liverpool there are such names as Roby, West Derby, Formby, Crosby, Kirkby, Litherland, and Uplitherland, Lathom, Lunt, North Meols, Skelmersdale, Ormskirk, Scarisbrick, Warbreck, Crossens, Croxteth, Toxteth. Of special importance is Thingwall, which must have been the meeting-place of a Scandinavian group. Here, also, is the Norse test-word *breck* (Scarisbrick, Warbreck); and an *erg* is found in a now lost name. Farther north, in Leyland H, there are found Becconsall (with an Irish-Norse personal name as first element), Croston, Hesketh and others. Names of minor places are also to a great extent Norse.

North of the Ribble, Lancashire is much narrower; and here Norse names are found, in varying frequency, all over Amounderness and in Lonsdale south of the Sands. Amounderness itself is a Scandinavian name. The Norse element is most pronounced in very low-lying parts and upon the rising land to the east. But even in the flat central parts not a few Norse names occur.[1] In the Lonsdale area, the majority of older villages have English names; probably there was not so much unoccupied land here. But there are some Scandinavian names, as Ireby, Hornby,[2] Skerton, Swainshead, Torrisholme, and the river-name Greta.[3] And as far as minor places are concerned, the Norse element is marked in Lonsdale as well as in Amounderness. Lonsdale north of the Sands is divided into Cartmel and Furness, both having Norse names. Old English names are found along the sea and in the river valleys, but they are interspersed with Norse names (Allithwaite, Holker, Kirkby Ireleth, Sowerby, Stainton). In the fell districts, Old English names are very few. Tilberthwaite north of Coniston contains an Old English name in -burg and Coniston itself may be a Scandinavianised OE *Cyningestūn*; the majority of place-names are here Norse, as Blawith, Hawkshead, Lowick, Torver.

Westmorland. The southern part, the old barony of Kendal, has a strong substratum of English names near the sea and in the river valleys, especially along the Lune.[4] Some English names are found even in the Lake District, as Rydal, Langdale, perhaps Grasmere; and Shap is high up in the fells. But there are many very typical Norse names interspersed with the English,[5] and the names of minor places, especially in the fell districts, are mostly Norse. In north Westmorland there must have been a considerable settlement of Danes.[6] But in the district round the Lowther, and around Ullswater and

---

[1] Sometimes townships have a composite name, consisting of one English and one Norse name, as Medlar with Wesham, Bispham with Norbreck, Layton with Warbreck. Strictly Scandinavian names are Ribby, Sowerby, Westby, Garstang, Larbrick, Norbreck, Warbreck (with Norse *breck*); and there are numerous names in -*erg*, as Goosnargh, Grimsargh, Kellamergh; Scandinavianised names are Bradkirk, Carleton, and Rawcliffe.

[2] On which see, however, p. 144.

[3] Examples of names in -*erg* are Docker, Salter and Winder.

[4] Barbon, Burton in Kendal, Casterton, Hutton, Helsington, Killington, Lupton, Fawcett and Sedgwick may be mentioned.

[5] The Norse names include Haverbrack, Howgill, Leasgill, Witherslack; names in -*erg*, as Mansergh, Skelsmergh, the inversion-compound Brigsteer; and many names of various kinds, such as Beetham, Grayrigg, Selside, Stangerthwaite, Wray, and the Scandinavianised Meathop. The only name in -*by* is Kirkby Lonsdale, but Kendal was formerly Kirkby. On the possibility of some place-names being Danish, see p. 144 above.   [6] See p. 143 above.

Hawes Water, the place-nomenclature has obvious Norse features. Here are Tirril and Winder (both names in -*erg*), and there seem to be inversion-compounds among minor names. Reagill and Sleagill are Norse names. It is evident that in this region there must have been considerable Norse immigration probably at a somewhat later period; and these Norsemen penetrated also into the Danish district on the Eden, and even into the North Riding.

Cumberland. It is sometimes stated that the Anglian settlement had hardly affected Cumberland (and Westmorland, too) at all, and that it was reserved for the Scandinavian settlers to Teutonise them. This is an exaggeration. There is a very marked substratum of old English names all along the coast, and also in the river valleys (especially that of the Eden with its tributaries), and there must have been an early and considerable settlement of Anglians here.[1] A name such as Whicham (from *Whitting-ham*) must date from a very early period. But though, in these areas, Old English village-names sometimes outnumber Norse village-names, the latter are also very common;[2] and names of minor places are preponderatingly Norse. All these Norse names are widely distributed, and no especially Scandinavianised districts stand out. Near the coast and in river areas there are many names in -*by*,[3] but many of these contain a Norman personal name (Aglionby, Botcherby, Johnby, Ponsonby, Robberby; Parsonby is "the parson's by"). Some of these personal names may have been borne by the Flemings who settled in Cumberland in the time of Henry I; Flimby means "the place of the Flemings". The element *by* was certainly in living use after the Norman Conquest, and names in -*by* in this district need not belong to an early stratum. In the Lake District, on the other hand, English names are very few. Keswick is a Scandinavianised name (OE *Cēse-wīc*, "cheese farm"). Sparket near Ullswater has an

---

[1] Old English names of villages are, for instance, from south to north along the sea: Millom, Bootle, Muncaster, Gosforth, Hale, Hensingham, Workington, Cockermouth, Ellenborough, Bolton, Wigton, Bampton, Dalston; on or near the Eden: Warwick, Wetheral, Plumpton, Hutton, Salkeld, Edenhall; on the Irthing: Brampton, Denton.

[2] There are many names of a typically Norse stamp, as Seascale, Winscales (*scale*), Hethersgill, Ivegill, Bowderdale (ON *būðar-*); many names in -*erg*, as Berrier, Cleator, Salter, Mosser; numerous inversion-compounds, as Aspatria, Bewaldeth, Kirkbride, Setmurthy; and a great number of other names, as Ainstable, Hesket, Stanwix (second element ON *veggr* "wall"), Whitehaven (first element an ON *Hvíta-hǫfuð*, "white headland"), Wreay. Many names of streams are Norse, as Bleng, Caldbeck, Whitbeck, Roe Beck. Scandinavianised names are, for instance, Stainburn, Rockcliff, Skelton.

[3] E.g. Dovenby, Melmerby (both with Irish-Norse personal names as first element), Arkleby, Birkby ("by of the Britons"), Gamblesby, Hornsby, Raby, Sowerby.

English name, and some names of hills and lakes are also English (Saddle-back, Buttermere, etc.). Most of these, however, are not found in early sources; yet Dodd occurs in an early thirteenth-century source.

In more remote districts, as in the Lake District proper, the place-nomenclature is almost exclusively Scandinavian (Norse). The colonisation in these tracts is evidently due chiefly to the Norse. Many names of hills and lakes are Norse, as Scafell, Skiddaw, Whinfell, Ullswater, Wast-water. All over Cumberland, Norse must have been very extensively spoken. One gets the impression that in the interior parts it was for a long time the only or chief language used.

Yorkshire. A Norse immigration into west Yorkshire from Lancashire and Westmorland, along the valley of the Lune and perhaps also that of the Ribble, is proved by many place-names. In the Craven district of the West Riding are numerous places with names in *gill*, *scale* (as Raygill, Wycongill, High Scale, Scaleber), and some names in *-erg* (as Battrix, Feizor). Farther south, in the Huddersfield district, is Golcar, another name in *-erg*; and Fixby may have an Irish-Norse personal name as its first element. But a certain amount of Norse immigration must be assumed in the other ridings. The element *erg* is found occasionally in the East Riding (Argam, Arram, Arras), and in the North Riding (as Airyholme in Cleveland and that in Ryedale; Arrathorne near Catterick; and Ery-holme on the Tees). Irish-Norse personal names occur in the place-names of these districts (e.g. Commondale, Duggleby, Lackenby, Melmerby), and some of the Domesday landholders have such names. Scorbrough, in the East Riding, is a Norse *skógar- búð* ("booth in the forest"). Names in *breck*, *gill* and *scale* are not rarely found in the North Riding (e.g. Breck near Whitby, Laskill in Ryedale, and Scargill), and also in the more eastern parts of the West Riding (e.g. Scholes in Barwick-in-Elmet). A Norse immigration into Yorkshire is easily explained as a result of the intimate connection between the Scandinavian kingdoms of York and Dublin; and, at the same time, these connections account for the Irish elements.

A certain number of Norsemen, too, may be supposed to have joined the Danish army in England; and some may have settled down with the Danes. There are several place-names such as Normanby, Normanton, which probably mean "the village of the Norwegians". There are Normanbys in Lincolnshire and Yorkshire; and Normantons are found in Derbyshire, Leicestershire, Lincolnshire, Nottinghamshire, Rutland and Yorkshire.

Wales. The Welsh chronicles certainly bear witness that the Welsh coast suffered from Scandinavian inroads during the ninth and tenth centuries.[1] But the brief notices of the chronicles do not hint at Norse colonisation. To place-names we must turn, but Welsh place-names in general have not been sufficiently investigated to make it possible to establish, with any precision, the extent of the Scandinavian settlement in Wales.[2] Some settlement certainly took place in Glamorgan, especially round Cardiff. Womanby, now a street in the town, is *Hundemanby* (c. 1270), identical with Hunmanby in Yorkshire. Not far from Cardiff are Lamby (*Langby*) to the east (in Monmouthshire), and Homri (*Hornby* fourteenth century) to the west near St Nicholas. Some street-names in Cardiff contain OScn *gata* "street" (as Millgate, now lost). Other settlements seem to have been made farther west. Laleston (*Lageleston* 1205) appears to contain a nickname formed from ON *lǫglauss* "lawless"; near it were formerly *Clakeston* (1205) and *Crokeston*. Swansea (*Sweynesse* c. 1175) is doubtless ON *Sveinsey*. In Pembrokeshire, Fishguard and Freysthrop look Scandinavian. Fishguard may be ON *fiskigarðr* "fishing-weir".

There are, however, several well-established Norse place-names along the coasts. Being seafarers and traders, the Norsemen naturally left vestiges of their activity upon coastal features. A number of islands off the Pembroke coast have Scandinavian names—Gateholm, Grassholm (*Gresholm*), Midland Isle (*Middelholm* 1326), Skokholm (*Scogholm* c. 1225, *Stokholm* 1275), the lost *Trellesholme* (1327), Ramsey, Skomer (*Schalmey* 1326), and Caldy near Tenby. Two skerries in St Bride's bay are called Black and Green Scar. Angle Bay at Milford Haven looks Scandinavian, as does Nab's Head (ON *nabbr* "point"). At Tenby there are Sker Rock and Gosker Rock. And in Glamorgan there are Sker Point and Tusker Rock (*Blachescerre*, *Skerra* twelfth century, cf. ON *sker* "skerry"). Of the same type are the names of the islands of Lundy (*Lundey* in the ON Orkneyinga Saga), Flat Holme and Steep Holme in the Bristol Channel. Anglesea in North Wales is *Ǫngulsey* in the Orkneyinga Saga and may be Scandinavian; Priestholm certainly is, and the same may be true of Orme's Head.[3]

---

[1] See B. G. Charles, *Old Norse Relations with Wales* (1934), for a study of the problem, particularly from the literary point of view.

[2] See the interesting papers by Dr D. R. Paterson in *Archaeologia Cambrensis*, "Scandinavian Influence in the Place-names and Early Personal Names of Glamorgan", 1920, p. 31; "The Scandinavian Settlement of Cardiff", 1921, p. 53; "The Pre-Norman Settlement of Glamorgan", 1922, p. 37: but these are apt to overrate the Scandinavian influence.

[3] See E. Ekwall in *Introd. Surv. Eng. Place-Names*, p. 79; and B. G. Charles, *op. cit.* p. 157.

CONCLUSION: THE DANISH SETTLERS

Place-names, while giving a general idea of the distribution of the Scandi-navian settlements, leave many questions unanswered, particularly those dealing with the numbers of the settlers and with the nature of their settlements.[1] It is often stated that the Scandinavian settlers in the Danelaw cannot have been very numerous, and that they soon became assimilated to the English population. Even place-names tell us that this cannot be correct. A small proportion of Danes could not have left such strong traces on the place-nomenclature. But more definite information is wanted. Some indication of the relative numbers of Danish settlers is given by personal names. Unfortunately, the earliest sources do not afford very much help here. Old English charters from the Danelaw are few;[2] and the Domesday Book is not detailed enough. All the more important, therefore, is the information yielded by the twelfth- and early thirteenth-century Danelaw charters that were issued by small land-holders. It is true this material must be used with caution. Personal names are much influenced by fashion; and Scandinavian names may have been adopted by English people and *vice versa*. Yet it is significant that in these charters the percentage of Danish names is very large. Professor Stenton writes:

Of the 507 Anglo-Scandinavian names recorded in the charters of the present collection, 266 may definitely be regarded as of northern origin or as including northern elements. The percentage lies between 52 and 53. It certainly represents an under-estimate, for all doubtful names have been referred to their possible Old English originals. A percentage which included all the names of possible Scandinavian descent would fall nearer to 60 than to 50, though it would not exceed the former figure. It is probable that these names are fairly representative of the general personal nomenclature of the Danelaw. An analysis of the native personal names occurring in the Lincolnshire Assize Rolls of 1202 shows 215 Scandinavian against 194 English forms....

But the character of these Scandinavian names is more important than their numerical preponderance. The 266 Scandinavian names recorded in these

---

[1] Upon these and kindred questions, the researches of Professor F. M. Stenton have thrown much light and some of his conclusions are utilised in the summary that follows. See the Bibliographical Note on p. 163.

[2] One important charter of 972–92 has been mentioned above; this gives an idea of the relative numbers of Danes in a district that does not seem to have been very strongly Scandinavianised. A later document (c. 1050) contains the names of Bishop Ælfric's festermen; forty-five out of about seventy-five have Danish names. The district is that of Snaith and Sherburn in Elmet in the West Riding of Yorkshire. See W. Farrer, *Early Yorkshire Charters* (1914), no. 9.

charters comprise 119 distinct name forms. They include many names whose general currency in England is already well established.... But the interest of the present series lies in the names of comparative rarity whose currency proves the strength of the Scandinavian tradition of nomenclature.... The northern stems were still producing compounds...the few diminutive forms which occur in these charters are really more significant than many compound names.[1]

An important criterion for the relative numbers of Danish settlers in the Danelaw is offered by the proportion of sokemen recorded in Domesday. A study of this class of peasant also throws some light on the status of the Danish population. A sokeman was mostly of a humble position economically, but he was a free man. "The sokemen of the Danelaw represent, as a class, the rank and file of the Scandinavian armies which had settled in this district in the ninth century";[2] while, presumably, the villeins and bordars represented the native peasant class. The number of sokemen varies somewhat, but in the districts where Danish place-names are numerous it is generally high. In Lincolnshire the sokemen are roughly from 73 to 20 per cent. of the peasant class; the fen wapentake of Elloe has the lowest number; there are here no Danish village-names. On an average, about half of the recorded rural population consisted of sokemen. The percentage is higher in the central parts (i.e. the area with the greatest number of Danish names) but it thins out towards the north, south and west. In Leicestershire, from 50 to 27 per cent. were sokemen, and they numbered nearly 2000 in the Domesday Book. For Nottinghamshire, more than 1500 are recorded. Derbyshire has few sokemen. As Professor Stenton writes elsewhere, "The Sokemen of Domesday were not a small body of men emergent through accidental circumstances, but a definite class which formed an integral and often a dominant part of the established order of rural society."[3] And if we assume that the proportions in the eleventh century correspond roughly to those of about 900, the general result is that in some regions the number of new settlers was about equal to, or even greater than, that of the native population.[4] It is possible,

[1] *Documents...of the Danelaw* (1920), pp. cxiv–cxvi.
[2] F. M. Stenton, "The Free Peasantry of the Northern Danelaw" (*Bulletin de la Société Royale des Lettres de Lund*, 1925–6), p. 79. The figures in this paragraph are chiefly from Professor Stenton's paper.
[3] *The Danes in England* (British Academy, 1927), p. 16.
[4] See the map showing the distribution of sokemen in F. Seebohm's *English Village Community* (1926 edition), p. 85. County averages hide the importance of sokemen locally. The average for Lincolnshire, for instance, is 45 per cent. of the total population.

however, that these figures should be somewhat reduced.[1] But although definite statistical evidence cannot be obtained, it is fairly clear that in the Danelaw the Danes must have formed a very considerable proportion of the population.

It has often been assumed that the Scandinavian settlements meant a displacement of the earlier population. There can be no doubt that some displacement took place, but it should not be too readily assumed that this was a common process. The long wars that preceded the settlements certainly killed many people. The fights for York alone, in 867, must have meant a serious decrease in the male population of Yorkshire; and there would be room for a considerable Scandinavian settlement without any displacement. In some cases, a Scandinavian leader with his followers could just step into the shoes of the English lord and his retainers. Nor would the position of the agricultural population be materially altered by the change of ownership. In other cases, it may well be that the Scandinavian settlements meant not so much a displacement of the English peasants, as a depression in their status. The settlement often may have taken the form of the superimposition of Scandinavian settlers on English villages.[2]

Further, that Scandinavians and English lived side by side in most districts is indicated by the mixed place-nomenclature. It is true that places with English names may have had a wholly, or partly, Scandinavian population; but the survival of such a great number of English place-

---

[1] If the land-holders with English names found in the twelfth- and thirteenth-century charters are generally to be looked upon as descendants of eleventh-century sokemen of English descent, then there would seem to have been not a few Englishmen among the sokemen of Domesday. This is not absolutely certain, as English names may have been adopted by Scandinavians in England. On the other hand, there were probably a good many people of Danish descent among the villeins and bordars.

The East Anglian material shows what caution must be exercised in using this criterion. The Feudal Book of Abbot Baldwin (in *Feudal Documents from the Abbey of Bury St Edmunds*, ed. D. C. Douglas, London, 1932), which is about contemporaneous with Domesday, gives the names of the free peasants of a number of Suffolk manors belonging to Bury. Of about 700 individuals only some sixty have Danish names. But there is some indication that a change of names had taken place. In a certain number of cases the father's name is given, and at least in ten cases the father's name is Danish, the son's English. The opposite change is very rare (at most, two cases). It looks as if the Danes began early to adopt English names.

[2] In such cases old villages very likely often got a new name. The common name Kirkby (Kirby), in most cases, was probably given to an old village with a church, for it is unlikely that all these places date from the time when the Scandinavians had been converted to Christianity and had begun to build churches.

names presupposes a strong local English population. Some place-names prove that there were free English villages in the most strongly Scandinavianised districts. "Ingleby", for example, means the village of the English; and there are three Inglebys in Yorkshire,[1] one in Derbyshire, and one in Lincolnshire. There are also a good many names in -by with an English personal name as first element, e.g. Atterby, Audleby, Barnetby, Kingerby, Worlaby, in Lincolnshire, Ratby in Leicestershire, Ellerby and the two Willerbys in Yorkshire. Such villages must have had English owners.[2]

Whatever the answers to these problems may be, it is clear that the Scandinavian settlement was no passing episode, but a lasting contribution to the development of English life. Scandinavian institutions, administrative divisions, and agricultural terms—all go to show that the Danelaw itself was far from being a geographical expression.

BIBLIOGRAPHICAL NOTE

This list does not include the numerous studies of the place-names of different counties where the Scandinavian element is important. Particular mention should be made however of the volumes of the English Place-Name Society, edited by A. Mawer and F. M. Stenton. These are now appearing and are invaluable.

ANDERSON, OLOF S. *The English Hundred-Names.* Lund, 1934.

BJÖRKMAN, ERIK. *Scandinavian Loan-Words in Middle English.* Halle, 1900–2.

—— *Nordische Personennamen in England.* Halle, 1910.

—— *Zur englischen Namenkunde.* Halle, 1912.

BUGGE, A. *Vikingerne.* Copenhagen and Christiania, 1904–6.

—— "The Norse Settlements in the British Islands", *Trans. Royal Hist. Soc.* 4th Series, IV, pp. 173–210 (1921).

CHARLES, B. G. *Old Norse Relations with Wales.* Cardiff, 1934.

COLLINGWOOD, W. G. *Scandinavian Britain.* London, 1908.

—— *Lake District History.* Kendal, 1925.

—— *Angles, Danes and Norse in the District of Huddersfield.* (The Tolson Memorial Museum Publications. Handbook 2.) Huddersfield, 1921.

EKWALL, E. *Scandinavians and Celts in the North-West of England.* Lund, 1918.

—— "The Scandinavian Element", in *Introduction to the Survey of English Place-Names,* edited by A. Mawer and F. M. Stenton. Cambridge, 1924.

—— *The Concise Oxford Dictionary of English Place-Names.* Oxford, 1936.

FERGUSON, R. *The Northmen in Cumberland and Westmoreland.* London and Carlisle, 1856.

---

[1] Two are in the Stokesley district, see footnote 1, p. 141 above.

[2] Relevant to these problems there is another. See E. Ekwall, "How long did the Scandinavian Language survive in England?" in *A Grammatical Miscellany offered to Otto Jespersen* (Copenhagen and London, 1930), p. 17.

HODGKIN, R. H. *A History of the Anglo-Saxons*, vol. II. Oxford, 1935.

KENDRICK, T. D. *The Vikings*. London, 1930.

LINDKVIST, H. *Middle English Place-Names of Scandinavian Origin*. Upsala, 1912. (This is very important.)

MAWER, A. *The Vikings*. Cambridge, 1913.

—— "The Scandinavian Settlements as reflected in English Place-Names", *Acta Philologica Scandinavica*, 1931–2.

STEENSTRUP, J. C. H. R. *Normannerne*. Copenhagen, 1876–82.

STENTON, F. M. *Types of Manorial Structure in the Northern Danelaw*. Oxford, 1910.

—— *Documents Illustrative of the Social and Economic History of the Danelaw*. London, 1920.

—— "The Free Peasantry of the Northern Danelaw", *Bulletin de la Société Royale des Lettres de Lund*, 1925–6.

—— "The Danes in England", *Proceedings of the British Academy*, XIII (1927).

STREATFEILD, G. S. *Lincolnshire and the Danes*. London, 1884.

WORSAAE, J. J. A. *Minder om de Danske og Nordmændene i England, Skotland og Irland*. Copenhagen, 1851.
  A pioneer book in demonstrating the importance of place-names. It was issued in English as *An Account of the Danes and Norwegians in England, Scotland, and Ireland*. London, 1852.

Chapter V

THE ECONOMIC GEOGRAPHY OF
ENGLAND, A.D. 1000–1250

H. C. DARBY, Litt. D.[1]

In an account of the economic geography of England between A.D. 1000 and 1250, the Domesday Book must loom large. Much evidence is provided by a variety of agricultural documents, by charters of various kinds, by the earlier Pipe Rolls, by the Anglo-Saxon Chronicle, and by other chronicles; but the bulk of the information that is available comes from the Great Survey itself, and from its satellites.[2] The story of its production is simple to tell. At the Gloucester Gemot, in the winter of 1085, the king "had deep speech with his witan....He then sent his men over all England into every shire" to collect the information now preserved in the two manuscript volumes that, within a century of their compilation, had become known as the Domesday Book. "So very narrowly did he cause the survey to be made", moans the Saxon Chronicler,[3] "that there was not a single hide nor a rood of land, nor—it is shameful to relate that which he thought no shame to do—was there an ox, or a cow, or a pig passed by, that was not set down in the accounts, and then all these writings were brought to him." And it is this material, in its summarised form, that provides the basis for any reconstruction of the contemporary geography.

But much statistical work will have to be done before the full geographical contribution of the Survey has been extracted. The pages that follow

[1] I want particularly to acknowledge my great debt to Professor F. M. Stenton who read through the manuscript of this chapter and made many valuable suggestions. I also owe much to the suggestions of Mr John Saltmarsh, Mr L. F. Salzman, Mr F. W. Morgan and Miss H. Cam.

[2] The edition of the Domesday Book used (and referred to as D.B.) in this chapter is *Domesday Book seu Liber censualis Wilhelmi Primi regis Angliae*, vols. I–II (ed. Abraham Farley), 1783, vols. III–IV (ed. Henry Ellis), 1816. Vol. I contains the greater part of England. Vol. II contains Norfolk, Suffolk, Essex. Vol. III contains Indexes and Introduction. Vol. IV includes the following supplementary documents— Exon Domesday, Inquisitio Eliensis, Liber Winton, and the Boldon Book.

[3] Under the year 1085. The Anglo-Saxon Chronicle is designated as the Chronicle in this chapter. The best edition is that of J. Earle and C. Plummer, 2 vols. (1892–99). The Rolls Series volume, with a translation, was edited by B. Thorpe, 2 vols. (1861).

can but outline the main features of the varied landscape that confronted the Norman invaders of the eleventh century, and can indicate but briefly the effects of Norman exploitation upon the geography of the time.

WASTE AND FOREST

The outstanding fact about the landscape of Norman England, compared with that of later centuries, was its uncultivated aspect. The Anglo-Saxon settlement, it is true, had pierced the wilderness and broken it up everywhere with "dens" and "leys" and "fields";[1] but, even so, almost every page of the Domesday Book shows that eight hundred years ago there was more woodland in England than there is to-day. A Domesday map of Sussex shows that only its southern part was closely occupied in 1086; at least one third of the county was part of the wild and wooded district that extended into Surrey, Kent and Hampshire. What was thus true of the Weald[2] was true in a varying degree of the whole of England. Although exact boundaries and quantities are sometimes difficult to ascertain, there can be no doubt about the wooded nature of large tracts of country.

We know as much as we do about these tracts because the wood was useful to the villagers in many ways.[3] And, despite all its imperfections, the Domesday Inquest does indicate some of the main features of the distribution of woodland in the eleventh century. On the one hand, large wooded areas, like that of Essex, stand out; on the other hand, the country-side generally seems to have carried local groves and small spinneys; and there were many intermediate stages. Such variation can be shown only on county maps drawn from the Survey.

Some of the waste country of eleventh-century England, marshland for instance, was the natural wilderness of primeval nature.[4] No comment

---

[1] See pp. 130–1 above.

[2] This is not to suggest that the Weald was devoid of settlement. Place-names suggest "that the settlement of the Wealden area took place at a quite early date". See A. Mawer, *Problems of Place-Name Study* (1929), p. 124. But the implications of Fig. 19 remain clear.

[3] See pp. 200–2 below.

[4] The marshes of the Somerset Levels were, for the most part, unmeasured—see R. W. Eyton, *Domesday Studies: Somerset* (1886), pp. 38 *et seq.* Some, however, were measured. "These were not many, nor large. Possibly they were parcels of moor-land reclaimed or half-reclaimed from the general swamp, their value arising in summer-pasturage, coarse fodder, rushes, or ozier growth" (p. 40). For the Fenland, see p. 180 below. See also J. H. Round in *V.C.H. Somerset*, I, 425, for an account of the Somerset moorlands.

is necessary here upon the physical background of the inhospitable regions of England. In Norman times, as in the Anglo-Saxon period, these were many.[1] Much of the waste, however, was devastated land; and for many counties, the Survey records a marked depreciation in values between 1066 and 1086.[2] Frequently, the devastated manors are named as containing so much "waste", much of which had been produced since the year 1066. Some areas had already been laid waste by armies on the march before the battle of Hastings.[3] In other areas there had been devastation earlier in the century;[4] the entry for 1006 in the Anglo-Saxon Chronicle speaks of "every shire in Wessex sadly marked by burning and by plundering"; under the year 1065, the Chronicle notes how earl Morcar led the men of the north of England to Northampton, where

they slew men and burned houses and corn; and took all the cattle which they might come at, that was many thousand: and many hundred men they took and led north with them; so that that shire, and the other shires which are nigh, were for many years the worse.

The great extent of this plundering may be seen from the large entries of waste recorded in a pre-Domesday Northamptonshire Geld Roll,[5] whose figures illustrate the considerable increase in the value of individual manors which the Domesday entries for Northamptonshire show between the years 1066 and 1086. The low Domesday values for 1066 suggest widespread devastation, but this had been well-repaired by 1086. Only Foxley was then completely waste,[6] although about a dozen other places in Northamptonshire contained wasted holdings.

To the west, the Welsh were active: in Herefordshire there were nine manors of nineteen hides entirely laid waste; and in eleven other manors there was land for thirty-six ploughs which was waste, and which had

---

[1] See chapter III, *passim*.

[2] See C. H. Pearson, *History of England during the Early and Middle Ages* (1867), pp. 661–70. Twelve out of twenty counties show an increase in values.

[3] See, for example, F. H. Baring, "Oxfordshire Traces of the Northern Insurgents of 1065", *Eng. Hist. Rev.* XIII, 295 (1898), for an account of the devastated manors in Oxfordshire.

[4] See the Chronicle, *sub anno* 1052, for devastation along the south coast; and, *sub anno* 1010, for devastation by the Danes in East Anglia.

[5] J. H. Round, "The Northamptonshire Geld Roll", in *Feudal England* (1909), pp. 147–56; Round assigned the date "ante 1075" to the roll. See also *V.C.H. Northants*, I, 259, for the amount of waste land in each hundred. "The total of the land returned as 'waste' is...one third of the county".

[6] D.B. I, 223b.

been so in the time of the Confessor, and which had never paid geld.[1] And a Cheshire entry noted land for sixteen ploughs, formerly held by eight freemen, and lying waste from a time before the Conquest.[2] Among the Shropshire entries, too, mention of devastation was frequent.[3]

Attention has been drawn to the connection between the wasted lands mentioned in the Survey and the movements of William's invading army. "It is obvious that a large army living as his did on the country it passed through must move on a wide front and leave a broad strip of ravaged country behind".[4] As the Chronicle puts it: "he plundered all that part which he overran, until he came to *Beorh-hamstede*".[5] Consequently, the great differences in the values of many manors just before and after the Conquest mark the footprints of the Conqueror's army in its march upon London. These depreciated manors extend from Hastings to Dover and then in a circuit around London—in Kent, Surrey, Hampshire, western Berkshire, eastern Hertfordshire, Buckinghamshire, and even into Bedfordshire and Cambridgeshire. "Outside the line of march the immediate effect of the Conquest in the value of land in the south-east seems to have been very slight".[6]

The south-eastern counties might have borne the first burden of the

---

[1] D.B. I, 181 *b*: "Rex habet in Herefordscire ix maneria wasta de xix hidis." D.B. I, 186 *b*: "In his xi maneriis est terra xxxvi carucis, sed wasta fuit et est. Nunquam geldauit, iacet in Marcha de Walis." See also Saxon Chronicle, *sub anno* 1067.

[2] D.B. I, 264: "Has terras tenuerunt viii liberi homines pro maneriis. Terra est xvi carucis inter totum. Wasta fuit et est tota." See J. Tait, *The Survey of Cheshire, Chetham Society Publications*, LXXV, 7–8 (1916).

[3] See D. Sylvester, "Rural Settlement in Domesday Shropshire", *Sociological Review*, XXV, 244 (1933). Professor Tait writes, "Many manors were lying waste at the death of Edward the Confessor, the results apparently of Welsh raids. Much wider destruction was wrought when the Normans entered into possession".—*V.C.H. Shropshire*, I, 281.

[4] F. H. Baring, *Domesday Tables* (1909), p. 214. See also "The Conqueror's Footprints in Domesday", *Eng. Hist. Review*, XVII (1898). For comments by J. H. Round see *V.C.H. Surrey*, I, 278–9. Round declared: "That the depreciation so strongly marked on certain Surrey manors in the early days of William was due to the march of his host through Surrey cannot well be doubted; but it is very hazardous to form too definite conclusions". For devastation in Bedfordshire see G. H. Fowler, (1) *Bedfordshire in 1086* (Beds. Hist. Rec. Soc., 1922); (2) "The Devastation of Bedfordshire and the Neighbouring Counties in 1065 and 1066", *Archaeologia*, LXXII, 41 (1922). This latter is a more scientific application of Baring's method.

[5] *Sub anno* 1066. This is sometimes identified with Little Berkhamsted in Hertfordshire, but the overwhelming probability is that Great Berkhamsted (usually called simply Berkhamsted) was the place implied.

[6] F. H. Baring, *op. cit.* p. 216.

Conquest, but they did not bear the heaviest brunt. In the following year, again according to the Chronicle, King William "let his men plunder all the countryside which they passed through". The effect of this and other plunderings can be seen in many counties. From Hemingford, in Huntingdonshire, comes the terse entry: "There is land for one plough. It is waste. Ralf the son of Osmund has it".[1] Of Beeston Green, in Bedfordshire, it is stated: "There is land for half a plough, but it is not there; this land has been laid waste, but when Turstin received it, it was worth 10 shillings. In the time of King Edward, 20 shillings ".[2] Entries like this could be repeated in great numbers. Occasionally, the waste was accredited with some small value; and it is only fair to say that the backward state of many manors may have been due not to devastation, but simply to the lack of capital and labour. In some cases the entry was very explicit: of Harbury, in Warwickshire, the Survey said, "wasta est per exercitum regis....Valuit x solid. Modo ii solid."[3] And in most places the inferences are only too clear. "Nearly all our authorities speak of widespread violence as the outcome of the revolts and the harrying to which they led".[4] A case in point is Staffordshire where there was a rebellion in 1069; when William arrived upon the scene he found nothing but anarchy, and the Domesday Survey gives very plain evidence of the results of his arrival. All over the country the commissioners reported isolated estates under the name "terrae vastae". Theirs was a sorry tale; in the hundreds of Pirehill (falsely rubricated as Cudolvestan) and Totmonslow (falsely rubricated as Pereolle) King William had fifteen and seventeen waste estates respectively. There was arable land in the county sufficient to employ $1225\frac{1}{4}$ teams of oxen, yet there were only $992\frac{1}{2}$ teams in stock.[5] This devastation extended, apparently, into the adjacent shire of Derby, where no less than 10 per cent. of the total number of places mentioned are recorded as waste.[6]

To the north, the host of the king made a desert with still more terrible vengeance. The tale of the "harrying of the north" is well known, although the account in the Chronicle is brief enough—the king "went northward

---

[1] D.B. i, 207.   [2] D.B. i, 216b.   [3] D.B. i, 239.

[4] For a discussion of the narratives see R. R. Darlington, "Æthelwig, Abbot of Evesham", Eng. Hist. Rev. XLVIII, 177 (1933).

[5] R. W. Eyton, Domesday Studies: the Staffordshire Survey (1881), p. 22. For these rebellions see Ordericus Vitalis, Historia ecclesiastica, in Patrologiae cursus completus, Ser. 2, tom. 188 (1855), lib. iv, cap. 5.

[6] F. M. Stenton in V.C.H. Derbyshire, i, 318. It should not be forgotten that the Conqueror was not the only person responsible for the depopulation. Malcolm Canmore invaded north Yorkshire after the departure of the Norman army, and his misdeeds loom very large in the pages of Simeon of Durham.

with all the troops he could collect and laid waste all the shire".[1] Seventeen years later, the marks of this devastation were to be seen on every page of the Domesday Survey; many pages present hardly anything else. A comparison of values tells its own tale:[2]

	1066			1086		
	£	s.	d.	£	s.	d.
West Riding	1122	13	2	597	9	5
North Riding	924	1	0	184	2	8
East Riding	1432	6	0	387	9	0

Thus, of the enormous soke centring around Earl Tosti's manor of Preston and containing sixty-one villages, only sixteen acres were still cultivated in 1086, and it was not known exactly who were the occupants even of these. All the rest was waste.[3] In the midland shires to the south, devastation and massacre were also let loose, although in a more spasmodic fashion; while farther south still, in Devonshire, terrible ravages were committed by the Irish pirates.[4]

Destruction was not restricted to the country-side. Urban centres were affected as well as rural communities. The table of the Domesday boroughs tells its own tale; reduced numbers of *mansiones* are evident in many boroughs. The plight of the burgesses of Norwich is instructive:

Twenty-two left, and stayed in Beccles, a village belonging to the Abbot of Bury St Edmunds. Those who fled, and others who remain, are entirely ruined (*vastatae*), partly through the forfeiture of the Earl R., partly through fires, partly by the King's geld and partly by Waleran [the sheriff].[5]

[1] *Sub anno* 1069.

[2] *V.C.H. Yorkshire*, II, 189. For an idea of the devastation, see the map of the West Riding given by J. Beddoe and J. H. Rowe, "The Ethnology of West Yorkshire" in *Yorks. Arch. Journ.* XIX, 31 (1906). For its subsequent reclamation see footnote 2, p. 187 below. For the activity of the Conqueror see also W. Edwards, *The Early History of the North Riding* (1924), pp. 62–9.

[3] D.B. I, 301 *b*: "In Prestune comes Tosti vi carucatas ad geldum. Ibi pertinent he terrae [61 names follow] Omnes hae ville iacent ad Prestune et iii ecclesie. Ex his xvi a paucis incoluntur sed quot sint habitantes ignoratur. Reliqua sunt wasta."

[4] D.B. IV, 301: "Hae ix predictae mansiones sunt vastatae per vilandinos homines." See the Chronicle, *sub anno* 1067: "One of Harold's sons came with a fleet from Ireland unexpectedly into the mouth of the River Avon, and soon plundered all the neighbourhood. They went to Bristol, and would have stormed the town (*burh*), but the inhabitants opposed them bravely. Seeing they could get nothing from the town (*burh*), they went to their ships with the booty they had got by plundering, and went to Somersetshire, and there landed." See also *sub anno* 1098.

[5] D.B. II, 117 *b*.

Besides the wear and tear of life, badly provided for and badly protected, there were other tragic happenings. One of the most disturbing influences in the life of a borough was the building of a castle. The simple description of Lincoln stands eloquent of the new order:

Of these waste houses there were destroyed for the sake of the castle one hundred and sixty-six; the rest, seventy-four, were laid waste outside the boundaries of the castle not on account of the oppression of sheriffs and bailiffs, but because of disasters, poverty and fires.[1]

Nor was Lincoln an isolated instance; many other towns suffered the same fate.[2] In all the Domesday Survey records fifty castles; and more were built during the twelfth century.

The story of spoliation did not end with the years immediately around the Norman Conquest.[3] And to devastation must be added famine.[4] In the twelfth century, with the anarchy of Stephen's reign (1135–54), the forces working for settlement and stability received a check once more. The Chronicle describes the year 1137 in words graphic enough:

Every powerful man built his castles and held them against him [i.e. the king]; and they filled the land full of castles. They cruelly oppressed the wretched people by making them work at these castles...and this state of affairs lasted the nineteen years that Stephen was king, and ever grew worse and worse. They were continually levying an exaction from the towns, which they called *tenserie*, and when the miserable inhabitants had no more to give, then they plundered and burnt all the towns so that you could even walk a whole day's journey without finding a man seated in a town, or its lands tilled.

[1] D.B. 1, 336*b*.

[2] In Gloucester sixteen houses had to be pulled down to make a site for the castle (D.B. 1, 162), in Cambridge twenty-seven houses (1, 189), in Stamford five (1, 336*b*). In Shrewsbury (1, 252), in Warwick (1, 238), in Northampton (1, 219), in York (1, 298), and in Wallingford (1, 56), there was "waste" owing to the same cause. These are only a few samples of the devastation. See footnote 4, p. 219 below.

[3] The Anglo-Saxon Chronicle for instance, under the year 1104, records that "wherever the king [Henry I] went, his train fell to plundering his wretched people, and withal there was much burning and manslaughter". The *Dialogus de Scaccario* (1177) tells us, of the same reign, how crowds of peasants came to court, or mobbed the king on his journeys, "offering him their [idle] ploughs in token of the decay of husbandry" (ed. Hughes, Crump and Johnson, 1902, p. 89). These of course may indicate nothing more than the normal operations of a somewhat harsh government.

[4] See p. 196. Under the year 1087, the Chronicle records: "from the badness of the weather which we have mentioned before, there was so great a famine throughout England, that many hundreds died of hunger. Oh, how disastrous, how rueful were those times!"

Then was corn dear, and flesh, and cheese, and butter, for there was none in the land. Wretched men died of hunger. Some sought alms who at one time had been rich; some fled out of the country. Never was there more misery.... The earth bore no corn; you might as well have tilled the sea, for the land was all ruined by such deeds; and it was said openly that Christ and his saints slept.

The garrisons of these castles maintained themselves by forays, in which they ranged farther and farther afield until at last they stood in the centre of a circle of desolated territory. Geoffrey of Mandeville, entrenched upon the isle of Ely, reduced the people of the fen country and those of the upland villages around to a condition of unspeakable misery.[1] And the writer of the *Gesta Stephani* used words no less strong than those of the Saxon Chronicler.[2]

It is true that the broad and sweeping statements of the chroniclers must be accepted with caution. But in this case Professor H. W. C. Davis[3] has shown how well founded they were. The pipe roll of 2 Henry II (1156) gives the figures for the danegeld levied in each shire, and these are revealing enough. The devastations were committed in different years, and the rates of recovery varied in different districts, but the figures of the table on the opposite page leave no doubt about the waste character of much of the country during the middle of the century.

Figures for a new danegeld are given in the pipe roll of 1162, and they are notable for the fact that in them the item of "waste" has practically disappeared. Six years of ordered government had done much to restore prosperity to the wasted land. This is an important point that agrees with the Domesday evidence; between 1066 and 1086 there had frequently been great recovery. Primitive agriculture did not suffer permanently by

[1] See J. H. Round, *Geoffrey de Mandeville* (1892), *passim*.

[2] *Gesta Stephani regis Anglorum*, ed. R. Howlett (Rolls Series 1884–89), p. 99: "In all the shires a part of the inhabitants wasted away and died in numbers from the stress of famine; while others, with their wives and their children, went dismally into a self-inflicted exile. You might behold villages of famous names standing empty because the country people, male and female, young and old, had left them; fields whitened with the harvest as the year verged upon autumn, but the cultivators had perished by famine and the ensuing pestilence."

[3] H. W. C. Davis, "The Anarchy of Stephen's Reign", *Eng. Hist. Rev.* XVIII, 630 (1903), which contains the table that follows based upon the *Great Rolls of the Pipe*, 2, 3, 4 *Henry II*, ed. J. Hunter (Record Commission 1844). Hampshire and Suffolk are omitted because of the nature of the evidence. Further, the list does not include the wasted boroughs paying *auxilium* or *donum*; they were Cambridge ($\frac{1}{4}$), Huntingdon ($\frac{1}{2}$), Oxford (over $\frac{1}{3}$), Hertford ($\frac{3}{5}$), London ($\frac{1}{2}$), Rochester ($\frac{1}{3}$), Winchcombe ($\frac{1}{10}$), Nottingham and Derby ($\frac{1}{2}$). The fractions indicate the proportion of waste.

war unless the plough teams and live stock were actually destroyed in large numbers. But, although it was sporadic, local and relatively ephemeral, this repeated devastation was an ingredient of no mean importance in the life of England during the earlier Middle Ages; and these periods of desolation were too frequent to be neglected in any reconstruction of the economic geography of the time.

The Great Pipe Roll of 2 Henry II

Shires	Danegeld due			In waste			Proportion of waste to total
	£	s.	d.	£	s.	d.	
Warwick	128	12	6	80	11	0	Nearly 2/3
Notts and Derby	112	1	11	58	11	6	Over 1/2
Leicester	99	19	11	51	8	2	Over 1/2
Oxford	249	6	5	96	2	10	About 2/5
Bucks and Beds	316	6	8	107	14	3	Over 1/3
Berks	205	11	4	77	16	7	Over 1/3
Cambridge	114	14	9	34	3	0	About 1/3
Gloucester	184	1	6	59	3	6	Nearly 1/3
Northampton	119	10	9	38	12	1	Nearly 1/3
Hereford	93	15	6	19	3	6	Over 1/4
Worcester	102	5	9	27	14	3	Over 1/4
Wilts	389	13	0	99	16	9	About 1/4
Essex	236	8	0	61	4	0	About 1/4
Herts	110	1	3	29	17	4	Nearly 3/11
Huntingdon	70	5	0	14	0	6	About 1/5
Stafford	44	1	0	8	8	0	Nearly 1/5
Somerset	277	10	4	54	5	0	Nearly 1/5
Surrey	184	16	0	30	12	9	Nearly 1/6
Middlesex	85	0	6	10	0	0	Nearly 1/8
Sussex	216	10	6	9	2	0	Nearly 1/23
Kent	105	16	10	0	8	0	Nearly 1/270

There was another factor increasing the waste land of the time. Not only the necessities but also the pleasures of the king took their toll of good land. With the Norman Conquest, the forest law and the forest courts of Normandy were introduced into England, and they resulted in a rapid and violent extension of "forest" land—that is, land outside (*foris*) the common law and subject to a special law, whose object was the preservation of the king's hunting.[1] The word "forest" was thus a legal and

[1] The best account of the early history of forests to 1217 is Felix Liebermann's *Ueber Pseudo-Cnuts constitutiones de foresta* (Halle, 1894). See also John Manwood, *A treatise and discourse of the laws of the forest* (1598, 4th ed. 1717); Percival Lewis, *Historical enquiries concerning forests and forest laws* (1811) is based largely on Manwood. Manwood's definition is: "A forest is a certain territory of woody grounds and fruitful pastures, privileged for wild beasts and fowls of forest, chase, and warren to rest, and abide there in the safe protection of the king, for his delight and pleasure; which

not a geographical term. At their widest extent, the forests must have included much land that was neither wooded nor waste. But although the term was a legal one, the consequences were geographical. Of the Conqueror, the chronicler wrote under the year 1087:

He made large forests for deer and enacted laws therewith, so that whoever killed a hart or a hind should be blinded. As he forbade killing the deer, so also the boars. And he loved the tall stags as though he were their father. He also appointed concerning the hares that they should go free. The rich complained and the poor murmured but he was so sturdy that he recked nought of them.

As these forests were not liable for geld, they are rarely mentioned in the Domesday Survey. The names of only five forests are given,[1] but other forests may be inferred;[2] and a number of parks are also mentioned. Further, in many places, particularly in Worcestershire, Herefordshire, Shropshire and Cheshire, there are references to *haiae*, which were enclosures for catching animals.[3] Thus, of *Bernoldune* in Herefordshire, it is said: "The wood there is large, but its extent has not been told. There is 1 *haia* in which he takes what he can".[4] The duty of making these

territory of ground so privileged is mered and bounded with unremovable marks, meres and boundaries, either known by matter of record or by prescription; and also replenished with wild beasts of venery or chase, and with great coverts of vert, for the succour of the said beasts there to abide: for the preservation and continuance of which said place, together with the vert and venison there are particular officers, laws, and privileges belonging to the same, requisite for that purpose, and proper only to a forest and to no other place."

[1] The New Forest in Hampshire (see below); Windsor Forest in Berkshire (D.B. 1, 58 b); Groveley Wood near Wilton in Wiltshire (1, 74); Wimborne in Dorsetshire (1, 78 b); some woods in Oxfordshire (1, 154 b). The Oxfordshire entry gave full particulars about Shotover, Stow Wood, Woodstock, Cornbury and Wychwood: "dominicae forestae regis sunt, habent ix leugas longo totidem lato. Ad has forestas pertinent iiii hidae et dimidia et ibi vi villani cum viii bordariis habent iii carucas et dimidiam. De his et omnibus ad forestam pertinentibus reddit Rainaldus x lib. per annum regi."

[2] Hampshire (D.B. 1, 38 b); Flintshire (1, 268 b); Sussex (1, 19 b); Wiltshire (1, 68 b); Gloucester (1, 166 b–67 b); Herefordshire (1, 179 b–86); Sussex (1, 19 b); Huntingdon (1, 208 b); Essex (ii, 5 b). Traces also occur in Worcester. One of the Hereford references runs "the wood of this manor is alienated to form part of the king's wood" (1, 180 b).

[3] See D.B. 1, 256 b of *Cortune* in Shropshire: "Ibi est haia capreolis capiendis"; 1, 176 b of *Chintune* in Worcester: "habebant i haiam in qua capiebantur ferae". See also G.B. Grundy, *Saxon Charters of Worcestershire* (1931), p. xi, for references to the term *haga* in the charters of Worcestershire and other southern counties. See also the *V.C.H.* articles on Forestry.

[4] D.B. 1, 187. The name seems to be lost.

deerhays was called "stabilatio", and this is frequently mentioned; the *venatores* in the Survey are also numerous.

About the New Forest, however, a wealth of detail was given; for it occupied a special section in the Hampshire Domesday. The actual amount of destruction caused in the making of the New Forest has long been a matter of controversy. The annalists of the twelfth century declared that the Conqueror reduced a flourishing district to a waste by evicting a large number of agriculturalists; even his own chaplain complains of the destruction of villages and churches. But many have declared this tale of wholesale destruction to be a calumny which agrees neither with the soils of the district, nor with the Domesday evidence itself. Certainly a familiar formula ran: "it is now in the forest"; but, by differentiating between the vills wholly destroyed and those only partly destroyed, Mr F. H. Baring has produced a most reasonable interpretation[1] which is summarised in the diagram on the next page (Fig. 21).

For the other royal forests there is no such detail. Their number and their size, in the eleventh and twelfth centuries, are alike uncertain; estimates range as high as one-third of the total acreage of the kingdom. Difficulties arise, however, in interpreting their distribution, because there were varying degrees of afforestation; and it is impossible to estimate with any precision the territorial extent of the forest reign by reign. But although exact measurements are unknown, and although a forest was not necessarily a waste place,[2] there can be no doubt about the large amount of land alienated from cultivation, or from further cultivation. The extension of arable land in the royal forests was discouraged by fines levied in the special forest courts. And this afforded a constant source of bickering between the king and his subjects. The forest land seems to

---

[1] *Domesday Tables* (1909), p. 202: "To sum up the evidence, the story which Domesday has to tell us of the forest is this. William found in a corner of Hampshire 75,000 acres practically uninhabited. Of these 75,000 acres he made a forest, if they were not a forest before, and he enlarged this forest by taking into it some twenty villages and a dozen hamlets, covering from 20,000 to 25,000 acres, more than half arable, ... It is not certain whether these additions date from the time when he first used as a forest the 75,000 uninhabited acres or whether they were made later, but from these 150 ploughlands he cleared off the population amounting to some 500 families or 2000 men, women and children. He thus formed what we have called the main forest, the limits of which corresponded roughly to the outer boundary of the present forest. To protect the deer there were further annexed on the borders of this main forest other 10,000 or 20,000 acres, mainly woodland, but including probably 500 inhabitants whose fate is doubtful." See also F. H. Baring, "The Making of the New Forest", *Eng. Hist. Rev.* XVI, 427 (1901).

[2] This is obvious from the fact that whole counties were subject to forest law.

have reached its widest extent during the middle years of the twelfth century. Not merely woods such as Sherwood, Selwood, Dean, Andred, Windsor, Arden, and such hill districts as the Chilterns, the Peak, Exmoor, Dartmoor and the Yorkshire Wolds, were subject to forest law, but whole counties—Devon, Cornwall, Essex, Rutland, Northampton, Leicestershire and Lancashire. Miss Bazeley believes that the reduction of the forest area was "already in progress when the Great Charter of 1215, followed by the Charter of the Forest in 1217, held out hopes of wider concessions".[1]

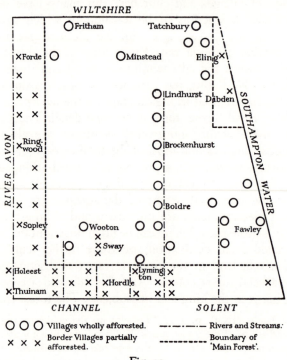

Fig. 21

Diagram of the New Forest district, re-drawn from F. H. Baring, "The Making of the New Forest", *Eng. Hist. Rev.* XVI, 427 (1901).

[1] M. L. Bazeley, "The Extent of the English Forest in the Thirteenth Century", *Trans. Roy. Hist. Soc.* IV, 4th series, 148 (1921). Further, "large districts in Devon, Cornwall, Staffordshire and Shropshire, Yorkshire and Essex were disafforested in 1204". Also, in the earlier part of the thirteenth century, "the counties of Sussex, Lincoln, Leicester and Middlesex obtained total exemption from the forest. In Derbyshire and Oxfordshire more than half; in Cumberland, Gloucestershire and Worcestershire more than a third; in Nottinghamshire, perhaps as much as two-thirds, and in

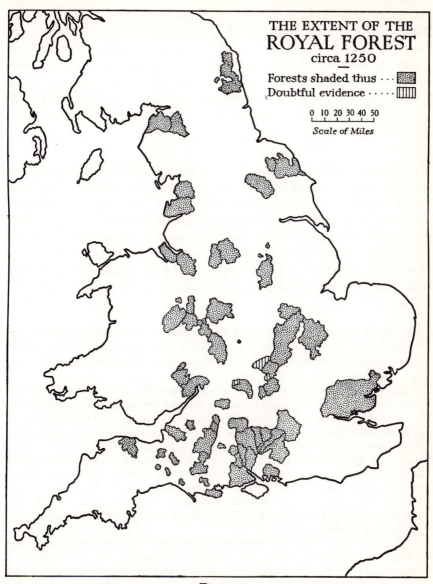

THE EXTENT OF THE
ROYAL FOREST
circa 1250

Forests shaded thus · · · ▓
Doubtful evidence · · · · · ▥

0 10 20 30 40 50
Scale of Miles

Fig. 22

Redrawn from M. L. Bazeley, "The Extent of the English Forest in the Thirteenth Century", *Trans. Roy. Hist. Soc.* IV, 4th series, 148 (1921).

Subsequent perambulations show that the disafforestments actually secured were considerable. In the later Middle Ages the disafforestation of district after district marked the economic progress of the country-side.[1] But substantial as these concessions were, very many royal forest districts remained untouched;[2] the accompanying map (Fig. 22) shows their distribution in the middle of the thirteenth century. Even as late as Tudor times, forest rights were still important in twenty counties of the kingdom.

COLONISATION

It is true that one of the immediate results of the Norman Conquest was the deliberate wasting of large tracts of country, but there were forces at work to redress something of this devastation. The colonising effort of Saxons and Scandinavians had not spent itself in the first impulse of their settlement. Despite the turmoil of the times, the encroachment of arable upon waste continued persistently, if slowly; and this expansion, gathering force with the centuries, was carried on under feudal leadership, both lay and clerical. A remarkable passage in a treatise by King Alfred (871–901) runs:

We wonder not that men should work in timber-felling and in carrying and building, for a man hopes that if he has built a cottage on the *laenland* of his lord, with his lord's help, he may be allowed to lie there awhile, and hunt and

Berkshire four-fifths of the forest area was surrendered by the Crown" (p. 152). The disafforestment was usually made by charter and a money payment exacted in return.

For the charters see Stubbs, *Select Charters* (9th ed. 1921), pp. 185 *et seq.* and pp. 344 *et seq.*; C. Petit-Dutaillis, *Studies and Notes Supplementary to Stubbs' Constitutional History*, vol. II (1915) of which deals largely with the forest; J. C. Cox, *The Royal Forests of England* (1905) deals with the later period; C. H. Pearson, *Historical Maps of England during the first thirteen centuries* (2nd ed. 1870) contains a list of forests. A most valuable treatise is C. J. Turner, *Select Pleas of the Forest, 1209–1334* (Selden Society, 1901). There are a number of histories of particular forests: the most valuable are W. R. Fisher, *The Forest of Essex* (1887); and E. J. Rawle, *Annals of the Forest of Exmoor* (1893).

[1] See p. 188 below.

[2] There is nothing like a complete list of forests before the year 1222, when a large number were mentioned in the Patent and Close Rolls for that year. A close letter, of June 27th, 1238, "to all the sheriffs in whose bailiwicks the king's forests are", names Yorkshire, Cumberland, Lancashire, Northumberland, Nottingham and Derby, Warwick, Northampton, Buckingham, Rutland, Huntingdon, Essex, Berkshire, Hampshire, Surrey, Somerset, Dorset, Devon, Wiltshire, Worcester, Stafford, Gloucester, Hereford. Oxfordshire and Shropshire may have been omitted by mistake. See M. L. Bazeley, *art. cit.* p. 153.

fowl and fish, and occupy the *laenland* as he likes, until through his lord's grace he may perhaps obtain some day *boc-land* and permanent inheritance.[1]

The thane, with his lord's permission, made a clearing in the forest or woodland, and lived there at first on game. In time, his log hut became one of a cluster; and this, in turn, became a group of homesteads. The cleared land was divided and tilled by the common plough, and there grew up a new hamlet in the outlying wood of some *ham* or *ton*.[2] The distribution of the "wood" place-name elements, "hurst", "ley" and the like not only provides an interesting summary of the process but also gives a commentary upon local soil conditions.[3]

The recovery of waste land also proceeded actively. In a curious document dating from the beginning of the tenth century,[4] Bishop Denewulf tells of his exertions to restore the wasted estate at Bedhampton granted him by the king; he succeeded in settling farmers on all its holdings, and he states explicitly that, after the last severe winter, there were 420 swine and 7 slaves and 90 acres sown, all apparently on the home farm. The same process of restoration following upon destruction can be seen in some other late Anglo-Saxon evidence during the struggle with the Danes.[5] Later still, the Chronicle itself provides a good example of recolonisation. Under the year 1092 it records how King William sent to the country around Carlisle "very many country folk, with their wives

---

[1] "Blossom Gatherings out of St Augustine", Brit. Museum, Vit. A, xv, f. 1. See F. Seebohm, *The English Village Community* (1883), p. 169. P. Vinogradoff in *Villainage in England* (1892), p. 333 wrote: "Everywhere in the world the advance of cultivation has been made the starting-point of privileged occupation and light taxation."

[2] Place-name evidence provides corroboration, and Professor Stenton writes: "Innumerable Domesday villages, as well as innumerable farms and hamlets bear names which resemble or are identical with the names of boundary-marks recorded in old English charters. The name Harston, which occurs in C and Lei, means literally 'boundary stone'. There are many parallels among old English boundary names to the Derbyshire village named Allestree, 'Æthelheard's tree', and the Warwickshire village named Austrey, 'Ealdwright's tree'. Names like these suggest movement outwards from the core of some earlier settlement to some point on the border of its territory." They suggest too that new villages were being formed throughout a long period.—*Introduction to the Survey of English Place Names* (ed. A. Mawer and F. M. Stenton), I, p. 40. [C = Cambridgeshire; Lei = Leicester.]

[3] See the distribution maps which are included in the more recent volumes of the Place-Name Society. See also pp. 130–1 above.

[4] W. de Gray Birch, *Cartularium Saxonicum* (1885–93), III, 653. See P. Vinogradoff, *Growth of the Manor* (1911), pp. 222 and 281.

[5] E.g. the activity of Bishop Oswald of Worcester in Heming's Cartulary (2 vols., ed. by T. Hearne, 1723).

and their cattle, to dwell there and cultivate the land". According to Florence of Worcester this area had remained desolate for 200 years after the devastations of the pagan Danes.[1] In other parts of the country the great ecclesiastical houses can be seen at work improving the dilapidated estates that had fallen into their hands.[2] An outstanding example of ecclesiastical activity comes from the great level of the Fens. "These marshes", wrote the chronicler of St Edmund, "afforded to not a few congregations of monks desirable havens of lonely life in which the solitude could not fail the hermits".[3] The records of the time tell how Crowland, Thorney, Peterborough, Ramsey, Ely and Chatteris arose in "the midst of a great solitude"; how others followed them; how they all established outlying cells; and how, in brief, the Fenland became famed as a home of monastic foundations.[4]

If we expect to find many specific references to this activity in the Domesday Book, we shall be disappointed. But one thing is certain—that clearings for the purpose of cultivation were already known as assarts, a word derived from the French *essarter*, to grub up or to clear land of bushes and trees, and so make it fit for tillage. But the allusions are very few in number, and we can only say that this is an example of the disparity between the real importance of phenomena and the notice taken of them by the Survey. The methods and the formulae of the Domesday commissioners have, apparently, hidden this silent revolution that was going on upon the English plain. There are four references to assarts in Herefordshire;[5] and in some counties on the Welsh march there were groups of *hospites*, who, wrote F. W. Maitland, "in fact or theory are colonists whom the lord has invited on to his land".[6] But the detailed entries for a number of vills in the Little Domesday throw, perhaps, a further sidelight

---

[1] *Chronicon ex Chronicis*, translated by T. Forester (1854). *Sub anno* 1092.

[2] On the restoration of the Abingdon estates see *Chronicon Monasterii de Abingdon* (ed. by J. Stevenson, Rolls Series, 1858), I, 486; and II, 1–2. Both references relate to the years immediately after the Conquest.

[3] *Memorials of S. Edmund's Abbey* (ed. by T. Arnold, Rolls Series, 1890–6), II, 13.

[4] Of course, these monastic houses did not attempt any large-scale reclamation in the Fen itself; but a certain amount of piecemeal reclamation seems likely.

[5] The references are: (1) Much Marcle: "In this manor there are 58 acres of land reclaimed (*proiecte*) from the wood". *Essarz*, is interlined as a gloss (D.B. 1, 179 *b*). (2) Leominster: "Of land reclaimed from the wood (*De exsartis silvae*) the profits are 17*s.* and 4*d.*" (I, 180). (3) Weobley: "There is a park there and assart land for one plough rendering 11*s.* and 9*d.*" (I, 184 *b*). (4) Fernhill: "The wood is half a league long and 4 furlongs broad, and (there is) assart land for 1 plough rendering 54*d*" (I, 184 *b*).

[6] F. W. Maitland, *Domesday Book and Beyond* (1897), p. 60. See D.B. 1, 259, 259 *b*.

upon this process. Some destruction of woodland may be evident in East
Anglia between 1066 and 1086. The clearing was associated with no one
area but was fairly general, and the following sample figures are, at any
rate, suggestive. They refer to the number of swine which the wood of
a village could support.[1]

		1066	1086
NORFOLK:	Cawston	1500 swine	1000 swine
	Baxton	1000 „	200 „
SUFFOLK:	Homersfield	600 „	200 „
	Coddenham	180 „	113 „
ESSEX:	Coggeshall	600 „	500 „
	Clavering	800 „	600 „

At one time it was believed "that this loss of woodland represents that extension of the cultivated area (*terra lucrabilis*) that was always in progress".[2]
But Mr Reginald Lennard has more recently shown that the destruction of
wood was not accompanied by expanding tillage.[3]

In some counties the distribution of wood in Domesday times seems
rather anomalous: it is absent from some claylands (where wood might
be expected) because, apparently, just these areas had been cleared at an
early date.[4]

The Weald is another area where colonising effort can be seen clearly
at work. Pre-Domesday charters show the intercommoning of the
villages in surrounding wealds. But the actual form in which this intercommoning took place in Kent may have been peculiar to the county.
Instead of turning out all their animals generally into the waste, the
Kentish villages came to have definite places in the wood peculiar to each
village; to these the swineherds drove their swine.[5] Some of these spots

[1] See pp. 200–2 below for the importance of swine in the economy of wooded areas.
[2] J. H. Round, *V.C.H. Essex*, I, 378.
[3] Reginald Lennard, "The Destruction of Woodland in the Eastern Counties under
William the Conqueror", *Econ. Hist. Rev.* XV, 36 (1945).
[4] See F. W. Morgan, "The Domesday Geography of Berkshire", *Scot. Geog. Mag.*
LI, 353 (1935).
[5] See N. Neilson, *V.C.H. Kent*, III, 81 *et seq.*; and N. Neilson, *Cartulary and
Survey of Bilsington, Kent* (1928). The latter deals in detail with "Custom in the
Weald". See also J. E. A. Jolliffe, *Pre-Feudal England: The Jutes* (1933), p. 56. The
thirteenth-century *Seneschaucie*, for instance, in describing the duties of the swineherd,
says: "if the swine can be kept with little sustenance from the grange during hard frost,
then must a pigsty be made in a marsh or wood, where the swine may be night and
day"—p. 113 of *Walter of Henley's Husbandry*, etc., ed. E. Lamond (1890).

13

became, in time, more or less permanently settled by men seeking pasture or timber; and each of these units was called a dene or denn. At Peckham in Kent, for instance, "the King has three denes where there are four villeins, and it is worth forty shillings".[1] The total number of denes mentioned in the Domesday Book is between fifty and sixty, and they are almost exclusively confined to Kent. From Saxon charters and thirteenth-century records, it is evident that by no means all the denes of the county are entered; those mentioned were probably inserted because of some particular circumstance, while many more lying in the great stretches of wood were included in the values of the villages to which they were attached. Distinction was made in the Survey between large denes and small denes; and, also (obviously cutting across the other division), between denes *de silva* defined, may be, by swine rents, and other denes which had progressed beyond the wooded state and had reached the stage of being cultivated by villeins and bordars, and of being measured by the ordinary measures of arable land. In one or two cases, Benenden and Newenden for example, places once rated as denes had come to be regarded as ordinary villages and were entered accordingly in the usual manner. But we do not know enough about the conditions under which the swarming-off from established villages was begun and conducted, and we cannot speculate with any certainty.[2]

Besides these scanty notices of newly reclaimed land, indications of the work of colonisation may be embodied in two of the most frequent Domesday formulae—*carucae* and *terra carucis*—the number of plough-teams (A) and ploughlands (B) respectively. The former presents comparatively little difficulty; in stating how many plough-teams were actually employed on an estate, it is only reasonable to suppose that the jurors were giving a rough estimate of the land actually cultivated. Frequently, sometimes usually, there is a difference between A and B, and we are driven to ask what does B really imply. Considerable controversy has taken place around this point. Some scholars believe it to be an estimate of land fit for tillage if the estate were properly cultivated. Variations in the formula suggest this: "land that can be ploughed by x teams" (*possunt arare x*

---

[1] D.B. 1, 7*b*.
[2] In some parts of the country, some of the outlying settlements may have had dairies denoted by the term "wick". See P. Vinogradoff, *English Society in the Eleventh Century* (1908), p. 369: "The derivation of 'herdwick' from a settlement of herdsmen making it the equivalent of *vacaria* seems tempting." But he also shows (*ibid.* p. 284) that the term may have been employed also in the sense of an agricultural farm.

carucae);[1] land where there could be x teams (*x carucae possunt fieri*);[2] "land for x ploughs" (*terra ad x carucas*).[3] Another variant is *x carucae plus possunt esse*;[4] here the contrast is not between A and B, but between A and the land which could be utilised by an increase in the number of teams. In the shires of Hertford, Middlesex and Buckingham, for example, both the existing teams are given and those which "could be made", the two together being in nine cases out of ten exactly equal to the number of teamlands.[5]

Difficulties arise, however, when an attempt is made to apply the interpretation generally. In Gloucester, Worcester and Hereford, there is no systematic mention of B at all, but only occasional reports which show that at certain places there might be more teams than there are. The Little Domesday likewise does not record information about B, but it gives information about "geld carucates" and about "plough teams"; the latter for the three dates—*tunc, post* and *modo*. When the existing teams are fewer than those that were ploughing in time past, it is sometimes added that a number of teams could be "restored".[6] Thus, of Rougham in Norfolk, it is said: "Then 3 ploughs on the demesne, afterwards and now, none; and 4 could be restored (*restaurari*)."[7] Egmere, also in Norfolk, is described as a berewick of "half a ploughland and there is nothing else there, but 1 plough could be [employed]".[8] That "*x* ploughs could be" (*fieri*) is a familiar formula; and there are many similar entries. Occasionally, the wording is more explicit; an entry for Thetford runs: "all this land is arable and 4 ploughs can till it".[9] But that is rare. Or,

---

[1] E.g. D.B. 1, 67: "Newetone...Totum manerium possunt arare x carucae" (S. Newton, Wilts); 1, 75: "Cerminstre....In ipso manerio habet episcopus tantum terrae quantum possunt arare ii carucae" (Charminster, Dorset).

[2] E.g. D.B. 1, 231: "Toniscote...Duae carucae possunt esse, et ibi sunt" (Cotes Deville, Leics.).

[3] E.g. 1, 29: "Alviet holds of William land for 1 plough (*terram ad unam carucam*); [it is part] of William's demesne and is not assessed in hides (*sine numero hidae*). There is 1 plough, and 1 mill yielding (*de*) 3 shillings. It was part of (*jacuit in*) Storgetune as (*in*) pasture. Now it has been lately brought under cultivation (*modo noviter est hospitata*). It is worth 10 shillings." This entry relating to the lands of William de Braiose, at Storrington in Sussex, is interesting because it contains one of the few explicit Domesday references to colonisation.

[4] E.g. 1, 174 a: "In omnibus his maneriis non possunt esse plus carucae quam dictum est" (Ecclesia Sanctae Mariae de Wirecestre).

[5] See F. H. Baring, *Domesday Tables*, p. 99. This type of entry is also found in Cambridgeshire.

[6] In any case the East Anglian assessment is peculiar. See p. 194.

[7] D.B. II, 120 b. [8] D.B. II, 113 a. [9] D.B. II, 118 b.

again for Flockthorpe (now in Hardingham) the entry is: "then 3 ploughs on the demesne, afterwards and now 2, and 1 could be restored".[1] And there are counties (e.g. Leicestershire) where *B* alternates with another formulae: "Ibi fuerunt T.R.E. *x* carucae". Occasionally, *nunc* or *unde* is added, which seems to include a contrast between 1066 and 1086. Many scholars have therefore believed that *B* refers to the number of teams in 1066, and not to the generally available arable land. The simple fact may be that the enquiry was executed differently in different counties according to the lights of the king's commissioners and of local juries.[2] But what, at any rate, is apparent from the contrast of *A* and *B* is the frequent prospect of further cultivation; and the commissioners, with the king's geld in mind, were not uninterested in this. Generally speaking it appears that agriculture reached its lowest ebb at the time when the lands were granted by King William; many estates had recovered since that date, and some were in an even better condition than they had been in 1066. This difference throws light upon the direction and aims, if not upon the initial spread, of reclamation.

The years that followed 1086 saw no cessation in the work of clearing. In France, on Palm Sunday, 21st March 1098, there was founded the monastery of Cîteaux, an establishment fraught with great consequences for English development.[3] The Cistercian movement, thus inaugurated, was but one of the great series of revivals within the Benedictine order; and in 1128 their first English foundation took place at Waverley in Surrey. Others soon followed. "In civitatibus, castellis, villis," ran one of the earliest regulations of the "first fathers", "nulla nostra construenda sunt cenobia, sed in locis a conversatione hominum semotis". And so the sphere of the Cistercian settlement was not confined to the southern and midland shires, but included also the wildest and most solitary districts of the land— among the dales and moorlands of the north, and in the deep valleys of

---

[1] D.B. II, 122.

[2] See F.W. Maitland, *Domesday Book*, pp. 422–3, for a discussion of the objects of the commissioners. But note P. Vinogradoff, *English Society in the Eleventh Century* (1908), p. 159: "It seems to me that it is quite out of the question to apply the T.R.E. number of the teams as a general clue for the *terra carucis* entries all through the Survey. In a very large number of cases the number of *terrae carucis* is a good deal larger than the number of teams (*carucae*), while the value of the estate T.R.E. turns out to be smaller —a most improbable contingency if *terra carucis* is taken to apply to the T.R.E. tillage."

[3] See A. M. Cooke, "The Settlement of the Cistercians in England", *Eng. Hist. Rev.* VIII, 625 (1893). Note, in particular, the map showing the distribution of Cistercian houses. See also Fig. 35, p. 243 below.

Wales. In 1131 Rievaulx was founded; in 1132 Fountains; and hence-forward the spread of the Order was particularly rapid. Daughter houses carried on the work of the *plantatio* to the third and fourth generations. And in this way was begun that great movement which transformed much of northern England from a wilderness into a sheep-run. Giraldus Cambrensis commented strongly upon their effective colonisation.[1] Indeed, as farmers, the Cistercians were held in high repute; their reputation for estate management was well known; and their skill in the breeding of horses and cattle was of no mean order. In Yorkshire, especially, they soon became famous for their wool production—in fact the export of wool by the monks was to become a great feature in the commerce of the realm.[2] The abbeys of the Order in England numbered about fifty by 1152, when the initial period of the Cistercian settlement came to an abrupt end, for a decree went forth from Cîteaux that no more Cistercian abbeys should be founded. Only a very few were established in England after this. Solitude proved itself but a poor defender of poverty; in England, by the end of the twelfth century, "as an element in the nation's spiritual life the order of Cîteaux, once its very soul, now counted for worse than nothing".[3] Failure on the one hand; success on the other hand; for, as a result of a generation's work between 1128 and 1152, many outlying tracts of south Britain had been made to share in the increasing prosperity of the later Middle Ages.

But the Cistercians were far from being the only monks set upon the improvement of dilapidated estates and upon the clearing of new land. Jocelin of Brakelond's *Chronica*[4] gives a graphic picture of the activity of Abbot Samson at the Benedictine foundation of Bury St Edmunds during the closing years of the twelfth century:

[1] *Giraldi Cambrensis Opera* (ed. J. S. Brewer, Rolls Series, 1873), IV, 113. At Revesby, in 1142, a reversal of the usual development took place. Revesby, "Scithesbi", and "Toresbi", seem to have been emptied of their inhabitants in order that the Cistercians might have an unencumbered site. See F. M. Stenton, *Facsimiles of Early Charters from Northamptonshire Collections* (1930), pp. 1–7. Professor Stenton writes: "What can be observed in safety is that unless the men of Revesby were very unlike their contemporaries, English villagers of the twelfth century must have been much less tightly rooted to their holdings than it is customary to suppose" (p. 5). Another example of the removal of inhabitants is provided by the foundation of the Carthusian monastery at Witham by Henry II.—*Magna Vita S. Hugonis Episcopi Lincolniensis* (ed. J. F. Dimock, Rolls Series, 1864), pp. 68–70.

[2] See p. 242 below.

[3] K. Norgate, *Angevin Kings* (1887), II, 345.

[4] See *The Chronicle of Jocelin of Brakelond* (ed. by L. C. Jane, 1925), pp. 43–64 *passim*.

The abbot caused inquisition to be made throughout each manor about the rents from the freemen, and the names of the villeins and their holdings, and the services due from each; and he caused all these details to be written down. Then he restored the old halls and ruined houses through which the kites and crows flew; and he built chapels and constructed inner rooms and upper stories in many places where there had been no buildings, except barns....He made many clearings and brought land into cultivation....He built barns and cattle-sheds, being anxious above all things to dress the land for tillage....The abbot further appeared to prefer the active to the contemplative life and praised good officials more than good monks.

To work indeed was to pray in the eyes of many monks; and from the other monasteries of the realm comes the same tale. There is but little point in giving detailed examples of monastic activity;[1] very many cartu-laries, chronicles and custumals record the progress of the advance of arable upon the surrounding pasture and waste. The phrases the "old assart" and the "new assart" register the stages in the process.

Nor was the movement confined to ecclesiastics. Laymen pursued the same policy with like energy. Charters, rentals, terriers and forest rolls provide the same evidence of new ploughing and enclosure, of which the cumulative effect must have been very great. Striking testimony of this activity comes from the counties of the north-west, where the Scandinavian element -thwaite (indicating a clearing) formed a characteristic feature of the local nomenclature. H. Lindkvist has summarised the situation:

Not one reliable instance seems to be on record in the O.[ld] E.[nglish] period. Domesday has a very few, belonging to Yo.[rkshire] and Norf.[olk]. Several others appear during the latter half of the 12th century in the monastic chartularies and registers, the Pipe rolls and in various other local documents. But the majority of these names are not met with until the 13th and 14th cen-turies, or even later. Their gradual appearance in literature reflects, to a certain extent, a progressive process of settlement, in the course of which a large proportion of them were coined, and the localities indicated by them were growing in importance. In many cases, when a þveit-name occurs for the first time in the records, it is expressly said to design a campus, or pastura, or cultura, or clausum, a meadow, vaccary, etc. It seems, in fact, that as far as the earlier

---

[1] Some good examples of reclaimed woodland in the shires of Leicester, Notting-ham and Lincoln are given by Professor F. M. Stenton in *Documents Illustrative of the Social and Economic History of the Danelaw* (1920), p. xli. In, or before, 1150 there were quite a number of grants of land "ad sartandam et arandam vel quicquid eis inde placuerit faciendam". One Sempringham charter, for instance, refers to a *cultura* at Hawthorpe "which in old time when it was a wood was called Simildehae".

half of the M.[iddle] E.[nglish] period is concerned, the bulk of these names are to be regarded as mere field-names, mostly designations of reclaimed land, or of out-of-the-way places, which, as time went on, were built upon and permanently inhabited; many of them formed some kind of dependencies of an old manor or demesne.[1]

The places subsequently designated by "thwaite" must, in 1086, have been waste and uninhabited, perhaps depopulated, at any rate of little importance. Yorkshire itself must have remained thinly populated for many generations after the devastation of 1069. "There was plenty of room for expansion; and we can observe the clearing of new land going on...". Mr A. T. M. Bishop,[2] who makes these remarks, describes in particular "the clearing which was carried on among groups of free tenants, to whom may be ascribed a comparatively unrestricted power of reclaiming the waste, and who were in any case the normal inheritors of the devastated parts of Yorkshire". The evidence suggests "that large tracts of waste lands were granted to isolated colonists". Already "by the thirteenth century there had been restored, in Yorkshire, a rival organisation resembling that which had survived the Conquest in the central Danelaw".

The documents show that this extension of the area under cultivation was effected in various ways; sometimes by a single settler with help from

[1] H. Lindkvist, *Middle English Place-Names of Scandinavian Origin* (Upsala, 1912), pp. 98–9. Lindkvist records the instances of *-thwaite* elements as follows: Yorkshire, 83; Cumberland, 52; Lancashire, 43; Westmorland, 23; Lincoln, 11; Norfolk, 7; Nottingham, 5; Leicester, 2; Suffolk, 2; Northumberland, 1. Although this list of names could be considerably increased, their relative distribution would not be much affected.

[2] "Assarting and the Growth of the Open Fields", in *Econ. Hist. Rev.* VI, 13 (1935). "That the settlers who recolonised large parts of Yorkshire tended to revert to tradition is shown by the fact that they did not found many new villages, or create a new type of village settlement...." (p. 17). Few thirteenth-century vills are not named, and fewer not identifiable, in Domesday. (See the English Place-Name Society's volume on the North Riding, ed. by A. H. Smith, 1928.)

This most valuable paper can be supplemented by another by the same author: "The Distribution of Manorial Demesne in the Vale of Yorkshire", *Eng. Hist. Rev.* XLIX, 386 (1934). This makes distinction between the manorialised vills (slightly less than half in number) that "had contained some rural population in 1086, and the non-manorial vills" which had been wholly waste in 1086. "The economic freedom of these vills must be attributed to the favourable conditions which have always been enjoyed by those who undertake the colonisation of waste land" (p. 406). For other information about post-Conquest improvement see W. Farrer, *Early Yorkshire Charters*, II, VII (1915).

For conditions in the adjoining Lancashire during the hundred years after the Conquest, see W. Farrer, *The Lancashire Pipe Rolls* (1902), p. xiv.

his lord; at other times by the entire village, or at least by a large group of peasants who joined together for the purpose. In the first case, there was no reason for bringing the reclaimed area into the scope of the compulsory communal practices, and there was an increasingly strong current towards individualistic husbandry throughout the period.[1] In the second case, the distribution of the acres and strips among the various tenants was proportioned to their holdings in the ancient lands of the village; this was the practice in the assarts recorded by the so-called Domesday of St Paul's in 1222. Assarts were duly dealt with in the early twelfth-century treatise, the *Leges Henrici Primi*, and they feature prominently in the *placita forestae* of the time. Many villages lay in, or near, forest land where the right to assart was disputed. Requests for permission "to assart"; grants of land "with leave to assart"; gifts of land "to be assarted as seems best"; records of "new land recently brought into cultivation"; disputes about the "tithes of assarts and clearings"—all these items are encountered again and again in the records of the time. Moreover, much of the work must have escaped mention in any document. The process was piecemeal in character—an occasional clearing, an enlargement of a field or "shot", an enclosure for pasture. And, in the main, it was a comparatively silent revolution; but its cumulative effect was to change the face of the countryside and to provide for an increasing population.

Besides being a reserve for colonisation, the waste land played an integral part in manorial economy because it provided common wood, common turbary and, above all, common pasture. Common rights constituted no incidental appurtenance, but were vital elements in the tenements of a village; and medieval husbandry was conducted on the principle of maintaining a balance between agricultural and pastoral pursuits.[2] Progressive encroachment upon the waste must therefore, sooner or later, reach a limit

---

[1] See P. Vinogradoff, *Growth of the Manor* (2nd edition, 1911), p. 330: "Enclosed plots of arable and private meadows, pastures and woods are also often to be found, and they occur more and more frequently as time goes on. We catch glimpses of the process of enclosure, and of the changes brought about by more intense and perfect husbandry. New sequences of crops are introduced, the soil of some portions of the demesne gets to be manured and cultivated more carefully, and, to protect these ameliorations, hedges have to be set up, 'intakes' are formed; and these intakes represent the most advanced technical progress of those times." For *inhoc facere* on the Malmesbury and Gloucester estates, see P. Vinogradoff, *Villainage in England* (1892), pp. 226–9. These changes served to complicate the ordinary three-course rotation; in some cases a new species of arable —the manured plot under *inhoc*—came into use. A. T. M. Bishop, in the paper cited above (*Econ. Hist. Rev.* 1935), discusses the relation between assarts and the open-field land.

[2] See pp. 197–8.

that could not but disturb the balance of rural economy. The arable benefited at the expense of the waste. When the lord enclosed successive portions of the waste for tillage, such enclosure naturally came to conflict very strongly with the interests of the community in general; and "approvement of the waste" (the right to enclose common land) developed into a burning question. The crisis came in the thirteenth century; and, in 1235, the Statute of Merton[1] recognised the lord's right to occupy the waste, provided he left sufficient pasture for his free tenants. This, as Vinogradoff has written, leaves

entirely in the shade a whole series of most interesting questions. What was the view of former generations on the rights of lords to approve? Were the processes of colonisation and reclaiming carried on entirely at random in former days or were they shaped by some order and custom, etc.?[2]

But this obscurity does not affect the fact that here may have been a turning point in the history of the village community. The implications of the Statute, both social and economic, were many. As far as clearing and settlement were concerned, it registered a stage in the history of open expansion. Of course, assarting and "new ploughings" continued long after 1235. Through the length of English history, heath and woodland continued to shrink, as new hamlets and new farms sprang up, and as the hunter-kings were forced to give up one forest jurisdiction after another.

THE ENGLISH VILLAGE

The unit of settlement upon the English plain was the village community. It was a remarkable association, because, despite all its inconveniences, it repeated itself not only over a great part of England but also over a great part of Europe; and it was prevalent throughout, at least, the span of a

---

[1] *The Statutes of the Realm*, 1, 2 (Record Commission, 1810).
[2] *Growth of the Manor*, p. 171. For a remarkable piece of evidence bearing on the approvement of woodland see the *Rolls of the Justices in Eyre for Lincolnshire 1218–9 and Worcestershire 1221*, ed. by D. M. Stenton (Selden Society, 1934), p. 449. When certain freeholders at Yardley, near Birmingham, complained in 1221 about approvement the reply was: "Such is the law in Arden that where there is a great pasture, he whose land it is may well make buildings and raise hedges and ditches within that pasture, provided that it is not in their [i.e. the freeholders'] exit or entry, or to their hurt." This is definite evidence that already by 1221 it was an accepted custom in one of the greatest midland forests for a lord to have the right of enclosing portions of woodland so long as it was not to the detriment of freeholders with pasture-rights. The Statute of Merton may have been essentially a definition of existing custom.

thousand years, from 500 to 1500, extending many of its features even into later times. Characteristic of agricultural plains that were easily settled, the village community was the systematic expression of a mode of life. "All such establishments, though different in form—a fact due to climatic differences or to different stages of social development—are the expression of the same need. That need is to centralise agricultural activity in some one place. A co-operative agreement as to dates in the agricultural calendar, and the time for certain tasks, is adopted for the advantage of all concerned."[1] Despite the variations of race and of history, this economic equation has a rough validity. It is true that the Norman occupation profoundly affected the English land system,[2] but that influence was legal and tenurial rather than agrarian. And, although much remains obscure, a comparison of documents immediately preceding 1066[3] with later extents and custumals, and with the agricultural treatises of the twelfth and thirteenth centuries,[4] establishes the essential continuity of English economic life notwithstanding the shock of the Norman Conquest.

The term "manor" came in with the Norman Conquest, but beneath the manorial organisation there seems to lie a more primitive institution— a community which in its essential economic needs, if not in its particular method of satisfying them, reaches back into a remote past.[5] Manorialism was but a shell and accretion around the village community.

In its purest form, the village community consisted of one cluster of houses (frequently there were two or more clusters) standing in the

[1] Vidal de la Blache, *Principles of Human Geography* (1926), pp. 299–300.

[2] In 1086 villeins appeared everywhere in place of the earlier sokemen, freemen and even the lesser thanes. The Survey really shows society under stress of change, change which was accelerated by the political events of twenty years. How great the change might be can be gauged from an entry relating to the Bedfordshire village of Crawley: "there are one villein and 7 bordars and a slave...9 thanes held this manor" (folio 158 *b*). F. W. Maitland, *Domesday Book*, pp. 62–3, pointed out that the sokemen of Cambridgeshire were reduced from 900 in 1066, to 213 twenty years later; and A. Ballard, *The Domesday Inquest* (1906), p. 100, declared that "not one per cent. of the land was owned (i.e. held of the king) in 1086 by the same men who had owned it in 1066, or by the sons or widows of the previous owners".

[3] The *Rectitudines Singularum Personarum* (c. A.D. 1000); the *Gerefa* (early eleventh century); and Aelfric's *Colloquium* (late tenth century).

[4] The Husbandry of Walter of Henley (thirteenth century); the Rules of Robert Grosseteste (mid-thirteenth century), and the associated anonymous Husbandry; and the anonymous *Seneschaucie* (not later than Edward I).

[5] Professor Stenton, for instance, in *Documents...of the Danelaw* (1920), p. lxi, writes: "It is, indeed, impossible to ignore the varied evidence which reveals the village and not the manor as the essential form of rural organisation in this region."

midst of its territory. The rights of a household in this territory involved an intricate mixture of claims, and necessitated constant co-operation between neighbours. The peasant's existence was really set against a double background, partly artificial and partly natural. The complicated tenurial relations between lord and peasant fitted into the agrarian framework provided by the physical conditions of the land itself. As Vinogradoff wrote: "the term which may best indicate its main characteristic would perhaps be that of a community of shareholders".[1] And the different aspects of the shareholding fell into three groups connected respectively with the cultivation of the arable, with the utilisation of meadow and pasture, and with the exploitation of the miscellaneous resources of wood and fen and river.

The arable land was the most important and conspicuous part of the village territory. It lay in strips, and every holding consisted of a number of these strips scattered about in different places, each strip separated from those of other tenements by fringes of unploughed turf or by unsown furrows. Each cultivator thus held a number of strips, but not in a compact bundle; and the intermixture of strips in this manner meant that the rules and methods of cultivation were the concern not of any individual but of the community as a whole. Such was the practice at a time when documents and descriptions were comparatively abundant, and there can be no doubt that the same arrangements were in existence right from the early stages of the settlement. Nor did the social status of the tenants make any difference. A half-servile community under a lord was not different in essence from a village of freemen, either in plan or in customary arrangements. It is true that the general method of the Domesday Book allows only very occasional glimpses of the system at work; but the Saxon charters contain numerous hints,[2] and the labours of modern enquirers have made clear the main features of the open-field system.

It was scarcely possible to cultivate the arable fields of the village year after year without exhausting the soil, especially in the absence of those

---

[1] *The Growth of the Manor*, p. 150.

[2] Occasionally, considerable detail is given. In a charter of King Edgar, conferring land at Avon, near Chippenham, in Wiltshire, the three hides given are said to be "singulis jugeribus mixtum in communi rure huc illuc dispersis loco" (*Cartularium Saxonicum*, 1120, ed. W. de Gray Birch, 1885–93). For other early examples see P. Vinogradoff, *Eng. Soc.* pp. 277–9. For a discussion of the data available see H. L. Gray, *English Field Systems* (1915), pp. 13–16; E. Nasse, *Agricultural Community* (1871), pp. 18–26; F. Seebohm, *English Village Community* (reprinted 1926), pp. 105–17.

important roots and seeds which were later to revolutionise farming practice. The land therefore needed periods of rest to recuperate; and recourse was made to regular fallowing by dividing the arable into large tracts, each of which was cultivated in turn. There came to be alternative arrangements.[1] Under the *two-field* system, the arable was arranged in two fields; of the field under cultivation one half was sown in autumn with winter corn, wheat or rye, and the other half in the beginning of the year with spring

	I Fallow	II Stubble of wheat or rye	III Stubble of barley or oats
October	Plough or sow with wheat or rye		
March		Plough and sow barley or oats	
June			Plough twice
August	Reap	Reap	
	Stubble of wheat or rye	Stubble of oats or barley	Fallow

[1] H. L. Gray, in *Eng. Field Syst.*, sums up the evidence (p. 82): "There is discernible in Anglo-Saxon England an open-field system, which first at the end of the twelfth century reveals itself as one of two or three fields;...it is certain that to some extent transition from two-field to three-field arrangements occurred during the thirteenth and early fourteenth centuries, and it is not improbable that the three-field system may have been altogether derived, arising from an improvement in agricultural method." Earlier on the same page he writes: "Hence it is not improbable that the predominantly three-field counties became such during the thirteenth and fourteenth centuries. If so, the system was a derived one, and midland England at the time of the Conquest was a region dominated by two fields." It is interesting to note that Walter of Henley (*op. cit.* pp. 8–9) in the thirteenth century, after describing three-field conditions, wrote: "And if your lands are divided into two, as in many places,...." This may be taken to imply that the two-field system was more uncommon, but this indication is hardly conclusive; and Professor Stenton informs me that "judging from charters—the earliest local evidence which exists—the two-field system seems to have been almost universal in the 12th century in regions where open-field cultivation of the normal sort prevailed."

For the transition in later times see p. 239 below.

crops, barley or oats;[1] thus was the labour of ploughing divided. The fallow field was ploughed twice at the beginning of summer. Under the *three-field* system, the land was divided into three portions—one with winter crops, one with spring crops and the third lying fallow—as shown on p. 192.[2]

Between seed-time and harvest, the land under crop was protected by temporary fences; when the harvest was over, the barriers were removed, and the village cattle were allowed to graze upon the stubble of the open fields. Under these arrangements, the merit of the three-field system was that it provided a greater acreage of crop for the same amount of ploughing;[3] but, on the other hand, the quality and the yield of the crop might suffer. Whichever variety prevailed more generally,[4] one thing, at any rate, is certain—the open-field system itself was not the only method of agricultural exploitation on the English plain. It was restricted to a large irregular area lying chiefly in the midlands, reaching northward as far as Durham and southward to the Channel, and extending from Cambridgeshire on the east to the Welsh border on the west (Fig. 23).

The Kentish system was the most divergent. Its feature, says Dr Gray, was "the *iugum*, the unit of villein tenure, which, compact and rectangular in shape, had its exact counterpart nowhere else in England".[5] While the Anglo-Saxons of the midlands developed the open-field system and the virgate holding, the Jutes who occupied Kent seem to have been "a nation of hamlet dwellers".[6] Further, the situation of the county resulting in an

[1] See E. Lipson, *The Economic History of England* (5th ed. 1929), p. 60. Walter of Henley (*op. cit.* p. 9) wrote: "...the one half sown with winter seed and spring seed, the other half fallow..."

[2] The diagram is that in W. Cunningham, *The Growth of English Industry and Commerce during the Early and Middle Ages* (5th ed. 1910), p. 74.

[3] This was shown in Walter of Henley's treatise on husbandry (*op. cit.* p. 8, note 1): 160 acres tilled on a two-field system meant that the plough would cover 240 acres to produce a crop 80 acres; 180 acres tilled on a three-field system meant 240 acres ploughing and 120 acres of crop. And although Walter's programme may have been ideal rather than actual (see Maitland, *op. cit.* p. 397) his argument remains valid enough.

[4] The choice between the two systems seems to have been simply a matter of agricultural opportunity. H. L. Gray (*op. cit.* p. 73) writes: "The regions which adhered to two-field husbandry were, on the whole, the bleak, chalky, unfertile uplands; those, on the contrary, which were possessed of better soil and better location came to be characterised by three fields. This can only mean that, whenever natural advantages permitted, men chose the three-field system by preference. The retention of two fields was usually a tacit recognition that nature had favored the township little."

[5] H. L. Gray, *op. cit.* p. 415.

[6] J. E. A. Jolliffe, *op. cit.* p. 1.

early money economy may have rendered obsolete the more rigid tenurial arrangements of the manor. The East Anglian *tenementum*[1] at times resembled the Kentish unit; at other times "it was usually less like the intact Kentish unit than like a Kentish holding after the *iuga* had for some

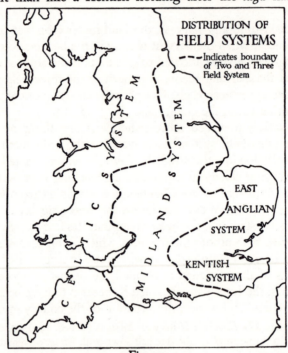

DISTRIBUTION OF
FIELD SYSTEMS
– – – Indicates boundary
of Two and Three
Field System

EAST ANGLIAN SYSTEM

KENTISH SYSTEM

Fig. 23

Drawn from H. L. Gray, *English Field Systems* (Harvard University Press, 1915).

generations been subdivided and a tenant had come to hold parcels in several neighboring *iuga*".[2] It suggests a pre-Domesday reorganisation of agrarian economy, associated probably with the influx of free Scandinavians.[3] The peculiar pasturage arrangements of Norfolk and Suffolk are also suggestive enough.[4]

[1] For a more detailed treatment of East Anglian tenurial peculiarities see D. C. Douglas, *Social Structure of Medieval East Anglia* (1927).

[2] H. L. Gray, *op. cit.* p. 416.

[3] H. L. Gray, *op. cit.* p. 353: "One naturally asks why incoming Danes brought into existence in East Anglia a unit different in aspect from the virgate and the bovate found elsewhere within the Danelaw.... In East Anglia, however, the Danes probably found a system divergent, then as later, from that of the midlands. To this they adapted themselves, being without doubt the minority of the population."

[4] See footnote 2, p. 197 below.

Between East Anglia and Kent lay Surrey and Essex. The peculiarities of their field arrangements may be explained by postulating a reorganisation of the disintegrating *iugum*, together with the adoption for the new unit of the name of the midland virgate. Finally, many of the characteristics of the midland manor were lacking in the lower Thames region of Middlesex, Hertford and the Chilterns. This really formed a borderland region. Moreover, it was fairly heavily wooded, and later irregularities may have arisen from the addition of assarted areas. "The Chiltern area should, therefore, be looked upon as a boundary region so influenced in its field system by its topography that its original affiliations cannot readily be discovered."[1] Whatever theory be adopted to explain these differences, the facts themselves show that no simple generalisation will cover the local variety in the agrarian topography of the English plain.[2]

As far as the Domesday Book itself is concerned, plough-lands and plough-teams are recorded with monotonous iteration in entry after entry. And on these plough-lands, and with these plough-teams, the peasant passed his life. The dialogue of Aelfric's *Colloquium* is vivid enough, with its conscious thraldom of a *theow* or serf:

"What sayest thou, ploughman; how dost thou do thy work?"

"Oh, my lord, hard do I work. I go out at daybreak driving the oxen to the field, and I yoke them to the plough. Nor is it ever so hard winter that I dare loiter at home, for fear of my lord; but, the oxen yoked, and the ploughshare and coulter fastened to the plough, every day I must plough a full acre or more."

"Hast thou any comrade?"

"I have a boy driving the oxen with an iron goad, who also is hoarse with cold and shouting."

"What more dost thou do in the day?"

"Verily then I do more. I must fill the bin of the oxen with hay, and water them, and carry out the dung. Ha! ha! hard work it is, hard work it is! because I am not free."

The *theows* are not described in the *Rectitudines*, but the services of the other peasants are to be found there at length. These, presumably, fared

[1] H. L. Gray, *op. cit.* p. 401.
[2] It is important to note in this connection that in the north the Scandinavian settlers did not, apparently, introduce any new methods of cultivation or any new field systems. "The field-names of Scandinavian origin, which are scattered in vast numbers over this country, illustrate the character of the settlement, but do not prove that it led to any material changes in agricultural practice" (F. M. Stenton, *The Danes in England* (British Academy Raleigh Lecture, 1927), p. 38).

better; but whether they were geburs or cottars or geneats or thegns, they could not escape the consequences of unseasonable weather, bad harvests, and "murrain among the cattle". In 1044, for instance, there was great scarcity in the land "so that the sester of corn went up to sixty pence and even further". And the Chronicle mentions a number of such disasters which, likely enough, depopulated many a village.[1] "Agrarian history", wrote F. W. Maitland, "becomes more catastrophic as we trace it backwards."[2]

To define the medieval tenement as simply a bundle of strips scattered in the open fields is to convey a totally inadequate impression. Earlier documents show that in pre-Norman times sheep and cattle were important factors in farming practice and that their milk was used for cheese and butter. The oxherd in Aelfric's *Colloquium* says:

I labour much: when the ploughman unyokes the oxen I lead them to the pasture, and all night long I stay by them to watch against thieves, and again first thing in the morning I commit them to the ploughman well fed and watered.

This sounds like a summer arrangement. In winter, no doubt, like Caedmon, he snatched a little sleep in the ox-stall. The shepherd too went abroad:[3]

First thing in the morning I drive my sheep to pasture and stand over them in heat and cold with dogs lest wolves should devour them, and I lead them back to their sheds and milk them twice a day and move their folds besides, and I make cheese and butter and am faithful to my lord.

And the *Rectitudines* lists not only the services of the oxherd and the shepherd but of the cowherd and the goatherd and the cheese-maker as well. But, apart from the ploughing oxen, the stock of a manor was passed over in silence by the Great Domesday Book. For eight counties only have we information about animals, and in these it was the stock in the

---

[1] Under the years 975, 976, 1039, 1044, 1046, 1054, 1087, 1095, 1097, 1098, 1105, 1110, 1115, 1124. The evidence of these years must not be taken as an indication of a generally bad climate.—The Domesday Book has thirty-eight entries of vineyards, measured usually by the "arpent", a French areal measure. For the especial abundance and quality of the vine in the vale of Gloucester see William of Malmesbury, *De Gestis Pontificum Anglorum* (ed. N. E. S. A. Hamilton, Rolls Series, 1870), p. 292.

[2] F. W. Maitland, *op. cit.* p. 365.

[3] There are indications that some herders may have lived in hamlets at some distance from the village adjacent to the pastures. The Domesday Book mentions colonies of herdsmen called *harduices*. Cow farms (*vaccaria*) were also located at some distance from the manor. See Andrews, *The Old English Manor* (1892), p. 224; and P. Vinogradoff, *Eng. Soc.* pp. 284 and 369, for a discussion of herdwicks.

demesne alone that was entered. But the inferences to be drawn from this incomplete material are instructive enough. Open the Little Domesday at hazard. On folio 77 is the entry of Aubrey de Vere's manor of Colne in Essex. On the demesne there were three ploughs in 1066 and five in 1086, while the ploughs of the men numbered three and four respectively: there was woodland for 400 swine, and there were 20 acres of meadow and two mills. The stock numbered:

	Cows	Beasts	Sheep	Swine	Goats	Horses
1066	20	19	120	60	60	3
1086	—	45	160	80	80	4

There were also in 1086 "6 asses and 20 mares"—altogether a varied collection.[1] In the following century, the Pipe Rolls and the Domesday of St Paul's throw light on the stocking and restocking of manors. And in the thirteenth century, Walter of Henley's advice was explicit enough: "if you have land on which you can have cattle, take pains to stock it as the land requires".

The animals were wretched little creatures, poor in size and quality alike. In the absence of root crops and artificial grasses, large numbers had to be slaughtered at Martinmas, and salted down for winter use. Some were fed upon the open fields themselves. After harvest, when the fences had been removed, the arable reverted to common pasture. Corn, reaped by the sickle, left behind much stubble; and, moreover, this use of the fallow constituted an important means of providing the soil with manure. It is significant that the freeman was characteristically described as "mootworthy", "fyrd-worthy" and "fold-worthy". A man who was *consuetus ad faldam* could hardly be considered free; for his sheep had to lie in the lord's fold and so manure it. Fold-soke (*soca faldae*) is the service that stands out most prominently in the Little Domesday.[2] The demand for manure played a large part in the economy of an agricultural community,

[1] E. Lamond, *op. cit.* pp. 9–10. Walter added, "And know for truth if you are duly stocked, and your cattle well guarded and managed, it shall yield three times the land by the extent." See also the thirteenth-century *Seneschaucie* for detailed description of the offices of cowherd, shepherd and dairymaid.

[2] H. L. Gray, *op. cit.*, points out (p. 48) that "in East Anglia different pasturage provisions deflected the field boundaries, and with them the field system, from the normal type.... In the East Anglian evidence there are references to pasturage arrangements of a sort not realisable under a two- or three-field system." Later, on p. 349, he states: "Arable farming was naturally better fertilised when sheep were folded regularly upon it than when the township herd and flock wandered aimlessly over it every second or third year, as they did in the midlands."

14

and so early medieval husbandry was conducted on the principle of maintaining a local balance between agricultural and pastoral pursuits.

But the arable in its fallow state was not the only support of the village
animals; there were meadows and pastures besides. The distinction was
clear. Meadow implied land bordering a stream, liable to floods, and
producing hay; pasture denoted land available all the year round for
feeding cattle and sheep; of course the two varieties of grassland merged
into one another, but they were usually entered separately. The quantity
of meadow in the average village was not great but it was important;
thirteenth-century evidence shows that an acre of meadow was frequently
considered to be twice or three times as valuable as the best acre of arable.[1]
The distribution of meadow land in the Thames basin is interesting
(Fig. 24). It lay chiefly along the bigger river courses, but its wide extension
in the clay vale running from north Berkshire, through Oxfordshire, to
mid-Buckinghamshire, is most noticeable. The usual method of exploiting
the village meadows was to divide them in strips among the households
by lot or rotation, and close them until Lammas Day (August 2nd). After
the grass had been mowed and gathered in, during July, the land was added
to the parish common, and became undivided pasture for the village beasts,
until it was shut up again in the spring.

The utilisation of the village pasture was marked by an array of common
rights which allowed the village cattle to wander freely "horn under horn"
over the common pastures.[2] These common rights were jealously guarded
against intrusion or encroachment, and the pasturing of outside cattle
was strictly forbidden. In later years, the number of animals which any
person could turn on to the commons was limited; this may have happened on some manors in the eleventh century also. Sometimes these
pasture arrangements included more than one village; pasture rights in
one village belonged to the inhabitants of another.[3] From Suffolk comes
the famous reference to "a certain pasture common to all the men of the
hundred" of Colneis;[4] and, for the Essex marshes, J. H. Round has demonstrated the relics of a system by which a number of vills had rights in
common pastures.[5] It was in the Fenland that this abnormal economy was

[1] See, for instance, F. W. Maitland, *Domesday Book*, p. 443; J. E. Thorold Rogers,
Six Centuries of Work and Labour (1906), p. 73; F. G. Davenport, *The Economic
Development of a Norfolk Manor, 1086–1565* (1906), p. 31.

[2] There is, however, a good deal of evidence for separate pasture, *pastura separalis*,
attached to the lords' demesne.

[3] D.B. iv, 108, of Haxon in Devon: "Haustona...communem pascuam Bratonae."

[4] D.B. ii, 339 b.

[5] J. H. Round, *V.C.H. Essex*, ii, 370.

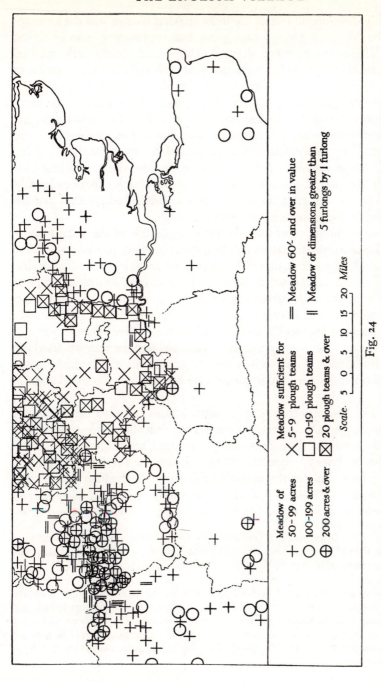

Fig. 24

The distribution of the larger Domesday meadows in the Thames Basin (constructed by Mr F. W. Morgan, M.A.). Note (1) It is impossible to equate the four systems of enumeration, and the groups of symbols are classified in an arbitrary fashion. (2) There was some meadow in almost every village—only the large meadows are shown.

most highly developed. There, the most striking feature was the arrangement of the villages in groups, each group intercommoning in an adjoining marsh.[1] These groups were sharply defined and clearly differentiated from one another. The villages claimed that their right to intercommon in the fen had existed time out of mind, and, says Miss Neilson, "the inference seems clear that the origin of such arrangements goes back to the early days of the settlement". In the twelfth and thirteenth centuries, rights of common were being differentiated, and boundaries established more accurately. Very many disputes were caused by these interlocking interests. In some instances these inter-village arrangements continued up to a much later period. Fossilised remains of the practice are occasionally preserved on the modern map in the mosaics of detached parishes that are to be found in some formerly marshy regions.[2]

Many other things went into the agricultural life of the village community. Later manorial extents mention quite frequent payments of geese and hens and eggs; and, in 1086, there must have been a considerable amount of *la petite culture* which the Survey passes over almost in complete silence. The rents of the shrievalty of Wiltshire included 480 hens, 1600 eggs and 16 sextars of honey[3]—but so explicit and full an item is rare. Hives of bees are frequently mentioned for the eastern counties; in the absence of sugar, their honey was most valuable, and the wax was used for candles. But among the miscellaneous appurtenances of the village community, two items stood out as particularly important—the wood and the fishery.

Terms like "silva infructuosa", "silva inutilis", "silva nil reddens", "silva sine pasnagio", do occur in the Survey, but such useless wood was rather the exception. For, in addition to minor perquisites like hawks' nests and wild honey, woodland provided material for fuel and for the repair of houses and for the making of fences;[4] the familiar phrases are

[1] See N. Neilson, *A Terrier of Fleet Lincolnshire* (1920), for an introductory essay on "Intercommoning in Fenland". For intercommoning in the northern shires, see footnote 1, p. 205 below.

[2] Particularly in Broadland and Fenland. On Tilney Smeeth, in the latter region, a space of less than 36 acres has been divided so that each of not less than six villages has a rectangular piece of it. See F. W. Maitland, *Domesday Book*, pp. 367–8.

[3] D.B. 1, 69.

[4] A good example of explicit declaration is afforded by an entry for the Wiltshire manor of South Newton: "Newentone...ad istum manerium pertinet habere per consuetudinem in silva Milcheti quater xx caretedes lignorum et paissonem quater xx porcorum et ad domos et sepes reemendandas quando opus fuerit." (D.B. 1, 68 *a*.) It is interesting to note here that the wood of Milchet was 15 miles away from the village of South Newton.

"silva ad clausuram", "nemus ad sepes reficiendas", "nemus ad domos curiae", and "silva ad faciendas domos". "Silva minuta" or "modica" was underwood or coppice; in a few entries, there occur "broca", brushwood; "grava", grove; and "spinetum", a spinney or thorny ground; and there are many minor variants. But the greater part of the woodland was recorded in one of two ways; sometimes quantitatively by linear or areal measures; at other times by recording its value for feeding swine. The most frequent quantitative formula is: "there is a wood x leagues by y leagues";[1] and the Domesday league may have comprised twelve furlongs or one and a half miles.[2] The whole question of Domesday admensuration raises many issues. In these cases, for instance, it is not certain that a definite geometrical figure was in the minds of the commissioners. They were, possibly, giving either the extreme diameters of irregularly-shaped woods, or the rough estimates of mean diameters; but any rigid interpretation becomes difficult. For some counties the extent of woodland was given in acres alone, occasionally even in hides. Although the normal wood estimates in the Little Domesday were in terms of swine,[3] the Norfolk hundred of Clacklose had its wood recorded in acres; the village of Stonham Aspall in Suffolk, too, had $1\frac{1}{2}$ acres of woodland and 28 acres "part woodland".

In many counties, the usual formula was worded in terms of swine, for these constituted an important element in the manorial economy of the time,[4] and they were fed normally upon acorns and beech-mast. Different phrases were employed to describe the values. Sometimes the phrase "silva de n porcis" stood for wood which could feed n swine; in Shropshire the usual formula is "wood for fattening (*incrassandis*) x swine", and there are many other variations.[5] In other counties, however, the number was

---

[1] For examples of this formula see (1) F. W. Morgan, "Woodland in Wiltshire at the Time of the Domesday Book", *Wilts. Arch. & Nat. Hist. Mag.* XLVII, 25 (1935). (2) H. C. Darby, "Domesday Woodland in Huntingdonshire", *Trans. Cambs. & Hunts. Archaeol. Soc.* V, 269 (1935).   [2] See F. W. Maitland, *op. cit.* p. 432.

[3] See H. C. Darby, (1) "Domesday Woodland in East Anglia", *Antiquity*, XIV, 211 (1934); (2) "The Domesday Geography of Norfolk and Suffolk", *Geog. Journ.* LXXXV, 432 (1935).

[4] Old English charters give similar information about cartloads of wood and pannage for pigs: e.g. Birch, *Cartularium Saxonicum*, no. 513 (866 A.D.): "70 porcis saginam in commone illa salvatica [silvatica?] taxatione...v plaustros plenas de virgis bonis et hunicuique anno i roborem ad ædificium". The *Rectitudines* mentions the swineherd, the serf-swineherd, and the woodward; and the *Seneschaucie* describes the duties of the swineherd at length.

[5] At Munden in Hertfordshire the phrase is expanded as "silva unde cc. porci pascerentur" (I, 137). See also II, 433 *b*.

much smaller and implied not a total, but an annual rent paid to the lord of the manor by villeins for pannage, the right of pasturing swine in the wood.[1] At Leominster, every villein paid one pig in ten to the lord;[2] on a side-note on folio 16*b*, one pig in seven is stated to be the rule "throughout Sussex". There were other proportions also, and consequently we cannot too certainly assume that the swine rent gave the real proportions between the woodland of different manors.[3] Further, the practice of intercommoning complicates the interpretation of the figures, but where there were a number of adjacent villages all responsible for wood profits, we may safely conclude that they lay in a wooded district.

The fishery provided another important source of gain. It was usually associated with the presence of a mill, without which few Domesday villages appear to have been complete. Indeed the right to erect a mill was a most valuable rural franchise. Many places had more than one mill; at Battersea in Surrey, for instance, there were four mills producing £42. 9s. 8d., or corn to that amount.[4] Not infrequently, the profits of the mill were stated in kind; and the mill-fisheries, scattered along the streams, were to be found everywhere in the English plain.

But the chief fisheries were in the fen counties, where the activity of the fishery attained the status of a characteristic industry. That of Doddington yielded 27,150 eels and a present of 24s. to the Abbot of Ely;[5] and very many pages of the Cambridgeshire Domesday bear witness to the importance of fisheries in that county. Renders of eels are repeated again and again. The implements are sometimes mentioned; at Swaffham the Abbot of Ely had 6s. "from the toll of the net";[6] at Soham the fisher had a "sagena", a fish-trap in the mere;[7] and, to the north, in Huntingdonshire, there were many boats on the mere of Whittlesea.[8] In the later fenland documents the weirs or "gurgites" become very frequent items. Early

---

[1] This applies particularly to those counties south of the Thames. The phrase is frequently expanded in the Domesday Book itself. At Kennington there is "tantum silvae unde exeunt de pasnagio 40 porci aut 54½d."; at Lewisham, "de silva 50 porci de pasnagio"; at Eastbridge, "silva de 3 porcis de pasnagio"; at Postling, "silva 40 porcorum"; at Monk's Horton, "de silva 6 porci". These examples, taken from Kent (D.B. 1, 12*b*–13*b*), all mean the same, i.e. swine rent.

[2] D.B. 1, 180*a*.

[3] A question that must be asked is: how accurate were the Domesday figures? In the case of the larger entries of a thousand or several hundred swine that occur for many counties, the round numbers may have been estimates rather than actual numbers; but, on the other hand, there are many instances of very detailed figures terminating in both odd and even numbers that suggest exactness.

[4] D.B. 1, 32.   [5] D.B. L, 191 *b*.   [6] D.B. 1, 190 *b*.
[7] D.B. 1, 192.   [8] D.B. L, 205.

in the twelfth century William of Malmesbury was declaring: "Here is such a quantity of fish as to cause astonishment in strangers, while the natives laugh at their surprise".[1] The fish were present not only in great quantity but also in great variety, and, as the *Liber Eliensis* records:

in the eddy at the sluices of these weirs, there are netted innumerable eels, large waterwolves (*lupi aquatici*), even pickerels, perch, roach, burbots, and lampreys which we call water-snakes. Indeed it is said by many that sometimes *isicii* (?) together with the royal fish, the sturgeon, are caught.[2]

But eels constituted the most abundant species and the etymologists of the time found at hand a ready explanation of the place-name of Ely. Finally, not only fish but a variety of resources—reeds and rushes, turbaries and salt-pans, and birds—combined to maintain a local habit of life quite different from the normal arable culture.

One further type of rural settlement remains to be discussed—the pastoral vill. The characteristic village community with its open fields was replaced in the north and west, as in the south and east, by another unit of settlement—but for different reasons. The difference between the western highlands and the rest of south Britain can be seen on the map to-day. In the west, wrote F. W. Maitland,

the houses which lie within the boundary of the parish are scattered about in small clusters; here two or three, there three or four. These clusters often have names of their own, and it seems a mere chance that the name borne by one of them should be also the name of the whole parish or vill.......As our eyes grow accustomed to the work we may arrive at some extremely important conclusions such as those which Meitzen has suggested.[3]

But, since Meitzen formulated his brilliant generalisations,[4] it has become apparent that too many complicated questions of an economic order are involved for any delimitation on either a racial or a geographical basis to go unchallenged. The one generalisation that emerges is that the isolated homestead is to be found characteristically in regions of predominant stock-raising, especially in an area where the arable land is disposed in small

[1] *De Gestis Pontificum Anglorum* (ed. N. E. S. A. Hamilton, Rolls Series, 1870), p. 322.
[2] *Liber Eliensis* (ed. D. J. Stewart, 1848), p. 231.
[3] *Domesday Book*, p. 15. On p. 368 Maitland wrote, "The science of village morphology is still very young...".
[4] *Siedelung und Agrarwesen der Westgermanen und Ostgermanen* ... (Berlin, 1895).

patches.[1] Here is nothing to require the varied services which life in a village community enforces. Habitations scatter.

Upon the highlands of Wales, the dispersed village developed most conspicuously—owing to the peculiarities of geography and of tribal custom. The laws attributed to Howel Dda,[2] together with their later accretions, and the twelfth-century descriptions of Giraldus Cambrensis,[3] indicate the pastoral basis of Welsh economy and the prevalence of hamlets rather than nucleated villages. These conditions seem to be reflected in the Domesday record, of Gloucester and Shropshire, by the enumeration of clusters of Welsh villages standing under one reeve. The groups were formed probably for the purpose of collecting rents in honey, milk, cheese, etc., and the units composing them seem to have been quite small, as one reeve was deemed sufficient to oversee from 7 to 14 of them.[4] As in Wales, so in the wilder and more rugged parts of northern England. There, climate and configuration emphasised the importance of pastoral activity. Sheep and cattle were important in the economy of much of Lancashire,[5] while many of the Durham vills had a very definite pastoral character.[6] "The Northumbrian charters indicate", writes Mr Jolliffe, "the use of summer

[1] Local studies would of course show the great importance of water-supply in influencing settlement.

[2] They are only known through medieval copies, but they may have a tenth-century basis. See J. E. Lloyd, *History of Wales*, 2 vols. (1911), for a discussion.

[3] *Itinerarium Kambriae*, and *Descriptio Kambriae*, translated with an Introduction by W. Ll. Williams in 1908. Giraldus wrote, "Almost all the people live upon the produce of their herds, with oats, milk, cheese, and butter; eating flesh in larger proportions than bread" (p. 166); and "they neither inhabit towns, villages nor castles, but lead a solitary life in the woods, on the borders of which they do not erect sumptuous palaces, nor lofty stone buildings, but content themselves with small huts made of the boughs of trees twisted together, constructed with little labour and expense, and sufficient to endure throughout the year. They have neither orchards nor gardens, but gladly eat the fruit of both when given to them. The greater part of their land is laid down to pasturage; little is cultivated..." (p. 184).

[4] D.B. 1, 162 (Gloucs): "Aluredus Hispaniensis habet in feodo ii carucatas terrae et ibi ii carucae in dominio. Isdem Aluredus habet in Wales vii uillas quae fuerunt Willelmi comitis et Rogeri filii eius in dominio. Hae reddunt vi mellis sextaria et vi porcos et x solidos." For food-rents in Hereford see D.B. 1, 179*b* and 1, 184. The districts bordering Wales were especially rich in food-rents of Celtic origin. For examples of small units see D.B. 1, 162: "Sub Wasuuic preposito sunt xiii uillae. Sub Elmui xiiii uillae. Sub Bleio xiii uillae. Sub Idhel sunt xiiii uillae. Hi reddunt xlvii sextaria mellis.... Sub eisdem prepositis sunt iiii uillae wastatae per regem Caraduech." Cf. also D.B. 1, 253*b*.: "Forde... cum xiiii berewichis" (Salop).

[5] A. Law, in *V.C.H. Lancs*, II, 263.

[6] G. T. Lapsley in *V.C.H. Durham*, I, 269. This is in a discussion of the evidence of the Boldon Book bearing on village types in Durham.

and winter grazings remote from the agricultural settlements,...by the month of May the cattle were being moved to the hill pastures, where the herds of many vills had their shielings upon the same ground."[1] Place-name evidence also provides indications of transhumance. "In some parts of the north," writes Professor Ekwall, "place-names tell us that the Scandinavians introduced the old Celtic and Scandinavian custom of sending cattle away to shielings in the summer. Names in -ergh, -booth and (some in) -set originally denoted shielings, many of which, however, at an early date developed into separate settlements".[2] But it is difficult to say, from the contemporary evidence, to what extent the dispersed village was associated with these conditions. There were numerous berewicks in many portions of Yorkshire,[3] and these may well be an indication of more than administrative arrangements. Some of the berewicks of Derbyshire seem to have presented the character of small hamlets,[4] and of them Professor Stenton has written:

These manors would seem to represent the most archaic type of agricultural estate to be found in the country; but we have to remember that the nature of the country in which they occur, consisting of great tracts of barren limestone rock with slender strips of cultivable soil along the watercourses, was not favourable to the development of the neat villar-manorial economy of the south of England. There is no doubt that the royal manors of north Derbyshire were largely the product of their geographical conditions. Under these circumstances it is only natural to find these hamlets on the royal lands grouped into large

[1] J. E. A. Jolliffe, "Northumbrian Institutions", *Eng. Hist. Rev.* XLI, 12 (1926). Mr Jolliffe adds: "Thus, much of the vill's pasturing was done, as the charter phrase runs, 'extra villam, prope et procul', and special grants were given to erect winter shelters beyond the bounds of the vill." Sometimes there was "a general practice of inter-commoning between neighbouring townships, but it was more usual to reserve the pasture of each vill to its own tenants and to form a central shire-moor out of the remaining waste", where vills could intercommon.

[2] *Introd. Surv. Eng. Place-Names*, p. 89. See E. Power in the *Cambridge Medieval History*, VII, 747, for another example of transhumance in the early thirteenth century on the estates of a Bishop of Lichfield and Coventry. See also p. 240 below.

[3] W. Farrer in *V.C.H. Yorks*, II, 144.

[4] P. Vinogradoff, *Eng. Soc.* p. 266: "A good example is afforded by the dale called Langendale in Derbyshire, on the border of Cheshire; it is unified by its name, and represents a vill and a manor in the Survey. Yet it consisted, as a matter of fact, of a string of small settlements of a couple of farms each, and these hamlets composing the administrative township were ranged along the dale at intervals of one or two miles, the whole dale covering a long, narrow range of some twelve miles in length." See D.B. I, 273. For other examples, see *ibid.* 272, "Neuubold cum vi Berewitis" (New-bold near Chesterfield); *ibid.* 272 b, "Aisseford cum [12] Berewitis" (Ashford); *ibid.* 273, "Vfre...3 Berewitae" (Mickleover).

manorial blocks for the sake of agricultural organisation, and this helps to account for the fact, otherwise strange, that the wildest part of the county is covered the most thickly with place-names on the Domesday map.[1]

In south-western England, conditions appear to have been analogous. Devon, with its moorlands, had a particularly low assessment;[2] while in Cornwall two distinct types of villages can be recognised. There were large vills with much plough-land, with English terminations like -ton, and located usually in the river valleys or along the coast. These were groups apparently similar to the typical large midland manors. Completely different from these were the small units, usually with Celtic names. They appear to have been small upland settlements. There are three instances where no plough-land at all is mentioned—'Rentin' (fol. 262 b), 'Trevill' (fol. 262), 'Witestan' (fol. 255 b); while half a plough-land is mentioned at 'Chori' (fol. 264 b) and 'Westcote' (fol. 239 b).[3]

There are 39 settlements with land for one plough; some of these are purely pastoral, and on them we find no mention of ploughs or ploughing oxen, e.g. 'Hela' (fol. 261 b), 'Languer' (fol. 236), 'Chilorgoret' (fol. 236 b), 'Telbrig' (fol. 245 b), 'Languer' (fol. 257 b), 'Trevillien' (fol. 229 b); here are six manors with a total population of seven bordars and four serfs; obviously these are small pastoral settlements of the scattered homestead type. Some of these one plough-land manors are small arable farms, e.g. 'Lanner' (fol. 236), 'Trehynoc' (fol. 256 b), 'Landrei' (fol. 261 b), 'Polescat' (fol. 230 b), 'Penquaro' and 'Torn' (fol. 233); these six manors contain in all 3 villeins, 15 bordars, and 11 serfs, and have among them 4 ploughs and a fraction; still, though they have an average of five households to a manor, we cannot call them anything but small hamlets. Sixty settlements have two plough-lands each; some, as before, are pastoral...others are agricultural....The names of most of these hamlets are strikingly un-English, and in sharp contrast to the English names which prevailed in the very large vills. Thus there are 104 manors, if so they can be called, each with two plough-lands or less; i.e. one-third of the whole number.[4]

[1] V.C.H. Derbyshire, 1, 385.   [2] V.C.H. Devon, 1, 385.

[3] The folio numbers refer to the Exon Domesday.

[4] In V.C.H. Cornwall, Part VIII, pp. 54–5 (published as a separate part in 1924). But note that there is no evidence of an extensive Celtic element in the adjoining county of Devon. The writers of the Devonshire Place-Name Society volume say 1931, (p. xiv): "As the overwhelming majority of the names in this county are of English origin, it is hard to believe that the settlements which they denote are, in any appreciable number, Celtic survivals." They are, indeed, less in evidence here than in Dorset and Somerset, forming less than 1 per cent. of the total place-names. The hamlets and scattered homesteads of the county cannot therefore be associated exclusively with Celtic influence.

As might be expected, the population of the county was sparse. Here also was a prevailing type of economy that had no relation to the three-field system.

Despite the uncertainty of the Domesday evidence, and despite the conjectural nature of any reconstruction of these past arrangements, one point emerges forcibly—there was no single village type. This was a time when rapid intercourse and the accumulation of capital did not easily modify the consequences of nature. Miss Neilson's survey of customary rents[1] is a revealing indication of the variety of agricultural services. Local variation was to be found everywhere, and it is evident that habits of settlement and cultivation were modified by peculiarities of local configuration.[2] A stretch of forest or a wold, a river or a marsh, might produce marked deviations. Only with that reservation does it become safe to generalise about the broad features of agricultural practice in the English plain. Contemporary descriptions, limited as they are, emphasise this point. The compiler of the *Rectitudines* (c. A.D. 1000) confessed: "land laws are diverse, as I said before, nor do we fix for all places these customs that we have before spoken of".[3] And the writer of the *Gerefa* (c. A.D. 1100) likewise brought out the same point: "in many districts the farm-work is earlier than in others; that is, ploughing-time is earlier, the season for mowing is earlier, and so likewise is the winter pasturing, and every other kind of husbandry".[4] And, in the thirteenth century, Walter of Henley was making other distinctions: "there are two kinds of land for spring seed which you must sow early, clay land and stony land"; and, again, "chalky ground and sandy ground need not be sown so early".[5] Nor were these the only differences, for there were variations not only in the methods of cultivation but also in the amount of the cultivated land.

The form and size of the settlements varied too. In many cases the nucleated character was ruled out by nature or by chance and the dwellings were disposed at random, *ut fons aut nemus placuit*, or constructed in regular rows, *connexis et coherentibus aedificiis*. Successive assarts and the progress of colonisation, too, left their mark upon the form of the village

---

[1] *Customary Rents* (Oxford Studies in Social and Legal History, 1910). For an emphasis on the factors producing variety see N. Neilson, "English Manorial Forms", *Am. Hist. Rev.* XXXIV, 725 (1929).

[2] See, for example, F. M. Stenton, *Documents...of the Danelaw* (1920), pp. xxxix *et seq.* for Lincolnshire tenements where "culture was based on meadow and marsh rather than on arable".

[3] *English Economic History, Select Documents* (ed. Bland, Brown and Tawney, 1930), p. 9.

[4] W. Cunningham, *op. cit.* p. 573.   [5] *Op. cit.* pp. 15–16.

and its fields. And as for size there was a good deal of variation even in the same locality. The total *recorded* populations for the villages of the hundred of Wetherley in Cambridgeshire illustrate that well enough:

Arrington	22	Haslingfield	82
Barrington	55	Orwell	24
Barton	33	Shepreth	31
Comberton	45	Whitwell	12
Grantchester	79	Wimpole	13
Harlton	22	Wratworth	36

Whatever factor must be used to obtain the *actual* populations from these figures, the fact of the variety is not affected. More general differences can also be observed. Vinogradoff's comparison[1] of the Domesday villages of Essex and Derbyshire is instructive enough when tabulated:

Sokemen and villein households per village	Derbyshire	Essex
Over 12	58 %	73 %
6–11	32 %	16·7 %
2–5	8 %	9·4 %

Some of the villages of eastern England were quite large. There were a number in Lincolnshire, for instance, with more than 30 sokemen and villein households. In this connection it is interesting to note the numerous Lincolnshire villages with considerable eleventh-century populations that had been deserted before the end of the Middle Ages.[2] Nor was Lincolnshire unique in this respect. In Norfolk, for example, many Domesday "thorpes" have disappeared as separate villages;[3] and of the Cambridgeshire villages listed above, Wratworth has vanished and Whitwell is remembered only as a farm name. As Professor Stenton has written, the

---

[1] P. Vinogradoff, *Eng. Soc.* pp. 269–73, and 495–506. For an explanation of these figures, see *ibid.* p. 270.

[2] Canon C. W. Foster has listed the Lincolnshire examples. See C. W. Foster and T. Longley, *The Lincolnshire Domesday and the Lindsey Survey* (1924), pp. xlvii–lxxii and lxxxvi–lxxxvii.

The Final Concords "also furnish information about parishes, vills, hamlets and manors which have become depopulated, or of which few traces remain". See C. W. Foster, *Final Concords of the County of Lincoln* (1920), II, l–lxv. Both of these volumes are published by the Lincoln Record Society.

For some Sussex villages "completely destroyed" by the Black Death see *V.C.H. Sussex*, II, 182. But, on the other hand, for a general estimate of the influence of the Black Death see p. 230 below.

[3] H. C. Darby, *art. cit. Geog. Journ.* LXXXV, 434.

work of topographical identification "is the foundation of detailed Domesday study".[1]

It is not always easy to determine local variations from summaries compiled over 800 years ago, for the summarised statistics of the Survey

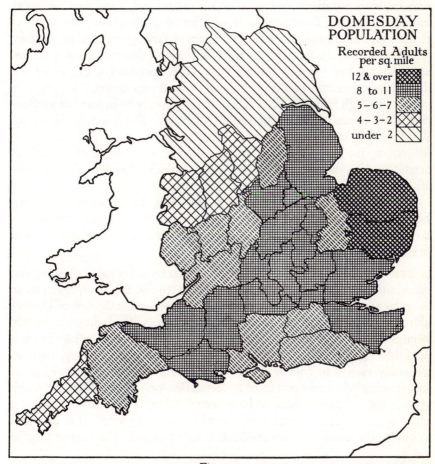

Fig. 25

For the many reservations to be remembered when looking at this map see p. 210.

present many difficulties and conflicting indications. The density of the recorded Domesday population may give as good a general indication as any of average conditions in England. This, taken county by county,

[1] In the Introduction to C. W. Foster and T. Longley, *op. cit.* p. xlvi.

shows a fair degree of uniformity upon the English plain.[1] But in interpreting Fig. 25 certain reservations have to be made:

(1) The Domesday figures present but an imperfect view of the population of the time. There must have been many people who, for various reasons, were not mentioned, and, on the other hand, some were recorded more than once in different connections.[2] But this imperfection was common to the whole realm and, as far as landowners and the agricultural population were concerned, the value of the map for regional comparison may not be seriously affected. In any case its figures are comparative and avoid the difficulties connected with any estimate of total population.

(2) The high figures for Norfolk and Suffolk may be in part but a reflection of the greater detail of the Little Domesday. But it is only fair to say that East Anglia was famed for its fertility during the early Middle Ages. Moreover, the two counties had escaped the worst evils of trampling armies in the years around the Conquest, nor did they contain any land afforested by the king.

(3) Allowance has been made, in the map, for changes in county boundaries, but the figures take no cognisance of variations within each county. The fens in Cambridgeshire, for example, brought down the average density for that county considerably.

This population map may be supplemented by other maps showing the distribution of Domesday place-names (Figs. 19 and 26). But, again, these must be accepted only with caution. As we have seen in the case of Derbyshire, a close density of names does not necessarily indicate a thickly occupied area.[3] On the other hand, "if, as often happens, a village bearing an Old English name is omitted from the Domesday Survey, there is generally reason to believe that it is included in the description of some larger estate".[4] Yet despite these uncertainties, one fact is clear: the wider gaps in the earlier distributions have been filled in, and most of the villages that appear in later records were already existing in 1066. It is true that considerable tracts still remained but thinly peopled. The barrier ring of

---

[1] The county totals are taken from Henry Ellis, *General Introduction to Domesday Book*, 2 vols., 1833 (Record Com.).

[2] See F. W. Maitland, *op. cit.* p. 20, footnote, for some of the difficulties. The difficult phrases "Alii ibi tenent" and "Plures ibi tenent" are discussed in *V.C.H. Suffolk*, I, 359.   [3] See p. 206 above.

[4] F. M. Stenton in *Introd. Surv. Eng. Place-Names*, p. 39: "In that part of Berkshire which lies between the Thames at Sonning and the Hampshire border at Sandhurst there are at least eight places, each bearing an Old English name, of which Domesday Book makes no mention."

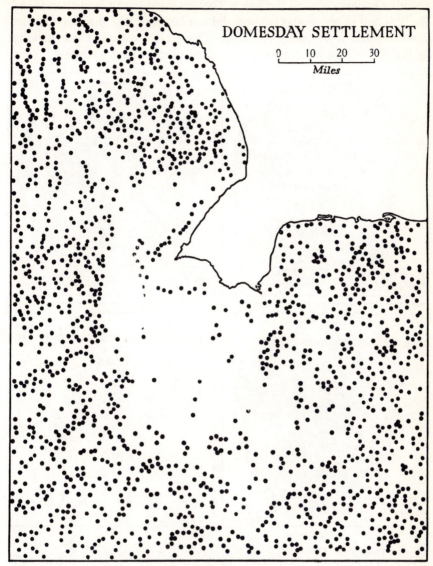

Fig. 26

The present-day coastline is shown; in all probability the Domesday coastline lay considerably behind this. The villages bordering the coast were situated upon the belt of silt that separated the peat-fen from the sea (see Fig. 73). The only settlements within this peat region were upon the islands. Adjacent to the middle portion of the Fenland, on the east, lay the Breckland which likewise had unoccupied tracts of country.

negative country around the lower Thames basin is still evident in 1086.[1] But this and other similar features did not begin to lose their significance until comparatively recent centuries. The general antiquity and stability of English village geography is one of its most striking features.

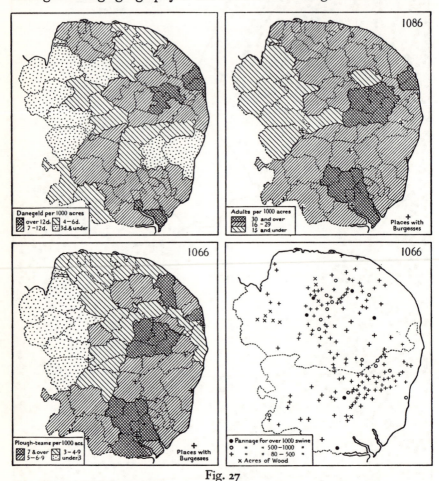

Fig. 27

Certain Domesday Distributions in East Anglia drawn from H. C. Darby, "The Domesday Geography of Norfolk and Suffolk", *Geog. Journ.* LXXXV, 432 (1935).

From these remarks it can be readily appreciated that the differences through the English plain can be made clear only by local studies. The accompanying sets of maps, relating to East Anglia (Fig. 27) and to Berk-

[1] See p. 132 above for a discussion of this point.

shire (Fig. 28), provide examples of regional variation. They reveal differences in intensity of settlement and in the productivity of the land; and these differences are fundamental in any discussion of the details of English medieval geography. This is not the place to consider the full implication of these maps;[1] they are just intended to show the kind of information yielded by the Survey when subjected to detailed examination. But there were other differences in Domesday England. Localism was

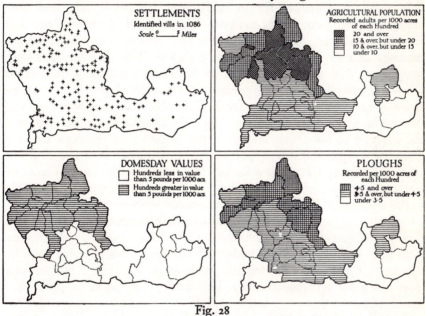

Fig. 28

Certain Domesday Distributions in Berkshire, drawn from F. W. Morgan, "The Domesday Geography of Berkshire", *Scot. Geog. Mag.* LI, 353 (1935).

The evidence of the Domesday statistics shows that in the northern clay vales of Berkshire there was closer settlement than in the south and east, and a greater density of agricultural population and activity; these facts were reflected in part in the higher values of the northern hundreds.

not confined to agricultural productivity and practice. Here it must be sufficient to say that the early Middle Ages witnessed the slow formation of those provincialisms which survived so tenaciously—variations in local architecture, local customs and beliefs and saints, local dialects and folklore; in short, all the mass of deep-rooted localism that gives to English life its variety and its sense of age-long contact with the soil.

[1] For a discussion see respectively (1) H. C. Darby, *art. cit. Geog. Journ.* LXXXV, 432 (1935); (2) F. W. Morgan, *art. cit. Scot. Geog. Mag.* LI, 353 (1935).

TOWNS AND COMMERCE

Any reference to urban origins in England and Wales must raise more questions than it can answer. The early history of towns in Britain is exceedingly obscure, and it is too much to hope that a general theory will tell the story of every or any particular town.[1] Municipal life, it is true, was a marked feature of Roman civilisation; but there was, apparently, little continuity between the towns of Roman Britain and those of later times. In some cases, the very sites of cities vanished from memory as they had vanished from the eye.[2] This cessation was only temporary. With the coming of the Scandinavians, community life received a new stimulus. The Danish supremacy in the midlands rested upon a confederate group of boroughs—Derby, Lincoln, Leicester, Stamford and Nottingham. And these were not the only Danish boroughs. Nor was the example of the Danes lost upon the English; the reconquest of the Danelaw was marked at every point by the creation of a stronghold. From the Anglo-Saxon Chronicle a list can be compiled of just over a score of boroughs established by Edward the Elder (901-25) to secure his conquests in the midlands.[3] "The exceedingly neat and artificial scheme of political geography that we find in the midlands, in the country of the true 'shires', forcibly suggests deliberate delimitation for military purposes. Each shire is to have its borough in its middle. Each shire takes its name from its borough."[4] And, as F. W. Maitland wrote on another page, "the borough does not grow up spontaneously; it is made; it is 'wrought'; it is timbered".[5]

Such a fortified establishment was called in Anglo-Saxon a *burh*, but the fact that the term "borough" came to mean town in our sense of the word does not necessarily justify the conclusion that this *burh*-building

[1] As will be seen, the uneven and fragmentary nature of the evidence prevents a general distributional treatment such as is possible in later times. For this reason, the material in this section must be less "geographical" than one could wish. In any case, local studies based upon large-scale topographical and geological maps can alone make clear the factors involved in the development of individual towns.

[2] For a discussion of survival see C. Petit-Dutaillis, *Studies Supplementary to Stubbs' Constitutional History* (1923) 1, 72 ff. For Roman towns see pp. 41 ff. above.

[3] See E. S. Armitage, *The Early Norman Castles of the British Isles* (1912), pp. 26 ff.

[4] F. W. Maitland, *Domesday Book*, pp. 187-8. See also H. J. Mackinder, *Britain and the British Seas* (2nd ed. 1915), pp. 206-7.

[5] *Ibid.* p. 219.

always meant the creation of urban settlements.[1] Some tenth-century boroughs would seem to have been essentially royal, military and administrative centres, whose inhabitants, for the most part, may have been socially indistinguishable from the country population, gaining a living directly or indirectly from agriculture. But even at the period of the creation of the military *burh*, the economic factor must have been important. The special peace in the king's *burh* gave all the more opportunity to utilise advantages of site and situation for the purposes of trade.[2] Long before the Norman Conquest a force had begun to operate which was ultimately to give to the English borough its most permanent characteristic —i.e. that of a trading centre.[3] The association of the early boroughs with markets and with mints, difficult as it may be to interpret, indicates the presence of a commercial element.[4] Altogether, in the tenth century, some 75 names can be drawn from the "Burghal Hidage" and the Saxon Chronicle; this list may be increased to 100 in various ways.[5]

In the eleventh century, the Domesday Book also supplies over 100 names of places containing burgesses.[6] The Domesday boroughs were a varied lot—some small, some large, some wealthy, some poor. Clearly the test for the classification was neither size nor general prosperity. "The word *burh* was an old word and for a long time past it had borne a legal meaning".[7] But although the commissioners appear to have been bound by traditional nomenclature, and although the Domesday Book shows that the salient features of the borough in 1066 may still have been military and agrarian, yet the Survey does provide many hints of the commercial

---

[1] F. W. Maitland, *ibid*. pp. 195–6; that a *burh* had a market-place "does not imply a resident population of buyers and sellers; it does not imply the existence of retailers".

[2] C. Petit-Dutaillis, *op. cit*. 1, 83: "And it is significant that, in the Anglo-Saxon laws, we sometimes find the town designated by the name of *port*, and that numerous charters tell us of a town's officer called port-reeve or port-gerefa. The *port* is the place of commerce; it is the old name for a town in Flanders, where civic origins have a clearly economic character."

[3] See F. W. Maitland, *op. cit*. p. 196: "We cannot analyse the borough population; we cannot weigh the commercial element implied by *port* or the military element implied by *burh*; but to all seeming the former had been rapidly getting the upper hand during the century which preceded the making of the Domesday Book."

[4] See, for example, A. Ballard, *The Domesday Boroughs* (1904), pp. 75–7, and 118–20 for a list of mints: pp. 115–18 are concerned with "the idea of the borough as a market"; but see p. 217 below.

[5] E. S. Armitage, *op. cit*. pp. 26 ff.; and A. Ballard, *The English Borough in the Twelfth Century* (1914), pp. 43 ff.

[6] A. Ballard, *The Domesday Boroughs* (1904), pp. 9–10.

[7] Sir F. Pollock and F. W. Maitland, *History of English Law* (1895), 1, 626.

activity of the time.[1] One indication is the establishment of new boroughs alongside the old, at Norwich, Nottingham and Northampton, settled mainly by Frenchmen. Colonies of *franci* are also reported at Hereford, Shrewsbury, Southampton, and at Wallingford.[2] Moreover, the tenants-in-chief of the Conqueror have been credited with a scheme of town colonisation chiefly in outlying regions; and, further, we are told that the privileges of these towns were modelled upon those of the Norman town of Breteuil.[3] Even when these views are subjected to criticism, the fact of burghal colonisation remains clear; and it would seem, too, that the new population "gave themselves up to industry and commerce".[4] But there is more definite evidence, too, in connection with some of the new castles built by the invaders. In Staffordshire, "Henry de Ferrers has the

---

[1] Professor Carl Stephenson believes that the Anglo-Saxon borough had no really urban character and that the origin of commercial towns (as distinct from fortified *burhs*) was essentially a post-Conquest development. These ideas he has set forth particularly in *Borough and Town—A Study of Urban Origins in England* (1933). An earlier summary can be found in his "The Origin of English Towns", *Am. Hist. Rev.* XXXII, 13 (1927). Essentially, Professor Stephenson's interpretation is an application to the English evidence of the general ideas of Professor Pirenne as outlined in "Les Origines des Constitutions urbaines au Moyen-âge", *Revue Historique*, LIII, 52 (1893), and as set out in his *Medieval Cities* (1925).

The validity of this parallel between England and the Continent has been disputed. For a critical estimate of Stephenson's ideas see the review of his book by Professor Stenton in *History*, XVIII, 256 ff. (1934): "The history of the coinage, to say the least, gives no support to the theory of a new urban concentration in England between 1066 and 1100, and it may certainly be doubted whether the statistics given by Domesday Book imply an increase in borough populations adequate to produce the change for which Mr Stephenson is arguing (p. 257)". See footnote 2, p. 223 below.

For another critical estimate see the review by Professor Tait in *Eng. Hist. Rev.* XLVIII, 642 (1933). Individual studies alone can illuminate the points involved, and in this connection see H. M. Cam, "The Origin of the Borough of Cambridge: a Consideration of Professor Carl Stephenson's Theories", *Proc. Camb. Antiq. Soc.* XXXV, 33 (1935). Miss Cam is unable to accept Professor Stephenson's interpretation as applied to Cambridge.

[2] D.B. II, 118; I, 280, 219, 179, 252, 52, 56.

[3] Mary Bateson, "The Laws of Breteuil", in *Eng. Hist. Rev.* XV, XVI and XVII (1900, 1901, 1902). Seventeen towns are given as certain: Hereford, Rhuddlan, Shrewsbury, Nether Weare, Bideford, Drogheda in Meath (and Drogheda Bridge), Ludlow, Rathmore, Dungarvan, Chipping Sodbury, Lichfield, Ellesmere, Burford, Ruyton, Welshpool, Llanvyllin, Preston. Eight are less certain: Stratford-on-Avon, Trim, Kells, Duleek, Old Leighlin, Cashel, Kilmaclenan, Kilmeadin.

[4] C. Petit-Dutaillis, *op. cit.* I, 89. See (1) C. Stephenson, *Borough and Town*, pp. 120 ff., for a critical estimate; (2) H. W. C. Davis, "The English Borough", *Quarterly Review*, CCVIII, 54 (1908), for another estimate; (3) A. Ballard, "The Laws of Breteuil", *Eng. Hist. Rev.* XXX (1915).

castle of Tutbury. In the borough about the castle are 42 men living only by their trading, and they render together with the market £42. 10s."[1] There were, apparently, similar foundations at Wigmore, Castle Clifford, Rhuddlan, Penwortham, Okehampton, Berkhamstead and Arundel.[2] At Clare, Louth, Bradford and Pershore, groups of new settlers called *burgenses* are found in close connection with a market;[3] at Eye there was, in 1086, a *mercatus* in which dwelt 25 *burgenses*;[4] at Ashwell there was a borough which rendered 49s. 4d. in toll and other customs.[5] At St Albans, the abbot had 46 burgesses worth £11. 4s. a year from toll and other dues.[6] But, on the other hand, it must be remembered that the right of holding a market was quite distinct from the legal essence of a borough.[7] There was only one borough (if that) in Cornwall, yet there were five market towns.[8] Of the 50–60 markets recorded in the Survey, fewer than 20 occurred in places which were boroughs;[9] at Berkeley, there was a *forum* in which 17 men lived who paid rent;[10] at Cheshunt, 10 merchants rendered 10s. *de consuetudine*;[11] at Abingdon, 10 merchants living before the gates of the church rendered 40d.[12] And the outstanding Domesday example of new commercial growth comes from what was recorded merely as the *villa* "where rests enshrined Saint Edmund King and Martyr of glorious memory". Accretion around the monastery at Bury St Edmunds had been considerable between 1066 and 1086:

For now the town is contained in a greater circle, including (*de*) land which then used to be ploughed and sown; whereon there are 30 priests, deacons, and clerks together (*inter*); 28 nuns and poor persons who daily utter prayers for the King and for all Christian people: 75 bakers, ale-brewers, tailors, washer-women, shoemakers, robe-makers (*parmentarii*), cooks, porters, stewards together. And all these daily wait upon the Saint, the Abbot, and the Brethren. Besides whom there are 13 reeves over the land, who have their houses in the said town, and under them 5 bordars. Now [there are 34 knights], French and English together, and under them 22 bordars. Now altogether (there are)

---

[1] D.B. 1, 248*b*.   [2] D.B. 1, 183*b*, 183; 269, 270, 105*b*, 136*b*, 23.

[3] D.B. 11, 389*b*; 1, 345, 67*b*, 174*b*.   [4] D.B. 11, 319*b*.   [5] D.B. 11, 135.

[6] D.B. 1, 135*b*. At Windsor there were 69 rent-paying houses, and 26 that were quit; but it was not called a borough.   [7] F. W. Maitland, *op. cit.* p. 193.

[8] A. Ballard, *The Domesday Boroughs*, p. 98. The markets were at St Germans, Bodmin, Launceston, Liskeard and Trematon. Ballard concludes, "that while a market was possibly an appurtenance of a borough, its establishment did not convert a village into a borough or confer borough rights". Bodmin may not have been a borough.

[9] See H. Ellis, *A General Introduction to the Domesday Book* (1833), 1, 248 ff.

[10] D.B. 1, 163.   [11] D.B. 1, 137*b*.

[12] D.B. 1, 58. See J. H. Round, in *V.C.H. Berkshire*, 1, 313.

342 houses in the demesne on the land of St Edmund which was under the plough in the time of King Edward.[1]

This comes from a place without the status of a borough, so that we cannot be sure that all the places so called in the Survey were actually towns, as contrasted with the many that were not so called. It is just as well to remember that we are dealing with two things, with the borough as a legal institution and with the "town" as an economic reality. Further it must be remembered that London, Winchester, Bristol, Southampton and a number of other places have no descriptions or, at best, incomplete descriptions, in the Survey. But despite the difficulties inherent in Domesday terminology, and although we may be dealing with legal entities rather than with geographical facts, the Domesday statistics do provide a great deal of information about urban centres in the eleventh century.[2] Of the Domesday boroughs in 1066, York had 1890 burgesses, four others had over 900 burgesses;[3] 16 others had over 200.[4] These figures may involve total populations of from about 1000 to over 9000; to which must be added a certain number of non-burgesses. "In a large proportion of these cases", writes Professor Tait, "we should feel sure that the burgesses had some other means of support than agriculture, even if the Domesday did not tell us that the 1320 burgesses of Norwich had only 120 acres of arable land, and the 538 of Ipswich (which had eight parish churches) only 40."[5]

But whatever the uncertainties of the Domesday evidence, and however disputed the interpretations may be, the fact remains that great changes were taking place. A. Ballard has listed charters for 114 boroughs up to the death of King John (1216);[6] some of these were confirmations of existing privileges; others inaugurated new creations. Comparison with the Domesday list shows about 50 names in common.[7] In the years that

[1] D.B. II, 272.

[2] See (1) J. Tait, "The Firma Burgi and the Commune in England", *Eng. Hist. Rev.* XLII, 360 (1927); (2) C. Stephenson, *Borough and Town*, p. 221, for the statistics.

[3] Norwich, Lincoln, Oxford, Thetford.

[4] Ipswich, Gloucester, Wallingford, Chester, Huntingdon, Stamford, Cambridge, Exeter, Sandwich, Northampton, Wareham, Canterbury, Shaftesbury, Shrewsbury, Derby, Torksey.

To these must be added those with over 200 burgesses in 1086—Leicester, Colchester, Warwick and Hythe, which were not reported for 1066. Also there was Dunwich which had grown remarkably between 1066 and 1086.

The doubtful meaning of some of the statistics makes it impossible to be quite certain about individual totals.

[5] *Eng. Hist. Rev.* XLVIII, 643–4 (1933).

[6] A. Ballard, *British Borough Charters, 1042–1216* (1913), pp. xxvi ff.

[7] See C. Stephenson, *Borough and Town*, pp. 74 ff.

followed 1216 the granting of charters went on apace.[1] Under the first
two Edwards 166 boroughs were summoned once or more often to send
representatives to parliament.[2] But the parliamentary line seems to have
been a most irregular one, and it is impossible to disentangle the degree of
exceptionality which made a borough.[3] Amongst this profusion of charters,
privileges, and legal identities, we grope almost in vain after the geographical
facts themselves, meeting everywhere difficulties and incompleteness.

There are however some symptoms which indicate the wider changes
of the time. The actual course by which any group of townsmen succeeded
in obtaining control over their fiscal, economic and judicial affairs is
written in the charters of their town.[4] The mercantile privileges of a town
were sometimes vested in a body known as the merchant gild. Its fundamental feature consisted in the exclusive right of its members to buy and
sell within the borough, on market days and at all other times without
payment of toll or custom. In many of the charters can be found a clause
similar to the following: "We grant a Gild Merchant with a hanse and
other customs belonging to the Gild, so that no one who is not of the Gild
may merchandise in the said town, except with the consent of the burgesses."[5] The relations between the merchant gild and the borough
community have been the subject of much controversy; but the importance
of the gild in the economic evolution of a town is quite evident. Dr Gross
has summed up the main facts thus:

The history of the Gild Merchant begins with the Norman Conquest....
Not until there was something of importance to protect, not until trade and
industry began to predominate over agriculture within the borough, would
a protective union like the Gild Merchant come into force. Its existence, in
short, presupposes a far greater mercantile and industrial development than
that which prevailed in England in the tenth century... it may be safely stated
that at least one third—and probably a much greater proportion—of the

[1] A. Ballard and J. Tait, *British Borough Charters, 1216–1307* (1923), pp. xxv ff.
[2] For a discussion see L. Riess, *Geschichte des Wahlrechts zum englischen Parlament*
(Leipzig, 1885), pp. 19–20. See Fig. 31 below for the later boroughs.
[3] "Could a borough be defined? We much doubt it"—Sir F. Pollock and F. W.
Maitland, *op. cit.* p. 653.
[4] The immediate effects of the Norman Conquest were frequently adverse;
houses were ruthlessly destroyed for one reason or another—frequently to make room
for a castle. Thus at Ipswich 538 burgesses paid custom to the king in 1066; while in
1086 there were only 210 burgesses of whom 100 were so poor that they could only
"render to the king's geld but one penny a head". Demolition and "waste" houses
are recorded at Cambridge, Canterbury, Northampton, Wallingford, Lincoln, Oxford,
Derby, Norwich, etc. See E. Lipson, *Economic History of England* (1929), I, 172–3.
See p. 171 above.   [5] C. Gross, *The Gild Merchant* (1890), I, 8.

boroughs of England were endowed with this gild in the thirteenth century; that in fact it was not an adventitious institution, but one of the most prevalent and characteristic features of English municipalities.[1]

Even so, the majority of boroughs may have been without merchant gilds and in many of the larger centres like Exeter, Norwich and Northampton there was, apparently, no gild; the name and formal organisation seems to have been absent from London.[2] So that despite its importance and despite its indications of increasing commercial prosperity, the merchant gild can provide no criteria by which to measure the comparative importance of twelfth-century towns in England.

The Pipe Rolls are of some assistance in this connection, and an indication of the relative rank of these towns may be obtained by setting forth the borough "aids" for 1130 and 1156. The list is that of Maitland:[3]

	Pipe Roll 31 Hen. I	Pipe Roll 2 Hen. II		Pipe Roll 31 Hen. I	Pipe Roll 2 Hen. II
	£	£		£	£
London	120	120	Wiltshire boroughs	17	
Winchester	80		Calne		1
Lincoln	60	60	Dorset boroughs	15	
York	40	40	Huntingdon	8	8
Norwich	30	33⅓	Ipswich	7	3⅓
Exeter		20	Guildford	5	5
Canterbury	20	13⅓	Southwark	5	5
Colchester	20[4]	12⅔[4]	Hertford	5	
Oxford	20	20	Stamford	5	
Gloucester	15	15	Bedford	5	6⅔
Wallingford	15		Shrewsbury		5
Worcester		15	Droitwich		5
Cambridge	12	12	Stafford	3⅓	3⅓
Hereford		10	Winchcombe	3	5
Thetford	10		Tamworth	2¾	1¼[5]
Northampton	10		Ilchester		2½
Rochester		10	Chichester[6]		
Nottingham ⎫ Derby ⎭	15	15			

[1] C. Gross, *op. cit.* These statements are taken from pp. 2–22. But the conclusions of Gross about the relations between gild and municipality have to be considered. See J. Tait, "The Borough Community in England", *Eng. Hist. Rev.* XLV, 530 ff. (1930). The earliest distinct reference to the gild merchant occurs in a charter granted by Robert Fitz-Hamon to the burgesses of Burford (1087–1107); see Gross, *op. cit.* I, 5. Note also the list of gilds in Gross, *op. cit.* 9–20; as a commentary on the borough lists this is of some interest.   [2] Gross, *op. cit.* I, 20.

[3] F. W. Maitland, *op. cit.* p. 175; see *ibid.* p. 176 for a discussion of this list. These lists should be compared (i) with those given by J. Tait, "The Firma Burgi and the Commune in England", *Eng. Hist. Rev.* XLII, 360 (1927); (ii) with those of C. Stephenson, *Borough and Town*, p. 225 (see *ibid.* pp. 161 ff. for his discussion of the figures).

[4] Nearly.   [5] This may come only from the Staffordshire part of Tamworth.
[6] Chichester pays in later years; but very little.

Unfortunately, the list has many omissions, but still it does show some of the leading towns of the twelfth century arranged in what may have been approximately their order of relative wealth. Comparison with the Domesday list reveals the essential validity of both lists. On the other hand, comparison with the list of borough charters show that the "aids" do not give a complete picture of the urban centres of the time. The Cinque Ports paid no aids; nor did those towns, like Bristol,[1] Leicester and Chester, that were not in the King's hands. After making allowances for exemptions, there are still surprises. Boston is a case in point. Lists of Domesday boroughs and of subsequent borough charters do not mention it; its gild merchant is doubtful; apparently it had no jewry. Yet the Pipe Roll of 1204 provides the following figures for a tax upon seaport merchants:[2]

	£	s.	d.
London	836	12	10
Boston	780	15	9
Southampton	712	3	$7\frac{1}{2}$
Lincoln	656	12	2
Lynn	651	11	6
Hull	344	14	$4\frac{1}{2}$
York	175	8	10
Newcastle	158	5	6
Grimsby	91	15	$0\frac{1}{2}$
Barton	33	6	9
Immingham	18	15	$10\frac{1}{2}$

The importance of the east coast parts is particularly noticeable.[3]

The information provided by the Domesday Survey, by the borough charters, by the merchant gilds and by the Pipe Rolls, may be supplemented with lists of jewries in thirteenth-century England.[4] These lists provide instructive comment upon certain aspects of economic progress in England during the thirteenth century, and they help to measure the distribution of

---

[1] A momentary glimpse of the importance of Bristol is provided by the Pipe Roll for 1194 owing to the temporary seizure of John's earldom of Gloucester in which Bristol was included; it was farmed at the rate of £145 per annum. Other farms which appear on the same roll are: Rochester, £25, Marlborough, £36, Gloucester, £65. See the Pipe Roll Society volume for 1194 (ed. D. M. Stenton, 1928), p. 240. For a twelfth-century description of Bristol, see p. 283 below.

[2] The "fifteenth from seaport merchants", Pipe Roll 6 John, m. 16 d.

[3] Comparison of Professor Stephenson's lists for Henry I and Henry II shows the rising importance of the east coast ports. See *Borough and Town*, p. 165.

[4] For a general discussion, see J. Jacobs, *The Jews of Angevin England* (1893), *passim*.

prosperity among some towns of the realm. The tallage of 1255[1] is instructive:

	marks		marks
London	472	Bristol	45
Lincoln	120	Exeter	36
Winchester	120	Wilton	36
Canterbury	90	Cambridge	36
Worcester	75	Stamford	30
Oxford	75	Gloucester	30
York	66	Hereford	30
Marlborough	45	Colchester	30
Northampton	45	Sudbury	15
Norwich	45	Nottingham	15
Bedford	45	Warwick	20s.

Despite the many difficulties that crop up in the interpretation of medieval statistics, no criticism can destroy one fact—the supremacy of London. From the latter part of the twelfth century comes the famous description of the city by William FitzStephen, written some time before 1183.[2] Independent evidence shows that it, although so highly coloured, was not the product merely of indiscriminate enthusiasm:

"Among the noble cities of the world that are celebrated by Fame, the City of London, seat of the Monarchy of England, is the one that spreads its fame wider, sends its wealth and wares further, and lifts its head higher than all others." It was great in size, containing within itself and its suburbs "thirteen greater Conventual churches, and a hundred and twenty-six lesser Parochial"; and, during the anarchy of Stephen's reign, mustering "twenty thousand armed horsemen and sixty thousand foot-soldiers". The citizens of London were renowned beyond all others for "their fine manners, raiment and table." "To this city, from every nation that is under heaven, merchants rejoice to bring their trade in ships." Within the city itself, "those that ply their several trades, the vendors of each several thing, the hirers out of their several sorts of labour are found each in their separate quarters and each engaged upon his own peculiar task". While about the city, "on all sides, lie the gardens of the citizens that dwell in the suburbs, planted with trees, spacious and fair, adjoining one another". To the north "are pasture lands and a pleasant space of flat meadows, intersected by

---

[1] *Cal. Patent Rolls*, 1247–58, pp. 441–4. For another tallage in the same year see *ibid.* pp. 439–40. See also (1) H. M. Chew "A Jewish Aid to Marry, A.D. 1221", *Trans. Jew. Hist. Soc. of Eng.* XI, 92 (1925); (2) I. Abrahams, "The Northampton Donum of 1194", *Miscellanies Jew. Hist. Soc. of Eng.* I, lix (1935).

[2] F. M. Stenton and others, *Norman London* (1934), pp. 26 ff. (H. E. Butler's translation). See also W. Page, *London: Its Origin and Early Development* (1929).

running waters, which turn revolving mill-wheels with a merry din. Hard by there stretches a great forest with wooded glades and lairs of wild beasts, deer both red and fallow, wild boar and bulls. The corn-fields are not of barren gravel, but rich Asian plains such as make glad the crops and fill the barns of their farmers with sheaves of Ceres stalk." There were plenty of amusements too —sports and athletics of all kinds. And in the twelfth century, as in the twentieth, many came out "to watch the contests of their juniors, and after their fashion are young again with the young".

It towered above all other towns, and to an increasing extent as time went on. Like Winchester, it is not surveyed in the Domesday Book, but its pre-eminence was no new thing. No other town had included so many moneyers striking pennies for Edward the Confessor or Cnut. And, writes Professor Stenton, "the idea that the Norman Conquest was followed by a revolutionary expansion of the trade of London receives little support from the evidence...the trade of Norman London was mainly due to geographical facts, which are not affected by foreign conquest".[1]

What was true of the trade of London, was true of the trade of the whole country generally. The Norman Conquest can hardly have meant a "revolution" in commerce. However pessimistic a view one takes of early trade, Charlemagne's letter (796) to Offa about interference with trade can hardly be explained away.[2] The evidence of Old English coinage, too,

[1] F. M. Stenton, *ibid.* p. 20. The biographer of Thomas Becket tells how "many natives of the chief Norman cities, Rouen and Caen, settled in London as the foremost town in England, because it was more suited for commerce, and better stored with the goods in which they were accustomed to trade". *Materials for the History of Thomas Becket* (ed. J. C. Robertson, Rolls Series, 1879), IV, 81.

[2] For a summary of the pre-Domesday trade see C. Stephenson, *Borough and Town*, pp. 71–2; E. Lipson, *op. cit.* I, 444 ff.; W. Cunningham, *Growth of English Industry and Commerce* (1922), I, 81 ff. C. Stephenson minimises early commerce but F. M. Stenton writes (*History*, XVIII, 258): "In the ninth century, king Burgred of Mercia is known to have employed more than fifty moneyers in a period of eight years. Moneyers bearing continental names produced much of the strange currency struck in England in honour of St Edmund, and this highly significant continental element persists in Wessex until the end of the Old English period. The direct references to English trade and trades in the written materials for Anglo-Saxon history are few and isolated, but they are certainly suggestive. Trade between England and Gaul must have been something more than the enterprise of a few adventurers if its temporary suspension marked a critical phase in the relations between Charlemagne and Offa. No doubt the foreign trade of England was slow to reach even its eleventh-century volume, but English ports looked to the north as well as to the south, and it would be very unwise to assume that they had no share in the commerce which created northern trading centres like Haithby near Sleswig in the ninth century."

cannot be gainsaid. Further, Danish contacts must have meant important developments in English commerce; and, certainly, from the time of Edgar (959–75) onwards, fragments of evidence relating to maritime trade begin to accumulate. There is the famous provision for the man who, with his own capital, should fare thrice across the sea and so become worthy of thane-right. The early chronicles add a few vague statements. But we are on surer ground with Aethelred (979–1016). His attention to the coinage, to weights and measures, and his commercial treaties with the Danes, are significant. More important still is his ordinance concerning tolls at London, where there were merchants from the Netherlands, from northern France, and from Germany, who brought to England lumber, fish, blubber, cloth, gloves, pepper, wine and vinegar, and who carried away wool, live stock and grease. There is, also, Abbot Aelfric's quaint description (*ante* 1051) of the merchant who tells the inquisitive schoolboy how he imported precious metals, gems, fine cloth, perfumes, drugs, ebony, glass, wine, oil, and other things, across the perilous sea. Details like these must have provided the dim beginnings of the sea-trade that is casually mentioned in the Domesday Book. These commercial contacts, developing in Anglo-Saxon and Danish times, and stimulated by the Conquest, showed their full force under the Angevin kings. The Conquest itself must have widened the horizon of the English merchant, and greatly stimulated internal trade and commerce.[1] Moreover, the factor of safety must not be forgotten. "Among other things", wrote the chronicler under the year 1087, "is not to be forgotten the good peace that William made in the land. It was such that a man, who was himself aught, might go over the kingdom unhurt with his bosom full of gold." "All ports and roads", William of Poitiers informs us, "he ordered to be open to merchants, and no injury to be done them."[2] The growth of markets and the rise of fairs into an essential part of the economic framework of English society registered the protective influence of the monarchy. In the charter of St Ives (1100) the king says: "I will and ordain that all who come to the fair, remain at it, and return from it, have my firm peace."[3]

[1] "In an age when reviving commerce was just gaining headway on the continent, England was brought into close union with a Continental state, and all traditional institutions that might have hampered the complete triumph of the mercantile class in the borough were swept aside" (C. Stephenson, *Borough and Town*, pp. 212–13).

[2] *Gesta Willelmi* (ed. F. Maseres, 1807), p. 149. It is a pity that the contemporary evidence relating to internal communications is so exceedingly scanty. For the Roman roads see pp. 78 ff. above; for later medieval communications see pp. 260 ff. below.

[3] *Cartularium Monasterii de Rameseia* (ed. H. W. Hart and P. A. Lyons, Rolls Series, 1884), I, 240.

The arrivals of foreign merchants at these fairs and in the ports indicates the spreading contacts of English commerce.[1] The Pipe Rolls, for example, contain much miscellaneous information which is very revealing. Detail can be added to detail: the presence of the "ships and merchandise of foreign merchants" at Boston Fair in 1196;[2] the small beginnings of a river police at the port of London in the same year;[3] the heavy amercements imposed, in 1198, upon trading in corn with Flanders;[4] the desire of some men to export corn from East Anglia to Norway in 1199;[5] these, and many others, are symptomatic items. Artificial factors stimulated these foreign connections. The French wars, the Crusades, the Irish campaigns —all helped the growth of that trade whose nature and contacts are analysed in later chapters of this book.[6]

Despite this increasing traffic, industrial activity in south Britain was still small in amount and restricted in extent. What activity there was comprised the working of salt, iron, lead and tin, together with the manufacture of cloth.

Salt was an indispensable item in the economy of domestic life in medieval times. "For five or six months in the year, our ancestors, at least the majority of people, lived on salted provisions."[7] Its importance can hardly be exaggerated. Further, rock salt was not worked in Britain until 1670, and so the commodity was obtained by evaporation either from sea-water or from brine-springs. The chief areas for maritime salt were the marshes of Lincolnshire, Norfolk,[8] Essex, Kent and Sussex[9]—but coastal salt-pans were far from being restricted to these counties. The inland centres were in Cheshire and Worcestershire—at Droitwich, Northwich, and at many other "wiches" as well.[10] The "laws and customs" of the Cheshire wiches were set out at great length by the Domesday Commissioners.

[1] See E. Lipson, op. cit. I, chapters VII and X; L. F. Salzman, English Trade in the Middle Ages (1931), passim; W. Cunningham, op. cit. pp. 182 ff. and 641 ff.

[2] The Great Roll of the Pipe, 1196, ed. D. M. Stenton (Pipe Roll Society, 1930), pp. xxiv and 248–9.

[3] Ibid. 1196, pp. xxv and 18.

[4] Ibid. 1198, ed. D. M. Stenton (Pipe Roll Society, 1932), pp. xv and xxv.

[5] Ibid. 1199, ed. D. M. Stenton (Pipe Roll Society, 1933), pp. xxiv and 289.

[6] Chapters VII and VIII.   [7] J. E. Thorold Rogers, op. cit. p. 95.

[8] For East Anglian salt-pans see H. C. Darby, art. cit., Geog. Journ. LXXXV, 442 (1935).

[9] See Fig. 29 for Sussex, Hampshire and part of Dorset.

[10] See J. Tait, "The Survey of Cheshire", op. cit. pp. 39 et seq.

Rights in some of these wiches were held by manors situated at a consider-able distance away. Rollright and Shipton-under-Wychwood (Oxford), and Risborough (Bucks), had salt-pans or salt workers in Droitwich. The manufacture was carried on in leaden vats (*plumbi*) and in small pits and reservoirs (*hocci, casuli*). The Worcestershire manor of Bromsgrove, for instance, had thirteen salt-pans in Droitwich, and three salt workers (*salinarii*) rendered for those pans 300 *mittas* of salt; in 1066 they had received 300 waggon-loads of wood from the woodwards.[1] The connection between the salt works and woodland (a source of fuel) was very naturally a close one.

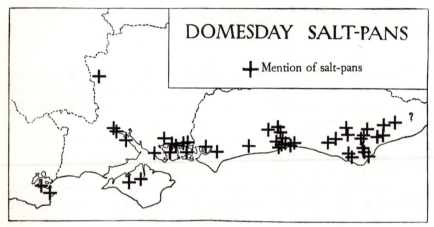

Fig. 29

Distribution of salt-pans along a section of the south coast. Note that this map does not give the actual territorial distribution, but only the distribution of holdings responsible for salt-pans—hence the inland pans.

Although iron was worked in Roman times,[2] there is remarkably little evidence of Saxon working. The Domesday Book records iron workers in a number of counties,[3] and during the twelfth century the industry seems to have expanded. In the north, the Survey mentions iron at Hessle in the West Riding, but twelfth-century grants name Egremont in Cumberland,[4] Denby and Kimberworth in Yorkshire,[5] Birley in Derbyshire.[6] But these isolated northern examples were surpassed by the activity in the Forest of Dean. The Survey had also mentioned iron working at Pucklechurch and

[1] D.B. 1, 172.   [2] See pp. 70 ff. above.
[3] Chester, Sussex, Northampton, Somerset, Gloucester, Hereford, West Riding.
[4] *V.C.H. Cumberland*, 11, 340.   [5] *V.C.H. Yorks*, 11, 342.
[6] *V.C.H. Derby*, 11, 356.

at Gloucester, and throughout the reign of Henry II the accounts of the sheriff of Gloucester tell of a constant output of medieval armaments, and iron bars, nails, pickaxes and hammers, despatched to Ireland, to France, and even to the lands of the Crusades.[1] "Throughout the thirteenth century the Forest of Dean retained its practical monopoly of the English iron trade, so far at least as the southern counties were concerned."[2] But this western area was soon to lose its priority. The industry of the Weald in Kent and Sussex, near to the London market, was expanding so rapidly during the thirteenth century that it ultimately came to eclipse the older centres.

The lead-mining centres of the country fell into four main groups:[3] (1) The "mines of Carlisle", that is to say of Alston Moor on the borders of Cumberland, Yorkshire and Northumberland, are mentioned in the Pipe Rolls of Henry II; and we have some details of the regulations in force there. Silver-bearing lead seems also to have been worked in the neighbouring district of Weardale.[4] (2) The Domesday Survey records a large number of lead mines in Derbyshire, and during the reign of Henry II quantities of lead were carried across to Boston and shipped to London and the Continent.[5] (3) The mines of the Mendips constituted another important source of lead and their workers seem to have been highly organised.[6] (4) Some lead was also mined in Shropshire, but the working in this county did not rival in importance the other three great lead-mining camps.

The evidence for tin mining in the south-west has been summarised by G. R. Lewis.[7] There is no mention of stannaries in the Domesday Book, and it is not until 1156 that the documentary evidence of the tin mines begins. Brief entries in the Pipe Rolls record the growth in production. In 1198 the mines were placed under the supervision of a warden. In 1200 the

[1] *V.C.H. Gloucester*, II, 217.
[2] L. F. Salzman, *English Industries of the Middle Ages* (1923), p. 24. On the use of coal, Mr Salzman writes (pp. 1–2): "with the departure of the Romans from Britain coal went out of use, and no trace of its employment can be found prior to the Norman Conquest, or indeed for more than a century after that date. It was not until quite the end of the twelfth century that coal was rediscovered, and the history of its use in England may be said for all practical purposes to begin with the reign of Henry III (1216)...."
[3] See L. F. Salzman, *ibid.*, chap. III. See pp. 72 ff. above for Roman lead mines.
[4] *V.C.H. Cumberland*, II, 339; *V.C.H. Durham*, II, 348.
[5] *V.C.H. Derby*, II, 316–17.
[6] *V.C.H. Somerset*, II, 36 ff. See J. W. Gough, *The Mines of Mendip* (1930), chap. III.
[7] G. R. Lewis, *The Stannaries* (1924), p. 34. See pp. 69 ff. above for Roman tin.

yield was 800 thousand-weight; and, in the following year, John issued the first charter of the stannaries, which "was followed somewhat tardily by a renewed interest in mining, and the output of tin, which from 1201 to 1209 had fallen to 500 thousand-weight per annum, now touched 800 in 1211, 1000 in 1212, and two years later made the record yield of 1200 thousand-weight, or about 600 long tons".[1] Statistics for the thirteenth century are not plentiful, but by the first half of the fourteenth century the tinners with their own customs and courts had achieved a definite status of their own under the English constitution.

Manufacturing activity was naturally less localised than mining, but, though on a small scale, was varied enough. Tile makers, pewterers, goldsmiths, glass makers, potters, tanners, founders, leather workers— these, and many other types of workers, are named here and there. Particular mention must be made of one activity—the manufacture of cloth. It is sometimes assumed that cloth making practically started with the introduction of Flemish weavers by Edward III. But the evidence shows that woollen manufacture was carried on in most parts of the realm at an earlier period.[2] Under Henry I and Henry II, weavers established gilds in London, Oxford, Lincoln, Nottingham, Huntingdon, Winchester and York. Nor were these the only centres of the industry. There were others at Stamford, Worcester, Gloucester and Darlington. In Yorkshire there was another group of towns—York, Beverley, Kirkby, Thirsk, Malton and Scarborough. To the twelfth century also belong the remarkable "laws of the weavers and fullers" of Winchester, Marlborough, Oxford and Beverley. It would seem as if the early English cloth industry was both extensive and widespread. Between 1233 and 1235 we find the king buying russets of Oxford and of Leicester, burnets, "powenacios" (?), and blues of Beverley, and blankets and haubergets of Stamford; while in 1265 English "Stamfords" were being imported into Venice. English cloth seems also to have been imported into Ireland and Spain. The idea that no cloth was manufactured for export needs to be modified. During the latter part of the thirteenth century, the making of cloth seems to have been on the decline. It remained in decline until the great revival under Edward III.

In summing up the commercial geography of medieval England before 1250, one point must be emphasised. When all is said, the fact remains that this expanding trade, this mining activity, this active manufacture,

[1] G. R. Lewis, *The Stannaries* (1908), pp. 36–7.
[2] See L. F. Salzman, *ibid.*, chap. IX, and E. Lipson, *op. cit.* chap. IX, for the references to the details that follow.

this rise of mercantile towns, were only the beginnings of things. They were simply hints of changes that were to come with the full tide of the "medieval renascence" in the thirteenth and fourteenth centuries.

BIBLIOGRAPHICAL NOTE

The footnote references are intended to give a guide to the sources. A. Ballard's *The Domesday Inquest* forms a useful introduction. Its second edition (1923) also contains "A Bibliography of Matter relating to Domesday Book published between the years 1906 and 1923". Most existing Domesday studies, however, have but little direct geographical interest, and, to repeat what has been said on p. 165, much statistical work will have to be done before the full geographical contribution of the Domesday Book has been extracted.

Chapter VI

FOURTEENTH-CENTURY ENGLAND

R. A. PELHAM, M.A., PH.D.[1]

The late Professor Unwin once cryptically summed up the main characteristics of the fourteenth century by saying that "...the contrast between the England of 1377 and that of 1327, whether in regard to its external relations, its constitutional development or its social and economic conditions, must have been scarcely less striking than the more familiar contrast between the beginning and the end of the Victorian era".[2] One might infer from this statement that the century should be studied not as one period but as two, separated perhaps by the brief economic and social interlude of the Black Death in 1348–9. But this inference would give undue prominence to the Black Death and its immediate consequences,[3] and would obscure the important and more fundamental changes that had not as yet found clear expression, though they had been operating for some time. Thus, although it is true that England, which had been an importer of cloth during the earlier part of the century, became an exporter of that commodity on a large scale in Richard II's reign (1377–99), the seeds of the change had been sown long before 1348, or even before 1331 when John Kempe and his friends came over as the first Flemish artisans enjoying royal protection. And although the freeing of the villein from his customary services, and the substitution of a money rent, was a marked feature of the latter part of the century, the death-knell of the manorial system had been sounded a long time previously. It is now generally appreciated that the Black Death merely hastened the changes that it was formerly thought to have originated.

[1] I wish to thank Professor Eileen Power and Mr M. M. Postan for many helpful suggestions during the preparation of this chapter and of chapter VIII.

[2] *Finance and Trade under Edward III* (ed. G. Unwin, 1918), p. xiii.

[3] See, for example, A. E. Levett and A. Ballard, *The Black Death on the Estates of the See of Winchester* (1916), p. 142: On the estates of the Bishop of Winchester "no period of anarchy follows upon the appearance of the Black Death, but there is evidence of severe evanescent effects and temporary changes, with a rapid return to the status quo of 1348". It should be noted, however, that poorer manors elsewhere were doubtless more severely disorganised than the more prosperous ones analysed by these authors.

THE DISTRIBUTION OF POPULATION

A rough estimate of the general distribution of population in the country can be obtained from the Poll Tax returns of 1377,[1] although these are not to be regarded as in any way comparable with modern census returns. When mapped (Fig. 30), they show that a line drawn from York to Exeter separated a sparsely populated north and west from a more densely peopled area to the south and east.[2] Within the latter region density was rather low in Surrey and Sussex[3] (owing to the Weald), and in Hampshire (where it was reduced by the New Forest), but groups of midland and eastern counties, separated by the fens of Cambridgeshire, had a density that was above the average,[4] the highest densities being in Holland (Lincs) with 54 per square mile, and Norfolk (48 per square mile). The grand total was 1,361,478.[5]

The three Yorkshire ridings formed the largest county total (131,000), Norfolk being next with 98,000, followed closely by Lincolnshire with 95,000. At the other end of the scale came Rutland with 6000, Westmorland with 7000 and Cumberland with 13,000. London stood out very

[1] E. Powell, *The Rising in East Anglia in 1381* (1896), Appendix I. It is only possible to map the returns for the lay population as given in the enrolled accounts, which show merely the totals for the counties and chief towns, as large numbers of the particulars accounts have not survived. Only known beggars were supposed to be exempt, although the "free miners" and other privileged groups may have evaded the tax. The tax was collected from the clerical population on a different regional basis and has therefore been omitted from the calculations.

[2] It is interesting to note that this boundary corresponds roughly with the line separating the north-western area in which villein services were seldom rendered before 1350 and a south-eastern area in which such services were the rule rather than the exception.—H. L. Gray, "Commutation of Villein Services before the Black Death", *Eng. Hist. Rev.* XXIX, 650 (1914).

[3] It must be emphasised that the density in each case is only the average for the whole county. Along the coastal plain of Sussex, for example, the density was almost certainly 40 per square mile.

[4] The method of grouping in Fig. 30 is arbitrary. Middlesex, with 39 per square mile, comes nearest to inclusion in the highest group. Including London it would have an average density of 122·5 per square mile. Both London and York are omitted from the county totals. Note the resemblance to the Domesday distribution of Fig. 25.

[5] This includes Plymouth and Winchester, which are given by J. Topham in *Archaeologia*, VII, 1785, but not by Powell. The other towns shown in brackets in Fig. 30 are not mentioned separately in the 1377 returns, being almost certainly included within the county totals for that year, but their population has been estimated in each case from the separate totals shown in the 1381 returns as given by E. Powell, *loc. cit.*

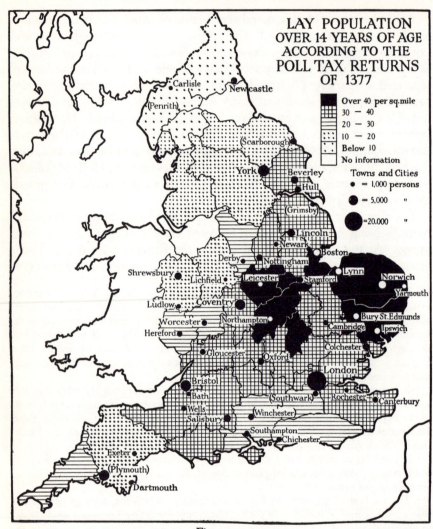

Fig. 30

The regions of greatest density were roughly the chief wheat-growing areas. More detailed statistics, if available, would show minor concentrations of population elsewhere, e.g. along the coastal fringes of Devon and Cornwall, the Vale of York and the coast plain of Sussex.

prominently, with a total of 23,000 which equalled the combined totals of York, Bristol, Plymouth and Coventry, the four largest provincial towns[1]. Even so, it accounted for less than 2 per cent. of the total for the whole country. If the formula used by Thorold Rogers[2] is adopted, the actual number of inhabitants in the whole country was about 2,000,000. Similarly, London's population was about 35,000, that of York, the next largest town, was about 11,000; Bristol probably had about 9500 inhabitants, Coventry and Plymouth a little over 7000 each, and Norwich nearly 6000.[3] Florence, in comparison, had a population of about 54,000 in 1351.[4]

Unfortunately, it is impossible to estimate the relative contributions of urban and rural settlements to the total population of the country. The towns and cities named in Fig. 30 only account for about 8 per cent. of this total, but there were undoubtedly others which, though smaller in size, had an urban population. There is much confusion as to what constituted a borough in the Middle Ages, and it is not safe to assume that a settlement which had a borough charter was necessarily urban in character.[5] Nevertheless, the general distribution of boroughs is of some interest. In Fig. 31,[6] which should be compared with Fig. 30, the parliamentary cities and boroughs are shown, and although they were not as numerous as the taxation boroughs of the period they were almost certainly urban settlements, whereas some of the taxation boroughs appear alternatively

[1] It is interesting to note that London's largest rivals lay at a distance of about a hundred miles at least along the western fringe of the more densely populated half of the country. York and Bristol were the only other towns in the fourteenth century that enjoyed county rank along with London (May McKisack, *Parliamentary Representation of the English Boroughs during the Middle Ages* (1932), p. 32).

[2] J. E. Thorold Rogers, *Six Centuries of Work and Wages* (1908), p. 118. His calculations are based on those of Topham, *loc. cit.*

[3] *Ibid.* p. 117. 50,000 was an extraordinarily high population for a medieval town according to Pirenne in *Econ. Hist. Rev.* II, 32 (1929–30).

[4] Pagnini, *Della Decima e di varie altre gravezza imposte dal comune di Firenze* (1765), I, 132.

[5] See e.g. M. McKisack, *op. cit.* p. 77 and A. Ballard and J. Tait, *British Borough Charters, 1216–1307* (1923), p. xlix. See also pp. 218 ff. above for a discussion of the borough.

[6] Drawn up from *Return of Members of Parliament, 1213–1702* (1878), which omits the parliament of 1275 mentioned by McKisack, *op. cit.* p. 5. This parliament, however, contained members from several "villatae" that were not represented in subsequent assemblies. It is impossible to be certain that a borough which made a return was actually represented (*ibid.* chap. IV) but this need not invalidate the general conclusions. Each borough normally was entitled to send two burgesses, but after 1355 London regularly sent four (*ibid.* p. 40).

as boroughs and "villatae" in the accounts.[1] In Fig. 31 the symbols are concentrated in the southern and south midland counties, in the middle Severn Basin and in the Vale of York,[2] but there is a comparative scarcity

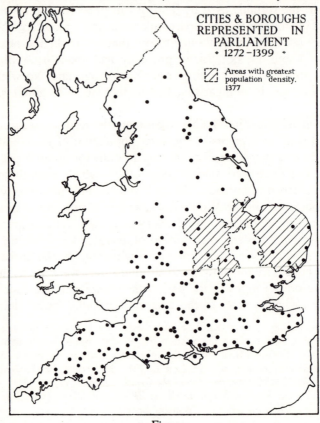

CITIES & BOROUGHS
REPRESENTED   IN
PARLIAMENT
• 1272-1399 •

Areas with greatest
population density.
1377

Fig. 31

Parliamentary representation even in the fourteenth century bore little
relation to the general distribution of population.

of symbols in the areas with the greatest density of population (over 40 per square mile).[3] These latter were the principal grain-producing

[1] As shown by J. F. Willard, "Taxation Boroughs and Parliamentary Boroughs, 1294–1336", *Historical Essays in Honour of James Tait* (1933), pp. 430–5.

[2] Many of the northern boroughs were only represented in the parliaments of Edward I, for Scottish raids during the fourteenth century impoverished the region and the boroughs were unable to pay the expenses of their members.

[3] Even the more generally numerous taxation boroughs were almost equally scarce in these areas.

regions,[1] over which the population was fairly evenly spread in villages rather than in towns. On an average, rather less than a half of the towns marked on the map were represented in any one parliament,[2] but even so the urban population appears to have had a much greater representation than its numerical strength justified.[3]

From the view point of occupation, the population of England in the fourteenth century may be grouped into three main categories associated with agriculture, industry and commerce, to which should be added mining, but unfortunately there are no means of estimating the size of the mining population since it was not subject to ordinary taxation. Broadly speaking, the richest agricultural areas (see Fig. 32) had the greatest density of population; the largest towns were those associated with the cloth industry (Fig. 39), whilst the population of the ports can be explained by their commercial activities. But despite the advent of industrialism, agriculture remained the chief occupation for the vast majority of the people and is the one that must be considered first.

AGRICULTURE

During the early centuries of feudal organisation, rural England was given over in large measure to subsistence farming. Manors were usually held not singly but in groups, often widely scattered, and intermanorial marketing was a common feature of the period. If necessary, the deficiency at one manor was made good from the surplus of another, but considerable quantities of produce were sold to the towns for local consumption or for eventual shipment abroad.

The carrying services, which formed part of the manorial tenants' obligations, often involved long and arduous journeys from one manor to another, but, during the twelfth and thirteenth centuries, many villages which possessed local advantages of position as route centres grew into towns, and thus changed their function. The substitution of money payments for the older services altered the conditions of production on the manor. The villein now became free to devote much more time to his own land, and, with a growing urban population ready to purchase his surplus produce, there was every inducement for a change

[1] See Fig. 32.
[2] The average during Edward I's reign was 75, and during Richard II's reign it was 83, out of a total of about 190.
[3] The rural interests were, of course, represented throughout the period by the Knights of the Shire, two of whom were elected for each county.

from subsistence farming to farming for profit.[1] This movement received a further impetus towards the end of the century, when many lords, finding that bailiff farming was no longer remunerative, began to lease out their demesnes for a money rent. Other lords retained their demesnes, but in areas that were suitable converted them into the sheep runs that

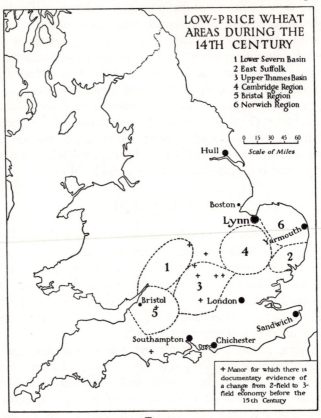

Fig. 32

The low-price areas were approximately the principal wheat-growing regions. The symbols for the ports roughly indicate the extent of the wheat trade of each during the century. For sources see p. 237, footnote 2, and p. 239, footnote 1.

became such a marked feature of the rural economy of the fifteenth century.

The rise of the local market did not necessarily mean that a town was supplied wholly from its immediate neighbourhood, for the corn middle-

[1] N. S. B. Gras, *Evolution of the English Corn Market* (1915), chap. I.

man soon made his appearance, and inter-urban trade on a considerable scale became general.[1] Nor did the much overrated transport difficulties of the period check its growth. In some years enough corn was grown not only to satisfy home demands but to leave a substantial surplus for exportation abroad. And although a good deal was imported in times of dearth, the balance seems to have been slightly in favour of exports. No statistics of production exist for the country as a whole at this early period, but there is some indirect evidence that is worth examining. Fig. 32 shows the regions in which, during the fourteenth century, the average price of wheat did

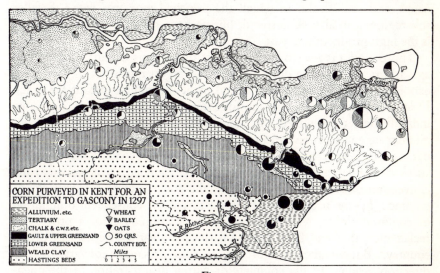

Fig. 33

Note the extensive wheat-growing to the north of the Chalk escarpment, especially in eastern Kent. The largest symbol, that for Wingham Hundred, represents a total of 455 quarters.

not rise above 6s. a quarter.[2] A low price meant no doubt that the supply was abundant; and since these were the areas from which most of the exported corn was drawn (as shown by the distribution of the ports engaged in the trade), it is probably safe to assume that these were the chief areas of production.

There was, of course, some local specialisation in crops at this time, but this is not easy to demonstrate because of the difficulty of obtaining detailed information for a group of contiguous manors covering a reasonably large area. But, for example, it is possible to reconstruct conditions

[1] N. S. B. Gras, *op. cit.* p. 163.   [2] N. S. B. Gras, *op. cit.* pp. 41, 47.

in Kent at the close of the thirteenth century, as shown in Fig. 33, which represents the amounts of grain supplied by each "hundred" in the county for an expedition to Gascony in 1297.[1] There are obvious and serious objections to accepting this by itself as an index of the quantitative distributions of the three principal crops; but, on the other hand, there is a considerable amount of evidence of a similar character, though unfortunately not complete enough for mapping, which confirms the conclusions that can be drawn from Fig. 33. The map reveals a striking contrast between the area north of the chalk escarpment, where wheat and barley predominated, and the Weald, in which the principal crop was oats.[2] Smaller sub-regions can also be detected, but perhaps the most important point is the high production in east Kent.[3]

The period 1250–1350 was the most prosperous century of medieval agriculture and the one during which the largest quantities of corn were exported. Whatever view may be taken of the significance of the Black Death, the fact was that in the following century a growing economic individualism undermined the conservatism of manorial life. The beginnings of enclosure constituted a symptom of changes to come. The term has been applied to a variety of processes: (1) the consolidation of scattered strips of arable land into compact plots surrounded by hedges, (2) the conversion of arable into pasture, (3) the concentrating (engrossing) of holdings, (4) the occupation of the waste and the consequent diminution of common rights, which was really the continuation of a process of reclamation that had been going on since the earliest settlement of the country. But whether new or old, all four processes converge in one direction, leading to the complete or partial disintegration of the open-

---

[1] Exchequer K.R. Accounts, 566/3 (P.R.O.).

[2] It is interesting to note that when this map is compared with those in *An Agricultural Atlas of England and Wales* (P. Howell, 1918), the resemblances between the two distributions are seen to be very marked, particularly in the case of the barley-growing areas. The decline in the cultivation of oats in Romney Marsh and the substitution of sheep pastures has been the only notable change. Of course, this correspondence could not be turned into a generalisation applicable to the whole country.

[3] East Kent is omitted from Fig. 32 because its average wheat price during the fourteenth century was slightly higher (6s. 1⅛d. per quarter in Gras' table, *op. cit.* p. 41) than the limit selected, but this was no doubt due to the great demand for Kentish wheat for shipment overseas (see *Cal. State Papers, Venetian, 1603–1607* (1900), p. 414). Although not apparent on the map, it should be noted that the average density of population in Kent (38 per square mile) was well above that of the whole country, so the correlation between considerable wheat production and high population density might be applied to the eastern part of the county. See also *Rot. Parl.* II, 194a, for reference to purveyance of corn in Kent.

field system. There were other changes, too, involving a transition from a two-field to a three-field economy.[1] These two systems or rather variations of the same system (for the rotation of crops in the common fields which differentiated them from the other land-holding systems was a feature of both) were confined to the somewhat narrow longitudinal area shown in Fig. 23. It is not necessary to discuss whether the two-field system, being simpler, was necessarily the older form, but it is important to note that the substitution of three fields meant a net gain of one-sixth in the annual area under crops.[1] Such a change must have involved a considerable amount of reorganisation and, in all probability, would only be undertaken in a progressive area or in one where the density of population was particularly high. It is significant that during the fourteenth century the available documentary evidence of the change relates almost entirely to manors within the low-price wheat areas of Fig. 32.

The undoubted importance of the wool trade has been rather over-emphasised at the expense of arable farming in the fourteenth century. The rich corn-growing areas may not possess such spectacular evidence to-day of their former prosperity, as do those that specialised in wool and cloth, but these corn areas certainly contributed handsomely towards the national income of the time.[3]

SHEEP FARMING AND WOOL PRODUCTION

Sheep rearing was by no means confined to the moorlands and to the chalk and limestone uplands, although these were no doubt the most favoured areas. Lowland regions generally, and marshlands in particular, supported large numbers in spite of the diseases that were more prevalent therein. During the winter months the animals were folded on the fallow,[4] or kept in sheep cotes.[5] Although there was a good deal of interchange of flocks

[1] H. L. Gray, *English Field Systems* (1915), chap. 11.

[2] See p. 193, footnote 3 for Walter of Henley's arguments on this point.

[3] It is difficult to illustrate this except in the case of Sussex, which is unfortunately not a typical county in this respect on account of the poor quality of its wool. An unusually detailed valuation of the wool and corn throughout the county in 1341 shows that corn accounted for more than half the agricultural wealth in each parish (*Inquisitiones Nonarum*, Record Commission, 1807); see also R. A. Pelham, "Some Mediaeval Sources for the Study of Historical Geography", *Geography*, XVII, 32 (1932).

[4] Wheat was cut high on the stalk. See p. 197 above.

[5] A. M. Melville, "The Pastoral Custom and Local Wool Trade of Sussex, 1085–1485" (unpublished thesis in University of London Library). Summary in *Bulletin of Institute of Historical Research*, X, 39 (1932). Crowland Abbey sheep were kept through the winter in the early fourteenth century, e.g. 8586 in 1314. See F. M. Page, "Bidentes Hoylandiae", *Economic History*, I, 603 (1929).

between different manors, there is little evidence of regular seasonal movements. Transhumance was, however, certainly practised in Wales,[1] and to some extent in England, judging by an order of the Bishop of Lichfield and Coventry in the early thirteenth century that tithes of wool taken by churches in his diocese should be divided "if the sheep be fed in one place in winter and another place in summer."[2]

It is possible to form a very rough estimate of the numbers of sheep in England in the middle of the fourteenth century by calculations based on the amount of wool exported and the quantity of English cloth exposed for sale. The former averaged about 30,000 sacks annually, the fiscal equivalent of which, on the basis of 240 wool fells[3] to the sack of 364 lb., would be 7,200,000 fleeces, but the 12,000 broadcloths subject to ulnage[4] would, on the basis of 4⅓ to the sack of wool,[5] account for a further 500,000 or so. If we then make a further addition for cloth not subject to ulnage[6] and for which, therefore, no statistics were drawn up, we reach a total of round about 8,000,000 fleeces, the produce of an equal number of sheep. Many of these sheep were no doubt killed off in the autumn owing to the comparative shortage of winter keep, but there is no evidence of wholesale slaughter.[7] The average number of wool fells exported annually between 1282 and 1290 was only about 125,000,[8] although this was increased to

---

[1] W. Rees, *South Wales and the March, 1284–1415* (1924), p. 218.

[2] E. Power, in the *Cambridge Medieval History*, VII, 747 (1932). See p. 205 above.

[3] Prior to 1368 the equivalent was 300 wool fells. R. J. Whitwell, in "English Monasteries and the Wool Trade in the Thirteenth Century" (*Vierteljahrschrift für Sozial- und Wirtschaftgeschichte*, II, 5, 1904), gives an average of 1 lb. 14½ ozs. for 1549 fleeces from a number of manors in Lincolnshire and Northamptonshire belonging to the Abbey of Peterborough. Since these fleeces were almost certainly of long wool, their weight would be greater than the average for the whole country. J. E. Thorold Rogers (*History of Agriculture and Prices in England* (7 vols, 1866–1902), I, 388–394) gives the weights of a number of fleeces which average about 1¾ lbs. These figures would only give about 200 fleeces per sack, but allowance must be made for wool thrown out in sorting, etc.

[4] See H. L. Gray, "The Production and Exportation of English Woollens in the Fourteenth Century", *Eng. Hist. Rev.* XXXIX, 34 (1924). The figures are: 1354–5, 10,665; 1355–6, 11,622; 1356–8 (av.), 15,610. By the end of the century these had increased to about 50,000.

[5] This is Gray's amended equivalent. His earlier estimate (*loc. cit.*) was 7 or 8.

[6] The Statute of 1353 did not apply to cloths equal to less than half a broadcloth.

[7] This did not occur until the great increase in stock following the enclosures. See H. E. Malden in *The Cely Papers* (1900), p. xii.

[8] Pipe Roll 133, m. 32 d, and 134, m. 3 (P.R.O.).

about 300,000[1] in the middle of the fourteenth century. Even when further allowance is made for the fells that were consumed in local manufacture, the total only accounts for a small proportion of the sheep in the country.

Unfortunately, only a general idea can be formed of the distribution of sheep throughout the country, except in the already-mentioned county of Sussex.[2] Here (Fig. 34) there was a marked concentration on the South Downs, especially towards the east, which was probably more open than the loam-covered area farther west.[3] The most striking feature was the large numbers that belonged to parishes on the rich arable area of the

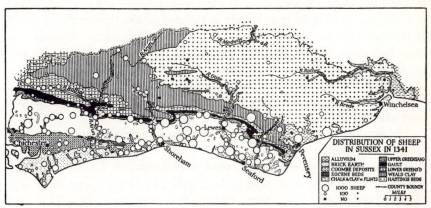

Fig. 34

The number of sheep in each parish has been calculated from the value of the fleeces. Note the concentration on the Downs east of Shoreham.

coastal plain around Chichester. Although there may have been a certain amount of intercommoning on the Downs, the stubble and fallow of the open fields no doubt played an important part in sustaining these flocks. The sheep were owned by all classes in the community. Battle Abbey possessed the largest flocks, but on the lay manors, the lords and free-holders, some of whom combined the occupation of wool merchant with

[1] Alice Beardwood, "Alien Merchants and the English Crown in the Later Fourteenth Century", *Econ. Hist. Rev.* II, 257 (1929–30).

[2] R. A. Pelham, "The Distribution of Sheep in Sussex in the Early Fourteenth Century", *Sussex Archaeological Collections*, LXXV, 130 (1934).

[3] At the beginning of the fifteenth century the merchants of Lewes petitioned Parliament for a renewal of their staple privileges on the ground that most of the wool produced in Sussex was grown within a fifteen-mile radius of the town (*Rot. Parl.* III, 497 (4 Henry IV)).

that of flockmaster, owned considerable numbers, whilst even the poorest cottar had a dozen or so.[1]

For sheep farming on a really large scale, however, we have to turn to Yorkshire and Lincolnshire, whose pre-eminence was due mainly to the activities of Cistercian abbeys.[2] Fig. 35[3] shows that most of the abbeys producing 30 or more sacks of wool annually were located within those two counties. On the basis of 200 fleeces to the sack, this implies flocks of at least 6000 sheep; and, in the case of the abbeys at Fountains and Rievaulx, which produced 76 and 60 sacks respectively, the flocks must have amounted to over 15,000 and 12,000. Evidence from another source shows that the flocks of the abbeys at Tintern and at Neath (Wales) amounted to 2364 and 4204 head respectively at the end of the thirteenth century.[4] It is important to notice, however, that although much emphasis has been laid upon the activities of the monks in the medieval wool trade, yet their contribution was much smaller than might be expected. The lists from which Fig. 35 has been drawn up are admittedly incomplete, but, even if the combined annual output of these monasteries be doubled, it would only account for about one-sixth of the national total.

The quality of the wool varied considerably in different parts of the country. A valuation made in 1343,[5] which is represented in Fig. 36, shows that the most expensive wool was grown in Lincolnshire and Shropshire, the least expensive in Cornwall and Devon. The latter was in fact so coarse that it was often referred to contemptuously as "Cornish hair".[6] The chalk lands of southern and south-eastern England, together with those of East Anglia, produced wool that was not at all distinguished for its quality, whilst the wool of the west midland counties approximated in value to that of Shropshire. Marsh wool was invariably lower in price than that of the neighbouring hill country, the counties specially mentioned in this connection being Lincolnshire, Essex, Middlesex, Kent and Sussex.

---

[1] A. M. Melville, *op. cit.* p. 17.   [2] See p. 185 above.

[3] Compiled from Pegalotti's lists given in W. Cunningham, *Growth of English Industry and Commerce*, I, Appendix (5th ed., 1922, pp. 628–41). See H. E. Wroot, "Yorkshire Abbeys and the Wool Trade", *Thoresby Soc.* XXXIII, Part I (1930); and R. J. Whitwell, "English Monasteries and the Wool Trade in the Thirteenth Century" *Vierteljahrschrift für Sozial- und Wirtschaftsgeschichte*, II, 1 (1904).

[4] *Taxatio Nicholai*, 1291 (Record Commission, 1807)—quoted by W. Rees, *op. cit.* p. 195.

[5] T. Rymer (ed.), *Foedera*, V, 369 (1708). A similar list in *Rot. Parl.* II, 138, gives nine marks for the Craven area of Yorkshire.

[6] *Studies in English Trade in the Fifteenth Century* (ed. E. Power and M. M. Postan, 1933), p. 49.

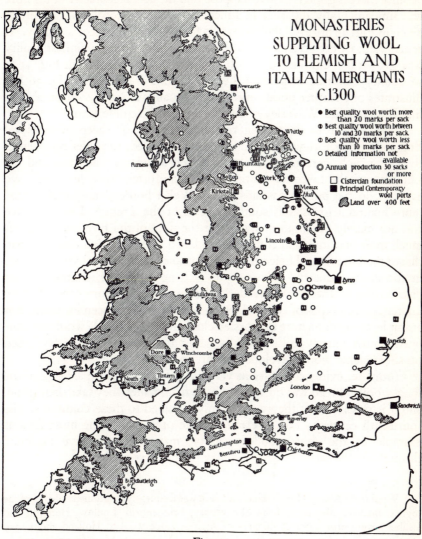

MONASTERIES
SUPPLYING WOOL
TO FLEMISH AND
ITALIAN MERCHANTS
C.1300

● Best quality wool worth more
than 20 marks per sack
◐ Best quality wool worth between
10 and 20 marks per sack
◑ Best quality wool worth less
than 10 marks per sack
○ Detailed information not
available
◎ Annual production 30 sacks
or more
☐ Cistercian foundation
■ Principal Contemporary
wool ports
▨ Land over 400 feet

Fig. 35

Pegalotti's list, from which this map has been compiled, is unfortunately incomplete.
Even so, important concentrations in Lincolnshire and Yorkshire, the two chief wool-
producing counties, are apparent.

The criterion of high quality seems to have been merely fineness of texture,[1] because there is good reason to suppose that the actual character of the fibre was fundamentally different in Lincolnshire and Shropshire. The prices given in Fig. 36 are essentially averages for a county or part of a county. If actual values were given it would be seen that wool of the highest quality was grown throughout the Welsh Marches.[2] The evidence in Fig. 35 as to quality confirms in general the conclusions drawn from Fig. 36 but illustrates one further point: even in low-quality areas, Cistercian abbeys like Beaulieu and Waverley were producing high-grade wool.

There appear to have been two main breeds of sheep in medieval England—the short-woolled and the long-woolled, the chief representatives of which were respectively the Ryeland on the one hand and the Lincoln and the Leicester on the other. The former type seems to have evolved on the poorer pastures of the Welsh border,[3] the latter on the more luxuriant grasslands south and east of the Trent.[4] The difference was not merely in length of staple, for the fibre of the Ryeland wool was much more serrated, and consequently had a greater felting capacity than that of the Lincoln and Leicester, a circumstance of considerable importance in the cloth industry.[5] There is, in fact, no means of ascertaining the distribution of these two breeds in the fourteenth century other than by inference from types of cloth which are known to have been localised. For example, the manufacture of worsted cloth from long-staple wool was confined almost entirely to East Anglia, whilst the broadcloth industry, based on short-staple wool, developed mainly in the west.[6]

From later evidence,[7] it seems very likely that the principal home of the long-woolled breed was the two counties from which they take their name, and it is possible that they were mainly confined to these counties in the fourteenth century. The Ryeland, on the other hand, the most famous example of the short-woolled variety, may formerly have extended throughout the sandstone hill country of the west midlands.[8]

---

[1] See Fig. 37.

[2] Wool from Abbey Dore in Herefordshire was frequently mentioned as being of the highest quality. Merchants from Shrewsbury, Bridgenorth, Ludlow, Hereford and Ledbury were prominent as shippers of highest grade wool to Holland about 1337 (Exchequer K.R. Accounts, 457/31).   [3] W. Youatt, *Sheep* (1837), p. 258.

[4] *Ibid.* p. 332.   [5] See p. 249 below.   [6] See below p. 254 for Suffolk broadcloths.

[7] See Fig. 38, and Youatt, *op. cit.* p. 344.

[8] Other short-woolled breeds extended over the counties noted for their coarse wool (see Fig. 37, and Youatt, *op. cit., passim*). The more wide-spread distribution today of long-woolled sheep is due mainly to the popularity of the New Leicester breed in the eighteenth century.

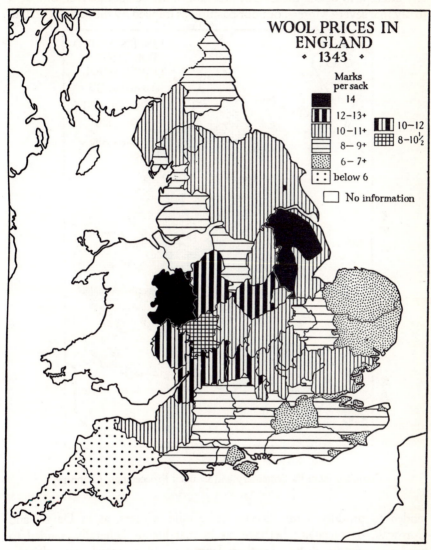

WOOL PRICES IN
ENGLAND
• 1343 •

Marks
per sack

14

12–13+

10–11+   10–12

8– 9+   8–10½

6– 7+

below 6

No information

Fig. 36

Some of the factors that have influenced the evolution and distribution of these two principal types of sheep are undoubtedly geographical, but any attempt to define them with precision must be speculative. Suffice it to say that severe winters, or poor pastures as implied by the name Ryeland,

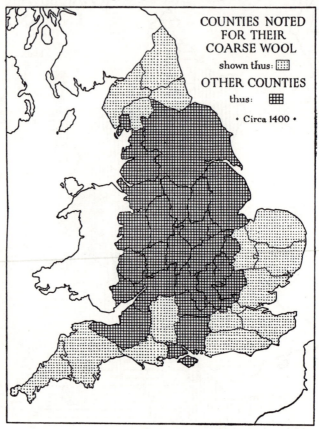

Fig. 37

Compiled from Parliamentary and Council Proceedings, Chancery, file 13, m. 1 (P.R.O.).

tended to produce a fine fleece; while mild winters, as in Devon and Cornwall, or succulent herbage, as in marshlands generally, tended to produce a coarse fleece. Further, length of fibre, which can also be roughly correlated with feeding, shows a tendency to increase with a super-abundance of nutriment.[1]

[1] W. Youatt, *op. cit.* p. 70.

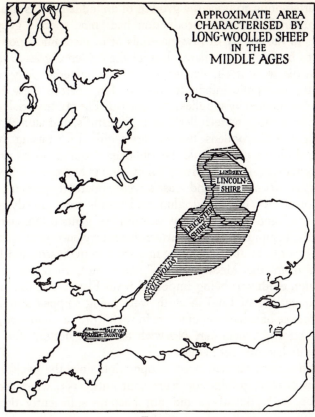

APPROXIMATE AREA
CHARACTERISED BY
LONG-WOOLLED SHEEP
IN THE
MIDDLE AGES

Fig. 38

It must be noted that this distribution is not based upon contemporary evidence
and can, therefore, only be tentative.

THE CLOTH INDUSTRY

During the Middle Ages the cloth industry passed through a number of
vicissitudes. In the twelfth and thirteenth centuries the manufacture of
cloth was very widespread; almost every important town was engaged
in it.[1] Then, largely as the result of gild restrictions, its development was
checked and many towns in the fourteenth century became seriously
impoverished. Finally, in Richard II's reign, the industry burst the bounds
of medieval town life and became established in rural areas, where it
steadily maintained itself until the great expansion of the Tudor period.

[1] See p. 228 above.

When attempting to trace the growth of an industry, it is usual to find that the original centres of production had some specific geographical advantages in the supply of raw materials or of mechanical power which were not to be found elsewhere. But all the evidence goes to show that, apart from the wool itself, whatever local raw materials may once have nurtured the economic infant, it was not long before these had to be heavily supplemented by supplies from abroad. The raw materials necessary for all types of cloth were fuller's earth, dyestuffs, and alum;[1] whilst, in the manufacture of woollens, teasels were required for raising the nap on the cloth. Fuller's earth could be obtained from deposits within the Jurassic formation,[2] in the Medway valley[3] and in Surrey;[4] some woad was grown on calcareous soils; and teasels were apparently grown on a large scale in Somerset.[5] But woad, alum and teasels were already articles of trade in 1228,[6] and when customs accounts of general merchandise first appear at the beginning of the fourteenth century enormous quantities of woad and other dyestuffs were being imported from Picardy and Toulouse, and alum from the Mediterranean.[7] It is not, therefore, surprising that although cloth was being imported from Flanders at this time, considerable quantities of English cloth were being shipped abroad.[8] This was the produce of urban manufacture in such towns as Stamford, Lincoln, London, Oxford, Colchester, Norwich and Winchester,[9] for the rural developments had not yet begun.

Although a cloth tax had been in operation since the twelfth century, no statistics of cloth production were kept until 1353, when the aulnage accounts began.[10] But these did not become sufficiently detailed for mapping purposes until the end of the century, when some of the rural areas had already started to produce on a large scale. Fig. 39 summarises

[1] Used as a mordant (L. F. Salzman, *English Industries of the Middle Ages* (1923), p. 208).

[2] It lies between the Inferior Oolite and Great Oolite, and is exposed on the valley sides.

[3] R. Furley, *History of the Weald of Kent* (1871–4), II, Part I, p. 330.

[4] *V.C.H. Surrey*, II, 279.

[5] *V.C.H. Somerset*, II, 357. For a history of woad see J. B. Hurry, *The Woad Plant and its Dye* (1930).

[6] N. S. B. Gras, *Early English Customs System* (1918), p. 157.

[7] E.g. Customs Accounts, 124/13 (Sandwich), 136/8 (Southampton); see N. S. B. Gras, *op. cit.* pp. 302 and 360.

[8] E.g. Customs Accounts, 5/7 (Boston). N. S. B. Gras, *op. cit.* p. 273.

[9] L. F. Salzman, *op. cit.* p. 194.

[10] H. L. Gray, "The Production and Exportation of English Woollens in the Fourteenth Century", *Eng. Hist. Rev.* XXXIX, 14 note (1924).

the information mainly for the latter part of Richard II's reign; the statistics for the towns in the counties of Norfolk, Suffolk, Dorset, Gloucester and Hereford relate to Edward IV's reign, but the county totals, in all cases save Norfolk, are for the earlier period. Statistical complications do not invalidate the conclusions, however, for there was little perceptible change in either the localisation or the output of the industry during the fifteenth century.[1] It is clear from this map that many of the older urban centres had ceased to count in the industry. Of the northern towns, only York remained, and even here a growing rural industry can be detected in the West Riding;[2] Coventry alone of the midland towns appears to have survived thus far the period of urban decay. Two main areas of activity are evident—the West Country and East Anglia. The former extended from north Devon to Hampshire, with a concentration on the slopes of the Mendips, whereas the East Anglian towns and villages were grouped in and around the Stour valley. London, the Weald of Kent, and the Steventon region of Berkshire formed isolated areas of minor importance.

Before discussing changes in the location of the industry during the century, certain technical points connected with cloth manufacture must be mentioned. Translated into terms of cloth, the distinction between long and short wool was a distinction between "worsteds" and "woollens". The long wool was combed, whereas the short wool was carded, before being spun; and the' latter underwent the process of fulling, to which worsted cloth was seldom submitted. To quote a contemporary:

> Cloth that cometh from the weaving is not comely to wear
> Till it be fulled under foot or in fulling stocks;
> Washen well with water, and with teasels cratched,
> Towked and teynted and under tailor's hands.[3]

Fulling had a two-fold purpose, it not only cleansed the cloth of dirt and grease, but it also thickened the material by inducing the fibres to "felt". Fuller's earth was indispensable for the cleansing, which was performed originally in a trough of water by a fuller, walker, or tucker, who trampled on the cloth until the necessary degree of cleanliness and thickness had been achieved. This was obviously a laborious but essential process; and, in the thirteenth century, an instrument known as "the

[1] See H. Heaton, *The Yorkshire Woollen and Worsted Industries* (1920), pp. 84–8. Ulnage accounts for the fifteenth century, however, are not reliable, being in many cases copies of earlier accounts. See E. Carus Wilson, "The Aulnage Accounts: a Criticism", *Econ. Hist. Rev.* II, 123 (1929–30).

[2] The centres were Ripon, Leeds, Wakefield and Pontefract.

[3] Langland, *Piers Plowman* (as quoted by L. F. Salzman, *op. cit.* p. 205).

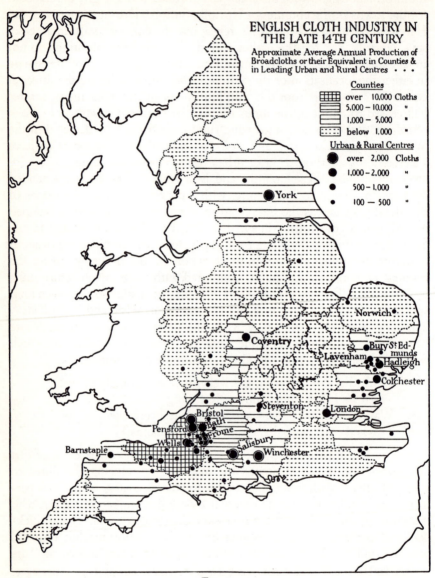

ENGLISH CLOTH INDUSTRY IN
THE LATE 14TH CENTURY

Approximate Average Annual Production of
Broadcloths or their Equivalent in Counties &
in Leading Urban and Rural Centres • • •

Counties
over 10,000 Cloths
5,000 – 10,000 "
1,000 – 5,000 "
below 1,000 "

Urban & Rural Centres
over 2,000 Cloths
1,000 – 2,000 "
500 – 1,000 "
100 – 500 "

York

Norwich

Coventry

Bury St Ed-
munds
Lavenham
Hadleigh
Colchester

Steventon
London

Bristol
Bath
Pensford
Frome
Wells
Barnstaple
Salisbury
Winchester

Fig. 39

This map is compiled from the Aulnage Accounts, which did not include worsteds.
The sizes of symbols must be regarded only as approximate because in many cases they
are based on the returns for only one or two years.

stocks" was introduced, consisting of an upright to which was hinged a wooden bar or flail with which the cloth was beaten.[1] This saved a good deal of labour, but a further improvement was made when water-power was applied to the process, and fulling mills enabled the work to be done much more expeditiously.[2]

The old established towns in which the industry had first grown up were situated for the most part on the lower and more mature stretches of large rivers where the current was sluggish. The town gilds resisted the mechanisation of the fulling process,[3] but whenever a mill was erected in such towns it clearly had to be worked on the "under-shot" principle because the "over-shot" principle was as yet little known and, in any case, the character of the lower valleys did not permit the construction of a dam, with its accompanying mill-race, which was necessary for its operation. The great increase in the output of cloth under royal stimulus during the second half of the fourteenth century led therefore to the multiplication of fulling mills on the upper reaches of rivers and in tributary valleys where the more efficient over-shot wheels could be employed. It was to these mills that the weavers in the towns began to send their cloth,[4] and around them, as time passed, that cloth workers began to settle. Thus the inability of the towns to keep pace with the development of the industry may have been due not merely to irksome gild regulations which drove large numbers of cloth workers into the country-side, but also to the fact that these towns were unable, by virtue of their location, to meet the demands of an increasing output from the weaver's looms. In this urban exodus the technical improvements in fulling may possibly have played an important part, so that what is usually referred to as the "period of decay of the towns" might be more appropriately re-named the "industrial revolution of the fourteenth century". On the other hand, these considerations affecting fulling in the felting sense did not apply to worsted cloth, which only needed scouring, so the question of water power did not affect its location.[5]

Fig. 39, it should be noted, only relates to woollen cloth, as worsted cloth was exempt from aulnage. It is fortunate that the accounts for Somerset[6] are more detailed than those for other counties, and, since this

[1] L. F. Salzman, op. cit. p. 221.
[2] This is illustrated in the building account of a fulling mill at Marlborough in 1237 (Exchequer K.R. Accounts, 501/18).
[3] E.g. fulling at mills was forbidden in London, 1298, 1376, 1391 (E. Lipson, Economic History of England (1915), I, 426).
[4] E.g. Bristol cloth was being fulled in the surrounding villages (W. Cunningham, op. cit. p. 426).
[5] R. A. Pelham, "The Distribution of Early Fulling Mills in England and Wales," Geography, XXIX, 52 (1944).
[6] Exchequer K.R. Accounts, 342/28, 30.

was the leading area of production, they merit closer study. The map (Fig. 40) shows very clearly that there were two distinct cloth-making districts within the county—the Mendips and the Parret valley, specialising in broadcloths and "straits" respectively. The narrow valleys of the Mendips could be easily dammed for water-power purposes, although the water supply may have been somewhat irregular; and it was in them that Bristol

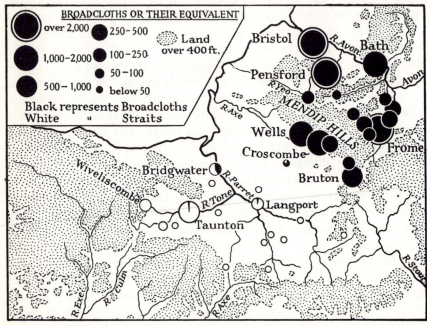

Fig. 40

This map of the Somerset area illustrates one of the earliest developments in the rural textile industry. It shows clearly the distinction between the hill country and the plain in the types of cloth produced. The map gives averages for years 1395-7.

cloth was being fulled, at this time, to the discomfiture of the Bristol fullers.[1] In the Parret valley, on the other hand, conditions were rather different. Water power was certainly available around the edge of the lowland, but since most of the cloth-making settlements were out in the plain it seems that this was not an essential factor. The medieval nomenclature of English cloths often hides the real nature of the material, and we cannot be certain of the precise meaning of the term "straits"[2] beyond the fact that such cloths were narrow, being only about half the width of a

[1] See footnote 4, p. 251 above.   [2] "*Pannus strictus.*"

broadcloth.[1] But we have already seen that the Vale of Taunton was probably inhabited by a long-woolled breed of sheep at this period,[2] so there is a strong possibility that the local cloth was made partly from long wool and partly from short wool,[3] the latter being grown on Exmoor, the Mendips and Dorset hills or obtained from farther afield. In that case, and in view also of the fact that the straits were made up in "dozens"[4] which were half the normal size, any fulling that was necessary could no doubt be done by the older method.

Towards the east, the rural character of the industry diminished and production was still concentrated in the towns. In East Anglia,[5] however, many rural settlements were making important contributions, and the industry had become established up the valleys where water power was more easily obtained. If the worsted area could be plotted in a similar way, we should find a concentration around Norwich possibly extending into north Suffolk, with a few centres in Lincolnshire, that is to say in areas lacking in water power, which was not, of course, necessary for that type of cloth.[6]

The distinction between the Norwich area and the Stour valley is clearly shown in Fig. 58, on the assumption that the exports of Yarmouth and Ipswich were mainly the products of their respective hinterlands. London was no doubt another centre of the worsted industry, but to what extent it is difficult to say. No long wool was grown in the vicinity, and the raw material must therefore have been brought from a distance. This was not difficult, for London had excellent communication with the rest of the country, and some of the cloth exported thence had certainly been made elsewhere.[7] London grew as a cloth market during the century, and

[1] The regulation size of a broadcloth was 24 yards by 1¼ yards (Statute 27, Edward III, I, cap. 4). [2] See Fig. 38.

[3] If it had been made entirely from long wool it would not have been liable to aulnage, as worsteds and serges were exempt. Serges were being made in Devonshire in Edward IV's reign, however (W. Youatt, *op. cit.* p. 213).

[4] I.e. 12-yard lengths.

[5] A detailed study of conditions in East Anglia is being made by Miss J. B. Mitchell, of Newnham College, Cambridge.

[6] Some worsted cloth was also made at York, apparently from Lincoln long wool; see, for example, *Cal. Inquisit. Miscell. Chancery*, II (1307–1349), No. 1628. Wroot (*op. cit.* p. 5) considers that long wool was grown in Yorkshire, but Youatt (*op. cit.* p. 302) attributes the present long-woolled breeds in Yorkshire to recent developments.

[7] E.g. in 1452 a London draper is recorded as having bargained with John Broke of Stoke Neyland in Suffolk for 100 Suffolk "straits" to be delivered to him at London for shipment abroad (W. T. Barbour, *History of Contract in Early English Equity* (1914), p. 211).

by the end of Richard II's reign one-third of all woollens exported from England were shipped from its quays.[1]

The so-called "broadcloths" exported from Ipswich (Fig. 58 and 59) were possibly not broadcloths in the full sense, and confusion frequently arose because cloths that were smaller and different in character were computed in terms of broadcloths for the purposes of taxation. The village of Kersey, for example, like the more renowned village of Worsted, had given its name to a type of cloth that was not being made exclusively in East Anglia during the fourteenth century. In Richard II's reign the chief areas of manufacture were Berkshire (around Steventon)[2] and Hampshire (mainly in the Isle of Wight),[3] and some were also made in the Lower Severn valley[4] and in Herefordshire[5] at a later date. These were all short-wool areas, and it can therefore be assumed that kerseys were made from short wool and were possibly milled. In that case they would tend to be as different from worsteds as the broadcloths of the Mendips were from the straits of the Parret valley.

The merchanting of wool and cloth led to the accumulation of wealth which found its most delightful expression in the rebuilding of parish churches throughout certain parts of England in the Perpendicular style. Many of these churches still contain the brasses commemorating their benefactors, and since the best examples are distributed over the northern Cotswolds that area is the one to which reference is most frequently made. A glance at Fig. 41[6] will show, however, that other areas are also well represented, and but for the destruction that has been meted out in the past to monuments of this kind there would be many more symbols to add to the map. For example, although Fig. 41 covers a much wider period (1391–1558)[7] than that of the fourteenth century, Yorkshire and the Marches are a complete blank, East Anglia and the southern counties nearly so. Nevertheless, in spite of the map's deficiencies we may note one or two points of interest. Only one brass is dated in the fourteenth century, that of Wimington in Bedfordshire (1391), but those dated up to 1430 probably relate to merchants who were operating within our period. These total six,[8] two of which are located at Lynwode, and one each at Northleach, Chipping Camden and St Albans. The Wimington brass

[1] H. L. Gray, *Eng. Hist. Rev.* XXXIX, 33 (1924).

[2] Exchequer K.R. Accounts, 343/24.   [3] *Ibid.* 344/11.

[4] *Ibid.* 346/25.   [5] *Ibid.* 347/4.

[6] Compiled from H. W. Macklin, *The Brasses of England* (1907); and M. Stephenson, *A List of Monumental Brasses in the British Isles* (1926).

[7] These dates of course refer to the deaths of the merchants.

[8] Including the Wimington brass.

commemorates a former mayor of the Staple and his wife, of whom it is stated: "Istam ecclesiam de novo construxerunt". The Chipping Camden merchant is referred to in a marginal inscription as: "Flos mercatorum lanarum totius Angliae".

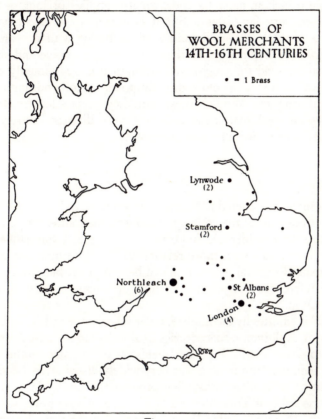

Fig. 41

Three of the London brasses are in the Church of All Hallows, Barking-by-the-Tower, near which lay the Wool Wharf in the Middle Ages (Stow's *Survey of London*, ed. W. J. Thomas, 1842, p. 17).

There is one further point of interest concerning the industry as a whole: it was not in the areas producing the highest quality wool that manufacturing first began on a large scale. There were obviously good reasons why developments should have taken place in Somersetshire, but conditions were almost as favourable, if not equally so, along the Welsh border. And if political instability be thought to have hindered industrial development in that region, why did the worsted industry not continue to grow in Lincolnshire instead of becoming centred around Norwich? The latter town appears to have been in a district characterised by poor quality short wool[1] which was totally unsuited for combing, although Lincolnshire, from which it must have drawn its supplies of long wool, was certainly not very far distant. But it is unwise to over-emphasise the importance of material considerations when dealing with the early stages of industrial developments. Human enterprise, initiative and skill were factors that counted for much, and although it is undoubtedly true that geographical factors were more directly powerful in conditioning men's activities during the Middle Ages than they are to-day, it cannot be necessarily assumed that those activities were dictated entirely by the circumstances of their immediate environment.

MINING

Although mainly restricted to Palaeozoic deposits, and therefore affecting only about one-half of the country, mining in the fourteenth century was an industry of considerable importance. There are, unfortunately, no statistics comparable with those relating to the cloth industry, and from which a map of mineral production can be compiled; but from numerous historical references it is possible to draw up a tentative map of the areas of exploitation.

Fig. 42[2] is admittedly incomplete, but it reveals one rather striking fact: the areas of exploitation six centuries ago[3] were almost co-extensive with those of to-day. Nearly all the coal-fields are represented on the map, and it will be seen that iron mining was very widely distributed. Lead, usually in association with silver, was being produced in Derbyshire, in some of the Pennine dales of Yorkshire and Durham, at Alston Moor in Cumberland,[4] in the Mendips,[5] and in Wales. One important, though rather small,

[1] E. Lipson, *op. cit.* p. 430, footnote 4, quoting J. James, *History of the Worsted Manufacture* (1857), says that Norfolk long wool was specially adapted for worsted cloth; but W. Youatt, *op. cit.* p. 326, shows that this long wool was from the New Leicester that had displaced the original Norfolk short-woolled variety.

[2] Compiled from various sources.

[3] The areas of "free miners" were: Forest of Dean—coal and iron; Devon and Cornwall—tin; Alston Moor (Cumberland), Derbyshire, Mendips—lead.

[4] Alston Moor and Upper Weardale are shown as one continuous area of exploitation on the map.

[5] Only one symbol is shown on the map as only scanty allusions to actual mines are to be found (see J. W. Gough, *The Mines of Mendip* (1930), p. 55).

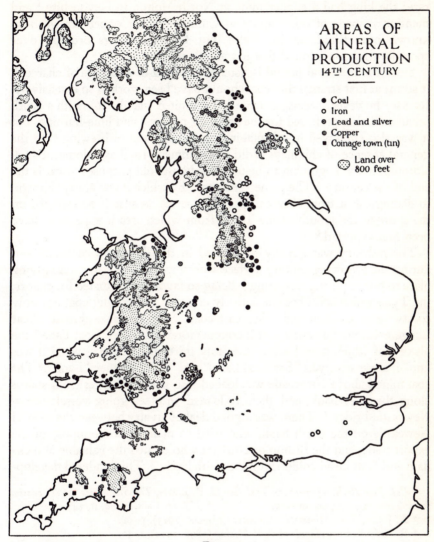

Fig. 42

It is interesting to note that Cistercian monasteries were associated
with much of the mining activity in the Pennines.

area was around Beer Alston in Devonshire. Tin streaming was, of course, localised in the south-western peninsula. Copper deposits were apparently rare; the king had a copper mine at North Molton in Devonshire,[1] and mention is made of other mines in Cumberland.[2] Copper ore was also known near Rhuddlan in Wales, but German miners brought over specially to smelt it found that it was valueless.[3]

Since the medieval period is usually regarded of as an age of charcoal, it seems at first strange that coal should have been worked so extensively. Its use, however, seems to have been limited mainly to smith's work (charcoal being preferred for the smelting process) and lime-burning, but it was also employed in salt making, in baking, and in brewing.[4] In the general absence of chimneys it did not commend itself as a domestic fuel, although it was beginning to be used for household purposes towards the end of the century.[5] The name "sea-coal" by which it was always known, to distinguish it from charcoal, probably arose because it outcropped on the shore at the mouth of the Tyne, from which area it appears to have been first exported.[6]

The industry was already established in the thirteenth century,[7] and during the following century surface workings in some parts were giving place to pits requiring timbering.[8] Being so favourably situated for export, the Tyne valley soon became actively engaged in shipping coal not only coastwise to London and other east coast ports but also abroad. Near Ludgate Circus this trade is still commemorated in "Sea Coal Lane" the history of which goes back to the early thirteenth century, when it was known alternatively as "Sea Coal Lane" and "Lime-burners' Lane".[9] The coal mined above Newcastle was loaded into keels or barges from staiths along the river bank, and then re-loaded into sea-going vessels below Newcastle bridge.[10] There was a good deal of rivalry between the men of Newcastle on the north bank, who tried to maintain a monopoly of the export trade, and the Bishop of Durham who owned the valuable Whickham and Gateshead collieries on the south bank.[11] This hindered develop-

[1] *Cal. Fine Rolls*, 1337–47, p. 454. See G. R. Lewis, *The Stannaries* (1908) *passim* for references to copper mining.   [2] *V.C.H. Cumberland*, II, 342.
[3] A. Jones, *Flint Ministers' Accounts, 1301–28* (1913), p. 96.
[4] R. L. Galloway, *Annals of Coal Mining* (1898–1904), I, 30.
[5] In 1389–90, a cargo of 31,000 chimney tiles was imported into London (Customs Accounts, 71/13).
[6] J. U. Nef, *The Rise of the British Coal Industry* (1932), II, Appendix D.
[7] L. F. Salzman, *op. cit.* p. 6.   [8] R. L. Galloway, *op. cit.* pp. 37, 45.
[9] *Ibid.* p. 29.   [10] *Ibid.* p. 38.
[11] *Ibid.* p. 43. The Whickham coal was specially suitable for smith's work (p. 44).

ment on the south side for some time, but by the end of the century the output on both sides of the river was considerable and the annual shipments to foreign countries alone amounted to several thousand tons.[1]

Of the iron-producing areas, the Forest of Dean was still probably the most important,[2] for the Wealden industry was as yet in its infancy although it had begun to supply London.[3] Local requirements in other parts of the country, however, seem to have been largely met by local supplies[4] or by imported Spanish or Swedish iron. The original method of smelting the ore was very simple; the "furnace" was erected on an exposed hill-top where the wind could create a natural draught. In the course of time bellows were introduced, worked at first by hand; but by the beginning of the fifteenth century, and possibly earlier, water power was utilised for the purpose.[5] The location of most of the workings on the edges of upland areas enabled them to take full advantage of this improvement when sufficient water was available, a foot-blast being used at other times.

In the south-western peninsula was the tin-mining industry, and here the first half of the fourteenth century was a period of prosperity, following upon the charter of 1305.[6] Production was concentrated mainly in Cornwall,[7] most, if not all, of the ore being obtained by streaming. This led to encroachments upon the alluvial land devoted to arable farming and resulted in numerous complaints of arable land being ruined and streams diverted.[8] Coal had already begun to take the place of charcoal as fuel, and was imported in large quantities,[9] possibly from south Wales, although it may also have been shipped coastwise from Newcastle.

[1] E.g. 5356 chaldrons (about 5000 tons) in 1380–1 (Customs Accounts, 106/4).
[2] L. F. Salzman, op. cit. p. 24.　　　　　[3] Ibid. p. 25.
[4] Sheffield was already famed for its cutlery, vide Chaucer "A Sheffield thwitel baar he (the miller) in his hose" (Reeve's Tale).
[5] L. F. Salzman, op. cit. p. 28.
[6] G. R. Lewis, op. cit. p. 39.
[7] The total output for Cornwall and Devon for the years in which details exist for both counties was:

Year	Cornwall	Devon
1301	560	53
1303	766	90
1355	497	0
1379	832	72
1400	1465	129

Numbers refer to the nearest thousandweight (1200 lb.). (G. R. Lewis, op. cit. Appendix J.)
[8] G. R. Lewis, op. cit. p. 94.
[9] E.g. Exchequer K.R. Accounts, 261/9.

TRANSPORT AND COMMUNICATIONS

It is customary to think of medieval transport in terms of pack-horses floundering in muddy lanes, and to emphasise the difficulties of communica-

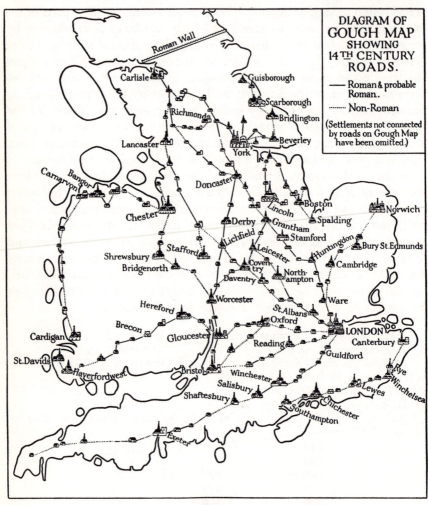

DIAGRAM OF
GOUGH MAP
SHOWING
14 TH CENTURY
ROADS.

— Roman & probable Roman.
...... Non-Roman

(Settlements not connected by roads on Gough Map have been omitted.)

Fig. 43

tion between various parts of the country except where water transport was available. But although water transport was the cheapest form of carriage, the complaints in contemporary records of dangers caused by weirs, kiddles and other obstructions in rivers[1] are so frequent that road transport must often have been preferred in spite of its greater expense. Nor was road transport confined to the pack-horse, which was limited in its usefulness by its inability to carry heavy and bulky produce; indeed, the outstanding feature of the period was the extensive use of carts and wagons for both local and long-distance work.[2] This suggests that road surfaces on the whole must have been much better than is generally supposed, although in bad weather conditions were often difficult.

Of the medieval road system, if indeed it can be so designated, but little can be known. The chief source of information is the Gough Map, which appears to date from the first half of the fourteenth century.[3] This shows a network of roads covering the whole of England and Wales, and upon analysis (Fig. 43) these are seen to be by no means wholly Roman. Although the Gough Map is not complete, it shows very clearly that London was the principal route focus. Among the provincial towns Coventry stands out as the most important post-Roman focus. The Great North Road, which is shown on one of the well-known Matthew Paris maps of Britain,[4] appears as a combination of Roman and medieval roads with a number of important junctions; an interesting confirmation of its accuracy is afforded by an account of the carriage of David Bruce's ransom to London towards the end of Edward III's reign.[5] The main line of penetration into Wales seems to have been the Roman roads that followed the Wye, Usk and Towy valleys, but the Edwardian towns on the north coast were also linked up, the road continuing southwards along the shore of Cardigan Bay. There is reason to believe that the Gough Map, as we know it to-day, is an incomplete copy of an earlier map, which may

[1] L. F. Salzman, *English Trade in the Middle Ages* (1931), chap. XI. The following disadvantages associated with water transport should also be considered:

(1) Full advantage of the cheaper rates of carriage could be gained only with large cargoes.

(2) In order to make up a cargo it was often necessary to bring goods from several places to a common centre. This involved break of bulk and storage which added to expenses and caused delay.

(3) Journeys by boat often took a considerable time owing to the circuitous courses of rivers.

[2] J. F. Willard, "Inland Transportation in England during the Fourteenth Century", *Speculum*, I, 368 (1926); J. F. Willard, "The Use of Carts in the Fourteenth Century", *History*, XVII, 248 (1932); L. F. Salzman, *English Trade*, p. 204.

[3] R. A. Pelham, "The Gough Map", *Geog. Journ.* LXXXI, 34 (1933).

[4] J. B. Mitchell, "The Matthew Paris Maps", *Geog. Journ.* LXXXI, 27 (1933).

[5] Exchequer K.R. Accounts, 598/11.

18

account for the absence of the London-Dover section of Watling Street and the road linking Southampton with Winchester.[1]

It is clear that the growth of towns and the development of trade during the early Middle Ages had led to the multiplication of highways throughout the country, and when the information on this map is supplemented by that derived from the numerous accounts dealing with the movements of produce (e.g. Figs. 44–7) it becomes apparent that scarcely any part of the country was sufficiently remote to be considered really isolated.

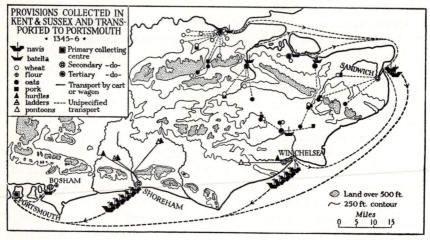

Fig. 44

Note the comparative absence of river transport. The term "primary collecting centre" in Figs. 44, 45, 46 and 47 refers to the final port of assembly in each case.

Some idea of the problems connected with the transportation of different kinds of produce may be gathered from Figs. 44,[2] 45,[3] 46[4] and 47[5]

[1] There is no evidence that the Pilgrim's Way, whatever may have been its function at an earlier period, was a commercial thoroughfare in the fourteenth century. It had serious disadvantages for wheeled traffic, and its use was no doubt largely limited to the pack horse. It may have been a route followed by pilgrims, though Chaucer's characters preferred the Watling Street.

[2] Compiled from Exchequer K.R. Accounts, 566/20; 588/17, 18, 22. The goods taken to Greenhithe on the Thames and to Sandwich were almost certainly conveyed in carts (although wheeled transport is not specified in the accounts), since payment was made for unloading carts at these two centres. The ladders carried to Shoreham were to be used in the siege of Calais.

[3] Compiled from Exchequer K.R. Accounts, 555/17, 18, 22.

[4] Compiled from Exchequer K.R. Accounts, 457/8, 14; 577/12.

[5] Compiled from Exchequer K.R. Accounts, 597/31.

which illustrate typical conditions in their respective areas. The contrasting conditions in south-east England and Devonshire are well shown by the fact that hurdles were conveyed on horse-back in the latter area, whereas

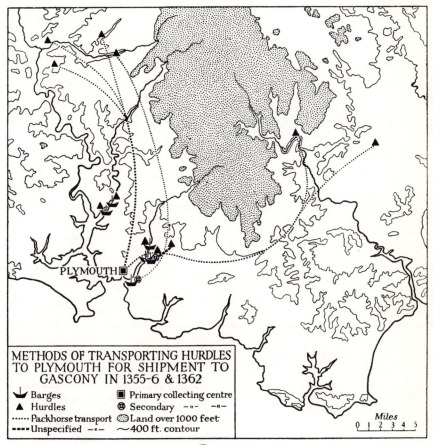

METHODS OF TRANSPORTING HURDLES
TO PLYMOUTH FOR SHIPMENT TO
GASCONY IN 1355-6 & 1362

Barges   ■ Primary collecting centre
▲ Hurdles   ⊕ Secondary – " –   – " –
······ Packhorse transport   Land over 1000 feet
▪▪▪▪ Unspecified –·– ⌇400 ft. contour

Miles
0 1 2 3 4 5

Fig. 45

The routes shown are conjectural, only the distances being given in the accounts. The avoidance of the Tamar is striking.

carts were used in the former, in spite of the fact that the routes traversed the Weald Clay vale. The general character of the movements in south-east England seems to have been centrifugal, the final stages being completed by boat, and this seems to have been true of all coastal regions. For inland areas there were usually three stages: first of all, carts or horses which conveyed the produce from where it was bought to a river port; here it

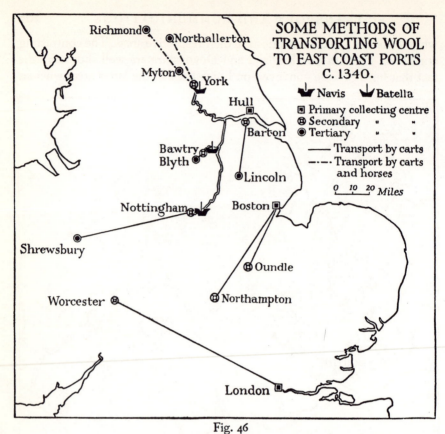

Fig. 46

The wool taken to Barton completed its journey across
the estuary to Hull in small boats.

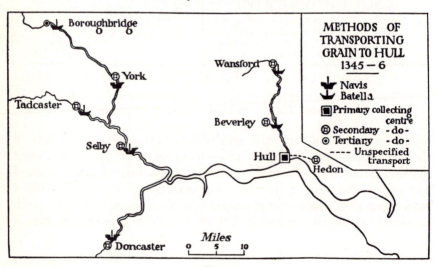

Fig. 47

was transferred to a small boat and carried downstream to a seaport, where it was loaded into a larger vessel for shipment overseas. But sometimes, as Fig. 46 shows, this scheme was short-circuited, and carts conveyed the goods direct to a seaport, although in one particular case it involved a journey of well over a hundred miles. Before the wool was taken from Shrewsbury and Worcester to the coast, cartloads of canvas in which to wrap it were brought from London, so it is evident that the difficulties of road transport at this period must not be exaggerated.

Rivers were used to a considerable extent in some parts of the country, especially in Yorkshire, for the conveyance of goods, in spite of certain disadvantages attached to this method of transport. Figs. 46 and 47 illustrate some of the ways in which the various streams converging upon the Humber were utilised. The Ouse and the Trent were able to accommodate sea-going vessels (*naves*), whilst smaller boats plied on these rivers and on tributary streams. Similar maps could be constructed to show the carriage of lead and timber, but they would merely serve to emphasise the important point which is already clear from a glance at Figs. 44 to 47, *viz.* that the full implications of these maps can only be considered in relation to the external trading activities of the country. But that is the theme of the next two chapters.

'BIBLIOGRAPHICAL NOTE

An historical background for a geographical survey of this period is to be found in those books by E. Lipson, J. E. Thorold Rogers and W. Cunningham referred to in the footnotes. Special mention must also be made of L. F. Salzman's *Medieval English Industries* (1923). The sections on Industries and on Social and Economic History in the Victoria County Histories are also valuable.

Chapter VII

MEDIEVAL FOREIGN TRADE: WESTERN PORTS

D. T. WILLIAMS, M.A., D.Sc.

In a study of the maritime activities of southern Britain during the Middle Ages it is imperative, first of all, to consider the changing types of vessel sailing the "narrow seas", the nature and perils of their navigation, and the general conditions affecting foreign trade during those centuries.

The earlier vessels were very small, averaging about 30 "tuns", but even in the thirteenth century ships of 100 tuns, or more, were occasionally mentioned.[1] The small boats of this century possessed neither bowsprit nor fore-and-aft sail, and had but imperfect steering appliances. Stern rudders were mentioned in 1328 but were not generally installed until much later. No pumps are shown in engravings of ships, or described as in use, before 1300. Thus, it is difficult to suppose that such vessels could have kept the wind within at least seven or eight points, and consequently they could only make progress when sailing large or before the wind. The compass was not yet a practical instrument, and masters of ships had to set their courses from headland to headland. Sailing within sight of land as far as possible was also made necessary by the frailty of their craft and by the frequent compulsion to seek shelter in a friendly port or roadstead to ride out a storm.[2]

The carrying capacity of most of the English vessels from the smaller ports remained well below 100 tuns throughout the medieval period. Even at the larger ports, such as London, Southampton and Bristol, it was not

---

[1] Sir N. H. Nicholas, *A History of the Royal Navy* (1847), I, 357–73; *Patent Rolls*, 1216–25, p. 540.

A "tun" was a cask of definite capacity, a measure of capacity for wine and other liquids, usually equivalent to 2 pipes or 4 hogsheads, containing 252 old wine-gallons. It came to be a unit used in measuring the carrying capacity or burden of a ship; originally, the space occupied by a tun cask of wine; now, for the purposes of registered tonnage, the space of 100 cubic feet. See *New English Dictionary* (Oxford, 1926), vol. x, Part I, pp. 124 and 464 under 'ton' and 'tun'.

[2] Off Ushant, a vessel bound for Southampton could choose one of two routes, either due north to the coast of Cornwall and thence along the English coast, or along the north coast of Brittany to Guernsey and thence due north to the Isle of Wight. In the first case more than a hundred miles of open sea had to be crossed; in the other case, less than sixty. Pilots could be taken on from the Ile de Batz and from Guernsey.

until the fifteenth century that larger cargo vessels appeared. The average cargo of the wine ships of Bristol increased from 88 tuns at the beginning of the fifteenth century to over 150 tuns in the middle of the century, and to some 200 tuns by 1500.[1] A marked improvement is also noticeable in the type of ship, showing, probably, Mediterranean influences in design. Instead of the earlier single-masted ship with its square sail, there appeared the two-masted or three-masted vessel with the three-cornered lateen sail rigged to the mizzen mast and the square sail fixed to the main mast. Such a rig enabled ships to manoeuvre close to the wind, and their pointed bows gave added strength and stability in a rough sea; the high castles at stem and stern gave increased shelter and passenger space, while the fighting top at the main mast acted also as a look-out, well needed in days of constant piracy. It was from the Mediterranean, too, that north-western Europe derived its first practical mariner's compass[2] which, however, did not come into general use in northern shipping until the fifteenth century. It was probably preceded for some two centuries by the occasional use of the lode-stone for fixing the position of magnetic north, but such a cumbersome and primitive instrument could scarcely be employed in holding a ship's course. Inevitably, it seems that the masters of ships in northern European waters must have become acquainted with the shipping routes by long service before the mast or else were dependent to a very large extent upon skilled local pilots.

The physical dangers to be surmounted were many. Ocean currents, tidal races, storms, fogs, unlit rocky headlands and islands—all lured the unwary shipmaster to destruction. It is no wonder that the Counts of Leon in north-west Brittany preferred to sacrifice to the Duke of Brittany the ports of Brest (1240) and Morlais (1275) rather than lose control over a rocky islet "fertile en naufrages", which they surnamed "their rock of gold".[3] Along most coasts, the rights of wreck were a rich perquisite eagerly sought and assiduously retained.

Piracy presented another danger. It seemed, indeed, to be a profitable occupation; Gascons, English, Bretons, Normans and Flemings were not at all particular in pillaging the ships of their own countrymen as well as those of their enemies.[4] Along the main trade routes of the period there

---

[1] *Studies in English Trade in the 15th Century* (ed. Eileen Power and M. M. Postan, 1933), pp. 239–40.

[2] Sir N. H. Nicholas, *op. cit.* I, 357–73; *Encycl. Brit.*, 14th ed., 1929, VI, 176–7.

[3] Charles B. de la Roncière, *Histoire de la marine française* (Paris, 1899), I, 463.

[4] E.g. *Rot. Litt. Claus.* I, 146, col. 2; 305, col. 1; 352, cols. 1 and 2; 653, col. 1, etc.; II, 48, col. 2, etc.; *Cal. Close Rolls*, 1234–7, pp. 194, 513; *Cal. Patent Rolls*, 1258–65, p. 651; 1292–1301, pp. 29–31, 34, 69, 376, 473.

were frequent nests of pirates. The political troubles between England and France and the many wars of these centuries intensified these dangers. It was mainly to counteract the ever-increasing dangers from pirates, and partly to prevent enemy attacks during the frequent wars with France, that the system of convoys was instituted. As early as 1226, Henry III had ordered the masters of all ships from England, Ireland, Wales and Gascony, bound for Gascony to sail in convoy, escorted by a larger ship under the command of an admiral, who was given complete charge of the convoy until its safe arrival in port.[1] This convoy system became very popular during the fourteenth century and was adopted by the fleets of England, Flanders and France.[2] Individual merchants were compelled to join the organised convoys whether they desired to do so or not. Friendly alliance with the large ships of Portugal, Castile, Bayonne, Genoa and Pisa was a valuable asset under such conditions and, therefore, political alliances with these countries were frequently arranged by French and English sovereigns.

Under such circumstances, groups of ports soon recognised the benefits to be derived from amalgamation and co-ordination of forces. At the commencement of the thirteenth century an association of Bayonne merchants completely controlled the commerce of the Bay of Biscay;[3] whilst, at the close of the century, the merchants of the Castilian ports of northern Spain entered into articles of association for their mutual benefit.[4] The Cinque Ports of south-east England, and the Hanseatic League, were somewhat similar unions. With the growth of national commerce and the organisation of centralised administration they disappeared.

Symptomatic of this co-operation was a uniform code of maritime laws that came into operation for the first time in north-west Europe, the famous

---

[1] *Patent Rolls*, 1225–32, p. 14.

[2] *Rot. Litt. Claus.* I, 69–70; *Nicholai Triveti Annales, Eng. Hist. Soc.* (ed. T. Hog), pp. 325–6; *Annales Monastici* (ed. H. T. Luard, 5 vols., 1864, Rolls Series), III, "Annals of Dunstable", p. 398; Bibl. Nationale, Paris, Pièces Originales, vol. 265, no. 12; Ancient Petitions, P.R.O., London, no. 5880.

The Norman wine fleet, for example, consisting of 100 large vessels and 100 smaller craft, was escorted by 22 Genoese galleys en route to La Rochelle in 1339. (Ancient Petitions, P.R.O., London, no. 5880. Cf. *Chronicon Henrici Knighton vel Cnitthon* (ed. Rev. J. R. Lumby, 2 vols., 1889, 1895), I, 336.) Indeed, the great sea battles of the thirteenth and fourteenth centuries were usually due to the accidental meeting of two opposing convoys.

[3] J. Balasque et E. Dulaurens, *Études historiques sur la ville de Bayonne* (1862, 3 vols), I, 439–49.

[4] C. Fernandez Duro, *La Marina de Castilla* (Madrid, 1894), pp. 391–6.

"Laws of Oleron".[1] These served to regulate and systematise traffic. They defined agreements between master and merchant, and between master and crew; they enacted the rules regarding jettisoning of cargo during storm; they specified the incidence of collision at sea and in harbour; and stated clearly the duties and liabilities of pilots in navigating their ships. Copies of these maritime laws, dating probably from the twelfth century, have been discovered in Castile, Brittany, London, Southampton, Bristol and Wisby.[2] As the "Sea Laws of Flanders" they soon reached the chief harbours of the Baltic; and, in the fifteenth century, they were incorporated bodily in the famous Saxon or Low German compilation of Wisby, known variously as the Gotland Sea Laws and the Supreme Sea Law of Gotland, which regulated the Baltic trade of the Hanseatic cities.

THE ENGLISH CHANNEL

The ports of Britain, from Southampton westwards, were for geographical reasons in closer contact with the trade of the western parts of Europe (and, later, of the Mediterranean area) than with the borderlands of the North and Baltic Seas. It must be borne in mind that in A.D. 1200, more than one-half of the land of France, including Normandy, Brittany, Maine, Anjou, Touraine, Poitou and Aquitaine, formed a part of the Angevin Empire, ruled over by the kings of England. But in 1204 England lost Normandy,[3] and political conflicts of the next 250 years reacted powerfully upon the commercial relations between England and the Continent.

Rouen and La Rochelle became the royal ports of France on the English Channel and the Bay of Biscay, respectively, set between the ports of Southampton and Bordeaux within the Angevin Empire. It is in its relation to the wine trade with Gascony, particularly with the port of Bordeaux, that we can trace the rise of England's medieval commerce. Bordeaux was the outlet of the fertile vine-growing province of Aquitaine, a factor which naturally favoured the growth of trade between its sunny vineyards and the cloudy meadow lands and cereal fields of England.

The nodal character of the radial streams of the Dordogne-Garonne

---

[1] Sir Travers Twiss, *The Black Book of the Admiralty* (4 vols., 1871, etc.), III, 4–33; P. Studer (1), *The Oak Book of Southampton* (2 vols., Southampton Record Soc., 1910–11), II, 76–9; see Introduction, pp. xxix–lxxi, for a critical dissertation upon the various publications of the Laws of Oleron.

[2] P. Studer (1), II, 51–5, xxxvii–xxxix, xliv–xlv.

[3] See F. M. Powicke, *The Loss of Normandy, 1189–1204* (1913), pp. 11–13, for the geographical affinities underlying the Angevin Empire. See Fig. 48.

system contributed to the pre-eminence of Bordeaux in the economic life of the entire region. And the town itself was a tidal port, sufficiently far upstream to prevent sudden piratical attacks, and set on firm ground where the width of the estuary first permitted the construction of a wooden

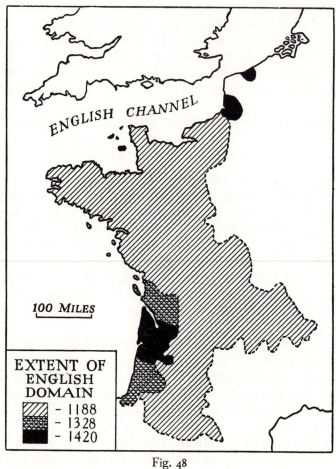

Fig. 48

Territorial changes which show the importance of the Gascony
wine area to England throughout this period.

or stone bridge. North-east, a dry routeway led towards Libourne, Angou-lême and Poitiers, avoiding the coastal swamps and marshes; and, likewise, to the south the pilgrim way extended towards Bayonne, Navarre and Compostella, keeping east of the barren sand dunes and marshy swamps of the Landes that border the rectilineal coast between the Gironde and

Adour. Camille Jullian, describing the condition of the city in the thirteenth and fourteenth centuries, goes so far as to say that "Quelle que soit sa condition, un Bordelais récolte, échange, ou vend du vin".[1] All were interested, in one way or another, in the traffic in wine. During these centuries, the cultivation of the vine in Gascony probably reached its maximum; and opinion was freely expressed that, in the midst of luxury, the city of Bordeaux ran the risk of famine, starvation and death through neglect of corn cultivation and the necessity for importing grain. This monoculture was a dangerous economic policy for Gascony. Any severance of the tie with England would lead to inevitable decay, and this fact was a most powerful incentive in maintaining the allegiance of the cities of the south-west of France to their overlords, the kings of England.[2]

During the thirteenth century the carrying trade between Bordeaux and English ports was practically a monopoly of the mariners of Bayonne. The Bordeaux custom rolls of 1308[3] illustrate the changing mercantile conditions; Bayonne vessels were still numerically supreme, but ships from London, Southampton, Bristol, Winchelsea, Yarmouth, Kingston-on-Hull, Sidmouth, Exmouth, Teignmouth, and from other English ports, participated in the Biscayan wine trade during this year. This tendency became more and more evident during the fourteenth century and, for 1372, Froissart[4] describes the arrival of 200 ships, owned by merchants from England, Wales and Scotland, to load with wines at the harbour of Bordeaux. This foreign trade was a valuable asset to the English kings. They encouraged, particularly, the wine trade to their Aquitaine province, for, in addition to the normal duties levied in peace time, the seizure of foreign ships with their cargoes was a simple means of obtaining supplies in case of war or to fill an empty treasury.

The maritime routeways from England and Flanders followed the coast of Brittany. Politically, the alliances of the Counts of Brittany had far-reaching consequences; commercially, the Breton mariners had ample

[1] C. Jullian, *Histoire de Bordeaux* (Bordeaux, 1895), p. 216.

[2] See F. B. Marsh, *English Rule in Gascony* (1912), pp. 152–3: "The extent of English rule on the Continent may roughly be defined as the radius within which the Bordeaux-Bayonne pressure was strongly felt. Outside that radius England never gained a durable hold. Within that radius her grasp was strong even in weakness.... A system of privileges which in its results closely approximates to the preferential tariff of today united the scattered realm of Henry." See also E. C. Lodge, *Gascony under English Rule* (1926).

[3] Exchequer Accounts, P.R.O., London, E. 162, no. 1, etc,

[4] *Chroniques de Froissart* (ed. S. Luce et G. Raynaud, Société de l'Histoire de France, Paris, 1869), VIII, 96. Cf. *Cal. Pat. Rolls*, 1385–9, p. 85.

opportunities to practice piracy. In a somewhat similar way the Channel Islands derived their medieval importance. They remained throughout these centuries a part of the Angevin Empire. Possessing in St Peter Port, Guernsey, a splendid roadstead and a sheltered anchorage, they became a possible asylum for English and Bayonne vessels *en route* to and from England.[1]

The activities of Southampton throughout the period were primarily concerned with maritime trade; it never became an industrial town. Its Gild Merchant was a powerful, closed corporation of burgess traders, and its municipal government was directed entirely by the Gild Merchant at the close of the thirteenth century.[2] The burgesses under their charter from King John (1199) were given "freedom from toll, passage and pontage money, both by land and water, both in fairs and markets throughout our dominions in England and on the continent";[3] and, in 1227, by the charter of Henry III, they received the customs of Southampton to farm for ever, as well as those of the port of Portsmouth, on the payment of £200 sterling, annually.[4] Southampton was the main outlet of the Hampshire basin during medieval times. Its immediate hinterland comprised the cities of Salisbury and Winchester, commanding not only the approaches into the Hampshire Basin from north-west and north-east respectively, but also the wool areas encircling it. The river Itchen had been made navigable as far as Winchester which lay on the direct route from Southampton to London. Through Salisbury the route from Southampton reached the woollen areas of the Cotswold Hills, and the Marches of Wales.[5] From the west, the products of Devon and Cornwall came in part by land to Southampton. Both Winchester and Salisbury became, as we shall see, important manufacturing centres[6] in marked contrast with their port, Southampton. They held important fairs,

[1] See D. T. Williams, "The Importance of the Channel Islands in British Relations with the Continent during the 13th and 14th Centuries", *Bulletin Soc. Jersiaise* (1928), pp. 1–89.

[2] P. Studer (1), I, Introduction; II, vi.

[3] H. W. Gidden, *The Charters of Southampton* (Southampton Record Soc., 1909–10), pp. 2–3.

[4] *Ibid.* pp. 4–5. Early in the following century the citizens of Salisbury and of Winchester were granted rights of trading at Southampton provided they paid the customary local dues. P. Studer (1), II, 18 and 53.

[5] For a description of the routes through Salisbury see pp. 339–40.

[6] See pp. 275–6.

St Giles' Fair at Winchester being especially famous. Here, too, there was a royal residence and a mint. It is in relation to these two hinterlands, both immediate and distant, that we must consider the character of the import and export trade of Southampton.

Protected by the Isle of Wight, the port was sheltered from sudden piratical attacks and even from foreign invasion. The double tides facilitated the entry and departure of the early oared and square-rigged ships. These could reach the quays and river wharfs of Southampton, but the larger carracks and galleys of the Genoese, Florentines and Venetians, in the fourteenth and fifteenth centuries, had, at times, to drop anchor in the roads and to discharge their cargoes into smaller boats, a splendid opportunity for smuggling[1]. Altogether, the port provided anchorage for the largest mercantile fleets of the Middle Ages while the roadsteads off St Helens and Yarmouth, on the sheltered leeward side of the Isle of Wight, became (together with Portsmouth) the recognised meeting-place of the English fleets in their frequent expeditions to northern France and Gascony.[2] With these advantages, Southampton became a main terminus for western trade. It derived much profit from the victualling of ships, fleets and armies; and it developed an important shipbuilding industry.[3] An indication of the extent of Southampton's trade early in the fifteenth century may be seen from the fact that Henry V obtained an advance of £14,000 from the Bishop of Winchester in 1417, and another advance of £14,000 in 1421, each on the security of the customs of Southampton and its dependent ports.[4]

Fortunately, Southampton and Portsmouth, together with their subsidiary ports on the Isle of Wight, were seldom attacked during the many wars of these centuries. Portsmouth was burnt by the French fleet of Behuchet during March 1338,[5] and a landing was made from the Genoese

[1] P. Studer (2), *The Port Books of Southampton* (Southampton Record Soc., 1913), p. xxviii.

[2] E.g. in 1346 and 1415. Cf. *Cal. Pat. Rolls*, 1225–32, pp. 259, 264; 1232–47, p. 44; 1247–58, p. 363, etc. Likewise, the convoys of wine ships to Gascony foregathered in the late summer off Portsmouth and the Isle of Wight during times of war and piracy. *Cal. Pat. Rolls*, 1338–40, p. 2; 1350–4, pp. 376, 543.

[3] *V.C.H. Hampshire*, v, 367. The royal ships of Henry V were built at the yards of Deptford and Southampton. It was at the latter port that the *Holigost*, the *Trinity Hall* (540 tons), the *Grace Dieu* (400 tons) and the *Gabriel* were constructed during the years 1414–18.

[4] *Ibid.* p. 369.

[5] *Ibid.* pp. 363, 366; *Adae Murimuth Continuatio Chronicarum* (ed. E. M. Thompson, 1889), pp. 87–90; B. B. Woodward, *A General History of Hampshire* (1863), ii, 174.

and Catalan ships in alliance with the French at Southampton in October of the same year, when the town was pillaged and set on fire. In 1339 and in 1340 the French fleets reappeared off Southampton, and off Yarmouth and St Helens Roads in the Isle of Wight. Portsmouth and the Isle of Wight were again attacked by the French in 1372. But, in general, the Solent ports were singularly free from those set-backs to commercial development that resulted from frequent enemy attacks.

The medieval trade of Southampton depended, primarily, upon two products, the import of wine and the export of wool (and, later, of cloth). The Patent and Close Rolls give some indication of the fact that it was England's principal wine port in the reign of King John.[1] Wines were brought in the ships of Bayonne and Bordeaux whose merchants remained in the town to sell their cargoes. Export through Rouen practically ceased after the loss of Normandy, and all imported wines at Southampton came from Anjou, Poitou and Gascony. The trade grew during the thirteenth century.[2] Return cargoes for these vessels were provided by the wool of the hinterland. Gascony became one of Southampton's chief markets for wool, wool-fells, and hides.

During the fourteenth century, the traffic at the port reflects the changing conditions in south-central England. We know that the chief shippers of wool from Southampton in the early part of this century were the Italians—Venetians, Genoese and Florentines.[3] It is for the year 1323 that the first record exists of the arrival of Venetian galleys at Southampton, but the Genoese carracks had frequented the port long before this date. Even when the staple for wool had been fixed at Calais in 1363, permission was granted by Richard II, in 1378, to all merchants of Genoa, Venice, Catalonia, Aragon and other countries "to the westward" to bring their ships to Southampton, and there to sell their merchandise and reload with wool, hides, wool-fells, lead, tin, and other staple com-

[1] *Rot. Litt. Claus.* I, 17, col. 2; p. 28, col. 2; p. 40, col. 2; p. 41, col. 1; p. 42, col. 1, etc.
[2] *V.C.H. Hampshire*, v, 361; N. S. B. Gras, *The Early English Customs System* (1918), p. 115; B. B. Woodward, *op. cit.* II, 210. In 1272, 3147 tuns of wine were imported at Southampton in comparison with 3799 tuns imported at the port of London. Of the imports of wine to Southampton in 1327–8 none was carried in Southampton vessels, whereas foreigners brought nearly 1500 tuns to the port. See Fig. 62.
[3] E. Power, "The English Wool Trade in the Reign of Edward IV", *Camb. Hist. Journ.* II, no. 1, p. 27. Cf. F. Miller, "The Middleburgh Staple, 1383–8", *Camb. Hist. Journ.* II, no. 1, pp. 63–5.

modities, on payment of the same subsidies and duties as were levied at Calais.[1] From 1297 onwards, whenever the fiscal policy of English kings determined staple ports and towns for the sale of the chief products of England, either Southampton or Winchester had been included in the list of wool staples.[2]

Southampton was one of the three nodal towns of the wool triangle in south-west England; Exeter and Bristol being the other two.[3] In spite of the growth of a local cloth industry during the second half of the fourteenth century, the fine wools, produced in the Cotswolds, continued to be bought by the Italian merchants, and were exported through Southampton.[4] Unlike the woollen exports from the other English ports in the fourteenth century, those from Southampton were never shipped in English vessels but remained in the hands of the privileged Italians.

The second half of the fourteenth century witnessed the development of the native cloth industry. This arose first of all in a few large cities and later spread into the surrounding rural villages and hamlets.[5] The cathedral cities of Winchester and Salisbury were two of the foremost industrial communities in England by 1350. Their output of broadcloths was sold partly in England itself, and partly exported through their port of Southampton.[6] At the close of the century the character of its export trade

[1] "Provided they sailed westwards, and not to a port east of Calais", see *Statutes of the Realm*, II, 8; J. S. Davies, *History of Southampton* (1883), pp. 250–1.

[2] *Cal. Pat. Rolls*, 1324–7, p. 269; 1327–30, p. 98; *Statutes of the Realm*, I, 332, 348. C. Gross, *Gild Merchant* (2 vols., 1890), I, 140–3.

[3] During the years 1333–6 the export of wool from Southampton averaged 3256 sacks (10·1 per cent. of the total English export of 32,307 sacks). At this time nearly 80 per cent. of the total English exports of wool passed through the staple ports of Hull (5326 sacks; 16·5 per cent.), Boston (7007 sacks; 21·7 per cent.), and London (13,334 sacks; 41·3 per cent.). Seldom, and then only in small quantities, was wool exported from Southampton *via* the staple at Calais. H. L. Gray, "The Production and Exportation of English Woollens in the 14th Century", *Eng. Hist. Rev.* XXXIX, 16 (1924). See Fig. 49.

[4] These Italian merchants shipped the following quantities of wool direct from Southampton when all other wools were sent either to Calais or to Middleburgh: 1380–1, 2416 sacks; 1384–5, 2818 sacks; 1388–9, 2495 sacks. These figures illustrate the decreasing quantities of wool exported during this century; the share of the Italian merchants decreased to between one-seventh and one-tenth of the total wool export of England. F. Miller, *op. cit.* p. 65. See Fig. 55.

[5] H. L. Gray, *op. cit.* pp. 30–2.

[6] During this time (1356–8), the trio of ports—Southampton, Exeter and Bristol—shipped 64 per cent. of all woollen cloths exported from England; H. L. Gray, *op. cit.* p. 33. See Fig. 58.

had altered considerably. The destination of the cloths exported from Southampton was very different from that of the wool. Gascony, Brittany and northern France were the chief areas purchasing the broadcloths exported from the ports of south-western England; during 1356–7, from Southampton, 3413 broadcloths were exported by English merchants, and only 349 by aliens (cf. Bristol, 2628 cloths by denizen merchants, none by aliens).[1] Towards the close of this century (1392–5), an interesting change in the exporters of cloth from Southampton becomes noticeable.[2] Not only had its exports of cloth increased by 266 per cent. in forty years (Salisbury was still the chief cloth manufacturing city in the whole of the country), but also alien merchants had become active participators in the trade. Hanseatic traders were never prominent at the port, but the merchants of France, Portugal, Spain and the Mediterranean countries frequented the port in ever increasing numbers during the fourteenth century.

The Port Books[3] show that Southampton had become an important entrepôt port by 1430. It was the place of arrival in England of the Venetian and Florentine galleys, of the Genoese carracks, and of Catalan vessels. These brought the produce of the Levant and of the Mediterranean.[4] The tonnage of these Mediterranean vessels must have been considerable, for each paid 3s. 4d. anchorage fee, a marked contrast with the fee of 4d. paid by the largest of the Hansa boats calling here, and the 1d. or 2d. paid by coasting vessels. These ships made the import trade of the port more and more a trade in luxuries. This is shown by the list of Mediterranean imports. The Venetians brought costly manufactured goods, and wearing apparel of silk, velvet, goldcloth and damask, gowns furred with sable or pole-cat, fruits such as dates and figs and raisins, spices, vegetable oils, woad, etc. The other Mediterranean traders, lacking the background of skilled Venetian manufacture, brought less costly articles and more raw materials (like woad, madder, scarlet dye and alum) useful in English textile industries. Sweet wines were also imported from Spain and the Mediterranean. Their ships loaded return cargoes of wool-fells, and hides

[1] H. L. Gray, *op. cit.* Appendix I.

[2] *Ibid.* Appendix III. 4399 cloths of assize were shipped by denizens, 12 by Hansards, and 5596 by aliens annually.

[3] P. Studer (2); see footnote 1, p. 273.

[4] *Ibid.* p. xxviii. Usually, some two to five Venetian galleys arrived at the port annually. Nineteen Genoese carracks arrived in 1428 and twenty-three in 1430, whilst one Savona carrack came in each year, three Florentine galleys in 1430 and three Catalan ships in 1428.

(especially to Venice), English cloth, pewter vessels, tin and lead. These merchants remained several months in the port and rented cellars and houses in the town. The oarsmen of the galleys had their own quarters and their own burial ground at North Stoneham. The re-export trade of Southampton was largely concerned with the distribution of Mediterranean products to other English ports and also to Flanders and northern Europe.

Other distant traders at the port in the fifteenth century included the Portuguese and Flemish. Although two Portuguese vessels in 1428 (and four in 1430) called at Southampton, their main port of call in England was Bristol.[1] The Portuguese merchants brought fruit, wine and wax, and returned with cloth and pewter. The Flemish trade[2] from Southampton was, on the other hand, rather important at this time, being evenly distributed between the vessels of the home port and those of Sluys and Middleburgh. In addition to re-exports of Mediterranean commodities, they carried cloth and pastoral products to Flanders, and brought back in exchange agricultural products, including onions, garlic, hops, German wines, corn, madder, woad and hemp, together with Flemish manufactures and fish.

In the fifteenth century the Gascony trade consisted for the most part of wine and cloth, the latter taking the place of wool that characterised the exports of the previous two centuries. An analysis of the Port Books and of the Enrolled Customs Accounts[3] illustrates clearly the reactions of the Gascon and Norman trade to political changes. During the first half of the century England finally lost her foothold in Aquitaine. It is, therefore, not surprising to find the amount of the reciprocal trade in wine and cloth progressively decreasing not to revive until Tudor times.[4] The carrying trade between Bordeaux and Southampton had also completely changed in character by this time. In 1428 only one Bayonne vessel unloaded wine at Southampton; no other French port was represented, whilst eight of the larger Southampton vessels were apparently shipping wine. The Mediterranean galleys and carracks that called at intermediate ports *en route* to Southampton may also have imported red wines from Bordeaux in addition to the sweet wines of Spain and the Mediterranean. The reciprocal trade in cloth from the ports of south-west England (Southampton to

[1] V. M. Shillington and A. B. Chapman, *The Commercial Relations of England and Portugal* (1907), pp. 15 *et seq.*, 58–9, 105, etc.

[2] Thirteen Flemish "busses" visited Southampton in 1428 and twenty-four came in 1430. P. Studer (2), p. xxiv.

[3] E. Power and M. M. Postan, *op. cit.* pp. 330–60. Enrolled Customs and Subsidy Accounts (1399–1482). Cf. Introduction to same by H. L. Gray, pp. 321–30.

[4] E. Power and M. M. Postan, *op. cit.* pp. 31–3 and 330–60.

Bristol, inclusive) decreased in volume by more than one-half in the years immediately following the loss of Guienne.[1]

Southampton was the most favourably situated English port for traffic with Rouen and Normandy. Political causes during the medieval centuries rendered such legitimate commerce impossible. The western ports of England had petitioned[2] Richard II in 1394 for permission to export wool direct to the ports of Normandy and Picardy *via* Southampton without calling at the staple of Calais. Their petition was not granted. But abundant evidence exists that the law was frequently broken,[3] and that there was a lucrative smuggling trade between Southampton and Rouen, utilising on occasion the favourably located Channel Islands as a "jumping-off" ground.[4] The extent of this traffic, which avoided cocket customs and infringed legal enactments, cannot be estimated.

Southampton had become the metropolis to which the merchants of southern England came in order to buy the goods brought to the port in foreign and native vessels. Their purchases were mainly distributed overland, but a certain amount of coastwise trans-shipment was involved. This was a monopoly of the craft from English ports outside Southampton. It was only in the inter-trade between Southampton and London that Southampton vessels participated to any extent.[5]

Ships from the ports of Suffolk, Norfolk, Sussex and Kent were well represented at Southampton. They brought herrings and stockfish from Suffolk and Norfolk; wheat, malt and iron from Sussex and Kent; while occasional ships from Kingston-on-Hull arrived with coal. The return cargoes consisted of wines, woad, alum, oil and fruits. There was also frequent interchange of products between Southampton and other Hampshire ports, including those of the Isle of Wight. Many of these smaller ports were under the jurisdiction of the major port.

The Dorset harbours of Poole, Wareham, Bridport, Melcombe Regis, Weymouth, Easton near Portland, and Lyme Regis had intimate maritime contacts with Southampton. Their maritime history illustrates both the changing values of geographical factors and the variable effects of

[1] E. Power and M. M. Postan, *op. cit.* pp. 28–9. It revived again after 1475. See Fig. 59.
[2] *Rot. Parl.* III, 332; J. S. Davies, *op. cit.* p. 254.
[3] *Cal. Pat. Rolls*, 1377–81, p. 79; 1446–52, pp. 442, etc.
[4] *Cal. Pat. Rolls*, 1343–5, p. 180. Cf. *Cal. Pat. Rolls*, 1361–4, p. 291.
[5] Studer (2), p. xix.

historical events. In Dorset the best flax and hemp in England was grown. Limestone was quarried at Lyme, Portland and Purbeck. Purbeck marble and stone were used in church building and for roofing purposes, whilst slates were also obtained from Lyme. Bridport, utilising local supplies of hemp and flax, was a flourishing manufacturing town for ropes, sails and cordage, which it supplied to the ships of Southampton and Plymouth. Purbeck marble was shipped from Corfe to Westminster in the thirteenth century.[1] Portland stone reached Exeter by sea in the fourteenth century but the quarrying of this stone does not seem to have survived the century.[2] Purbeck stone was exported from Poole to the Low Countries and to other parts of the continent in the fifteenth century.[3] Slates from Lyme were also sent by sea to Southampton.

These Dorset ports were well-placed to participate in piracy and smuggling. Their mariners captured and looted merchant ships coasting along Lyme Bay, and, at times, crossed to the Channel Islands and the Bay of St Malo to intercept English and foreign ships using the southern routeway.[4] They also engaged in the smuggling of wool, cloth and other merchandise into Normandy.[5] It was no wonder that the French retaliated and attacked the Dorset ports when the opportunity presented itself. Melcombe Regis, the precursor of Weymouth, was virtually destroyed by the French in 1377. In the years preceding this French attack, the port of Melcombe possessed twenty-four sea-going vessels and forty fishing boats; its customs and subsidies totalled a sum of £1000 annually.[6] During the fifteenth century this trade of Melcombe migrated to Poole.

Local physical conditions were unfavourable to the survival of such ports as Lyme and Bridport. They were open to the westerly winds that swept into the eastern half of Lyme Bay; Bridport river was liable to be blocked by bars of sand and pebble, and attempts to improve the harbour during the years 1385–96 met with little success. The port of Lyme suffered a like fate; its harbour, or cobb, was washed away by the sea during the gales of November, 1377.

Although they had prospered during the favourable circumstances of medieval commerce, serving a rich hinterland, acting as ports of supply and distribution to neighbouring Southampton, and deriving wealth from piracy and smuggling, Poole, Lyme, Bridport and Weymouth were included amongst the decayed towns of the time of Henry VIII. In less than thirty years (1440–68), foreign or alien settlers in Dorset had decreased

[1] *V.C.H. Dorset*, II, 333. [2] *Ibid.* p. 339. [3] *Ibid.* p. 336.
[4] *Ibid.* pp. 182–3; *Cal. Pat. Rolls*, 1334–8, p. 359; cf. pp. 367, 377.
[5] *V.C.H. Dorset*, II, 241. [6] *Ibid.* p. 188.

from several hundreds to five only. The ever-increasing size of ships made survival impossible for open ports like Lyme Regis, or for ports, like Poole, with shallow bars at their harbour entrances. Moreover, the sheltered and deep harbours of Weymouth and Portland displaced them in subsequent centuries.

It was, however, with the ports of the south-west (Devon and Cornwall) that a flourishing coastwise trade from Southampton was maintained at this time.[1] These western ports not only exported the products of their own region, but their vessels also took part in the wine trade from Bordeaux to Southampton. The small ports of Lyme Bay, Otterton and Sidmouth, sent fish—herring, ling, mulwell and hake; and from Seaton there came local boats carrying building stone. The larger ports of Devon— Dartmouth, Kingswear, Plymouth and Exeter—were engaged in general carrying trade and their merchants arrived at Southampton not only with such home products as slates, herrings and canvas, but also with wines, madder and woad, most probably obtained from Gascony and south-west France *via* Bordeaux. The "cogs" of Cornwall brought fish (ling, mulwell, hake, pollock), tin and slates.[2]

Along this coast of Devon and Cornwall lay one of the important trade routes of medieval times, and one of the most productive fishing grounds of north-west Europe. A barren agricultural hinterland, virtually surrounded by an important sea and provided with excellent natural harbours, encouraged the development of a supplementary industry in fishing, piracy and trade. The ships of Cornwall and Devon participated in the carrying trade from Gascony and Brittany to Irish ports (Drogheda and Dublin), to Welsh ports (Beaumaris, Carnarvon, Milford and Tenby), and to Chester, Bristol and Southampton.[3] It is impossible to estimate the

[1] During the years 1428 and 1430 there were seventy-two merchants and captains from Cornwall and Devon engaged in the trade of Southampton. Fifty-one vessels from Devon ports, and twenty-nine from Cornwall, arrived at Southampton during these two years.

[2] The fish came from the ports of Penzance, Looe, Perranporth and St Ives; the tin and slates came from Fowey and Falmouth.

[3] Francisque-Michel, *Histoire du commerce et de la navigation à Bordeaux* (2 vols. Bordeaux, 1867–71), I, 123–5. Michel wrote of the thirteenth and fourteenth centuries: "Après Londres et Hull, on peut, ce nous semble, mentionner Exeter et Dartmouth comme des ports en correspondance fréquente avec la France et en particulier avec Bordeaux." E. A. Lewis (1), "A Contribution to the Commercial History of Medieval Wales", *Y Cymmrodor*, XXIV, 108, 110, 113, etc. (1913). Wines, alum and woad from Bordeaux *via* Toulouse and Narbonne, and salt

volume of the local trade of the separate ports, but indirect evidence suggests its nature and existence. The chief ports of Cornwall were situated on its southern coast, only the inferior fishing ports of St Ives and Padstow were located on the open, northern coastline.

However, the two counties had their special commodities to contribute to general trade. They sent their salted and dried fish to Southampton and to Gascony; their tin, lead, silver and stone went to several destinations. In addition, Devon exported wool and broadcloths.[1] The tin of Cornwall and Devon was a valuable commodity throughout the Middle Ages, but it is impossible to estimate the relative importance of the amount exported by sea and that transported by land to London, Bristol and Southampton, for manufacture or re-export. The centre of tin mining gravitated westwards from Dartmoor to Land's End as the years progressed.[2] The output of the alluvial deposits of Devon during the twelfth century was replaced by the tin ores of Cornwall during the thirteenth century; and the output of the mines of eastern Cornwall around Bodmin and Lostwithiel were, in turn, surpassed by that of Helston, Truro and the western half of Cornwall. The industry was closely regulated, and the coinage was confined to a small number of towns.

Tin was in demand not only in England itself but also in the Mediterranean countries and in Flanders. In the thirteenth and fourteenth centuries ships sailing from Fowey, Plymouth and Dartmouth took the

from the Bay were the most important commodities. Salt was in great demand for curing purposes in England as well as in Normandy, Picardy and Flanders. The coastal marshes of Brittany, Poitou, Saintonge and Gascony, where the salt water was evaporated by the heat of the sun, were the chief sources of supply to the western parts of England and Wales. The fish diet, so important in the economy of Roman Catholic peoples, taxed the ingenuities of the age to provide an adequate supply, and was a direct stimulus not only to the salt trade itself, but also to the development of the fisheries along the shores of England, Wales and Ireland. See *Rot. Litt. Claus.* II, 48, 129, etc.; *Cal. Pat. Rolls*, 1232–47, p. 311; *Cal. Close Rolls*, 1337–9, pp. 455–6.

For some twenty years after the loss of Guienne in 1453 the reciprocal trade with Gascony languished only to revive rapidly after the Treaty of Picquigny in 1475.

[1] During the second half of the fourteenth century east Devon became a cloth manufacturing area, and from its three southern ports, Exeter, Dartmouth and Plymouth, large quantities of broadcloths were exported annually. These cloths were, at first, carried almost entirely by native merchants, but in the year 1481–2 4149 broadcloths were shipped by denizens and 2165 by alien merchants from these three ports. E. Power and M. M. Postan, *op. cit.* pp. 337–9, 352–4. See Fig. 59. Exeter exported through its outports, Topsham and Exmouth, owing to the closure of the Exe by weirs and dams in 1312.

[2] G. R. Lewis, *The Stannaries* (1908), pp. 43–4.

tin to La Rochelle and Bordeaux[1] whence it was sent *via* Toulouse to Narbonne and Marseilles for Mediterranean distribution by the merchants of Genoa, Florence and Venice. In the year 1198 came a decree[2] that "No man or woman, Christian or Jew, is to export tin by land or by sea without the licence of the chief warden of the stanneries". Moreover, "honest and lawful men are to be appointed in the ports of Devon and Cornwall who are to take the oath of all pilots and mariners of ships there anchored that they will not transport tin or allow it to be transported in their ships unless it is weighed and stamped according to the king's regulations and unless they have the licence of the warden of the stanneries". A lucrative trade in the smuggling of tin continued throughout the period. Calais became the legal staple for the export of tin and remained the sole legal staple until 1492.[3] But some tin, however, was exported from other ports by royal licence, and, in addition, large quantities were shipped abroad illicitly. The merchants of the west had petitioned Richard II to add Southampton to the list of staples, but this was refused.[4] Southampton became the port of export of tin by Mediterranean merchants, but we do not know what proportion of the re-exported tin reached the port of Southampton from Cornwall by sea.[5] Mariners from Cornish ports must also have continued their shipments of tin coastwise to Calais, Flanders and Gascony, from Fowey, Penzance, Falmouth, Plymouth, Dartmouth, etc., although the actual amounts from each port are not available.

THE BRISTOL CHANNEL

The physical character of the Bristol Channel and Severn Estuary encouraged the growth of medieval maritime commerce; navigation within sight of land was facilitated, and the tidal rise and fall were valuable assets to all types of shipping. Although there were many small ports in medieval times, trade tended to gravitate towards a few favourably located centres—Bridgwater, Bristol, Gloucester and Chepstow. Each of these ports served its own particular region, but the rapid growth of Bristol soon dominated the distant commerce of the other ports, and the products of their respective hinterlands found their chief market in the town and port of Bristol.[6]

[1] *Cal. Pat. Rolls*, 1216–25, pp. 256, 273; *Rôles Gascons* (ed. Michel and Bemont (Paris, 1885), 3 vols. and Supplement), nos. 495, 2091, 4480.
[2] G. R. Lewis, *op. cit.* p. 237, Appendix A.
[3] *Rot. Parl.* v, 334 *b*.   [4] See above, footnote 2, p. 278.
[5] P. Studer (2), pp. xxi, 126, Table V; E. Power and M. M. Postan, *op. cit.* p. 323.
[6] *V.C.H. Somerset*, II, 245.

The advantages of the site of Bristol as a port were recognised in the twelfth century:[1]

A certain part of the province of Gloucester, being narrowed into the form of a tongue and extended a long way, forms the city, two rivers washing two of its sides and meeting together in a great abundance of waters on the lower side, where the land itself is defective. Moreover, a quick and strong tide, ebbing and flowing abundantly night and day, causes the rivers on both sides of the city to run back upon themselves into a wide and deep sea, and forming a port very fit and very safe for a thousand vessels, it binds the circuit of the city so nearly and so closely that the whole city seems to swim on the water and sit on the banks. But on one side, where it is esteemed more exposed to a siege and more assailable, a castle raised on a considerable mound, fortified with a wall and bulwarks, towers and various machines, prevents the approach of assailants.

The port site at the confluence of the Frome and Avon, some seven miles from the sea, was improved in 1239 by the construction of a new channel for the Frome across St Augustine marsh which gave a better harbour and which overcame to some extent the tidal influence of the Avon.[2] The defence of the growing town was strengthened by a wall, c. 1255,[3] whilst a stone bridge across the Avon linked together north and south Bristol.

Its maritime and commercial development was fostered by the Norman kings of England. A trading community, enjoying the rights of burgage tenure, must have existed at Bristol before A.D. 1188 and, possibly, even a hundred years earlier.[4] Its citizens were granted freedom from toll and passage and all customs throughout England,[5] and the control of trade was vested in the burgesses of the city.[6] Unfortunately, the records of the Gild Merchant of the city have been mislaid, and the valuable details of the trade relations between Bristol, France and other countries, which they probably contained, are thus lost. Towns, which claimed by their charters freedom from tolls, etc., throughout the lands of the kings of England, were recorded in the "Little Red Book of Bristol",[7] the first

[1] *Chronicles of the Reigns of Stephen, Henry II and Richard I* (ed. R. Howlett, 1884–6, Rolls Series, 3 vols.), III, 36–7; E. W. W. Veale, *The Great Red Book of Bristol* (1931), p. 6.

[2] N. D. Harding, *Bristol Charters* (1155–1373) (Bristol Record Society, 1930), pp. 18–19. See Fig. 79 for the site of Bristol.

[3] *Ibid.* p. 31; its cost was defrayed by a grant of the customs of the port from Easter, 1255 to Michaelmas, 1257.

[4] E. W. W. Veale, *op. cit.* p. 7.   [5] N. D. Harding, *op. cit.* pp. 10–11.

[6] E. W. W. Veale, *op. cit.* p. 7.

[7] F. B. Bickley, *The Little Red Book of Bristol* (Bristol and London, 1900, 2 vols.), II, 199, 215, 235, etc.

entry of which is dated c. 1344. In it are the confirmations of the charters of Liverpool, Tenby, Totnes, Cork, Barnstaple, Lampeter, Tintern Abbey, etc. The Little Red Book also contains the Laws of Oleron.

The trade with south Ireland was, without doubt, one of the earliest maritime contacts of the port of Bristol; and it remained throughout the medieval period of great importance. From the Irish ports of Waterford, Cork, Kinsale, Limerick, Galway and Sligo came fish (herring, cod and salmon especially), hides, skins of wild animals, horses, timber, flax and linen. Return cargoes of salt, iron, wine, leather and cloth (fifteenth century) were taken from Bristol. The Irish vessels engaged in the trade outnumbered those from Bristol. It appears, indeed, that the carrying trade to and from Ireland fell into the hands of Irish traders and of merchants from the smaller Bristol Channel ports, e.g. Minehead, Chepstow, Tenby, etc.[1] Although Bristol merchants were granted freedom from toll at the ports of Drogheda and Dublin, the trade of this part of Ireland was mainly with Chester and Liverpool during the thirteenth and fourteenth centuries.

Trade with Gascony and Brittany formed another artery of Bristol commerce in the Middle Ages. The wine trade was important in John's reign,[2] and, by the close of the thirteenth century, Bristol had become one of the chief centres of the Gascony wine trade.[3] In the thirteenth century, the ships of Bayonne were the great transporters of wine to the ports of western England, but the ships of Cornwall and Devon became particularly active later on. The inter-port trade of the Bristol Channel during the thirteenth and fourteenth centuries has interesting geographical aspects, too. Welsh merchants and mariners were always prominent at Bristol.[4] They took part in the Irish trade and brought hides, fish, leather, wool, and later, cloth, from their home ports (Tenby, Carmarthen and Kidwelly). Chepstow, at the mouth of the Wye, was the western outlet of the Forest of Dean coal and iron area: timber, iron and coal were sent thence to Bristol and Gloucester, while quantities of wool were shipped from Tintern Abbey to Bristol.

Gloucester was the outlet for the Severn valley and the Marches of Hereford and Shropshire. Its site controlled the ford and, later, the bridge

[1] E. Power and M. M. Postan, *op. cit.* p. 192.
[2] *Rot. Litt. Claus.* I, p. 24, col. 1; p. 40, col. 2, etc.
[3] F. Michel, *op. cit.* p. 122; *Rôles Gascons*, nos. 1257, 2441, 2589, etc.
[4] At the "Welsh Back" quay Welsh and Irish ships loaded and discharged their cargoes. Such names as Vaughan, Lloyd, ap Rhys and Lewis are frequent in the population of fifteenth-century Bristol. See E. Power and M. M. Postan, *op. cit.* p. 234.

across the Severn nearest the sea; it was on the direct land routeway from England into south Wales and thence to Ireland; it had river navigation up the Severn and its tributaries; and it was one of the strategic fortresses of the Marchland. The charter of Henry II to Gloucester (1163–74)[1] indicated its commercial importance in the twelfth century:

> We command that the men of Gloucester and all those who wish to go by the river Severn shall have their way and passage by the Severn with wood, coals and timber and all their merchandise, freely and quietly, and we forbid anyone from vexing or disturbing them in aught hereupon.

Gloucester had previously been granted freedom from tolls and customs throughout the lands of Henry II,[2] and it was as a market and fair town that the city derived much of its prosperity; its fair lasted for five days commencing on the Eve of St John the Baptist's Day. Iron and coal from the Forest of Dean mines[3] were sold at the fair, and agricultural implements were bought by the surrounding village communities. The wool and hides of the Marches formed another important commodity. Gloucester also manufactured iron nails for shipping, and horseshoes and weapons for the armies of medieval England. Transport by water was much superior to poor road transport in the Severn area, especially in the Forest of Dean, during these centuries. A list of taxes on cartloads and horseloads of goods brought to Gloucester by road, and on salted meat, salt, cloth, wines, coals, etc., brought by sea, in aid of the paving of the town (1335) suggests the nature of the commerce of Gloucester in the fourteenth century.[4]

A petition[5] of the commonalty of Bristol in 1363 states that corn used to be imported to that city from the counties of Gloucester, Worcester and Hereford; and it was probably shipped from Gloucester and Tewkesbury.[6] But the city of Gloucester could not rival the growth of the neighbouring port of Bristol; indeed, in 1487–8, the petition[7] of the mayor and burgesses of Gloucester to the king, requesting the amelioration of the fee-farm of £65 per annum due to the king, was granted. It was pointed out that the number of dwelling-places in the town had decreased by over three hundred in a few years; that the burgesses had to bear the

---

[1] W. H. Stevenson, *Calendar of Records of the Corporation of Gloucester* (1893), p. 4.
[2] *Ibid.* p. 3.
[3] H. G. Nicholls, *Iron Making in the Olden Times* (1866), p. 71.
[4] W. H. Stevenson, *op. cit.* pp. 50–1.
[5] *Cal. Pat. Rolls*, 1361–4, p. 409.
[6] *The Great Red Book of Bristol*, Text (Part I) (Bristol Record Society, 1933), pp. 132–3.
[7] W. H. Stevenson, *op. cit.* p. 61.

cost of maintaining the walls, gates and towers of the town, repair and maintain the great bridge over the Severn; and that the town was the chief defence of a great part of the coast adjoining the Marches of Wales.

Bristol increasingly dominated the whole region. Towards the middle of the fourteenth century it was becoming an industrial town in addition to being a large port. It had gilds of skinners, cordwainers, tanners, etc., but its principal industry was the manufacture of broadcloths.[1] The Cotswold hinterland itself became a manufacturing district in the following century, and Bristol became the outlet and marketing centre of a prosperous industrial region.[2] Its actual wool exports must have been small in comparison with those of London, Southampton, Boston and Hull.[3] But figures show that Bristol surpassed both London and Southampton in the exportation of cloths by English merchants.[4] Hansards and Italians seldom visited the port of Bristol during the fourteenth and fifteenth centuries.

This industrial and commercial prosperity of Bristol raised the problem of its food supply. The corn of its immediate agricultural district had become insufficient for the needs of the growing population. Supplies were obtained from the counties of Gloucester, Worcester and Hereford.[5] A petition of the citizens of Bristol, in the late fourteenth century, stated[6] that all the corn produced within ten leagues around Bristol was insufficient for the sustenance of the city's population. And it was even suggested that the corn vessels from Gloucester, Tewkesbury and other Severn ports

---

[1] F. B. Bickley, *op. cit. passim.*

[2] During the years 1353–60, 32·3 per cent. of the total broadcloths sent abroad from England (15,848 from a total of 49,088 broadcloths) were exported from Bristol. Of these exports, 98·4 per cent. were by denizen merchants; only 259 broadcloths were exported by aliens during these seven years. In forty years (1356–96), its own manufacture of broadcloths had practically doubled. In the last decade of the fourteenth century Bristol merchants retained a monopoly of its cloth exports. Between 1392 and 1395 the native merchants exported 4924 cloths, alien merchants only 157 cloths, annually. H. L. Gray, *Eng. Hist. Rev.* XXXIX (1924), 13–35, Appendix I; Appendix II.

[3] See p. 275 above; cf. Fig. 49.

[4] At the two latter ports, Italians, Hansards and other aliens controlled more than 50 per cent. of their export trade in cloth (London: denizens 4197, Hansards 4373, other aliens 5353; Southampton: denizens 4399, Hansards 12, other aliens 5596 cloths, annually, during the years 1392–5). Cf. Fig. 59.

[5] See footnote 5, p. 285. Cf. *Cal. Close Rolls,* 1374–7, p. 324. There seem to have been certain restrictions and hindrances (commercial rivalry with Gloucester, and the increasing industrial population in the north Cotswold zone).

[6] *Cal. Close Rolls,* 1374–7, p. 324.

discharged their cargoes at places other than at Bristol. The king gave orders that the import of corn and other victuals to Bristol should be resumed under his protection.

The development of the port during the fifteenth century is clearly a natural growth from the previous century. Its hinterland extended ever outwards and reached into the cloth area of the Cotswolds, the midlands (Warwickshire and Coventry, especially), and north and east Devon; into the lead-mining area of the Mendips and the tin-producing counties of Devon and Cornwall. Bristol became the collecting and distributing centre for much of the overseas and inland trade of a hinterland covering the west of England from Coventry to Land's End. Its maritime contacts likewise extended in area. The Irish trade developed into the Icelandic fishing trade,[1] a prelude to the Atlantic trade of Tudor times; the Biscayan trade in wine and salt grew into the Castilian and Portuguese traffic: attempts were even made by Bristol merchants to break the Mediterranean monopoly possessed by the Italians and Catalans.[2] From Gascony, Bristol received, in addition to wine and salt, woad, alum and cineres essential to its cloth manufactures. The principal return cargoes were cloth and hides. It has been said that more than one-half of the imports of wine from Gascony to Bristol during the first decade of the fifteenth century arrived in Bristol ships;[3] but the evidence suggests that the vessels of Fowey, Plymouth, Dartmouth and of the other south-west ports, as well as the vessels of Bayonne, were also active participants in this trade. The Gascony commerce of Bristol suffered considerably by the loss of that province in 1453 at the close of the Hundred Years' War, and it did not fully recover its former prosperity until after the Treaty of Picquigny in 1475.[4]

During the fifteenth century, too, the merchants of Bristol captured the trade of Portugal and of the Castilian ports of the Bay of Biscay which, formerly, had been in the hands of the merchants of those areas. Wines and cork were the principal imports from Portugal, and return cargoes of cloth were sent from Bristol to Lisbon and Oporto.[5] Until 1460 the export of broadcloths from Bristol by native merchants was 200 to 300 times greater than those by aliens.[6] Bristol became in this century the

---

[1] E. Power and M. M. Postan, *op. cit.* pp. 192–201.

[2] *Ibid.* pp. 224–30. W. Hunt, *Bristol Historic Towns* (ed. E. A. Freeman and Rev. W. Hunt, 1877), p. 93; T. Rymer (ed.), *Foedera*, etc. (10 vols., The Hague, 1739–45), v, ii, 67.

[3] E. Power and M. M. Postan, *op. cit.* p. 205.

[4] See p. 278 above.

[5] V. M. Shillington and A. B. Chapman, *op. cit.* p. 54, footnote 1, 58–9, 105.

[6] E. Power and M. M. Postan, *op. cit.* "Enrolled Customs Accounts", pp. 333–5.

chief western port of England. Although it entered into competition with Southampton, Hull and Chester, in the commerce to and from Gascony, Iceland and Ireland, its trade grew on lines which foreshadowed pre-eminence and even monopolies during the succeeding Tudor period.

It is difficult to assess the relative importance of the south Wales ports on the Bristol Channel at this period. Few records dealing with their maritime trade are extant. There are no customs records available for the lordship-marcher areas but, fortunately, we possess fragments of the records dealing with the administration of those parts of south Wales under the English crown,[1] and it thus becomes possible to interpret the commercial activities of such ports as Haverfordwest, Milford, Pembroke, Tenby and Carmarthen. The commercial development of the ports of Glamorgan and Gwent has to be inferred from the references contained in their charters, inquisitions and surveys,[2] and, indirectly, from the prosperity of their respective hinterlands and their contacts with other ports.

The military conquest of south Wales from the close of the eleventh century onwards resulted in the creation of several lordships-marcher directly related to the many strategic castles. Mixed agriculture, around the castles and manors, was practised by the Anglo-Normans, but the production of cereals in medieval Wales was quite insufficient to provide an item in its exports.[3] The rearing of sheep and cattle remained through-out the period the dominant occupation of the upland dwellers, and pastoralism was also important in the lowland areas.[4] Coal mining seems to date from the early fourteenth century in south Wales and there is the possibility of the mining of lead in the neighbourhood of Llantrisant and Llanharan in Glamorgan.[5] The sea and river fisheries[6] were important supplements to agriculture throughout south Wales, and the construction

[1] See E. A. Lewis (1), pp. 107–88 for "Tabulated Accounts, 1301–1547"; H. Owen, *Calendar of the Public Records relating to Pembrokeshire* (1911), 1 (Haverfordwest).

[2] J. H. Matthews, *Cardiff Records* (1898), 1, chap. I–III; G. G. Francis, *Charters granted to Swansea* (1867); G. T. Clark, *Cartae et alia Munimenta quae ad Dominium de Glamorgancia pertinent* (6 vols., 1910).

[3] *Cal. Pat. Rolls*, 1330–4, p. 80. Licence granted to the burgesses of Haverfordwest to buy 1000 quarters of corn in England for sale in Wales.

[4] J. H. Matthews, *op. cit.* chaps. II and III.

[5] *Ibid.* pp. 173, 274; G. G. Francis, *op. cit.* pp. 7, 233; G. T. Clark, *op. cit.* pp. 990–1000, 1067.

[6] J. H. Matthews, *op. cit.* pp. 132, 137, etc.; A. Ballard, *British Borough Charters, 1042–1216* (1913), p. 63.

of weirs and dams along the rivers Taff and Teifi may have retarded the maritime development of Cardiff and Cardigan during the Middle Ages.[1]

There is no evidence to show the existence of a single large port in south Wales at this time; there were several small ports serving restricted hinterlands or even a single castle or religious centre.[2] It seems that the trade of south Wales gravitated mainly to Bristol and remained in the hands of foreign merchants.[3] It was not until the fourteenth and fifteenth centuries that Welsh ships and Welsh merchants came to compete with the English and foreign merchants from Bristol, Plymouth, Dartmouth and Bayonne, both in home and in distant trade.[4] There must have been a good deal of coastwise traffic between the ports of south-west Wales during this period,[5] and pilots were available for the navigation of certain rivers, e.g. pilots for the navigation of the Towy to Carmarthen could be obtained at Llanstephan.

The principal imports[6] during the period either supplemented those home commodities which were not produced in sufficient quantities to satisfy the native demand (iron, lead, coal, wax, honey), or they were essential items unobtainable at home (salt, wine, oil, pitch, rosin). In exchange, butter, cheese, wool, skins, hides and cloth formed the staple exports of south Wales.[7] Welsh wool and, later, Welsh cloth were exported by sea, mainly to Bristol.[8] It is interesting to note that the staple of wool in Wales during the fourteenth century was confined to two or three royal towns. After 1284, legislation regarding the export of wool

[1] *Cal. Pat. Rolls*, 1313–7, p. 99.

[2] *West Wales Historical Records*, IX, 84, 86, 87. Porthclais, port of St David's Cathedral. Cf. J. H. Matthews, *op. cit.* p. 124; ports of Aberthaw, Barry and Ogmore.

[3] A. Ballard, *op. cit.* p. 184. Bristol burgesses were free from the payment of tolls and customs in Wales according to the terms of their charter of 1164. H. Owen, *op. cit.* pp. 145, 146, 148. See p. 284 above. In B. G. Charles, *Old Norse Relations with Wales* (1934), pp. 135–163, the possible influence of Norse trading stations upon medieval Welsh maritime trade in general is discussed, and emphasis laid upon Welsh participation in Bristol slave trade to Ireland.

[4] E. A. Lewis (2), *Medieval Boroughs of Snowdonia* (1912), pp. 107–33; E. A. Lewis (1), *passim; Cal. Close Rolls*, 1313–8, p. 227; *Cal. Pat. Rolls*, 1391–6, p. 442; 1385–9, p. 324.

[5] H. Owen, *op. cit.* pp. 84, 162, 164; *West Wales Historical Records*, IV, 6, 29; IX, 84, 86, 87.

[6] E. A. Lewis (1), pp. 107–33; J. H. Matthews, *op. cit.* pp. 134, 140, etc.

[7] E. A. Lewis (1), pp. 107–47; J. H. Matthews, *op. cit.* pp. 113, 114, 124, 133–4, 140.

[8] E. A. Lewis (1), pp. 128, 141, 146.

from Wales was identical with that from England. Haverfordwest was the only Welsh staple town included in the list of staples for 1326 from which wool and wool-fells could be exported to parts beyond the sea;[1] in 1327, Shrewsbury, Carmarthen and Cardiff were staple towns,[2] but a few years later, in 1332, Cardiff, a marcher town, had ceased to be a staple for wool.[3] Shrewsbury was thus, at this time, the chief wool market for mid-Wales, and Carmarthen for south Wales. Carmarthen, in 1353, became the sole staple for Wales.[4] From this royal town, after taxation had been exacted, the wool either was sent "across the mountains" to England, or it was exported by sea. This policy of one staple town for the whole of Wales proved unsatisfactory, and the Black Prince allowed his loyal Welshmen to take their wool to the English staples direct. One must bear in mind that with the exception of the wool obtained in the border region of Hereford, Shropshire and Radnor, Welsh wools were of the worst quality. They were described, in 1341, as "coarse and of little value". We find, however, that from the ports of Carmarthen and Haverfordwest, in the fourteenth century, wool was exported by sea in moderate quantities.[5] The extant returns of the Staple of Carmarthen for the period 1354–1461 have been published, as well as abstracts from the custom accounts of Haverfordwest, Milford and the adjacent ports of Pembroke-shire for certain years.

These records show that native and English vessels from Tenby, Haver-fordwest, Carmarthen, Bristol, Dartmouth, Teignmouth, etc., in addition to occasional boats from Flanders, transported the wools, and that there were representatives of English, Flemish and Lombard merchants at these ports. Moreover, the Welsh monastic houses of Strata Florida, Margam and Tintern consigned their wools to Flemish markets, too. In the four-teenth century, Pembrokeshire was an important wool-producing county apparently, but by the fifteenth century the export of Welsh wool had decreased considerably. It had been replaced by the export of Welsh russet cloths. Fulling mills had spread from Pembrokeshire throughout the marcher areas of Wales—the region of the feudal manors—with the centralisation of production displacing the old homestead industries. As in the case of wool, Welsh cloth was taken not only to the border towns of Chester, Oswestry, Shrewsbury, Ludlow and Hereford, but also by land

---

[1] *Cal. Close Rolls*, 1323–7, p. 585; but see *Cal. Pat. Rolls*, 1324–7, p. 274.
[2] *Cal. Pat. Rolls*, 1327–30, p. 98.
[3] *Cal. Pat. Rolls*, 1330–4, p. 363. Cf. *Cal. Close Rolls*, 1330–3, p. 525.
[4] *Statutes of the Realm*, I, 332, 348.
[5] E. A. Lewis (1), pp. 134–47, 149–61.

and sea to Bristol.[1] On occasions, Welsh russets were even taken to the great St Bartholomew's Fair, in London. But Bristol was really the chief centre for the sale of Welsh cloths, and to it came the cloths of south Wales to be sold at home and abroad, in Flanders, Brittany and Gascony. In 1494, the total value of Welsh cloth sold at Bristol amounted to £815. 17s. 6d.[2]

Associated with wool and cloth were exports of hides, especially salted hides, wool-fells, and skins, from the ports of Carmarthen, Haverfordwest and Milford.[3] There are no records of exports from other south Wales ports in Glamorgan and Gwent, but the arrival of south Wales ships at Bristol points to a similar type of trade. South Wales vessels participated in the Irish trade with Bristol. They brought fish—especially herrings, hake, salmon and salted fish—hides and Irish cloths from Irish ports to Bristol and took in return salt, wine, iron, English cloths and general merchandise. Indeed, in the fourteenth century Welsh boats from Milford, Haverfordwest, Tenby, Carmarthen, Chepstow, etc., engaged in the wine and salt trade from Bordeaux and the Bay to Bristol.

The growth of towns following the Norman settlement of parts of south Wales encouraged the foreign trade in wine, iron and salt. During the thirteenth and fourteenth centuries, alien vessels from Bayonne, Brittany and Spain, and English boats from Cornish and Devon ports, and especially from the port of Bristol, arrived at Carmarthen, at Haverfordwest and at Milford with cargoes of salt and wine.[4] The virtual monopoly of the Bristol and Chester merchants in the Welsh foreign trade of the thirteenth century had largely disappeared during the following century. Chester boats traded principally to the north-Wales ports of Beaumaris, Carnarvon and Conway. It was seldom that they put into a south-Wales port, such as Milford.

The maritime and trading activities of the Pembrokeshire ports around Milford Haven had been well regulated as early as the twelfth century. Pembroke possessed its merchant gild,[5] its fair lasted for eight days,[6] and its burgesses were quit of toll, pontage, havenage and all customs at Bristol, Gloucester, Winchester, and in Devon and Cornwall, and Rhuddlan.[7] This provides further indication of the direction of trade

---

[1] E. A. Lewis (1); and (3), "The Development of Industry and Commerce in Wales during the Middle Ages", *Trans. Roy. Hist. Soc.* (new ser.), XVII, 159 (1903).

[2] E. A. Lewis (3), p. 160. Cf. E. Power and M. M. Postan, *op. cit.* p. 334, footnotes 4 and 5. Welsh cloth to the value of £198 in 1431–2 and £248 in 1434–5 exported from Bristol by the English, "Bristol Enrolled Customs".

[3] E. A. Lewis (1), pp. 134–47, 149–61.   [4] *Ibid.* pp. 127, 129, 156, etc.

[5] A. Ballard, *op. cit.* p. 205.   [6] *Ibid.* p. 172.   [7] *Ibid.* p. 192.

from south-west Wales. A twelfth-century charter (1154–89) decreed that

all ships which enter with merchandise the port of Milford (Haven?) shall come to Pembroke bridge if they wish to buy or sell in the land, and shall there buy and sell. But, if they do not wish to trade they shall wait their breeze at the cross, paying their lawful customs. Moreover, all merchandise which is bought in the county of Pembroke to be exported to England must be shipped at Pembroke bridge or at Tenby on payment of their customs.[1]

But some years later (1219–31), ships entering Milford Haven were compelled to call at either Pembroke or Haverfordwest to sell their cargoes.[2]

The Welsh ports were, apparently, not represented by any ships at the great sea battles of the fourteenth century, such as Sluys (1340), Winchelsea (1350), La Rochelle (1372); but, on occasion (e.g. in 1335, and in 1342) certain merchant ships at Carmarthen (one of 80 tons, and another of 100 tons), at Swansea (of 100 tons, and 60 tons), at Tenby (one of 200 tons, and another "magna navis"), at Milford (one of 100 tons, and the second of 80 tons), were arrested and converted into men-of-war. The details of their reconstruction are extant;[3] it included the fitting of fore- and top-castles and, in one case, of an aft-castle, or poop. This was the work of special artificers, called castlewrights. The complements of the ships and the payments made to the crews are also given. It is an indication of the presence of fairly large merchant ships at south-Wales ports in the mid-fourteenth century and of the contribution of the mercantile marine towards an impoverished navy when the exigencies of war demanded.

CARDIGAN BAY AND THE IRISH SEA

An understanding of the commercial and maritime activities of north Wales is not possible without a knowledge of its political evolution which defined, above all else, the growth of the urban centres that, in turn, made possible maritime contacts on any scale. In the "patria" of north Wales the tribal, pastoral life of the Welsh people continued throughout the period. The towns, with some few exceptions, were alien settlements, political in origin, and intended to facilitate and consolidate the English conquest of north Wales. Their charters of incorporation were all based upon, or affiliated with, the charter granted to the border city of Hereford. They claimed the valued privilege of a gild merchant which gave to the

[1] A. Ballard, op. cit. p. 169.
[2] A. Ballard and J. Tait, British Borough Charters, 1216–1307 (1923), p. 243.
[3] Cymmrodorion Transactions (1925–6), p. 67.

burgesses a monopoly of trade regulation and of trading within the borough.[1] Their alien origin and character might have retarded their economic progress; and to overcome this possibility Edward I decreed that:

No markets, no fairs, nor any other places of trade forsooth, for the buying and selling of oxen, cows, horses, etc., excepting all articles of food such as butter, milk, and cheese, shall be held elsewhere in North Wales than in the towns of Conway, Beaumaris, Newborough, Carnarvon, Criccieth, Harlech and Bala.[2]

This centralised trade in a few royal English towns led to conflict with the smaller Welsh towns which were gradually acquiring for themselves a commercial character. It did not prevent, however, the holding of the old Welsh fairs in the rural parts of north Wales.

The growth of trade was encouraged not only by legislation but also by useful works. Open ways were cleared through thick woods for commercial and political purposes; timber was used for building and for fuel in the towns and was frequently transported by river to the ports; fishing weirs that blocked the passage of navigable waterways were removed or reconstructed to permit the passage of boats; quays were built at the ports, and their maintenance in good repair was assured by local dues upon shipping frequenting the ports for the victualling of the castles. English methods of trading and the use of English weights and measures aroused, for two centuries, the hostility of the native Welsh. And it is easily perceived that maritime trade at the ports of north Wales was essentially alien in character. The Welsh can hardly be called in medieval times a maritime people. The ships calling at the major ports of north Wales belonged to the merchants of Chester, occasionally to those of Bristol and to those of other south-west ports, and sometimes to Breton and Bayonne traders.[3] The nature of their trade was dictated, primarily, by the needs of the castle and of its small associated communities, and, secondarily, by the harvest of the sea and rivers, and the profits of the land, whose native population was practically self-sufficing in its needs.

A regional survey of the north-Wales ports substantiates the truth of the preceding analysis. Along the shores of Cardigan Bay and the peninsula of Lleyn, there were small townships such as Aberystwyth, Aberdovey, Towyn, Barmouth, Harlech, Portmadoc, Criccieth, Pwllheli and Nevin. All of these, in spite of the castles at Aberystwyth, Harlech and Criccieth, were essentially fishing and local trading ports, peopled by

---

[1] E. A. Lewis (2), p. 167.   [2] *Ibid.* p. 174.
[3] E. A. Lewis (1), *passim*; *Cal. Close Rolls*, 1318–23, pp. 453, etc.

20

communities of farmer-fishermen.[1] They possessed good harbours and, in all probability, the waters of the estuary of the Dwyryd river flowed much nearer to the walls of Harlech Castle than they do to-day. The herring fisheries of Cardigan Bay gave employment to the local boats, but few records have survived to indicate the arrival of merchant vessels at these ports or of the engagement of local ships in their trade. The fishing ports of Nevin and Pwllheli were created borough towns by the Black Prince, but they remained throughout the period distinctly Welsh in population.

Along the coast of north Wales were the ports of Beaumaris, Carnarvon and Conway—castellated and walled towns—of which Beaumaris was by far the most important. They held yearly fairs and weekly markets; but, legally, it was impossible for a Welshman to become a burgess of one of these castellated boroughs. Commercial wealth centred upon these three boroughs and they became the main arteries of foreign trade. Their castles had to be repaired and victualled and for these purposes vessels arrived at their ports from distant parts. The pastoral districts of Snowdonia could contribute but little to foreign export trade and there were few industries in being at this time. Carnarvon, the centre of English political administration, was not the chief port. As at Conway, local fishing boats were the principal callers, and upon them a local custom was exacted for the maintenance of the new quay which was built in Edward I's reign to supply the castle. Ferry boats and royal boats were also maintained at both ports.

Beaumaris, on the opposite coast of Anglesea across the Menai Strait, became the chief centre of commercial importance,[2]—less liable to attack and siege from the mountains of Snowdonia, and more accessible from seaward. It was sheltered from westerly gales, possessed good anchorage facilities, and was well situated to supply the needs of the Edwardian castles of north Wales. Fishing boats made it a favourite port of call, and it became the centre of foreign trade from which coasting vessels supplied the other ports of north Wales with provisions and stores. It was a town of merchants and traders. Its quay and warehouses were improved by the proceeds of customs levied upon the herring boats and merchant ships that called at the port, and it was at Beaumaris that the royal customs were collected during the Middle Ages. Edward I, in his settlement of north Wales, had decreed that all ships approaching the coast of Anglesey should call at Beaumaris and there display their goods and merchandise for sale,

[1] E. A. Lewis (3), p. 150.
[2] E. A. Lewis (2), pp. 206–7; (1), passim.

and not elsewhere.[1] An analysis of the custom revenues of the north Welsh ports shows that Chester merchants were the chief participators in their trade.[2] Vessels also arrived from Ireland, Brittany and Gascony, from the ports of Liverpool, Fowey, St Ives and Plymouth, and, in the fifteenth century, from Portugal. Imports comprised mainly wine, some iron and lead, and occasional cargoes of corn and peas from Ireland and from the south-west peninsula of England. In the fifteenth century the port of Beaumaris exported wool and cloth, shipped in Breton vessels from St Malo and St Pol de Leon.[3] It foreshadowed the prosperity of this port in Tudor times. Apparently, most of the wool from north Wales found its way to the markets of the border towns; but from the Rhuddlan staple small cargoes of wool were very occasionally sent by sea in ships chartered mainly by Chester merchants.[4] But a merchant, John de Banham, a burgess of Newport-on-Usk, who traded at the Carmarthen staple in December 1396, visited the Rhuddlan staple in the following year.[5] These are details. The statistics available warn us against exaggeration. In northern Wales, foreign trade was small in amount and spasmodic in character.

There were many points of similarity between Chester and Gloucester. Both were fortresses commanding approaches into Wales; both were bridge-head towns on navigable rivers; and both were destined to lose their maritime trade to neighbouring ports, Liverpool and Bristol respectively. It is possible to reconstruct the maritime activities of Chester from references contained in its several charters.[6] In its earliest charter, granted by Henry II (c. 1175),[7] reference was made to the trade with Ireland from Chester in the time of Henry I. Permission was given by Henry II to the Chester merchants to buy and sell at Dublin as freely as in the time of his grandfather. They were licensed to traffic in corn and grain at Drogheda and in all other ports. This Irish trade from Chester to Dublin and Drogheda remained important throughout the medieval period.[8] Vessels of all three ports engaged in the reciprocal trade, and, during the fifteenth century, the ships of Dublin became the most

[1] E. A. Lewis (2), p. 206. [2] E. A. Lewis (1), p. 115.
[3] *Ibid.* p. 129. [4] *Ibid.* pp. 127–8. [5] *Ibid.* p. 127.
[6] For a discussion of the site and trade of Chester see H. J. Hewitt, *Medieval Cheshire* (1929), chap. VIII.
[7] R. H. Morris, *Chester in the Plantagenet and Tudor Reigns* (Chester, 1893), pp. 457–8, 480, for the Charter of Henry II.
[8] *Ibid.* p. 573. List of Irish ships entering the port of Chester in the fifteenth century.

frequent arrivals at the port of Chester. The Irish exports from Dublin and Drogheda included skins (marten, hare, squirrel, etc.) and hides, fish (salmon, herring and hake) and linen.

It has already been pointed out that the Chester merchants were frequent visitors to the ports of north Wales.[1] Their ships also visited Gascony and Brittany for wine and salt, and it was they, apparently, that revictualled the castles of north Wales and supplied the needs of the garrison and castle of Chester.[2] The evidence shows that, although the ships were freighted by Chester merchants, they were not necessarily vessels of that port nor of Bordeaux. Many ships from Ottermouth, Portsmouth, Lynn, Drogheda, Dartmouth, Kingswear, etc., left Bordeaux for Beaumaris or Chester.[3]

But in the fifteenth century the decay of Chester as a commercial port was evident, and its replacement by more favourably situated ports such as Beaumaris and Liverpool soon became apparent. In 1445 the mayor and burgesses of Chester had their fee-farm of £100, annually, reduced to one-half.[4] The grounds for their request were two-fold: (1) the closure of the harbour by the accumulation of silt so that no merchant ship could approach within twelve miles and more of the city; and (2) the loss of trade sustained by its merchants during the revolt of Owen Glyndwr, and the refusal of the Welsh to visit Chester after this revolt. Forty years later, in 1484, the whole of the fee-farm was remitted[5] by Richard III, but, in 1486, £80 of the fee-farm were remitted by Henry VII owing to the "influx of sand and the silting up of gravel".[6] Chester merchants were thus compelled to use neighbouring ports in spite of the fact that an attempt was made to recover some of the former prosperity of the port by the construction of a new quay or haven on the Wirral peninsula lower down the estuary of the Dee. This quay, some six miles below Chester, through procrastination took 59 years to construct (1541–1600).[7]

From the scanty evidence available it is difficult to postulate any large

---

[1] See p. 295 above.

[2] F. Michel, op. cit. I, 124. The freightage rates between Bordeaux and the Red Bank of Chester were recorded at Bordeaux in the fourteenth century (Chester or Beaumaris 18s.; Dublin 15s.; Drogheda 16s. per tun). The freightage of the ship had to be paid within ten days of its arrival in port; and towage and pilotage was to be paid by the merchant, except that the master could take a pilot at his own expense at Milford or at Dalkey when bound for Beaumaris or to the Red Bank of Chester.

[3] Ibid. pp. 124–5.

[4] R. H. Morris, op. cit. p. 511. Charter of Henry IV, 4th August 1445.

[5] Ibid. pp. 518–21.   [6] Ibid. pp. 521–4.   [7] Ibid. p. 459.

amount of trade for the ports of Lancashire and north-west England at any time during these centuries.[1] This area suffered frequent invasions from both Scotland and Wales, and its political instability reacted adversely upon its agricultural and industrial development. The wool of Cumberland, Westmorland and north Lancashire (Furness Abbey) was of inferior quality; the area had little else to export with the possible exception of small quantities of iron, salt and fish. The records show that even the port of Liverpool does not date its maritime prosperity from these centuries.

It had been founded as a free, royal borough by King John, in 1207, and settlement was encouraged therein by granting to its burgesses "all the liberties and free customs enjoyed by any borough on the sea-coast".[2] He apparently fixed upon its sheltered pool as a suitable point for embarkation of men and supplies from northern England for service in Ireland. Liverpool had its ferries across the Mersey to Birkenhead and Runcorn as early as the thirteenth century, but it was in connection with the transport of armies and materials to Scotland and Ireland that the port derived some maritime importance during the fourteenth and fifteenth centuries.[3] Industry did not flourish in the new settlement, and although it had been granted its gild merchant at an early date there is no trace of craft gilds at Liverpool until the sixteenth century.[4] Liverpool, unlike Bristol or Southampton, remained moribund during the fifteenth century and it was still described as a "decayed town" some fifty years later.

BIBLIOGRAPHICAL NOTE TO CHAPTERS VII AND VIII

Unfortunately there is no comprehensive work dealing with English maritime trade during the Middle Ages. A general survey of commercial relations is given in the following:

L. F. Salzman, *English Trade in the Middle Ages* (1931).

W. Cunningham, *Growth of English Industry and Commerce during the Early and Middle Ages* (5th ed. 1910).

E. Lipson, *Economic History of England* (5th ed. 1929).

The chapter footnotes provide a bibliography of published and unpublished sources, but particular mention must be made of *Studies in English Trade in the Fifteenth Century* (1933), edited by E. Power and M. M. Postan. Some of the Victoria County Histories contain sections on the maritime relations of the counties during the Middle Ages.

[1] See *V.C.H. Lancaster*, II, 265, etc., for meagre evidence of medieval maritime trade at Preston and Lancaster.

[2] *V.C.H. Lancaster*, II, 265; IV, 2.

[3] *Cal. Pat. Rolls*, 1377–81, p. 385; 1385–9, p. 163; 1388–92, pp. 134, 385, 405; 1399–1401, pp. 164, etc.; *V.C.H. Lancaster*, IV, 3.

[4] *V.C.H. Lancaster*, IV, 3, 9. See p. 352 below.

Chapter VIII

MEDIEVAL FOREIGN TRADE: EASTERN PORTS

R. A. PELHAM, M.A., Ph.D.

During the Middle Ages English society was organised mainly on a two-class basis: a land-owning aristocracy on the one hand, and an agricultural population concerned primarily with the means of subsistence on the other. The feudal system was not elastic enough to contain the rapid growth of a middle class devoting itself to commerce, and it is not surprising, therefore, that the internal trade of the country was for some centuries partly in the hands of Jews,[1] whilst foreign trade was largely controlled by groups of aliens who had become established in most of the east coast ports.[2] Since both the merchanting of goods and their shipment to the Continent was mainly in the hands of aliens, native enterprise was at first limited to coasting trade,[3] or to trade with south-west France, where alien competition was less severe. The breakdown of the manorial system, however, and the growth of an industrial population was accompanied by an increase in the commercial activities of Englishmen, with the result that the merchant staplers and the merchant adventurers, engaged chiefly in the exportation of wool and cloth respectively, secured much of the trade that aliens had formerly enjoyed. For merchants operating on the east coast this was a struggle against Hansards, Dutch, Brabantines, Flemings, French and Italians, and, although Englishmen failed to break the monopoly of Hansards in the Baltic and of Italians in the Mediterranean, they finally gained control of a great deal of the traffic across the narrow seas.

[1] For thirteenth century jewries see pp. 221–2 above.

[2] There were Hanseatic "factories" at Hull, Boston, Lynn, Yarmouth, Ipswich and London (W. Vogel, *Geschichte der deutschen Seeschiffahrt* (1915), 1, 267), although only London, Lynn, Ipswich and Boston were apparently functioning in the fifteenth century. (E. Power and M. M. Postan, *Studies in English Trade in the Fifteenth Century* (1933), p. 148.) A Flemish Hanse existed in London, the members of which represented three distinct groups of towns, viz. those immediately around Bruges, those connected with Ypres, and those of Walloon or French Flanders.—L. M. Seckler, *Anglo-Flemish Trading Relations during the Early Middle Ages* (1932), p. 34 (unpublished thesis in University of London Library). London ranked as one of the four great centres of Hanseatic activity in Northern Europe, the others being Bruges, Bergen and Novgorod (Vogel, *loc. cit.*).

[3] Prior to 1549 only occasional records of coast shipments are found (N. S. B. Gras, *Early English Customs System* (1918), p. 145, note).

THE EAST COAST PORTS

The physical configuration of the eastern half of England, with its rivers converging upon a series of navigable estuaries, and the fact that the country's trade associations throughout the Middle Ages were mainly with North-west Europe, naturally led to a great concentration of commercial activity along the east coast. The finest of the estuaries led inland to the greatest of the ports, and London dominated the overseas trade, just as it did the internal trade, of the country. The prosperity of a port must always be largely determined by the character and extent of its hinterland, but it is not true to assume that at this early period the hinterland of a port was restricted to the region with which it was in immediate contact. The transport facilities of medieval England[1] enabled an east coast port to extend its influence at times over a very considerable area, and the economic watershed of the country was thrust far to the west, so that cloth from Wales[2] and wool from the Marches[3] gravitated towards the North Sea coast, as did lead from the Pennines[4] and even metal from Cornwall.[5] Furthermore, the extensive use of credit in both home and foreign trade implied much less uncertainty in commercial relationships than the political and social history of the period would indicate.[6]

It is not easy to summarise the trade of the Middle Ages as a whole over the vast area with which these ports had direct or indirect trading relations. Changes were many, some brought about by arbitrary restrictions demanded by the political exigencies of the moment, as in the case of the transference of the wool staple; other changes were produced by the growth of native industry, such as the English textile revolution of the fourteenth century, which within a few decades reversed the flow of cloth along the old-established routes between England and the Low Countries;

[1] See pp. 260–5 above.

[2] E.g. Welsh frieze was shipped from Lynn by Hanseatic merchants in 1467 (Customs Accounts, 97/8; N. S. B. Gras, *op. cit.* p. 615).

[3] Wool from the Abbeys of Stanlaw and Combermere was sent to Boston and from the Abbey of Darnhall to London in Edward I's reign (H. J. Hewitt, *Mediaeval Cheshire* (1929), p. 48). See also Fig. 46 for transport of wool from Shrewsbury to Hull.

[4] Lead from Nidderdale (Exchequer K.R. Accounts, 598/9, P.R.O.) and Chesterfield (*ibid.* 580/36,) was taken by road and river to Hull, in the mid-fourteenth century, for shipment to London.

[5] E.g. a consignment of silver was carried on pack-horses from Calstock to London in 1357 (Exchequer K.R. Accounts, 261/9).

[6] M. M. Postan, "Credit in Mediaeval Trade", *Econ. Hist. Rev.* 1, 234 (1927–8).

others, again, were caused by the opening up of new markets such as that of Prussia, the corn supplies from which minimised the risk of famine in this country during the later Middle Ages, and helped to maintain the wine trade of Gascony.[1] Isolated events like the rebuilding of the nave of Canterbury Cathedral in Richard II's reign, which led to a temporary revival of the old trade in stone between Caen and Sandwich, and the appearance of new commodities, such as hop beer,[2] at the end of the four-teenth century—all help to complicate a general picture. If to these be added the more disturbing factors of wars and rumours of wars, piracy and plague, it will readily be seen that economic conditions in the overseas trade of the country were subject to countless interruptions. To the mer-chant class, the Middle Ages were essentially a period of adjustment to these changing conditions, and no geographical account of the period must fail to recognise the part which their energy and resourcefulness played in transforming the local economy of the earlier centuries into the complex systems of international trade which characterised the later Middle Ages.

There was considerable variation between individual ports. Physically, Hull, Boston, Yarmouth and Sandwich had certain features in common; each was situated at the mouth of a river on which, a few miles upstream, stood an important manufacturing city. Between city and port there grew up a close association, the commercial operations of the one being com-plementary to the industrial activities of the other.[3] But each port also served a much wider hinterland, and just as wool from the Marches and lead from the Pennines found their way to Hull and Boston,[4] so goods from all parts of the country which had been brought to London were some-times taken along the Watling Street to Sandwich before being shipped.[5] Within a narrower radius of each port there lay a rich arable area, on the fringes of which were pastures for thousands of sheep. For their rather more circumscribed, though industrialised, hinterlands, Yarmouth was compensated by the valuable herring fisheries,[6] and Ipswich by the estuary

---

[1] E. Power and M. M. Postan, op. cit. p. 140.

[2] L. F. Salzman, English Trade in the Middle Ages (1931), p. 361, note.

[3] Sea-going vessels were able to penetrate up-stream as far as York (see Fig. 46 and 47) and Norwich however.

[4] The Fosse Dyke, a canal linking the R. Witham and the R. Trent had been cut at least as early as 1121, though frequent complaints of its having become choked were made during the fourteenth century. W. Dugdale, History of Imbanking and Drayning (1662), p. 167; V.C.H. Lincs. II, 383.

[5] This applied particularly to wool in the fifteenth century.

[6] The Yarmouth monopoly was challenged by Lowestoft (see Parliamentary and Council Proceedings, Chancery, File 12, P.R.O.).

of the Orwell, which was a renowned place of assembly for vessels sailing to the Continent.[1]

Lynn was in a somewhat different category. Its water communications[2] were more extensive than those of the other ports excepting Hull, and it was not over-shadowed by a large manufacturing centre. Behind it lay a particularly rich arable region supporting a dense population,[3] and it is not surprising therefore, that it became a great emporium.[4] At the two extremities of the region under consideration were Newcastle and Winchelsea, both largely concerned with shipbuilding and the exportation of fuel. The former had also developed as the commercial focus of the northern counties, and was the staple for wool grown north of the Tees.[5]

The actual sites of Hull and Winchelsea are of special interest because these two towns were re-established by Edward I in the last decade of the thirteenth century.[6] Hull, lying in the marshes near the confluence of the River Hull and the Humber, had belonged to the Abbey of Meaux near by, and had served its apprenticeship as a wool port before becoming a royal borough.[7] Change of ownership was indicated by change of name, and henceforth it was known as Kingston-on-Hull. With an excellent harbour and defensive site,[8] it served as an admirable base for operations

---

[1] The first recorded use of Orwell Haven as a strategical base for men-of-war occurs in 1326 (*V.C.H. Suffolk*, II, 205). Two hundred vessels sailed thence for the Battle of Sluys in 1340 (*ibid.* p. 206). The Orwell-Zealand route was no doubt safer for vessels shipping wool to the Continent when piracy was rife in the narrow seas. Compare Chaucer's Merchant:

"He wolde the see were kept for anything
Betwixte Middelburgh and Orwelle".

Wool ships from London (Exchequer K.R. Accounts, 457/7) and Hull (*ibid.* 457/8) *circa* 1337 were piloted to Dordrecht *via* Orwell, Sluys and Middelburgh.

[2] *Cartularium Monasterii de Rameseia* (Rolls Series, 1893), III, pp. 141–57; N. S. B. Gras, *Evolution of the English Corn Market* (1915), p. 174.   [3] See p. 231 above.

[4] It was the nearest port to Stourbridge Fair, and was constituted a staple port in 1373 because of its cheap and easy water communications with the counties of Warwick, Leicester, Northampton, Rutland, Bedford, Buckingham, Huntingdon and Cambridge (*Rot. Parl.* II, 319); L. F. Salzman, *op. cit.* p. 209.

[5] Owing to its poor quality (see Fig. 37) this wool was allowed to be exported free of duty in the fifteenth century, with the result that the more expensive wools of Yorkshire and Lincolnshire were sometimes carried to Newcastle for shipment in order to evade the customs (L. F. Salzman, *op. cit.* p. 307).

[6] T. F. Tout, *Mediaeval Town Planning* (1934), pp. 22–6.

[7] Over 4000 sacks of wool were exported in 1275 (N. S. B. Gras, *Early English Customs System*, p. 225).

[8] The encircling of the town by water was completed by the cutting of a channel from the River Hull to the Humber.

against the Scots,[1] and for commercial purposes its water communications left little to be desired.[2]

The hinterland of Winchelsea was much more limited in area, and considerably poorer in resources. But the town could be easily defended[3] and it commanded an estuary in which large fleets gathered.[4] It stood at the confluence of three rivers[5] which gave it access into the heart of the Weald whence supplies of timber for shipbuilding could be easily obtained.[6]

The Cinque Ports Confederacy, of which Winchelsea was a member, appears to have reached the zenith of its prosperity in the early fourteenth century,[7] and the process of decay that continued throughout the later Middle Ages has left many traces in both local and national records, for the men of the ports sought by the plunder of friend and foe alike to make good the losses that changing geographical circumstances were imposing upon them. The creeks that had served as nurseries for seamanship in earlier centuries became blocked with sand or shingle, or their protecting headlands were eroded away by the fierce storms that were a feature of the time.[8] Hastings was the first to go,[9] Old Winchelsea soon followed,[10] Rye,

[1] Kingston-on-Hull replaced Ravenser, at the mouth of the Humber, which fell a victim to coastal changes in the north as did Old Winchelsea in the south (see *Chronica Monasterii de Melsa* (Rolls Series, 1868), III, 120–1; *V.C.H. Yorks*, III, 402. For map see J. R. Boyle, *The Lost Towns of the Humber* (1889), frontispiece).

[2] In the middle of the fourteenth century the River Ouse below York was described as "a highway and the greatest of all the king's rivers within the kingdom of England" (*Public Works in Mediaeval Law*, ed. C. T. Flower, Selden Soc. (1923), XL, p. 253). The Ouse was tidal as far as York and eight miles beyond (*The York Mercers and Merchant Adventurers, 1356–1917*, ed. M. Sellers, Surtees Soc. (1918), CXXIX, p. ii), whilst the River Hull was tidal as far as Beverley (*Beverley Town Documents*, ed. A. F. Leach Selden Soc. (1900), XIV, p. 22). See Fig. 47. The rise of Hull affected neighbouring ports (*The History of Scarborough*, ed. A. Rowntree (1931), p. 166).

[3] It was built on a steep-sided hill of sandstone that was occasionally surrounded by water (W. Dugdale, *op. cit.* p. 91).

[4] E.g. in 1297 (*V.C.H. Sussex*, II, 132). See also *Cal. Pat. Rolls*, 1324–7 (1904), p. 26, for reference to the spaciousness of Winchelsea Haven.

[5] The River Rother is believed to have changed its course in 1287 after a very severe storm. Formerly it entered the sea at Romney (M. Burrows, *The Cinque Ports* (1888), p. 16).

[6] Winchelsea was one of the three ports commissioned to build two galleys each in 1294, the others being Bristol and Shoreham (with Seaford). No other port was required to build more than one. See R. C. Anderson, "English Galleys in 1295", *Mariner's Mirror*, XIV, 220 (1928).   [7] *V.C.H. Kent*, II, 264.

[8] See J. A. Williamson, "Geographical History of the Cinque Ports", *History*, XI, 97, note (1926).   [9] M. Burrows, *op. cit.* p. 8.

[10] It was finally overwhelmed in the storm of 1287 (M. Burrows, *op. cit.* p. 114).

Romney and Hythe remained insignificant ports throughout the four-teenth and fifteenth centuries; even Dover never rose above mediocrity, though it clung tenaciously to its monopoly of the cross-channel passenger traffic, which seems to have been the only legitimate activity of any con-sequence that remained.[1] Sandwich alone survived, but although it re-tained its position as a staple port it was overshadowed by London on the one hand and Southampton on the other. Nevertheless, its position in the narrow seas enabled it to function as an important entrêpot whence Italian produce was re-exported to the continent,[2] whilst its excellent communica-tions with London enabled it to act as an outport for the capital, especially during the fifteenth century, when wool destined for Italy was often taken by road or water to Sandwich before being loaded into Venetian galleys.[3]

The position of outstanding importance held by London in the eco-nomic life of the country has been referred to several times, but two further points need to be mentioned. In the first place, whereas the Channel ports and Bristol carried on most of their trade with south-west Europe, and the east coast ports mainly with the Baltic and the Low Countries, London maintained important connections with all parts of the Continent throughout the Middle Ages. And in the second place, as a corollary of this, although the merchants of Newcastle, for example, were interested chiefly in coal, and the merchants of Lynn in corn, those of London were concerned with every branch of trade, each mistery having a well-organised body of men whose activities were not necessarily confined to the city.[4] In the Lay Subsidy of 1332 the wealthiest group of merchants—those with possessions worth £60 or more—included four pepperers (later known as grocers), two mercers, two vintners, two woolmongers, a draper, blader and butcher, and lower in the social scale are to be found fish-mongers, cordwainers, goldsmiths, pewterers and a host of others.[5] Though remote from the areas yielding supplies of tin and lead, London also became

[1] J. B. Jones, *Annals of Dover* (1916), p. 146.
[2] Irish wool and hides were re-shipped from Sandwich in the mid-fourteenth century (Enrolled Customs Accounts, 5, P.R.O.). See L. F. Salzman, *op. cit.* p. 293, note.
[3] E.g. in 1463, when a cargo of wool was taken in a Dutch vessel from London and transferred to a galley at Sandwich (Customs Accounts, 73/35; N. S. B. Gras, *Early English Customs System*, p. 604).
[4] G. Unwin (ed.), *Finance and Trade under Edward III* (1918), p. 238. Their mem-bers were drawn from all parts of the country (see C. L. Kingsford, *Prejudice and Promise in Fifteenth-century England* (1925), pp. 120–1, 126, and J. J. Lambert, *Records of the Skinners of London* (1933), p. 18. See also the chapter by S. Thrupp, "The Grocers of London", in E. Power and M. M. Postan, *op. cit.* pp. 247 *et seq.*
[5] G. Unwin, *op. cit.* p. 45.

the chief centre of the pewter industry, and considerable quantities of pewter ware, tin and lead were shipped from its quays throughout the fifteenth century, mainly by Hansards.[1] In fact, long before the close of the Middle Ages the trade of London was all-embracing, for every branch of foreign commerce and every aspect of English trade and industry were represented in its activities. It is pertinent to ask how far this phenomenal growth at such an early stage in the economic development of the country was due to special local circumstances. That London had certain geographical advantages denied to the other ports will be readily conceded, but whether they were sufficient to justify a growth of this magnitude is less certain. It is abundantly clear, however, that any attempt to explain these developments solely in terms of material considerations would constitute a grave injustice to the merchant class whose enterprise and ability undoubtedly constituted two of the most potent factors in the evolution of medieval England.

THE WOOL TRADE

From the viewpoint of commodities, the foreign trade of all these ports, in spite of certain well-marked differences in their geographical settings, was more uniform in character than might be expected. All were easily accessible from the Continent, and, thanks to inland transport facilities, the hinterland of each was widely shared.[2] It was, therefore, very unusual for a port to have the monopoly of either the importation or the exportation of any particular commodity. The main contrasts between the ports were due rather to differences in the quantities of goods passing through each and to the varying proportions of English and alien merchants engaged. For wool and cloth, the existing documents are particularly detailed and numerous, and since they make it possible to form an idea of the relative importance of the various ports engaged in the shipment of these commodities they merit first consideration.

It is a commonplace to say that wool was the leading export of medieval England, and the commodity which determined much of the political as well as the economic activities of the Middle Ages. Its history as an article of commerce goes back a long way, but the earliest systematic accounts of its shipment abroad date only from Edward I's reign (1272–1307).[3] These

[1] Power and Postan, *op. cit.* p. 142.
[2] See footnote 5, p. 301 and Fig. 46.
[3] With the institution of the "Ancient Custom" of 1275 (see N. S. B. Gras, *Early English Customs System*, chap. II).

enable conditions at the close of the thirteenth century to be reconstructed
in some detail. Taken together, the material falls into four groups: (*a*) the

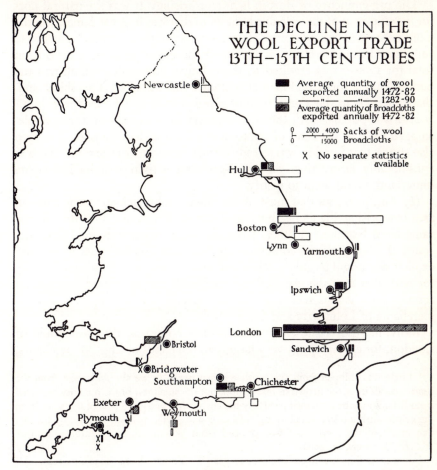

Fig. 49

The details for the earlier period are taken from Pipe Roll 133, m. 32 d, and 134, m. 3
(P.R.O.) and for the fifteenth century from tables of Enrolled Customs Accounts in
E. Power and M. M. Postan, *op. cit.* pp. 330–60. It should be noted that worsteds
are not included in the symbols for cloth.

relative amounts of wool exported from the various ports, (*b*) the merchants
engaged in the trade, (*c*) the vessels employed by them, (*d*) foreign markets.

    (*a*) Fig. 49 shows, among other things, the average quantity of wool
shipped annually from each of the leading ports of England during the

period 1282–90. Boston was then the chief exporter, followed closely by London, and then by Hull and Southampton.[1]

The pre-eminence of Boston is not surprising. It was the natural port for the shipment of the valuable long wool that was in great demand on the Continent,[2] and behind it were clustered many of the monasteries that were supplying Italian merchants at this time.[3] Indeed, 15 per cent. of the Boston total for 1296–7 was credited to three Italian companies.[4] Hull was the natural outlet for a region that included the whole of Yorkshire[5] and, thanks largely to the Trent, its hinterland stretched as far as Shropshire.[6] It also was popular with Italian merchants who, in 1275–6, for example, shipped over 50 per cent. of the total amount that left the port. The map also shows that London was already attracting a share of the trade that was considerably larger than the wool-growing activities of its immediate hinterland would seem to justify.[7]

(b) Fig. 50 shows that most of the alien wool merchants came from an area comprising Flanders, Brabant, Zealand, Artois and Picardy. A few came from Normandy; others from Germany, from Italy, and from the south of France. If the contemporary Southampton accounts were more explicit we should be able to add a few more symbols to Southern Europe, for with that area the port did a considerable trade. Indeed, less than a century later it was recognised as the special port of shipment for the merchants of Lombardy and Catalonia.[8]

Individual ports show some interesting contrasts; the wool trade of London, for example, was very largely in the hands of English merchants,

---

[1] Little was shipped from the south-western ports, as we should expect from the coarseness of the wool grown in that region, but the smallness of the quantity exported from Bristol may be explained partly by the absorption of a good deal of the wool from the neighbourhood in local cloth manufacture, and partly, perhaps mainly, by the fact that much of the local wool was shipped from Weymouth and Southampton (see *V.C.H., Somerset*, 11, 406).

[2] It was used in the manufacture of serges (W. Youatt, *Sheep* (1837), p. 312).

[3] See Fig. 35.

[4] No Italian vessels were engaged in this trade, most of the ships employed being either Dutch or Flemish (see Fig. 51).

[5] Scarborough and Yarm were subsidiary ports through which wool from the North York Moors was exported, often uncustomed (see A. Rowntree, *op. cit.* p. 168). See C. Frost, *Notices relative to the Early History of the Town and Port of Hull* (1827) chap. VI for details of the early wool trade.   [6] See Fig. 46.

[7] See A. Schaube, "Die Wollausfuhr Englands vom Jahre 1273", *Vierteljahrschrift für Sozial- und Wirtschaftsgeschichte*, VI (1908), p. 54. *Cal. Pat. Rolls*, 1340–3 (1900), p. 274, mentions carriage of 825 sacks of wool to London from Hertford, Essex and Sussex.

[8] D. Macpherson, *Annals of Commerce* (1805), I, 587.

who were drawn from a wide area. At other ports, alien merchants were in a substantial majority, especially at Hull, which was visited by merchants from Lübeck, the Netherlands, France and Italy.

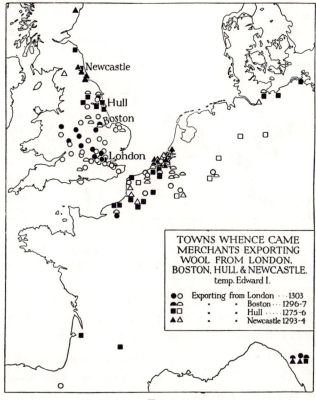

TOWNS WHENCE CAME
MERCHANTS EXPORTING
WOOL FROM LONDON.
BOSTON. HULL & NEWCASTLE.
temp. Edward I.

●○   Exporting from London ···1303
▲△   ·   ·   Boston ··· 1296-7
■□   ·   ·   Hull ····.1275-6
▲△   ·   ·   Newcastle 1293-4

Fig. 50

This map and Fig. 51 are compiled from Customs Accounts as follows: London, 68/8 (March–April, 1303) and 68/9 (April–Michaelmas, 1303); Boston, 5/5 (May 1296–July 1297); Hull, 55/1 (June 1275–April 1276); Newcastle, 148/4 (Easter 1293–Easter 1294). The black symbol refers in each case to a merchant whose town is given in addition to his surname, e.g. John Fadirson of Middelburgh; the white symbol if there is no surname, e.g. William of Middelburgh. The towns with 3 symbols are Cologne and Dinant, and the Italian towns are Lucca and Florence. See footnote 7, p. 329.

(c) Fig. 51 shows very clearly that the carrying trade in wool was almost the monopoly of ports lying near the mouth of the Rhine, except in the case of London which seems to have relied mainly upon English vessels.[1]

[1] Note the comparison with Fig. 52, which shows the extent to which local shipmasters maintained their hold over the carrying trade in wool from London.

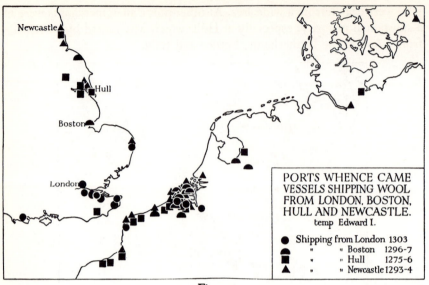

Fig. 51

There were actually many more vessels from Zealand than even the thickly clustered symbols in that region would suggest, owing to the large numbers of ships from Middelburgh and neighbouring ports. For sources see Fig. 50. See footnote 7, p. 329.

The relative importance of English and foreign vessels at the four chief ports in the years covered by Fig. 51 can be seen in the following table which includes all vessels whose nationality can be identified.[1]

| Year | Port | Approximate numbers of | |
		English vessels	Alien vessels
1303–4	London	19	7
1296–7	Boston	5	34
1275–6	Hull	7	55
1293–4	Newcastle	10	34

Conditions at London were in many ways a large-scale version of those obtaining in the western ports. This can be illustrated by reference to Kent and Sussex, for which the surviving accounts are particularly informative. The details for a particular year are summarised in Figs. 53[2] and 54 and it will be seen that conditions at Seaford and Sandwich were com-

[1] Although these figures, especially in the case of London, do not represent the total numbers of ships engaged, they undoubtedly show the character of the carrying trade at these ports.

[2] Compare Fig. 34 which shows the wool-growing areas of Sussex.

pletely different. It is important to notice that although the Sussex ports were within easy reach of Flanders, their wool trade was mainly in the hands of local merchants and shipmasters, whereas alien shipping had secured a monopoly of the carrying trade at Sandwich. This distinction

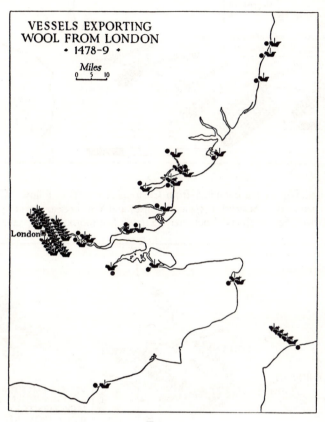

VESSELS EXPORTING
WOOL FROM LONDON
• 1478-9 •

Miles
0   5   10

London

Fig. 52

Compiled from Customs Accounts, 73/40. The absence of alien vessels is very marked. See footnote 7, p. 329.

between the two parts of the region shown on the map is really one between the eastern and western ports generally, for the conditions at Shoreham and Seaford correspond much more closely to those at ports farther west than they do to those on the east coast, excepting London.

(d) The two most important overseas markets for English wool during the Middle Ages were the Netherlands and Italy. Prior to the fourteenth

21

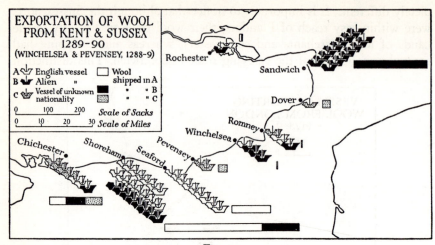

Fig. 53

This map and Fig. 54 are compiled from Customs Accounts as follows: Chichester, 32/2; Shoreham, 135/4; Seaford, 135/4A; Pevensey and Winchelsea, 147/11B; Romney, 147/11A; Dover, Sandwich and Rochester, 124/1A. See footnote 7, p. 329.

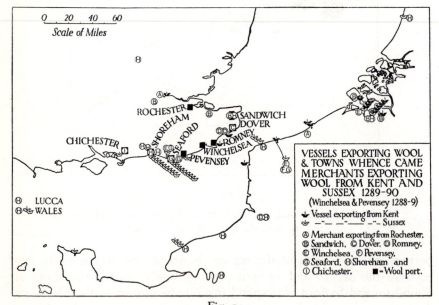

Fig. 54

The sources are the same as for Fig. 53. See footnote 7, p. 329. Note the preponderance of local merchants and vessels in the trade of Sussex.

century, as far as sea transport was concerned, they counted as one;[1] wool destined for Italy was shipped across the narrow seas to marts in the Netherlands or northern France, and thence carried overland (see Fig. 55).

Space will not permit a similar analysis of the wool trade at later periods, but certain tendencies during the two following centuries must be noted. Fig. 45 shows that by the end of the fifteenth century[2] the amount of wool shipped abroad had declined very considerably. At Newcastle, Hull, Boston, Lynn and Chichester the annual total had dropped to a mere fraction of what it had been at the end of the thirteenth century; at London (which now took the lead),[3] Sandwich and Southampton the decline was appreciable though less pronounced; whilst in only one instance, that of Ipswich,[4] was there an increase. The smaller totals were, of course, due mainly to the expansion of the English cloth industry, but it is interesting to notice that the drop in the wool exports at Newcastle, Hull, Boston, Lynn and Chichester was not compensated by a corresponding increase in the amounts of cloth shipped abroad. Since the Chichester totals can be ignored on account of their smallness, one possible inference to be drawn from the map seems to be that wool grown in the midlands and the north, that had previously been shipped from east coast ports, was being diverted to the cloth-making areas in East Anglia and southern England, whence it was exported in a manufactured state through London and the south-western ports.[5] At all events, it is scarcely probable that there had been a decrease in the amount of wool grown in the midlands and the north comparable with the shrinkage in the quantities shipped from these east coast ports, and Fig. 39 shows that there was no large-scale industry in these areas to absorb it.

An important change also took place in the Italian wool trade. During the fourteenth century, a sea route to Italy *via* the Strait of Gibraltar

[1] The "Flanders galleys" first sailed from Venice in 1317, but Genoese vessels had been following the sea route in the thirteenth century, e.g. a Genoese galley exported 22 sacks of wool from Sandwich in 1287 (Customs Accounts, 124/1).

[2] The decline had set in long before 1472. It really began with the growth of the textile industry in the latter part of the fourteenth century.

[3] This lead had been established in the fourteenth century (see for example, footnote 3, p. 275 above).   [4] See p. 301, footnote 1.

[5] Bristol, Exeter and Weymouth, but not Southampton. Much of the cloth manufactured in the hinterland of that port possibly found its way to London before being shipped abroad.

began to be preferred;[1] and it is not, therefore, surprising that by the beginning of the fifteenth century this section of the trade had become concentrated at Southampton and Sandwich, and to a lesser extent at London, all of which were more convenient ports of call for Italian vessels than those on the east coast.[2] Fig. 55 summarises the changes in the Italian section of the trade throughout the period.

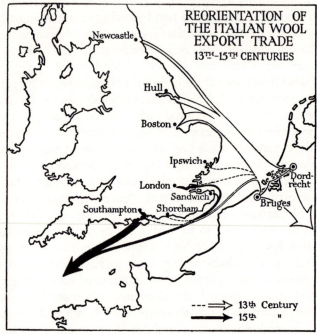

REORIENTATION OF
THE ITALIAN WOOL
EXPORT TRADE
13TH-15TH CENTURIES

Newcastle
Hull
Boston
Ipswich
London
Sandwich
Southampton
Shoreham
Dordrecht
Bruges

- - -> 13th Century
——> 15th   "

Fig. 55

The location of the staple abroad was frequently changed during the reigns of Edward I (1272–1307) and Edward II (1307–27), but it was fixed at Calais for the greater part of the two centuries during which that port

[1] The Champagne Fairs were flourishing in the thirteenth century, but declined during the fourteenth century, when political conflict in Flanders and the outbreak of the Hundred Years' War combined to make the overland route unsafe. (See A. Allix, "The Geography of Fairs", *American Geog. Rev.* XII, 536 (map); C. Walford, *Fairs, Past and Present* (1883) chap. XIX. For routes followed see F. Bourquelot, "Études sur les Foires de Champagne", *Mémoires Présentés...à l'Académie des Inscriptions et Belles Lettres*, 2nd series, vol. V (1865), Part I, chap. VIII.

[2] Italian wool from east coast ports was shipped almost entirely in Flemish and Dutch vessels in the thirteenth century.

was under English rule (see Fig. 56).[1] The only wool that did not pass through Calais was that shipped to Italy *via* the Strait of Gibraltar; the coarse wool exported direct to the Netherlands from Newcastle and Berwick; the wool shipped to non-staple ports by special licence; and, of course, the not inconsiderable amounts that were smuggled abroad.[2]

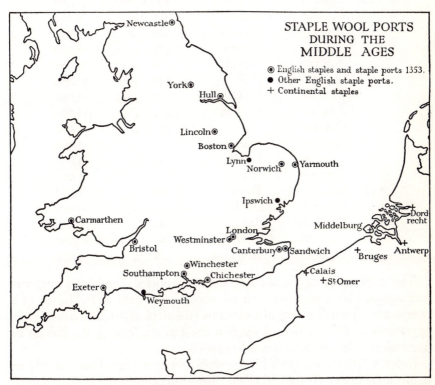

Fig. 56

Although the sea passage to the Continent was short, the risks of loss from official and unofficial enemies were great, so that ships often sailed in fleets, the convoy system being frequently used throughout the Middle

[1] A. L. Jenckes, *The Origin, the Organization and the Location of the Staple of England* (1908), p. 79. Hull, Boston, Yarmouth, London, Sandwich and Southampton were at first only the outports for the staple towns but in some cases the port tended to overshadow its neighbour (see, for example, the conflict between Boston and Lincoln in P. Thompson, *The History and Antiquities of Boston* (1856), p. 55). The staple was actually transferred from Lincoln to Boston in 1369.

[2] E. Power and M. M. Postan, *op. cit.* p. 47.

Ages.[1] Some idea of this may be gathered from the following table, which summarises the shipments of wool and wool-fells from various ports in 1470–1.[2]

Date		Port	Wool	Wool-fells
1470	April 11	Ipswich	591 sacks	18,000
	„ 12	Boston	259 „	9,000
	May 4	Ipswich	242 „	7,000
	„ 9	Boston	1,374 „	18,000
	„ 15	Ipswich	134 „	13,000
	„ 26	Hull	841 „	34,000
	June 10	Ipswich	103 „	12,000
	Aug. 4	London	2,114 „	256,000
	Oct. 29	„	491 „	27,000
	Nov. 12	Boston	37 „	1,000
	„ 28	London	940 „	2,000
	Dec. 22	„	238 „	5,000
1471	Jan. 29	Boston	146 „	1,000
	Feb. 12	London	1,022 „	6,000
	March 10	Hull	184 „	8,000
	„ 28	London	1,161 „	81,000
	various	Sandwich	6 „	1,000

The diminution in the annual amounts exported after about 1370 was accompanied by a concentration of most of the trade in the hands of a comparatively small group of merchants (the staplers)[3] who confined their operations to four or five of the east coast ports, leaving the Italians in possession of Southampton and Sandwich.

Hides, which were included in the wool accounts, played but a small part in medieval trade. At the end of the thirteenth century the average annual

[1] See p. 268, footnote 2; E. Power and M. M. Postan, *op. cit.* p. 75; *Black Book of the Admiralty* (Rolls Series), 1876, IV, p. 27; Exchequer K.R. Accounts, 55/8. The following sailings, between 5 July 1478 and Mich. 1479 (Customs Accounts, 73/40), refer to London:

24 July, 1478: 41 vessels		13 Feb., 1479: 5 vessels		
25 Sept. „	3 „	27 Mar. „	24 „
31 Oct. „	2 „	18 May „	14 „
13 Nov. „	8 „	9 Sept. „	18 „
14 Dec. „	5 „		

[2] E. Power and M. M. Postan, *op. cit.* p. 42.

[3] Possibly as a result of the negotiations between Edward III and the wool merchants and the regulation of the wool trade after 1336 (see "The Estate of Merchants", in G. Unwin, *op. cit.* pp. 179–255).

total for the whole country was only about 45,000;[1] and two centuries later it had dwindled to less than 2000.[2] But just as the decline in the shipments of wool was a measure of the development of the cloth industry, so the diminishing exports of hides may perhaps be regarded as an indication of the growth of leather working. During the period 1282–90 wool-fells

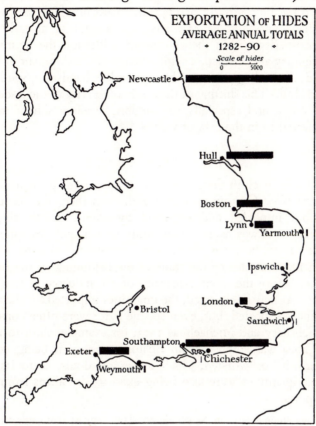

Fig. 57

The figures on which this map are based are taken from Pipe Roll 133, m. 32 d and 134, m. 3 (P.R.O.).

exported outnumbered hides by about 3 to 1; by the end of the fifteenth century the proportion had risen to something like 200 to 1. But the substitution of sheep for grain in the rural economy can only be a partial explanation of the change, and there must still have been many thousands of cattle in the country in the fifteenth century.

[1] Excluding Bristol.   [2] E. Power and M. M. Postan, *op. cit.* p. 323.

Fig. 57 shows that Newcastle, Southampton, Hull, Boston and Lynn were the leading ports of shipment during the early period;[1] London was of little account, possibly because the leather industry there was sufficiently developed to absorb most of the hides that were brought into the city. Although London cordwainers do not figure among the wealthiest contributors to the Lay Subsidy of 1332, the ward in which they were concentrated had a very high *per capita* average, suggesting an early development of prosperity within their mistery.[2] Although the figures for the fifteenth century are so small, a slight change in the shipment of hides can be detected; the leading ports now being London, Sandwich, Southampton, Bristol and Hull.[3] This shifting of the centre of gravity of the trade towards the south coast, and especially to London, recalls a similar movement already referred to in the case of wool.[4]

THE CLOTH TRADE

Of the trade in cloth there are no detailed records prior to the imposition of the New Custom in 1303,[5] and, even then, the accounts only include cloth imported and exported by aliens. Cloth shipped by English merchants was exempt from duty until 1347.[6] In spite of their incomplete nature, however, these early accounts throw some interesting light upon the character of the trade at the beginning of the fourteenth century, i.e. before the great expansion of the industry in England had taken place. As may be expected, the quantities imported by aliens greatly exceeded those exported, but, even so, the latter were often considerable. Between February and Michaelmas 1303, for example, cloth to the value of over £2000 was shipped from Boston, most of it being designated "English cloth" or worsted, which leaves no doubt as to its origin.[7] Appreciable quantities were also being exported by aliens from most of the other ports.

[1] The heavy shipments from Newcastle and Hull suggest that cattle were kept in large numbers in the north. Vaccaries (cattle farms) were numerous in Lancashire (G. H. Tupling, *Economic History of Rossendale* (Chetham Soc. LXXXVI (1927), 17–27) and Yorkshire (P. A. Lyons, "Compoti of the Yorkshire Estates of Henry de Lacy, Earl of Lincoln", *Yorkshire Archaeological and Topographical Journal*, VIII, 351 (1884)). See also R. J. Whitwell, "English Monasteries and the Wool Trade in the Thirteenth Century", in *Vierteljahrschrift für Sozial- und Wirtschaftsgeschichte*, II, 20 (1904), for the substitution of sheep for cattle on the estates of Meaux Abbey.

[2] G. Unwin, *op. cit.* p. 41.   [3] E. Power and M. M. Postan, *loc. cit.*

[4] See p. 311 above.

[5] See N. S. B. Gras, *Early English Customs System*, p. 66.   [6] *Ibid.* p. 72.

[7] Customs Accounts, 5/7; N. S. B. Gras, *op. cit.* p. 273.

Most of the imported cloth came from Flanders and Brabant, but Italian merchants often appear, though less frequently than those from the

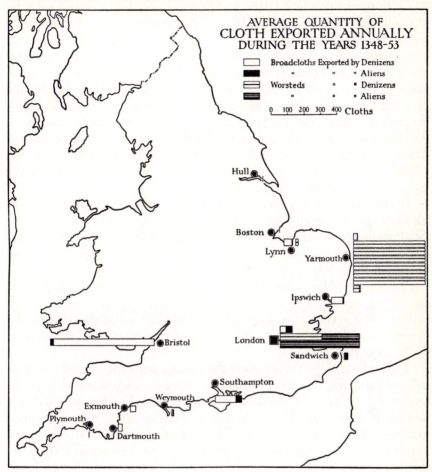

Fig. 58

This map is compiled from Customs Accounts, 158/15 and Exchequer K.R. Accounts, 457/19, 20, 22, 23. The period covered includes the Black Death, and the totals for Hull and Boston were disproportionately lower, whilst those from Bristol were somewhat higher, than those of other ports for the following decade.

Low Countries, in the accounts. In the Sandwich returns for 1304–5,[1] for example, merchants from Florence, Milan and Parma were listed, but, since most of the Italian trade with England came overland at this period, there

[1] Customs Accounts, 124/13; N. S. B. Gras, *op. cit.* p. 302.

is no means of telling whether they were importing Italian cloth or cloth purchased at the great marts in the Low Countries.

Cloth production in England had become localised in certain areas by the end of the fourteenth century, and this localisation is reflected in the export figures. A comparison between Fig. 58 and Fig. 39 will show that these areas were, in fact, already exporting cloth on a fairly large scale by the middle of the century; Bristol and Southampton served as the outlets for the west country broadcloth region, Yarmouth and London for the worsted areas. Fig. 58 shows also that, at the three provincial ports, English merchants had the trade largely in their own hands, whereas at London alien merchants had secured a firm footing. If the figures for Hanseatic merchants were included they would increase still further the aliens' share of the total at London in particular, and to some extent at the other east coast ports as well.

During the years 1353–60 there was a great increase in the amounts of cloth exported from all parts of the country. In 1359–60,[1] for example, the Southampton total of broadcloths was nearly four times as great as that for the whole period 1348–53 shown in Fig. 58, whilst that of London had increased three-fold. The Black Death, no doubt, reduced the totals for the earlier period below the normal level; in which case the figures provide interesting sidelights upon the rapid recovery made by the cloth industry within a few years of that catastrophe. Although worsteds shared in the general recovery, the amounts of worsted cloth exported did not increase in the same ratio.[2]

Throughout these twelve years the leading ports maintained the same relative positions, but, by the end of the century, a significant change had occurred. As far as broadcloths were concerned, London now led in a remarkable way (see Fig. 59), and its predominance was maintained throughout the fifteenth century. The remaining ports still stood in very much the same relation to each other as they had done about forty years before, but the gap between Bristol and Southampton on the one hand,

[1] H. L. Gray, "The Production and Exportation of English Woollens in the Fourteenth Century", *Eng. Hist. Rev.* XXXIX, 34 (1924). The figures for 1359–60 are as follows (broadcloths only):

Newcastle	19	Sandwich	75
Hull	447	Chichester	109
Boston	1022	Southampton	3060
Yarmouth	592	Exeter	1498
London	1225	Bristol	3554

[2] The export of worsted cloth steadily declined during the fifteenth century (E. Power and M. M. Postan, *op. cit.* p. 325).

and the lesser ports on the other, had narrowed very considerably. The shrunken totals for these two ports are no doubt to be accounted for not by any decrease in the amount of cloth manufactured in their immediate hinterlands, but by the deflection of much of the west country cloth to

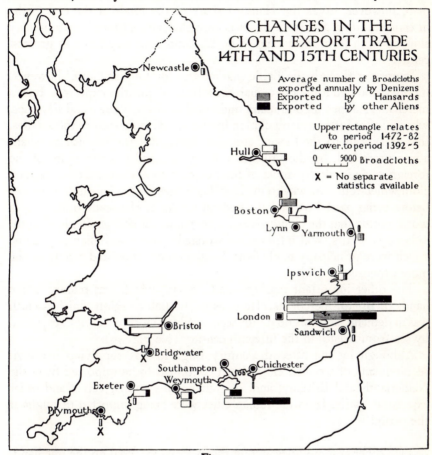

CHANGES IN THE
CLOTH EXPORT TRADE
14TH AND 15TH CENTURIES

Average number of Broadcloths exported annually by Denizens
Exported by Hansards
Exported by other Aliens

Upper rectangle relates to period 1472-82
Lower, to period 1392-5
0   5000 Broadcloths
X = No separate statistics available

Fig. 59

The averages for the earlier period are taken from Gray's table (*op. cit.* p. 35) and for the later period they are compiled from the Enrolled Customs Accounts in E. Power and M. M. Postan, *op. cit.* pp. 330–60. "Other aliens" were mostly Italians (exporting from Southampton) and Dutch (exporting from London).

London for export. In other words, the increased shipments from London are almost certainly an index of greater commercial, rather than industrial, activity in the city, and mark one further stage in its development as the economic capital of the country.

It would be interesting to know how far the increase in the London cloth exports was due to greater activity on the part of Hanseatic merchants. This cannot be estimated, for Hansards were apparently exempt from the Cloth Custom of 1347, on the returns for which Fig. 59 is based. This map, however, gives the Hansard share of the totals at the end of the fifteenth century, and from this it is clear that, although they occasionally shipped cloth from most English ports, London was by far their greatest centre.

Important though the Baltic market was for English cloth,[1] however, it only accounted for about a quarter of the total quantity exported. Italian merchants shipped a good many undressed cloths to Venice and Florence, whence, after further manipulation by the skilled craftsmen of those cities, they were sent to the Levant and elsewhere. But the English merchant adventurers had, by the fifteenth century, secured a major share of the foreign trade in cloth;[2] most of their exports took the form of undressed cloth which found its way to the Low Countries for the finishing processes before being passed on into the general stream of Hanseatic traffic.[3] It is worth noting that the Hansards exported mainly the cheaper kinds of cloth[4] —kerseys, straits, Welsh friezes and worsteds—whereas Italian shipments, which were principally made from Southampton, consisted chiefly of the more expensive qualities.

The other important market was Gascony, which was mainly in the hands of Bristol merchants. Here, as in the Baltic region, finished cloth was in demand, and this no doubt accounted for most of the cloth exported by sea from Bristol in the fifteenth century (see Fig. 59).

Although by the fifteenth century England was supplying her own requirements of woollen cloth, linen continued to be imported from the Netherlands and Brittany, and, of course, silks, satins, velvet and other expensive textiles from the Italian cities were being imported throughout the period.

Two other commodities among the exports, viz. coal on the one hand,

[1] The woollen cloth industry had not become established in this region, so finished cloths were in demand (E. Power and M. M. Postan, *op. cit.* p. 144).

[2] E. Power and M. M. Postan, *op. cit.* pp. 139, 151.

[3] *Ibid.* pp. 143, 145.

[4] Possibly owing to the lighter taxation of these types of cloth (*loc. cit.*). The area of distribution was very extensive; Cologners carried English cloth along the Rhine to Frankfort, whence it was distributed throughout the upper Danube basin and Galicia, whilst Prussian merchants supplied the markets of Russia, Poland, Hungary and Wallachia.

and timber, together with wood fuel and oak-bark on the other, deserve mention because they differed markedly in character from wool, hides and cloth in being the virtual monopolies of Newcastle and Winchelsea respectively.

The earliest accounts of Newcastle reveal a well-established export trade in coal,[1] amounting to several thousand tons a year and serving a wide area. Fig. 60 gives an idea of its extent at the beginning of Richard II's reign and

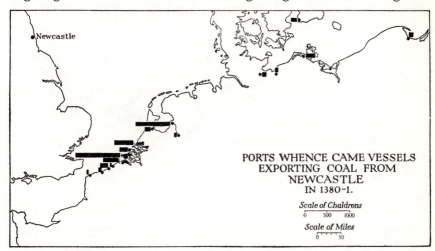

PORTS WHENCE CAME VESSELS
EXPORTING COAL FROM
NEWCASTLE
IN 1380-1.

Scale of Chaldrons
0   500   1000

Scale of Miles
0   50

Fig. 60

The precise destination of the coal is not given in the accounts, the amounts shown on the map being merely placed opposite the ports from which the vessels came. (Compiled from Customs Accounts, 106/4.) See footnote 7, p. 329.

incidentally illustrates the two main branches of trade that affected all the east coast ports. Many of the vessels shipping coal do not appear on the imports side of the account, and since coal was usually the only cargo that these particular vessels took on board it is possible that they had called at another port farther south, discharged their goods and shipped a small return cargo, in which case coal would have been taken on board partly as ballast. It is interesting to notice small quantities of this coal reshipped from a continental port and landed at such places as Winchelsea.[2] The coal exported from Sandwich in 1303[3] and Southampton in 1309[4] had

---

[1] See J. Brand, *History and Antiquities of Newcastle-upon-Tyne* (1789) II, pp. 241 *et seq.*; L. F. Salzman, *op. cit.* pp. 281–2.

[2] Customs Accounts, 32/6.

[3] Customs Accounts, 124/11; N. S. B. Gras, *Early English Customs System*, p. 273.

[4] Customs Accounts, 136/8; N. S. B. Gras, *op. cit.* p. 361.

in all probability been shipped coastwise from Newcastle in the first place, although it may have been brought originally from Newcastle *via* a continental port.[1]

Winchelsea at no time compared favourably in size and importance with the other east coast ports, but its position at the natural outlet of the Weald made it the greatest exporter of timber and fuel in England throughout the Middle Ages.[2] There is abundant evidence of the shipment of these commodities coastwise to London and other English ports,[3] but the bulk

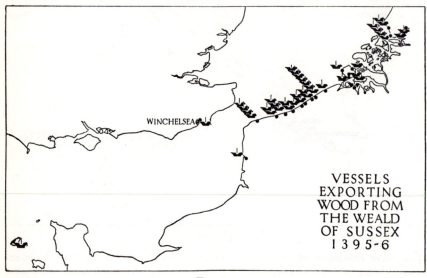

VESSELS
EXPORTING
WOOD FROM
THE WEALD
OF SUSSEX
1395-6

Fig. 61

Large quantities of oak-bark were included in the cargoes exported by these vessels (compiled from Customs Accounts, 33/28). See footnote 7, p. 329.

of the trade was done with merchants from the Netherlands, Picardy and Normandy. Their distribution in a typical year is shown in Fig. 61. The Weald in general, however, was a region of comparative poverty, and the purchasing power of its inhabitants was low. Consequently, the alien vessels seldom brought any large cargoes to exchange for the forest products, and most of them came over in ballast. This was tipped over-board in the mouth of the Rother, and it is rather ironical that the

[1] If brought direct from Newcastle, it would not have been listed in the account.

[2] Small quantities were also exported from the Medway (Customs Accounts, 125/1).

[3] E.g. in 1358 when timber was taken to Boston for bridge repairs (*Cal. Pat. Rolls*, 1358–61, p. 119). London was supplied with firewood from this area—J. F. Willard, "Inland Transportation in England during the Fourteenth Century," *Speculum*, I, 371 (1926).

very trade on which the early development of the port was so largely
built should have hastened in this way the silting up of its haven and the
ultimate destruction of its prosperity.[1]

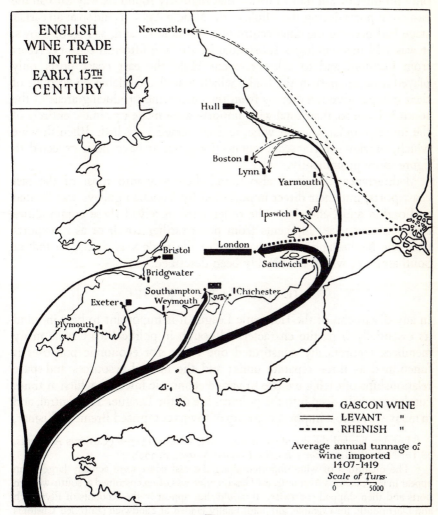

ENGLISH
WINE TRADE
IN THE
EARLY 15TH
CENTURY

Newcastle

Hull

Boston
Lynn
Yarmouth

Ipswich

Bristol   London
Bridgwater   Sandwich
Southampton   Chichester
Exeter Weymouth
Plymouth

━━━ GASCON WINE
═══ LEVANT   "
------ RHENISH   "

Average annual tunnage of
wine imported
1407-1419

Scale of Tuns.
0   5000

Fig. 62

Small quantities of wine were also imported at Chester and other western ports which
do not appear regularly in the accounts.

[1] R. A. Pelham, "The Foreign Trade of the Cinque Ports in the year 1307–8",
Studies in Regional Consciousness and Environment (ed. I. C. Peate, 1930), p. 145.

The imported commodities that were common to all the North Sea ports can be divided into three categories: (*a*) Gascon and Rhenish wine, (*b*) Baltic produce, and (*c*) Low Country produce. Gascon wine was the only product of Southern Europe that regularly found its way into all the east coast ports during the whole of the Middle Ages. London at an earlier stage had become the chief centre of importation;[1] and, as Fig. 62 shows, it was still maintaining a substantial lead in the fifteenth century. Apart from London, and to a lesser extent Hull, the east coast ports only played a minor part in the trade, which is understandable on account of their comparative remoteness from the chief wine-producing areas of the south.[2] Even so, their total wine imports were made up almost entirely of the more popular Gascon wine, and contained very little Rhenish wine which, in view of Hanseatic connections, might have been expected to figure more prominently.

Mediterranean products also found their way into most of the east coast ports, but their direct importation by Venetian galleys was limited to London and Sandwich. The other ports received these commodities either as coastwise shipments from ports farther south or as re-exports from the Netherlands. The character of the trade was similar to that at Southampton, which has already been described.[3]

THE HANSEATIC TRADE[4]

In any discussion of the Hanseatic League it is important to bear in mind its essentially tripartite character. Although in political matters its many members theoretically constituted one body, for economic purposes it functioned as three separate units; and the natural resources and space relationships of each gave rise to varied economic interests which at times cut dangerously deep into the political life of the League. The central, and in many ways most important, group of towns comprised Bremen, Wismar,

---

[1] London headed the list of ports importing wine in 1327–8 (Customs Accounts, 78/3*a*; N. S. B. Gras, *Early English Customs System*, p. 399).

[2] The quantities of wine imported along the east coast were actually larger than appear in the Customs Accounts, for Gascon wine was often customed at south-western ports and then shipped coastwise. It would thus appear in the accounts of Plymouth and Dartmouth, for example, although finally landed at Sandwich (Enrolled Customs Accounts, 5 (P.R.O.)—the year being 1350–1). Southampton had the monopoly of the sweet wine trade after Henry IV's reign, but the local dues charged were so excessive that importation elsewhere was permitted unofficially (A. L. Simon, *The History of the Wine Trade in England* (1907), II, pp. 105–7).   [3] See pp. 272 *et seq.* above.

[4] This section is based upon the chapter by M. M. Postan, "Anglo-Hanseatic Economic Relations", in E. Power and M. M. Postan, *op. cit.* pp. 91–153.

Rostock, Hamburg and Lubeck, the last two commanding the narrow land route across the base of the Jutland peninsula;[1] the western group included the towns of the Low Countries and the Rhineland, especially Bruges and Cologne; away to the east, and extending to the confines of Muscovy, lay a third area with its two great commercial foci of Danzig and Novgorod.

Commercial contacts between England and the Hanseatic towns date back to the early Middle Ages, but by the beginning of the fourteenth century exploration and adventure around the fringes of the Baltic region were giving place to economic consolidation, and Hanseatic merchants were establishing control over the mineral wealth of Sweden, over the fisheries of Skania, and over the fish and fur trade of Norway. Some of that produce was brought direct to England by sea, but much was carried to the great fairs of the Low Countries for disposal. The bulk of this trade was in the hands of merchants from the central and eastern groups of towns who may be designated collectively as the Baltic Hansards, to distinguish them from the merchants of the western group of Hanseatic towns, who were only indirectly concerned with the Baltic trade, their main activities being confined to the trade between England and the Low Countries. Before the mid-fourteenth century, these western Hansards had an important share of the wool trade,[2] but during Edward III's reign they were displaced by merchants of the staple, though they continued to import miscellaneous goods into England, together with a certain amount of Baltic produce which came *via* the marts of the Netherlands.[3]

During the course of the century the expansion of the English cloth industry led numbers of English merchants to seek fresh markets abroad for their cloth. They gained a footing in the trade of the Low Countries,[4] and began to penetrate into the Baltic (at the same time that Hanseatic and Dutch vessels from the western area were attempting to establish an all-sea route into the Baltic), to the discomfiture of Lübeck whose monopoly of the transit trade across the peninsula was thereby threatened. During the second half of the century many English merchants settled in Danzig, and, for a time, shared in the Hanseatic trade with this country.

In many ways that trade was ideal in character, for the two areas concerned were complementary, and many of the economic needs of each

[1] See W. Vogel, *op. cit.* p. 184.

[2] K. Kunze, "Hanseakten aus England, 1275–1412", *Hansische Geschichtsquellen*, VI, 331 *et seq.* (1891).

[3] No doubt taking advantage of the route behind the dunes (W. Vogel, *op. cit.* p. 233).

[4] Especially after 1363, when the wool staple was transferred to Calais (Salzman, *op. cit.* p. 332).

22

could with little difficulty be met by the surplus produce of the other. English cloth found a ready welcome in the cooler Baltic lands and was distributed over a widespread area, while in return for this, timber in various forms, and other forest products, were eagerly accepted in England,[1] together with corn, in times of dearth, from the manorial estates of the newly settled region east of the Elbe. The westward cargoes of timber and corn, however, were much more bulky than the cargoes of cloth from England, and additional transport was often needed for return journeys. Vessels could be purchased at Danzig, which was noted for its shipbuilding, and considerable additions to our merchant shipping must have been made in this way during the period of English participation in the Baltic trade.

The only articles of importance, apart from cloth, shipped from England to countries around the Baltic were tin, pewter ware and coal. The imports, however, were very varied.[2] These latter can be classified broadly into (a) materials for shipbuilding, (b) various articles of wood, (c) raw materials for cloth manufacture, (d) metals and metal goods, (e) fish, (f) corn, (g) other commodities. These are not listed in any particular order, but the first-named was of outstanding interest to a maritime nation like England, and comprised timber, various kinds of boards,[3] masts, oars, pitch, tar, hemp for ropes, canvas and sailcloth. Prominent in the second category were bowstaves. Of the materials for cloth-making, ashes and litmus came from the Baltic region proper, madder from Westphalia. The chief metals imported were high-grade osmund iron from Sweden and copper from Hungary. The commonest fish were stockfish[4] and herrings, but salmon and sturgeon also occur quite frequently in the accounts. Among the unclassified commodities may be mentioned wax and furs from the eastern Baltic, and beer from Bremen and Hamburg. A direct sea route also grew up between Danzig and Gascony, wine from the latter, together with fruit and salt, being exchanged for the corn of the former, thus reducing the burden of supplying Gascony with corn from England.

It was a great misfortune for English merchants that their privileged position in the Baltic was so short-lived, but in the fifteenth century they

---

[1] E.g. on the north-east coast as early as the thirteenth century boards were purchased from German or Dutch merchants (R. J. Whitwell and C. Johnson, "The Newcastle Galley, A.D. 1294", *Archaeologia Aeliana*, 4th series, II, 148 (1926)).

[2] See *The Libelle of Englyshe Polycye*, written in 1436 (ed. Sir G. Warner, 1926), pp. 16, 17; L. F. Salzman, *op. cit.* chap. XVII.

[3] These included wainscot (oak boards), righolts (apparently from Riga), Eastland boards, "barelholts" and "deles" (L. F. Salzman, *op. cit.* p. 363).

[4] These were dried fish of various kinds, including cod.

had to contend against the efforts of the Hanse to regain its monopoly. Although successful in the early stages of the conflict, the English merchants were finally forced to retire from the trade, and, by the end of the fifteenth century, Hanseatic supremacy was complete. But the volume of trade was probably affected very little. The Hansards continued to stimulate the textile industry in England by exporting several thousand cloths each year. And, in any case, a new outlet for commercial activity was found in Iceland,[1] which became a suitable alternative goal more particularly for those merchants who had lost their connection with Bergen as a result of Hanseatic pressure.[2]

THE TRADE WITH THE LOW COUNTRIES

The geographical setting for trade between England and the Netherlands was excellent. The latter lay at the point of intersection of the sea routes between southern Europe and the Baltic, on the one hand, and the overland routes linking England with the Mediterranean and the Danubian lands on the other. Thus to the advantages of a short sea passage was added the ease with which goods from any part of the commercial world could be imported direct from great marts like Bruges and Lille, whilst the dependence of the Flemish industrial population upon foodstuffs[3] and raw materials from England ensured reciprocity. Unfortunately, the wealth to be gained from this trade, and the vulnerability of the routes followed, made it a prey to pirates and political intriguers alike, but in spite of frequent interruptions its general character remained very much the same throughout the Middle Ages.

There were marked regional differences within the rather ill-defined limits of the Netherlands. Flanders was essentially an industrial area, whereas the ports that had grown up in the delta of the Rhine were engaged mainly in commerce. But there were important contrasts within Flanders itself; manufactures were concentrated in the Walloon towns around Douai during the twelfth and thirteenth centuries, whereas Bruges was a centre of commercial activities.[4] Political events at the end of the thirteenth century, however, caused a migration of industry towards the

---

[1] See E. Carus-Wilson, "The Iceland Trade", in E. Power and M. M. Postan, *op. cit.* p. 155.

[2] The English settlement at Bergen was destroyed in 1402 (L. F. Salzman *op. cit.* p. 331).

[3] These were very varied and included grain, ale, cheese, bacon, beans, honey, fish, butter, swine, cattle, lard and meat (L. M. Seckler, *op. cit.* pp. 212–13). Most of the grain and dairy produce came from East Anglian ports, especially Lynn, and the ports of south-eastern England.　　　　　　　　　　[4] *Ibid.* p. 4.

North, and Bruges became not only one of the four leading Hanseatic "factories" of northern Europe but an important manufacturing centre as well. However, in the fifteenth century, the silting up of the Swin, combined with local gild restrictions, helped to bring about a further change of focus, and Antwerp superseded Bruges.[1]

The Dutch, who had been engaged to some extent in the carrying trade during the early Middle Ages, completed the main part of their defensive works against the sea by the beginning of the fourteenth century,[2] and began henceforth to take a more active part in commerce. Throughout the later Middle Ages, Dutch and English seamen gradually weakened the supremacy which Flemings and Hansards had formerly enjoyed, and thus prepared the way for the great naval conflict of the seventeenth century.

Wool was the commodity of outstanding importance that crossed the narrow seas. It made its way by slightly different routes at different periods, according to the location of the staple,[3] and its final destination changed with the northward spread of industry in the Netherlands, but the zone of traffic remained narrow. After 1317 the bulk of the wool exported from England to Italy was sent *via* the Strait of Gibraltar, although some was re-shipped from Calais or Bruges. English wool was of vital importance to Flanders mainly because there was no effective substitute, for Spanish wool, which was imported in considerable quantities, was of no value unless mixed with the finer varieties from England.[4]

The author of *The Libelle of Englyshe Polycye* was unfortunately prejudiced against the Flemings,[5] and his description of their activities is somewhat misleading. He recognised that their chief commodity was cloth:

> Fyne clothe of Ipre, that named is better than oures,
> Cloothe of Curtryke,[6] fyne cloothe of all colours,
> Muche fustyane and also lynen cloothe.[7]

but his statement that Flanders produced nothing "But a lytell madere and Flemmyshe cloothe"[8] ignored the important leather industry, for

---

[1] The Merchant Adventurers settled in Antwerp in 1407. Henceforth this city became their main centre (L. F. Salzman, *op. cit.* p. 332).

[2] E. Power and M. M. Postan, *op. cit.* p. 94.

[3] See A. L. Jenckes, *loc. cit.*

[4] *Libelle*, p. 6; L. F. Salzman, *op. cit.* p. 357.

[5] Because it was primarily a region of commercial exchange:
"For the lytell londe of Flaundres is
But a staple to other londes iwys." (p. 7.)

[6] Courtrai.   [7] *Libelle*, p. 5.   [8] *Ibid.* p. 7.

which large quantities of oak bark were exported from Winchelsea and the Medway estuary, and also the manufacture of hats,[1] hose, metal and wooden articles, bricks and tiles, of which there is abundant evidence. The commodities of Brabant and Zealand, according to *The Libelle*,

> Be madre and woade, that dyers take on hande
> To dyen wythe, garleke and onyons,
> And saltfysche als for husbond and comons.[2]

But to these might be added earthenware from Zealand,[3] onion seed, beer, which made steady progress as an import in spite of opposition[4] and hops, which were imported at least as early as 1420.[5]

There were, besides, a host of manufactured products, for the most part metal goods, the raw materials for practically all of which had been imported into Flanders in the first place from northern and central Europe.[6] It was this dependence upon foreign supplies that made the Flemings particularly sensitive to political disturbances—to anything, in fact, which endangered their livelihood by threatening to cut their lines of communication. To what extent the tension here, as elsewhere, was relieved by the partial re-orientation of European interests towards the East Indies and the Americas after 1500 is a matter for discussion, but the impression gained from a survey of the trade[7] of the later Middle Ages is that some important economic readjustments were long overdue.

For Bibliographical Note see p. 297.

[1] Chaucer's merchant wore "a Flaundrish bever hat".   [2] *Libelle*, p. 28.
[3] From the area that later specialised in Delft ware (L. F. Salzman, *op. cit.* p. 361).
[4] L. F. Salzman, *English Industries of the Middle Ages* (1923), p. 295.
[5] Customs Accounts, 72/17 (London); N. S. B. Gras, *op. cit.* p. 501, etc. For details of Dutch trade see H. J. Smit (ed.), *Bronnen tot de Geschiedenis van den Handel met Engeland, Schotland en Ierland 1150–1485* (2 vols, 1928), *passim*.
[6] J. G. Van Dillen, *Het Economisch Karakter der Middeleeuwsche Stad* (1914), p. 69; L. M. Seckler, *op. cit.* pp. 220–1.
[7] It should be pointed out that Figs. 50, 51, 52, 53, 54, 60 and 61 are not as complete as could be wished, owing to the fact that (*a*) a town or port is not always given against the name of a merchant or vessel in the accounts, and (*b*) that some towns and ports, even when mentioned, cannot be identified. In Figs. 52, 53 and 54, however, the omissions are negligible, amounting to only two or three symbols in each case, but although there are more omissions from the other maps, especially Figs. 50 and 51, there is good reason to believe that the conclusions drawn from them are not thereby invalidated.

It has been thought desirable to map the details for individual years rather than attempt to show average conditions over a period of years, owing to the varied nature of the documents concerned, but the year illustrated is representative of its period in each case.

Chapter IX

LELAND'S ENGLAND

E. G. R. TAYLOR, D.Sc.

"Within living memory" is a phrase which may perhaps be stretched backwards to cover a hundred years, but no more. Three-quarters of a century is bridged by the direct memories of the octogenarians, a further twenty-five years by the dimmed recollections of what these heard from their elders. Hence it is natural that every hundred years or so some inquiring spirit has felt impelled to take stock after his own fashion of a changed and ever changing England.

John Leland was a pioneer of the method of direct inquiry and observation, and, although his primary objective had been a survey of the libraries of the doomed monasteries, he became (as he related to King Henry VIII)

totally enflamed with a love to see thoroughly al those Partes of this your opulente and ample Realme that I had reade of yn the aforesaide writers yn so muche that al my other Occupations intermitted, I have so traveled yn your Dominions boothe by the Se Costes and the midle Partes, sparing nother Labor nor Costes, by the space of these VI Yeres paste, that there is almost nother Cape, nor Bay, Haven, Creke or Peere, River or Confluence of Rivers, Breches, Waschis, Lakes, Meres, Fenny Waters, Montagnes, Valleis, Mores, Hethes, Forestes, Chases, Wooddes, Cities, Burges, Castelles, principal Manor Placis, Monasteries and Colleges, but I have seene them; and noted yn so doing a hole Worlde of Thinges very memorable.

Unfortunately a mental disorder prevented this diligent observer from writing his proposed *Description of the Realm of England*; and his voluminous notes passed at his death in 1552 into the hands of Edward VI's tutor, Sir John Cheke, to be dispersed subsequently, and not to be reassembled until the eighteenth century.[1] But unco-ordinated and incomplete though they remain, they afford us a picture of Early Tudor England which is as revealing by what it omits as by what it emphasises.

The geographical situation of medieval England had been neatly summed up in the *Libelle of Englyshe Polycye* so far as it affected policy, strategy and overseas trade.[2] Britain interposes a 600-mile barrier aligned from south-

[1] By Thomas Hearne, from whose 3rd edn. (1768) of the *Itinerary* Leland is quoted.
[2] This appeared in 1436 and was written by an anonymous author who probably was Adam Moleyns, afterwards bishop of Chester. It was included in Hakluyt's *Voyages* (1589), and it has been edited by Sir George Warner (1926).

east to north-west, and stretching from Dover to the Outer Hebrides, athwart the peripheral sea-way between the Baltic and Mediterranean Seas. Hence the merchant ships of the Hanse and of Italy alike were forced into the gut between Dover and Calais. In the very Narrows, as it chanced, lay Flanders, the goal alike of the sea-ways from north-east and south-west, and of the axial land-way from the south-east across Europe. England itself offered a natural prolongation of this land-way, that of the

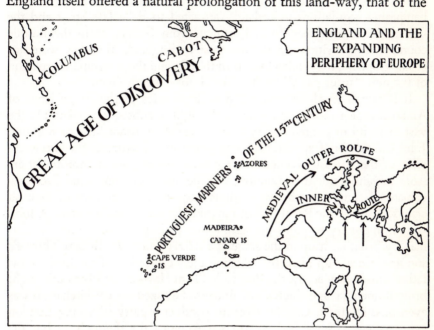

Fig. 63

Dover-London-Chester line, whence was a final outlook towards northern Ireland. The ocean boundary therefore, the true "utter margin" of Europe, was not lowland England, but embraced Ireland, the islands and highlands of Scotland, and the High Fjeld and Skerry Guard of Norway. Lisbon, Galway, Lerwick, Bergen formed a chain of ports along an outer route the use of which familiarised sailors and merchants with the grey and stormy Atlantic itself. To widen this periphery, and to reach the fisheries of Finmark, Iceland, and even Greenland, made no new demands upon courage, skill or nautical equipment. In the fifteenth century the line of the "utter parts" ran from Arctic waters through Iceland to the Azores and thence to the Madeira group, the Canaries, Cape Verde Islands and Guinea. Bristol men as well as Portuguese were made free of the Eastern

Atlantic, and dreamt of what might lie still farther afield. It was, indeed, as inevitable as the widening of the circle made by a stone cast into the water that a Portuguese-trained Columbus, and a John Cabot supported by two merchant shippers of Bristol,[1] should, by the close of the fifteenth century, have extended the radius of Europe to embrace the Antilles and the New Found Land. The report of cod-fishing grounds of incredible richness, far from the domains of the king of Denmark (who interposed his hampering regulations in Iceland), attracted to Belle Isle Strait and the Grand Bank, fishermen from Portugal, from Galicia, from the Basque country, from Brittany, and from western England. But they were not cosmographers, they sailed by rule of thumb, and the geographical relation of the new "utter parts" to the old did not concern them.

It is worth emphasis that the first discoveries of the fringe lands of America appear momentous only in the light of subsequent events: to the vast majority of contemporary Englishmen they meant little or nothing. John Cabot's brief appearance swaggering in silk attire as the successful pilot of the Bristol adventurers provoked one or two humorous letters from Italians resident in London, and the interpretation of his discovery demanded some interchange of diplomatic notes in respect of Spanish claims, based on the Papal Bull of 1493: but he was soon forgotten. A John Rastall might try to point the significance to England of the new world-map by bringing it upon the stage in a *Merry Interlude*; Robert Thorne's son might draw up plans for reaching the "backside" of the lands of his father's discovery across the Pole; a William Hawkins of Plymouth might bring home a savage chieftain of Brazil to be gazed at in Whitehall: it was even possible for a Captain Hore to conduct a party of young London gentlemen across the Atlantic to see America for themselves; but the stories they had to tell on their return were of danger and horror. Two or three projects for English exploration were put forward during the reign of Henry VIII, but they fell through for lack of intelligent support.[2] The Merchant Staplers had their interests at Calais, the Merchant Adventurers at Antwerp; while the Mediterranean and Levant trade based on Seville was expanding. In Leland's day the essentially south-eastern orientation of the English outlook was still unchallenged. He pictured, it is

[1] Hugh Eliot and Robert Thorne the elder.
[2] The City Companies were approached at Wolsey's instance in 1521 to furnish ships for an expedition to the north-west, with a promise of very substantial privileges in the new trades to be opened up. It is clear from their reply that they were unable to envisage the meaning of the proposal, and saw in it only a wanton hazarding of men's lives. J. A. Williamson, *Voyages of the Cabots* (1929), p. 97.

true, a changing England; in one place there was rapid growth, and in another there was decay, but only rarely or indirectly can such changes be linked to the discovery of the New World.

Exception must be made of the disappearance of the Italian spice merchant from Southampton and London consequent upon the Portuguese discovery of the sea route to India. "Hampton is a goodlie towne", wrote Roger Barlow, "and of grete trat and spetiallie when the venetians had the trat of the spices in ther handes, for then thei made there ther chief scale, but sence that the king of portugal had the trat from calicout he removed the scale into flandres at antwerpe".[1] So too, Stow remarks that the Galley Key in London had been long deserted and was built over with dwelling-houses and storehouses, while the old warehouses of the Italian merchants were falling in ruins and were let out as stables or as ale-houses and eating-houses for sailors.[2] Leland, it is true, wrote[3] of a Cornish harbour as fit to hold the great carracks of Genoa, and mentions the great "Ragusais" which sheltered in the downs off Broadstairs, but such ships were already a memory only. Nevertheless, even in Southampton, the disappearance of the merchant strangers could have had little effect upon general prosperity. Not only was the French wine and woad trade expanding, alongside that of Mediterranean produce handled by the Anglo-Spanish merchants, but as London grew, so Southampton, as its inevitable out-port, grew likewise. The south-west winds which brought ships up the Channel and up the Solent did not serve for the tedious doubling of the North Foreland, and goods were often landed at Southampton for transport by road to London, as well as to the nearer markets of Winchester and Salisbury.[4]

[1] A Brief Summe of Geographie (ed. E. G. R. Taylor, Hakluyt Society, 1931), p. 46.
[2] The Key did not, however, fall completely out of use. It was being used for landing general merchandise in 1582.
[3] "Falmuth is a Haven very notable and famose, and yn a maner the most principale of al Britayne. For the Chanel of the Entre hath be spece of ii Myles ynto the Land xiiii Fadum of Depes, which communely ys cawlled Caryke Rood by cawse yt ys a sure Herboro for the greatest Shyppes that travayle be the Ocean" (Itinerary, VII, 121).
[4] "Ther be 3 principall Streates in Hampton, where of that that goithe from the Barre Gate to the Watar Gate is one of the fayrest Streates that is in al England, and it is welle buildid for Timbre Building. There ys a fair House buildid in the Midle of this Streat for Accomptes to be made yn. There cummith fresch Water into Hampton by a Conduct of Leade, and there be certen Castelletes onto this Conduct withyn the Town. There be many very fair Marchants Houses in Hampton; But the chefest is the house that Huttoft, late Custumer of Hampton, buildid in the West side of the Town. The House that Master Lightster, chief Bane of the King's Escheker, dwellith yn is very fair. The House that Master Mylles the Recorder dwellith yn is fair. And so be the houses of Nicoline and Guidote Italians" (Itinerary, III, 77).

On the east coast, more varied fortunes had befallen the sea and river ports, all of which in the past had owed much to the Hansards, and to the export of English wool. Hampered by taxes and restrictions, as well as by English competition, the merchants of the Hanse had gradually withdrawn, while the expansion of the home cloth industry had changed the direction of the movement of wool. To Newcastle, this was of little consequence, for there the wealthy coal merchants had already appeared. Yarmouth had the great herring fair; Norwich (though sadly decayed) still had its worsted industry; Hull, on the major river estuary, had plenty of business to do with the vales of York and Trent. But Boston and Lynn, with Lincoln and Stamford behind them, all pre-eminent under the earlier régime, had found no new activities to recompense them for the disappearance of the old stapling days. They still (especially Lynn) had their functions, for behind the Wash lay a fertile hinterland, but there was a definite shrinkage of prosperity.[1]

To the modifications of geographical values which followed upon changes of external trade and expansion of industry, must be added those which were the result of the acceleration during the early part of the sixteenth century of what has been termed the "agrarian revolution"—a movement of which "enclosures" of very varied types were the outward symbol.[2] From the geographical standpoint a leading feature of the revolution was the accompanying improvement in agricultural production (and especially the development of animal husbandry), which brought changes in the distribution of population, and additional internal traffic in its train. Finally, there must be mentioned the more violent and sudden alterations in the face of England brought about by the suppression of the religious houses and the wholesale transference of property into new hands. The

[1] "Mr Paynel a Gentilman of Boston [actually he was Customer of the port] tolde me that syns that Boston of old time at the great famose Fair there kept was brent, that scant syns it ever cam to the old Glory and Riches that it had; yet syns it hath beene manyfold richer then it is now. The Staple and the Stiliard Houses yet there remayne; but the Stiliard is little or nothing at alle occupied. There were iiii Colleges of Freres. Marchauntes of the Stiliard cumming by all Partes by Est were wont greatly to haunt Boston: and the Gray Freres took them yn a manor for Founders of their House, and many Esterlinges were buried there" (*Itinerary*, VI, 59).

Brackley, standing at the south-western end of the limestone belt running through Northampton, afforded an example of a decayed inland town. "The toune of Brakeley by Estimation of old Ruines hath had many Stretes in it, and that large....It...was a Staple for Wolle, privileged with a Major, the which Honor yet remaynethe to this pore Towne." And in the old Chapel of Ease Leland saw "Stone Imagis beringe Woll Sakks in theyr Hands, in token that it was of the Stapelers Makyng" (*Itinerary*, VII, 11).

[2] See p. 238 above, and pp. 345–6 and 397–9 below.

great abbeys were interested, like other landowners and capitalists, in wool and the woollen industries, and were large-scale employers of labour: their disappearance spelt loss and decay to the towns and villages in which they stood. "Mr Blake the last Abbate [says Leland] buildid two Fulling Milles at Cirencestre that cost a 700 Markes of Mony. They be wonderfully necessary, by cause the Town standith alle by Clothing."[1]

An anonymous writer quoted by Stow makes an interesting attempt to explain the decay of provincial towns and ports, which was increasingly apparent as the century advanced.

But here...I have shortly to answer the accusation of those men, which charge London with the loss and decay of many (or most) of the ancient cities, corporate towns, and markets within this realm, by drawing from them to herself alone, say they, both all trade of traffic by sea, and the retailing of wares and exercise of manual art also. Touching navigation, which I must confess is apparently decayed in many port towns, and flourisheth only or chiefly at London, I impute that partly to the fall of the Staple, the which being long since a great trade, and bestowed sometimes at one town and sometimes at another within the realm, did much enrich the place where it was, and being now not only diminished in force, but also translated over the seas, cannot but bring some decay with it; partly to the impairing of havens, which in many places have impoverished those towns, whose estate doth ebb and flow with them; and partly to the dissolution of religious houses by whose wealth and haunt many of those places were chiefly fed and nourished.[2]

As far as the impairing of havens was concerned, Leland met with abundant examples of disastrous silting, especially round the Devon-Cornwall coast, which he attributed in many cases to the stream-washing of tin which was carried on in those counties.[3] The choking of small

[1] *Itinerary*, II, 24.
[2] A *Discourse* written circ. 1578 and published in John Stow's *Survey of London*, 1598.
[3] "Now Sandwich be not celebrated by cawse of Goodwine Sands, and the Decay of the Haven.... Hithe hath bene a very great Towne in lenght, and conteyned iiii Paroches that now be clene destroied.... And yt may be well supposed that after the Haven of Lymne and the great old Toun there fayled, that Hythe strayte thereby encreesed and was yn Price. Finally to cownt from Westhythe to the Place wher the Substans of the Towne ys now ys ii good Myles yn lenght al along on the Shore to the which the Se cam ful sometyme, but now by Banking of Woose and great casting up of Shyngel the Se is sumtyme a Quarter sumtyme *dim.* [½] a Myle fro the old Shore. In the tyme of King Edward the 2 there were burned by Casualte xviii Score Houses and mo, and strayt followed great Pestilens, and these ii things minished the Towne.... The Havyn is a pretty Rode, and lieth Meatly strayt for Passage owt of Boleyn [Boulogne]. Yt croketh yn so by the Shore a long, and is so bakked fro the mayn Se with casting

harbours is, in fact, an inevitable process during a period of stillstand of the shore line, but it is a process which many factors may tend to accelerate or retard at particular periods. It is possible that the post-Roman or "Flandrian" depression which deepened inlets in the Middle Ages had been compensated by accretion and refill by the sixteenth century. It must be recalled, however, that the substitution for the medieval cog of ships of larger burden had reacted unfavourably upon the smaller ports not merely because of the need for greater depth of water and longer wharfs, but because the larger cargoes could only be disposed of at more populous centres.

The fact that London was already three times as large as York, the next most important English city, had the effect of stimulating agriculture in the surrounding counties from which it drew provisions, and hence the best lands of Suffolk, Essex, Kent and Surrey were more densely peopled than any other part of rural England. Many statesmen, officials, courtiers, merchants, lawyers and other people whose duties kept them in London also had their country seats in these counties. The geographically-minded writer who has already been cited was able to explain why the site of London was the "inevitable" one for the metropolitan city.

This realm [he wrote] hath only three principal rivers, whereon a royal city may well be situated. Trent in the north, Severn in the south-west, and Thames in the south-east, of which the Thames both for the straight course in length reacheth furthest into the belly of the land, and for the breadth and stillness of the water[1] is most navigable up and down the stream; by reason whereof London, standing almost in the middle of that course,[2] is more commodiously served with provision of necessaries than any town standing upon the other two rivers can be, and doth also more easily communicate to the rest of the realm the commodities of her own intercourse and traffic. The river openeth indifferently upon France and Flanders, our mightiest neighbours, to whose doings we ought to have a bent eye and special regard: and this city standeth thereon in such convenient distance from the sea, as it is not only near enough for intelligence of those princes, and for the resistance of their attempts, but also sufficiently removed from the fear of any sudden dangers that may be offered by them; whereas for the prince of this realm to dwell upon Trent were to turn his back

of Shinggil, that smaul Shippes may cum up a large Myle toward Folkeston as yn a sure Gut" (*Itinerary*, VII, 138–141).
    Thus Leland writes of the shore of Kent, and, elsewhere, describing Sandwich, he says: "The caryke that was sonke yn the Haven yn Pope Paulus tyme did much Hurt to the Haven, and gether a great Banke." See also pp. 322–323 above.
    [1] Both Severn and Trent have tidal bores.
    [2] I.e. the east-west section to Reading.

or blind side to his most dangerous borderers; and for him to rest and dwell upon Severn were to be shut up in a cumbersome corner, which openeth but upon Ireland only, a place of much less importance. Neither could London be pitched so commodiously upon any other part of the same river of Thames as where it now standeth; for if it were removed more to the west it should lose the benefit of the ebbing and flowing, and if it were seated more towards the east it should be nearer to danger of the enemy, and further both from the good air and from doing good to the inner parts of the realm; neither may I omit that none other place is so plentifully watered with springs as London is. And whereas amongst other things, corn and cattle, hay and fuel, be of great necessity; of the which cattle may be driven from afar, and corn may easily be transported. But hay and fuel being of greater bulk and burthen must be at hand; only London, by the benefit of this situation and river may be served therewith.[1]

The writer might have added that not only hay and fuel, but malt, bricks, tiles, building stone, and lime were commodities that moved preferably by water, but even were direct evidence lacking, it is sufficiently clear from the minute attention that Leland pays to bridges and causeways[2] over the country generally that road traffic by cart and waggon as well as by pack-horse had by the early sixteenth century reached a considerable volume. Such traffic took place not merely from farm, mine or quarry, to the nearest shipping point or market town: goods were taken by road over long distances, as for example from Southampton to London and from London to Bristol. For the time being, the highways were able to bear the demands made upon them; the passage of wheeled vehicles and of droves of cattle was not such as to ruin them past repair as was to happen a century later; and, indeed, later outcries about the roads have reference almost

---

[1] From the *Discourse* published by Stow cited above.
[2] He thus catalogued the bridges over the river Nidd: "From Patley Bridge [of wood] and Village, a Membar of Ripon Paroche, to New Bridge of Tymber 3 Miles. Thens to Killinghal Bridge of one great Arch of Stone 3 Miles, and 3 Miles to Gnaresbrughe, where first is the West Bridge of 3 Arches of Stone, and then a little lower Marche Bridge of 3 Arches. Both these Bridges serve the towne of Knaresborow. Gribolol-bridge [Goldsborough] is about a Mile beneath Marche Bridge, and is of one very greate Bridge for one Bowe. Then to Washeford Bridge a 4 Miles. It is of 4 Arches; Then to Catalle Bridge of Tymbar a 2 Miles, to Skipbridge of Tymbar and a great Causey. The last and lowest Bridge on Nidde is this Skipbridge. This Cawsey by Skipbridge towards Yorke hath a 19 small Bridges on it for avoiding and overpassing Carres [pools] cumming out of the Mores therby. One Blake that was twyse Maior of Yorke made this Cawsey and another without one of the suburbs of Yorke." Thus there were nine bridges between Pateley Bridge and York, a distance of some 30 miles as the crow flies. Skipbridge and its Causey carried the main road between York and Boroughbridge (*Itinerary*, VIII, 68).

entirely to those sections which cross the heavy clay belts. As will presently be shown, the main roads followed as far as possible the outcrops of permeable, and therefore naturally drained, limestones and sandstones. Actually with the disease, nature herself had often supplied the remedy, for difficult clay vale and navigable longitudinal waterway are causally linked.

The value of particular waterways may be judged from the extent to which they were kept free from obstructions in the form of weirs, mills, and of all but large and well-constructed bridges. But obstructions were the rule rather than the exception, and they combined with the inconvenience of the many river windings to push traffic on to the roads. The Thames as far as Reading was a water-route of the first rank, and towns like Staines and Maidenhead (the latter a shipping point for timber) had some importance as river ports. Navigability by small boats and barges extended to Lechlade, but owing to the changing direction of the river, and to the many obstructions, this upper section was of much less value than the lower. Ware River (the Lea) afforded an important link with the corn counties, and there were grist-mills on its lower reaches which perhaps explain the bakers' carts which came daily into London from Stratford at Bow. From the south, the Medway below Yalding served to carry to the Thames Wealden iron and Wealden timber, the latter in billets ready for burning, or partly shaped for shipbuilding and other purposes. To Londoners, however, fuel already meant Newcastle coal carried coastwise by sea.

To York, standing as it does in an alluvial plain, and off the main north road (which kept to the better drained land to the west), the Humber-Ouse waterway from Hull to the city, and thence up to Boroughbridge to meet the highway, was very important. York was the lowest bridge-point. The lower Derwent was similarly useful, and had no bridge across it below Sutton, 15 miles above the confluence with the Ouse. Yet after Sutton bridge was reached, there was a succession of bridges across the Derwent every 3 miles or so, for in this section the river formed a barrier between the market of York and the corn-growing areas of the limestone hills and wolds to the east. Save for the important crossing at Newark, the Trent was kept free from bridges below Nottingham, and there were ferries at fixed points. The Idle below Bawtry was a feeder of the Trent route, but the Fosse Ditch which linked it with Lincoln, and thence by the Witham with Boston, had been allowed to fall into disuse.

The Severn together with its estuary linked up the chief cities of the west, and small sea-going vessels passing Gloucester went up as far as

Tewkesbury. Bewdley was essentially a river port, better placed than its older rival Bridgnorth, and above it the shallow-draft Severn "trows" went right up to Shrewsbury. Other rivers were of local importance only, notably the Exe serving Exeter, the Yare serving Norwich and the Parret serving Bridgwater. On the Exe, sea-going ships could come up no farther than Topsham, and Leland, while noticing the business of this little out-port, remarks that the Exeter men were anxious to extend the haven up to the city.

The "London way" or high road from Exeter, which Leland himself followed as far as Honiton only,[1] crossed the successive Lias, Oxford, and Kimmeridge clay belts at their narrowest, and reached the Upper Greensand at Shaftesbury. Thence it ran across the chalk through Salisbury and Andover as far as Basingstoke, where it was necessary to pass over the London Clay. The Bagshot Sands, however, afforded a dry surface until within a few miles of the Thames crossing which was made at Staines bridge. Beyond Staines, the road ran over the Terrace Gravels to Hounslow and New Brentford, which, by Leland's reckoning in old English miles, was 8 miles from London. The highway from London to Bristol ran across Hounslow Heath, and, after crossing the alluvial meadows of the Colne "at rages of rayne by rising of the river much overflown", took advantage of the Taplow Terrace until the Thames was crossed by the timber bridge at Maidenhead. For two or three miles beyond Maidenhead the subsoil was chalk, but the road then encountered a patch of London Clay which was forested, and beyond this the alluvial vale of the Loddon had to be passed at Twyford before Reading was reached. The Loddon, like the Colne, was divided into several arms, and Leland reported that he rode across them by a succession of four bridges. Beyond Reading (which could, of course, be reached alternatively by river wherry or barge) the road to Bristol lay entirely upon permeable rocks, save for the short crossing of the Oxford Clay outcrop between Calne and Chippenham. This section, it may be remarked, was the first to be turnpiked, although this did not take place until the reign of Queen Anne, by which date, of course, a much heavier traffic had come into existence.

It was a fortunate coincidence that the lighter soils, which produced the finest and best wools, were also those over which movement was the easiest. From the famous wool market of Chipping Campden, in the northern Cotswolds, the road to Salisbury ran all the way over permeable rocks save for the section which crossed the Oxford Clay and Thames alluvium

---

[1] He made extensive use of by-roads.

(by the shortest possible route) at Lechlade. Beyond Salisbury, Poole Harbour could be reached by a road which followed outcrops of chalk and Tertiary sands. The routes to Christchurch and to Southampton too, across the Hampshire basin, were mainly over sandy rather than clay beds.

The important highway to Gloucester diverged from the Bristol way just beyond Maidenhead, and recrossed the river at Henley, whence it ran across the richly wooded Chilterns by Nettlebed and so down to the Chalk Platform. Here, some five or six miles of heavy Gault clay could not be avoided, but between Dorchester and Culham Hythe the way lay over the Lower Greensand, and once the Thames was crossed at Abingdon, the belt of Corallian Limestone afforded dry going as far as Faringdon. Between Faringdon and Fairford lay the Oxford Clay[1] and thenceforward the road lay over the oolites as far as the foot of Birdlip Hill. Yet another heavy clay belt was here encountered, and crossed as directly as possible, and the road reached Gloucester Cross at a level well above that of the Severn floods. Thence it dropped some 50 feet to the Severn Bridge and the quay, and it was carried by a causeway over Alney Island to the firm Triassic sandstones on the west bank of the river.

Leland rarely complains of the roads, but he makes clear that the belt of Lias clays lying under the Cotswold escarpment made bad going, and he himself made a detour when journeying from Evesham to Gloucester which enabled him to take advantage of higher and drier ground, as well as to pursue his enquiries concerning the old cloth-making town of Winchcombe and its environs.

According to common report, the passage of the Thames at Abingdon, which was made by the Gloucester road in Leland's day, represented a diversion of an older route by Wallingford, consequent upon the substitution of a bridge for the former boat crossing at Culham Hythe. This bridge was built about 1416, and in 1459 a tablet was put up to Geoffrey Barbour, the rich Abingdon merchant and sometime bailiff of Bristol, who had supplied the funds. His benefaction was praised[2] in verse:

> Of alle Werkys in this Worlde that ever were wrought
> Holy Chirche is chefe....
> Another Blissed besines is brigges to make,

---

[1] "From Farington onto S. John's Bridge of 3 Arches of Stone and a Causey a 3 Miles *dim*. [3½ miles] al by low ground, and subject to overflowinges of Isis.... From S. John's Bridge to Lechlade about half a mile....From Lechlade to Fairford about a 4 Miles al by low ground, in a maner in a levelle, most apt for grasse, but very barein of Woodde. Fairford is a praty uplandisch Toune..." (*Itinerary*, II, 22).

[2] According to Hearne (*Itinerary* VII, Pt. 2, 64 *b*, footnote).

There that the pepul may nor passe after greet showres...
For cartis with cariage may goo and come clere,
That many Wynters afore were mareed in the myre...
For now is Culham Hithe i com to an ende,
An al the contre the better and no man the worse.
Few folke there were coude that way wende,
But they waged a wed or payed of her purse.
And if it were a begger had breed in his bagge,
He schulde be ryght soone i bid for to goo aboute,
And of the pore penyles the hiereward wold habbe
A hood or a girdel, and let them goo withoute.
Many moo myscheves there weren I say.
Culham hithe hath causid many a curse.

Wallingford certainly showed signs of a much greater importance in the past when Leland visited it; and he was told that it had been desolated by the great plague in the days of Edward III. But, it may be noted, the way between Wallingford and Faringdon ran over a broad outcrop of Gault clay, and once the bridge was built, the route by Abingdon was to be preferred.[1] Leland cited Wilton as another town which suffered by the diversion of the highway, in this case after Old Sarum was deserted for the new town in the valley.

It was impossible, of course, for the roads to avoid the clay and alluvial belts entirely, and recourse was sometimes had to building a raised causeway as in the approach to York already described. This was done also in a section of the important highway between London and Banbury, which left the Chalk for the Gault at Wendover. "There is a Causey made almost [says Leland] to pass betwixt Alesbury and it, els the way in wett tyme as in a lowe stiffe Claye were tedious and ill to passe."[2] Other notable causeways were that from Oxford to Hinksey Ferry; the two of sixteen and thirty arches respectively by which Leland approached and left Oundle as he was travelling down the Nene valley; the long causeway linking Nottingham with the bridge across the Trent; those serving as approaches to Bideford and Barnstaple bridges; and many others besides. Merchants

[1] Abingdon lay also on an important north-south road, coming from Banbury and Oxford, and leading south (across the Kimmeridge and Gault beds at their narrowest) to the chalk country, across which it passed either by Hungerford and Salisbury and thence to Poole, Christchurch or Southampton, or alternatively to Newbury and Winchester and thence to Southampton or Portsmouth. Leland relates that a weaver was Bailey of Abingdon in the late fifteenth century and took part in the Lollard rising: it was thus early an industrial centre.

[2] *Itinerary*, IV, 192.

23

who rode about the country in the course of their business were often the
donors of bridges and causeways—George Monox, the London draper,

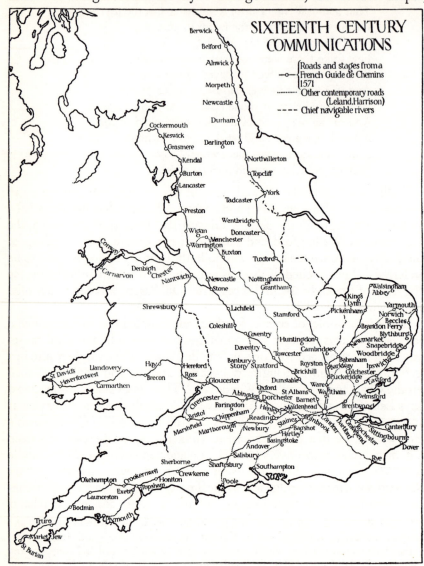

SIXTEENTH CENTURY COMMUNICATIONS

Roads and stages from a
French Guide de Chemins
1571
Other contemporary roads
(Leland.Harrison)
Chief navigable rivers

Fig. 64

erected the timber causeway across the Lea marshes between Walthamstow
and the Lock bridge, on the way to London, and Hugh Clopton gave the
handsome bridge at Stratford-on-Avon.

The benefactions of wealthy men reflect the needs of the day; and the growth of the towns, particularly London, had the result that many donations and legacies were made for the provision of conduits of spring water. Leland always mentioned a town thus provided, for this was an amenity that could not be taken for granted. It is clear, too, from his references that the great abbeys, long before the towns, had taken steps to provide themselves not only with water, but also with drainage, usually by the diversion of an arm of a river through the offices. Occasionally, the lay-out and position of a town was such that running water could be diverted through the kennels, and it may be inferred from Leland's comments on the dirtiness of Droitwich and Nantwich, that uncleanliness was exceptional so far as main streets were concerned. The streets of the principal towns were always paved, and were overlooked by handsome well-built houses and inns, beautiful churches and almshouses. Free schools and grammar schools were already to be found in many country towns, especially in those with rich citizens (usually wool staplers or drapers), and the open pillared and roofed markets, which still adorn so many of the Cotswold towns where the woollen industry once flourished, were being built in the late fifteenth and early sixteenth centuries.[1]

Building materials, owing to their bulk and weight, were rarely sought more than ten or twelve miles away, so that from Leland's descriptions of domestic architecture it is easy to follow the broad swathe of handsome stone-built and stone-roofed houses that coincided with the belt of oolitic limestone running from Dorset to the North Riding. South-east of this belt, bricks and tiles were increasingly displacing the characteristic timber and thatch, especially in the London and Hampshire basins. At Hull and Southampton, Leland particularly noticed the general use of brick (here of course of long standing), while he mentioned (although infrequently) the flint used in East Anglia and elsewhere in the chalk belt. In the well-wooded midlands, beyond the Jurassic escarpment, it was to the handsome timber and plaster houses that he called attention, while in the Pennines stone houses of a less attractive type and stone-walled parks reappeared.

So precious were shaped stones and roofing lead, that an abandoned building was soon plundered; and, once unroofed, the interior quickly fell into ruins. By the beginning of the sixteenth century, castles in England had had their day, except for those on the Border or for those newly built

---

[1] For example at Bruton, a clothing town in east Somerset: "Ther is in the Market Place...a new Crosse of 6 Arches, and a piller yn the midle for Market folkes to stand yn, begon and brought up to *fornix* by Ely laste Abbate of Bruton. The Abbay ther was afore the Conqueste a place of Monkes..." (*Itinerary*, II, 45).

as Channel defences. In his search for relics of antiquity Leland made particular enquiry for castles, and had often to report that nothing remained of a former stronghold save a mean farmer's house or a cattle fold within the fallen walls. Barmouth Castle, Rockingham Castle, Cottingham Castle, Malton Castle, Torrington Castle, were all in ruins, and Belvoir Castle had been unroofed to serve a new building at Ashby-de-la-Zouch. In not a few cases, however, a noble owner had remodelled his ancestral castle in accordance with the changing ideas of the day, adding more comfortable rooms and more commodious offices, while leaving the unwanted dungeon or keep to decay. Such was the case of Warwick Castle, while Wressel Castle on the Derwent (built of squared French stone, as it was easily carried by water) was beautifully equipped without and within, the adjustable reading desks in the study called "Paradise" particularly delighting the scholarly Leland. Town walls, now fulfilling no necessary function, were also allowed to crumble away, and nearly every town had "pretty suburbs" without the old gates; these suburbs often took shape as a "ribbon development" along the main roads.

The passing of the castle and the town wall had taken place by slow degrees. The passing of the magnificent abbeys, priories and nunneries, and of the humbler buildings of the white, black and grey friars in the towns was more abrupt. Sometimes these buildings were turned to secular uses: at Gloucester, the Grey Friars' College was transformed into a brewhouse, the Black Friars' into a drapering house; and at Malmesbury the abbey buildings were used as weaving sheds.[1] A rich citizen of Dorchester (Oxon.) bought part of the abbey church from the king and gave it to augment the parish church; the townsfolk themselves bought the abbey church at prosperous Malmesbury. The stones of Aulcester Priory were taken to repair Bitford Bridge, and the Groom Porter of the Court went to live in Cookfield Nunnery. In very many cases the religious houses were remodelled or rebuilt to serve as manor houses or country mansions.

All over England, as Leland noted, there were springing up those handsome dwellings, beautifully situated, which a later generation was to know as "gentlemen's seats", and which were typical of the new era of more luxurious living, and more widespread wealth associated with the Renaissance. On the site of the monastery of Tichefield, Mr Wriatherly had built a stately house with battlements and a handsome gate, having conduit water laid on to the middle of the courtyard. The Lord Russell had partly rebuilt Chenies in brick and timber; the owner of the manor

---

[1] William Stumpe, the master clothier, who put his workpeople into Malmesbury Abbey, rented Osney Abbey also in 1546.

house at Ewelme had shown his modern spirit by using iron bars instead
of cross-beams in his hall, and had added a "fair and lightsome parlour".
Around these houses were elaborately planned flower gardens, walks,
orchards, ponds and plantations.[1]

Every considerable landowner had his paled park, and often two, one
for red and one for fallow deer, and although there are indications that
disparking had begun,[2] the total acreage thus removed from cultivation in
each parish must have been considerable. Deer were not confined to the
parks (even could the park palings or walls always have kept them within).
Fallow deer ran wild in the lowland woods and fens, and in the unfenced
chases and fern brakes; red deer were numerous on the moorlands in the
Pennines and in the south-western peninsula. Where the arable fields lay
open and unfenced, as they did in the greater part of the midlands, the
depredations of deer meant a heavy loss to the husbandman.[3] Nor can it
be doubted, in view of their geographical distribution, that many of the
enclosures that Leland reports were made against deer. Besides these
there were, of course, the enclosures of pasturage for sheep, against which
such an outcry was raised by the publicists of the day. More than a
beginning had been made of the movement (accelerated later in the century)
for laying together a man's formerly scattered strips or quillets of plough-
land, which naturally led to their enclosure in accordance with a more
enlightened system of husbandry.[4] In the west country [Somerset, Devon

[1] "Morle (in Darbyshire) Mr. Lelandes place is buildid, saving the Fundation of
Stone squared that riseth within a great Moote a vi foote above the water, al of tymbre
after the commune sorte of building of houses of Gentlemen for most parte of Lancastreshire. Ther is as much pleasur of Orchardes of great varite of Frute and fair made
Walkes and Gardines as ther is in any Place of Lancastreshire" (*Itinerary*, v, 83).

[2] Radley Park was disparked "by reason that the scollars of Oxford much resorted
thethar to hunt".

[3] "In the Forest [of Delamere] I saw but little corn, bicause of the Deere." The
Bishop of Durham had "wild bulls and kine" in his park at Auckland as well as deer
(*Itinerary*, v, 82).

[4] Tusser, in his famous *Five Hundred Points of Good Husbandry*, written in 1573,
made a "Comparison between Champion Countrie and Severall" very much to the
disadvantage of the former:

"The countrie enclosed I praise
The tother delighteth not me..."
"The champion liveth full bare
The woodland full merrie doth fare."

He cited Suffolk and Essex as counties affording examples of the superiority of enclosed
fields. For references to enclosures see p. 334 above, footnote 3. Fig. 66 (Mr E. C. K.
Gonner's map) sums up the situation at the end of the century.

and Cornwall] the thick hedges were of old standing, as they were also in Lancashire and in the Weald, for in these regions the open-field system had never been characteristic.

Leland's descriptions of the landscape from the farming standpoint are at first sight disappointingly meagre. It is either champaign (open and unenclosed) or "metely wooded", it is either rich, or moderate, or scant of corn, of pasture, and of meadow. With subsistence agriculture in many places, a certain uniformity necessarily resulted, but that climate and soil imposed a degree of regional differentiation is nevertheless quite clear. The term "corn" included not only the grains—wheat, barley, oats and rye—but also beans, peas and vetches, that is to say all crops sown for food for man or beast. Wheat was noticeably scanty on the poor sandy soils of the Lower Trias in the midlands, and in the northern counties beyond Tees and Mersey, where oats took its place. On the other hand, certain favoured areas produced wheat far beyond their requirements; the part of Warwickshire between the Avon and Edge Hill was described as the "very granary" of the whole county. Beans were a special feature of the vale of Aylesbury and the plain of Somerset, and were used to victual ships when grain was dear. The grass country of east Leicestershire and Northampton was already famous, as were the sweet pastures for cattle on the Mountain Limestone in the Pennine dales and in Staffordshire near Uttoxeter.

Marling was the chief method of soil amelioration practised, and Leland mentioned it especially in relation to the poor sandy soils of Shropshire, Cheshire and Lancashire. The results were sometimes amazingly good, and he noted the "marvelous good corne and pastures" round Mr Spurstow's house in Cheshire; in the same county was Mr Bouth's place where "by good culture is made veri good corne ground, wher somtime was very ferny and commune ground". It is possible to follow very accurately Leland's passage across the varying geological formations; the journey from Poole to Salisbury is typical.

From Pole to Winburn 4 Miles, whereof 3 and an Half be Morish and Hethey Ground [i.e. the Bracklesham beds]. The soile about Winburn Ministre is very good for Corne, Grasse and Woodde [i.e. on the London clay].... From Winborne to Horton [still on the clay] 4 Miles by much woddy Ground.... From Horton to Cranbourn [thenceforward on the chalk] a 3 Miles al by Champain Ground having neither Closure nor Wood. Cranbourn is a praty thorough Fair, and for one Streat meatly welle buildid. There rennith a fleting bek thorough it.... From Craneburn I passed a 2 Miles or more al by playne Champain Ground...thens a 6 Miles by like Ground to Honington... and so to Salisbury al Champayn Ground a 2 Miles.... The site of the very Town of Saresbyre and much Ground therabout is playne and low, and is a Pan or Receyvor of most parte of the Water of Wyleshire.[1]

The north of England was much less densely peopled, and included much more waste land, than the south, but the impression of poverty left by Leland's description of Durham is in part due to the fact that his route lay along the marginal hill spurs of the Pennines, carved from the Coal Measures, where natural mountain pasture alternated with woody vale. "From Duresme over Framagate Bridge to Chester in the Streate partly by a litle Corn Ground, but most by Mountaineose Pasture and sum Mores and Firres.... Thense to Gateshed vij [long] Miles by Mountaniouse Ground with Pasture, Heth, More and Fyrres. And a little a this side Geteshed is a great Cole pit." Riding back from Durham to Auckland, the landscape was the same. "I rode through a great Wod standing on a Hille, and so came by hilly, morish and hethy Ground to S. Andres Akeland." Thence he rode southwards to "Raby Castle 5 Miles part by Arable but more by Pastures and Morish hilly ground baren of Wood".[2] In Lancashire, mosses were a feature of the landscape, and afforded peat which was the commonly used fuel.

Mr. Leland of Morle [whose house had already been described] brennith al Turfes and Petes for the Commoditie of Mosses and Mores at hand.... And yet by Morle as in Hegge Rowes and Grovettes is meately good Plenti of Wood, but good Husbandes keepe hit for a Jewell.... Riding a mile and more beyond Morle I saw on the right hand a Place nere by of Mr. Aderton, and so a ii Miles of to Lidiate Moss, in the right side whereof my Gide said that ther were Rootes of Fyrre Wood. About this Mosse I began to se a Hille or Hilles on the right hand that stil continuid on the same hand as a mighty long bank ontil I came to Lancastre....[3] The Soile about Lancastre is veri fair, plentiful of Wood, Pasture, Medow and Corne.... A ii Mile from Lancastre the Cuntri began to be stony, and a litle to wax Montanius.... The ground beyond Warton and about is veri Hilly and mervilus Rokky onto Bytham a v Miles of. In the Rokkes I saw

[1] *Itinerary*, III, 54. [2] *Ibid.* I, 91.

[3] *Ibid.* V, 83–5. The Pennines. "Up toward the Hilles by Greenshaugh Castle be iii Forests of redde Deere, Wyredale, Bowland and Blestale. They be partly woody, partly hethye. The ground betwixt Morle and Preston enclosid for Pasture and Corne, but were the vaste Mores and Mosses be, whereby as in Hegges Rowes beside Grovettes ther is reasonable Woodde for building, and sum for Fier, yet al the People ther for the most part burne Turfes. Likewise is the Soile bytwixt Preston and Garstang: but alway the most part of Enclosures be for Pasturages. Whete is not veri communely sowid in these partes aforesaid."

Herdes of Gotes. . .. Thens I roode over a great Bek called Staunton Bekke and so ridding a ii Miles farther cam to a Soile less Stony and more frutiful of Corne, as sum Whete, much Ootes and Barle or Bigge, and so to Kendale. ...

Industrial crops were few in early sixteenth-century England. A people who knew what it was to experience scarcity of corn (and at times the poorest were even driven to make bread of acorns) were jealous of using the soil for the production of any but food and fodder crops. Little patches of hemp and flax were grown by individual farmers and cottagers for their own use: the hemp for halters, cords, nets, traces, and even shirts and shifts for the very poor; the flax for linen cloth "without which", wrote Leland's contemporary, Dr Wm. Bulleyn,[1] "we shalbe like beastes both at boorde and bedde. Linen is comely both for men and specially for women, and more commendable in the temple than holly." "Barley", said the same writer, "is the principal vine grape of Englande, that our Malt is made upon. And mother of our Beer and Ale, whiche Malt in the time of our extreme nede, have been sold awail into forrain realmes: by certain old known theves, called Humber and Linhaven, with his braunches. From Cambridge, St Ives and Mildenhaule, Thetford and Brandon Ferie: and also Yarmouth in Norffolke."

Dyer's madder and fuller's teazles were grown in gardens, saffron in the fields of Cambridge, Essex and Norfolk. Kent was already famous for cherries, Worcestershire for pears, and Hereford for cider apples. Rape seed was a speciality of Marshland in Norfolk. Radishes and other salads were very generally grown in gardens, but Bulleyn complained of the import from abroad of cabbages, onions, and hops, all of which might easily have been grown at home. Animal husbandry was beginning to be specialised, but, in general, sheep and cattle and to a less extent horses were indispensable in all current farming practice.[2] It was only on the Cotswolds, the Dorset Heights, and the Lincoln Edge, that Leland specially noticed large flocks of sheep. He might have seen them also, had he gone that way, on the saltings which fringe Essex and Kent, where the ewes were milked, and upon Romney Marsh too. The problem of winter fodder for cattle remained a serious one for another century and more. Hence the available acreage of water-meadows yielding hay was an important consideration, and the idea of draining these meadows and rendering them fit for the

[1] Wm. Bulleyn, *Bulwark of Defence*, 1562.

[2] Horses were bred in great numbers for riding and pack animals, and were most important in the Fens, in the Hampshire Basin and wherever common pasture was abundant. The king's horses were bred in the Severn valley, at Upton and Tewkesbury.

plough was not to be entertained. The lay-out of parish boundaries shows the anxiety of the village community to participate in meadow land, and Leland noted that there were as many as fifty-eight villages lying in, or "butting of", Walling Fen in Howdenshire, Yorks.

The different ways in which the alternation of belts of heavy wet clay with belts of lighter and better drained limestone and sandstones has always affected the human geography of England, have already been illustrated. In the sixteenth century it was still possible to make out clearly the original contrast of natural vegetation, for remains of the old oak forests were most abundant in the clay vales. It is true that the areas of legal "forest"[1] were neither wholly under timber nor devoid of settlements, yet they were still heavily wooded. Epping Forest, the Forest of Bere, Woolmer Forest, Rockingham Forest, Selwood Forest, Kingswood Forest, the Forest of Wyre, Galtres Forest, Needwood Forest, were all on heavy soils.

As far as mining was concerned, there was little that was new in Leland's day. The old-established lead mines in Derbyshire and the Mendips, and the tin mines of Cornwall and Devon, provided metal for export as well as for home use, the tin being partly exported in the form of pewter made in London, to which the tin was carried coastwise. Every surface-outcrop coal-field was worked for the market afforded by the immediate neighbourhood,[2] but only from Newcastle was there as yet any considerable output or considerable movement of coal, London being the principal consumer. Iron was smelted with charcoal in the Weald, in the Birmingham district, in the West Riding, in the Forest of Dean, and in the Clee Hills. Smiths made use of coal in their forges; at Sheffield and Rotherham there were long-established cutlers; and at Birmingham edge-tool makers, loriners and nailers; while Ross was well known for its smiths, and Gloucester for its nailmakers.

Leland's careful mention of innumerable quarries is evidence of their value. From the far famed free-stone quarry at Hamden, near Crewkerne in Dorset, for example, he noted that Mr Strangways took 3000 loads of

[1] See above pp. 174-5, and below p. 356.
[2] "Though betwixt Cawoode and Rotherham be good plenti of Wood, yet the People burn much yerthe Cole, bycause hit is plentifully found ther and sold good chepe. A mile from Rotherham be veri good Pittes of Cole. In Rotherham be veri good Smithes for all cutting tools" (*Itinerary*, v, 92). "Halamshire hath plenti of Woodde, and yet ther is burned much Se Cole" (*Itinerary*, v, 97). Ling and furze were the only fuel in some treeless moorland areas, for example in Cornwall, and the people of the Isle of Portland burned dried cow-dung.

stone for his new house, twelve miles away. Burford quarry was also famous, and so was a quarry of very hard sandstone close to Manchester, while limestone from a quarry in the Cotswolds behind Prestbury was used nine miles away in Tewkesbury. Roofing slates were quarried in Cornwall, north Wales and Charnwood Forest; marble in Teesdale and in Derbyshire; alabaster (gypsum) from the Triassic rocks in the Isle of Axholme, near Leicester and near Burton-on-Trent; grindstones and mill-stones were obtained from the Pennines, fuller's earth from the Great Oolite and from Woburn.

Salt was made from the brine springs of the New Red rocks, principally at Droitwich and Nantwich. As the evaporation was carried out in lead pans, coal could not be used, and the neighbourhood was denuded of wood. The salt-makers of Droitwich, for example, had to fetch their fuel from the borders of Warwickshire, twelve miles away. Leland described[1] Northwich as

a praty market toune but fowle, and by the Salters Howses be great Stakkes of smaul Clovyn Woode to sethe the Salt Water that thei make white Salt of.... Ther be ii Salt Springges at Middlewich, that standith, as I remember, upon Dane River, and one at Nantwich, the which yeldith more Salt Water than the other iii. Wherefore ther be at Nantwich a iii hunderith Salters. The Pittes be so set abowte with Canales that the Salt Water is facile derivid to every Mannes Howse. And at the Nantwiche very many Canales go over the Wyver River for the Commodite of deriving the Water to the Salters Troughs. They sethe the Salt Water in Furnesses of Lede.

A coarser salt was made at many points round the sea-coast, notably in Northumberland, Lincolnshire and Lancashire, and a salt pan was a valuable property. Dr Bulleyn, who has already been mentioned, owned one near Tynemouth. Leland described the method of extraction used at Cockerham (Lancs), where he rode across the sands "not without sum Feere of quikkesandes. At the end of the Sands I saw divers Salt Cootes, wher were divers Hepes of Sandis taken of Salt Strandys, owt of the wich by often weting with Water they pike owt the Saltnes, and so the Water is derivid into a Pit, and after sodde." But the home production of salt did not meet the needs of the fishing industry either as to quantity or quality, the strong bay-salt from France, Portugal and Spain being considered the best for curing fish. The Anglo-Spanish merchants therefore dealt in salt, and Nicholas Thorne, a prominent Bristol merchant (brother of Robert Thorne), left gifts of salt to the poor of the city.

[1] *Itinerary*, v, 82.

The most widespread and important English industry of the period, and that which most constantly came under Leland's notice, was the weaving and finishing of cloth. In the belt of limestone country behind Bristol, running through Gloucester into the borders of Somerset and Wiltshire, every town and townlet suitably situated on a clear running stream had its prosperity based on "clothing". Malmesbury was held to be the most prosperous among many prosperous communities, but where so much depended upon the skill and reputation of a few individual "clothiers" who directed the industry, a town might "decay" (as Bath had done in Leland's time) through a succession of deaths or personal misfortunes. In the Severn basin, where the industry included the draping or finishing of the Welsh woollens, the industry was concentrated in the large towns, notably at Worcester, Shrewsbury and Gloucester, and, on a smaller scale, at Hereford, Oswestry and Kidderminster.[1] Bridgnorth had formerly a cloth industry, but it had decayed, perhaps for the same reason as that affecting Leominster:

The Towne of Lemster by reason of their principall Wool use great Draping of Cloth, and thereby it flourished. Since of latter dayes it chanced that the Cityes of Hereford and Worcester complained of the frequency of People that came to Lemster, in prejudice of both theyr Marketts in the Shyre Tounes, and also in hindringe their Drapinge. Whereupon the Satturday Markett was removid from Lemster and a Markett on Friday was newly assigned unto it. Since that tyme the Towne of Lemster hath decayed.[2]

The Devonshire kersey industry was carried on in nearly every market town in that county, and the local wool was supplemented by an import of Irish yarn at Minehead. Irish yarn came into Liverpool, too, whence it reached Manchester, and was utilised in the rug, frieze, fustian and "cotton" industries which were growing up on the west Pennine rivers. "Bolton [says Leland] standith most by Cottons and Cowrse yarne. Divers villages in the Mores about Bolton do make Cottons."[3] They included Blackburn, Rochdale and Bury.

Kendal cloth was held in very high esteem but, so far as volume was concerned, the West Riding industry, carried on at Wakefield, Bradford, Halifax and Leeds, and turning out rather coarse products, was more

[1] It was stated in an Act of Parliament of 1565 that there were 600 shearmen and frizers at Shrewsbury who lived by draping Welsh cottons. They were complaining of the competition of outsiders.

[2] *Itinerary*, IV, 178.

[3] *Ibid.* VII, 57.

important.[1] Woollen manufactures were also to be found in several towns far from the regional centres but standing on the main highways to London; such were Newbury, Reading, Abingdon and Coventry. The last-named made the cloth caps which at one time were worn by high and low alike, but fashions change, and in Leland's day Coventry was suffering because such caps were no longer in demand. Salisbury and Winchester, since they handled so much wool, had their clothiers, and there was a small industry scattered about in Hampshire, Surrey, Sussex and Kent, for example at Farnham, Godalming, Guildford and Tenterden. Leland did not visit East Anglia and so has nothing to tell us about the worsteds of Norfolk, the narrow cloths of Suffolk, or the handywarps of Essex, made at Coggeshall, Braintree and Bocking.[2]

Various Acts of Parliament of the mid-sixteenth century show how the conservative elements in the country strove to stem what appeared to them the flood of change. The rules and regulations of the medieval gilds were proving too narrow for the new industries springing up, and one way to evade them was to set up business in a new locality which had no traditions or prescriptive rights. Liverpool grew at the expense of Chester because trade there was less hampered,[3] and, for similar reasons, the cloth industry was deserting the cities for the villages. The Act of 1557, already cited, declared that workers had set themselves up in villages and husband towns, thus not only emptying the cities but displacing the husbandmen, and causing the decay of husbandry and tillage by taking over farms and pastures into their own hands. It was therefore enacted that they should no longer do so, but exception was made of all the areas where cloth-making was either a new industry, or traditionally a cottage or a village industry, as in Cornwall, in Wales, and in the northern counties, by Stroudwater (the only exception in the broadcloth belt), in the Godalming district, and the West Riding. But such a law was as futile as that laying down that no dyes should be used save those customary for the last twenty years: progress was inevitable.

The quiet persistence of men like Roger Barlow was to bear fruit. One of the last acts of Henry VIII had been to recall Sebastian Cabot to England, and he was brought over by his son-in-law, an English-Spanish merchant. Within a year or two, Cabot's World Map hung on the walls of scores of

---

[1] In an Act of Parliament of 1557 the West Riding manufactures were said to consist of broadcloths and the cloths called pewkes, tawnies, violets or greens.

[2] See G. Unwin, "The History of the Cloth Industry in Suffolk", *V.C.H. Suffolk*, II, 254–71.

[3] See also p. 296 above for the silting of the Dee estuary and the decay of Chester.

counting-houses, scholars' closets, and noblemen's libraries. As a result the first national effort to seek out undiscovered markets was made in 1553.[1] England was awake to a New World.

BIBLIOGRAPHICAL NOTE TO CHAPTERS IX AND X

The sources of material for chapters IX and X are sufficiently indicated in the footnotes. Leland's description has been edited by L. T. Smith as *The Itinerary of John Leland in or about the years 1535–1543*, 5 vols. (1906–10).

Although written from a different standpoint, the works of the economic historians on this period will be found to supply much basic material for the reconstruction of sixteenth-century geography, and will serve also as a guide to original sources. The following should be consulted: (1) the relevant sections of C. Read, *Bibliography of British History, Tudor Period* (1933); (2) R. H. Tawney, "Studies in Bibliography: Modern Capitalism", *Econ. Hist. Rev.* IV, 336–56 (1933); (3) the copious bibliography in J. U. Nef, *Rise of the British Coal Industry*, 2 vols. (1932); (4) the bibliographies of contemporary works in E. G. R. Taylor, *Tudor Geography* (1930), and *Late Tudor and Early Stuart Geography* (1933); (5) the current bibliographies to be found in each issue of the *Economic History Review*.

[1] The search for a north-east passage to Cathay found wide support, as it was believed that "civil people" would be met with *en route*, in Scythia and northern Tartary, who would purchase English woollens. No such hopes were possible if the journey was made by the north-west passage, of which Cabot claimed to hold the secret; in America there were only half-naked savages.

Chapter X

CAMDEN'S ENGLAND

E. G. R. TAYLOR, D.Sc.

William Camden was born about the time that Leland died, and his journeys through England were spread over the last thirty years of the reign of Queen Elizabeth. He plundered his predecessor's manuscripts (as did Harrison and Lambarde) for topographical detail, no doubt because his own interest lay specially in the reconstruction of Roman Britain, in "Britannia" rather than in England. The contemporary landscape, contemporary industries, and contemporary changes he regarded and recorded somewhat perfunctorily. Nevertheless, it is possible to extract from his pages an outline geography of Elizabethan England, and to discern especially the elements of permanency in the rural scene.[1]

In Cornwall (where, following classical example, he commenced his description), he could not pass over the tin mining, and described both lode- and stream-washing. The pilchard fisheries, too, called for remark; the cured and barrelled fish were esteemed as a delicacy not only in England but in Mediterranean countries, where they were known as *fumadoes*. Another local peculiarity to which his attention was drawn was the way in which the Cornish husbandmen improved their sterile soil by laying upon it both sea-wrack and sea-sand. The latter was carried inland by barge and pack-horse for distances up to ten miles from the coast. In Devonshire, tin mining had now ceased to be of any importance; but (although Camden omitted to say so) copper mines had been newly opened, and there were silver-lead mines at Combe Martin, which were worked by Bevis Bulmer, who had charge of the Mines Royal.[2] The

[1] The citations from Camden's *Britannia* are made from Bishop Edmund Gibson's translation of the Latin edition of 1607. Gibson carefully distinguishes his own *addenda* from the original. His work was first published in 1695, and again in 1722. For the various editions of Camden see footnote 2, page 389 below.

[2] Exeter held only antiquarian interest for Camden, but judging from the customs receipts it had become one of the busiest of English ports outside London, and Dartmouth was not far behind it. The wealthy Exeter merchants and clothiers, like those of Bristol, were interested in Atlantic trade, alike to the south, north and west. The Devonshire ports were also favourably situated for the important cross-Channel trade in the linens, buckrams, sailcloth and other manufactures of flax and hemp from Rouen, Moriaix and St Malo.

steady output of lead sows from the Mendips provided a staple export from Bristol, but the Mendip mining was taken for granted, and he did not even mention the recently exploited zinc (calamine) mines there, although these formed the basis, with copper, of brass founding, and manufactures of English brass were being exported to Ireland.[1] The major contrasts of Somerset farming, however, struck his eye: the great summer beast fair at Somerton on the borders of the cattle-rearing plain of Sedgemoor, and the excellent, although rather light, grain crops on the limestone hills. At Warminster, just over the Wiltshire border (situated where Chalk, Greensand and Gault outcrop side by side in quick succession), there was a great corn market at which what he termed "incredible" quantities of grain were sold weekly. Warminster lay on the main road between Southampton and Bristol, the latter

so populous, that, next to London and York, it may justly claim the pre-eminence over all the cities in Britain; for the trade of many nations is drawn hither by the conveniency of commerce, and of the harbour, which receives vessels under sail into the very heart of the city. And the Avon swells so high by the coming in of the tide, that ships upon the shallows are borne up eleven or twelve fathoms. The citizens derive a rich trade through Europe and make voyages to the remotest parts of America.

There was naturally a brisk movement of grain to provision such a port.

In Dorset and Hampshire, Camden found nothing to describe of interest to the geographer,[2] but he gave a clear picture of the productive and prosperous Isle of Wight, which was typical in its "scarpland" agriculture of the English plain at large.

To say nothing of the abundance of fish in this sea, the soil is very fruitful and enables the inhabitants to export corn to other parts; there is everywhere plenty of rabbits, hares, partridge and pheasants; and it has besides a forest and two

---

[1] The Mendip zinc mines were discovered in 1566 by Wm. Humfrey, the Say Master and founder of the "Company of Mineral and Battery Works". At about the same time he fixed upon Tintern, near the Forest of Dean, as a suitable site for iron-wire drawing works. He introduced German miners from Saxony as technical experts in his various enterprises.

Lord Burghley's papers, preserved among the Lansdowne MSS., throw much light upon the economic geography of the day, and have been freely drawn upon here.

[2] The growth of Southampton, side by side with other of the larger Channel ports, did not interest him. The following quotation from an annalist's notes is significant: "4 June 1567. By a letter written in Southampton reports that certaine merchants of Antwerp fled thether for religion have viewed xx of the fayrest howses in Southampton and mean theere to inhabit and use their Trade, so that they may have licence" (Harleian MSS. 253, fol. 112).

parks, which are well stocked with deer for the diversion of hunting. Through the middle of the island, there runs a long ridge of hills, where is plenty of pasture for sheep; whose wool, next to that of Lemster and Cotswold, is reckoned the best, and is in such request with the clothiers, that the inhabitants make great advantage of it. In the north part there is very good meadow-ground, pasture and wood; the south part is all in a manner a corn country, inclosed with ditches and hedges.

He went on to add that on the island were thirty-six towns, villages and castles, and that 4000 good soldiers could be raised in time of war.

The Thames valley in Berkshire was likewise fruitful, for it was "chequered with cornfields and green meadows, clothed on each side with groves", and the towns of Abingdon and Wallingford grew rich by their trade in malt. Towards the east and south, near the best sheep pastures, were the clothing towns of Newbury, Reading and Wokingham, but between the latter and Windsor lay great expanses of chase and forest that prompted Camden to reflect on this wasteful aspect of land utilisation. He first explained the term "forest" in its legal sense: "A forest is a safe harbour for beasts: not of every sort, but for such only as are wild, not in every place, but in some certain places fit for that purpose.... And it is incredible [he continued] how much ground the kings of England have suffered every-where to lie waste and have set apart and enclosed for deer." Whereupon he proceeded to quote John of Salisbury: "The husbandmen are debarred their fallows whilst the deer have liberty to stray abroad: and that their feeding may be enlarged the farmer is cut short of the use of his own grounds. What is sown or planted they keep from the husbandmen, and pasturage from the graziers."

Passing into Surrey, Camden found it chiefly notable for its magnificent noblemen's seats, and for its royal palaces: already it had a "dormitory" aspect. Neighbouring Sussex lent itself more easily to geographical characterisation. "The sea coast of this county has very high green hills, called the Downs, which, consisting of a fat chalky soil, are very fruitful. The middle part, chequered with meadows, pastures, cornfields and groves, makes a very fine show. The hithermost and north side is shaded pleasantly with woods, as anciently the whole county was, which made it unpassable...." These three belts, the chalk country (under the plough until the last decades of the nineteenth century), the mingled sands and clays of the Weald, and the timbered forest ridges, are recognisable to-day, and there is little doubt that the "chequer-board" of small fields now to be seen in the second belt was already marked out by the tall hedges when Camden saw it. The beautiful contemporary maps of the Buckhurst estates in Kent and Sussex, surveyed in 1595, show the identical field boundaries of the ordnance map of to-day, with no trace of any open-field system.

This county [observed Camden] is full of iron mines, all over it; for the casting of which there are furnaces up and down the country, and abundance of wood is yearly spent; many streams are drawn into one channel, and a great deal of meadow-ground is turned into ponds and pools for the driving of mills by the flashes, which, beating with hammers upon the iron, fill the neighbourhood round about it, night and day, with continual noise. But the iron, wrought here, is not everywhere of the same goodness, yet generally more brittle than the Spanish, whether it be from its nature, or tincture and temper. Nevertheless, the proprietors of the mines, by casting of cannon and other things, make them turn to good account; but whether the nation is in any way advantaged by them is a doubt which the next Age will be better able to resolve. Neither does this county want glass-houses: but the glass (by reason of the matter or making I know not which) is not so transparent and clear and therefore is only used by the ordinary sort of people.

The armament industry of the Weald grew to such dimensions in Elizabeth's reign that mariners complained bitterly that their enemies destroyed them with English guns and English shot. In 1574 forty iron-mill proprietors entered into bond under penalty of £2000 apiece not to found or sell iron ordnance without the queen's licence; but such licences were not withheld, and, in 1591 for example, the Earl of Cumberland had permission to export guns up to a weight of 100 tons. Illegal export, too, was rife, in spite of proclamations and penalties, for it was difficult to check. In 1601 a certain Edward Peake made an eloquent appeal to the House of Commons to forbid any export at all, even under licence. As he pointed out, although the queen might make £3000 a year out of her custom of £4 a ton on exported ordnance, she lost more than that sum on cargoes captured and destroyed at sea by enemy ships armed with English guns. The Swedish and Spanish iron founders (he said) could not make cast pieces equal in quality to English, and hence if the English were withheld foreigners would be compelled to use the more expensive brass (cast in Spain), and costing £6 the hundredweight against 10s. the hundredweight for cast iron. "It appeareth", he added piously, "to be a particular blessinge of God given onlie to England, for the defence there-of, for albeit most Countreys have their Iron, yet none of them all have Iron of that toughnes and validitie to make such Iron Ordnance of."[1] The

[1] Harleian MSS. 253, fol. 73.

glass industry was introduced into England in 1567 when continental wars cut off the supply from the Vosges mountains; carried on at first by foreign workmen from Lorraine and Normandy, it gradually came, in part at least, into English hands, and besides bottles, utensils, and drinking glasses, window glass and imitation Venetian glass were turned out at the glass-works in the Weald, in Hampshire, in London and in Staffordshire.

The rapid growth of industry put the queen and her advisers in a terrible dilemma, for it meant an equally rapid decrease of timber supplies. That great timber, especially oak, was necessary for the navy, and that a strong navy was England's only safeguard were propositions held to be axiomatic: yet equally was it necessary to have iron and lead and glass and saltpetre, all of which consumed wood in the shape of charcoal. Thomas Proctor's contention in 1589 that he had discovered a way to smelt ore with pit coal or peat was hailed with delight, and he was granted an exclusive patent: but it was premature. The depletion of the forests went on, especially in the Weald, and the Surveyor of the Royal Parks suggested substituting quick-set hedges for park palings to enclose them!

The growth of the navy brought with it changes along the Kentish shore of the Thames, which became lined with towns. Chatham was the dock-yard for the queen's ships in the Medway; at Woolwich and Deptford were shipbuilding and repairing yards; and at Deptford also stood the great naval stores and the Trinity House. Gravesend, where the continental route left the river for the Dover Road, grew with the increasing number of travellers, for Dover Harbour (in spite of the repeated need for costly harbour-works over a long period of time) continued to be the most frequented port for France. Away from the towns, Camden saw Kent as already "the garden of England", not only everywhere rich in the normal trio of corn and meadow and pasture, but abounding "with apples beyond measure, as also with cherries.... They thrive exceeding well in those parts and take up great quantities of ground, by reason they are planted square and stand one against another which way soever you look". Making a strong contrast with the orchard country were the marshes of accretion in the south. "Some places which within the memories of our grandfathers stood upon the sea-shore are now a mile or two from it. How fruitful the soil is, what herds of cattle it feeds that are sent hither from the remotest parts of England to be fatted, and with what art they raise walls to fence it against the incursions of the sea are things which one can hardly believe that has not seen them." Not only horned cattle, but tens of thousands of sheep grazed on the marshes, and the occasions when the pastures were flooded by unusual storms (as in 1570) were disastrous. Rye Harbour was ruined by the recent rapid accretion[1] although it had been a frequented port for Normandy in Leland's day. Situated behind it lay a little group of clothing towns, Tenterden, Benenden and Cranbrook, where the industry had been introduced by the Flemings invited over by Edward III. The Wars of Religion had latterly brought many foreign refugees into Kent, but Camden mentioned neither the Flemings who were now making "bays and says" in Sandwich, nor the Walloons who were carrying on textile and other industries in Canterbury. He spoke, however, of a new industry germane to clothing, that set up in Sheppey Island where "a certain Brabanter lately undertook to make brimstone and copperas of stones found upon the shore by boiling them in a furnace". The "stones" were pyrites and the copperas a sulphate of iron used as a mordant in dyeing.[2]

Going down the estuary from Sheppey, past the busy port of Faversham,[3] with its neighbouring oyster beds, and past the twin towers of Reculver, always a famous sea mark, Camden reached Thanet, and recorded of the inhabitants "who live near the roads or harbours of Margate, Ramsgate and Brodsteare" that (like the Norwegians) they were both husbandmen and mariners.

According to the several seasons they make nets, fish for cod, herring, mackeral, etc., go to sea themselves and export their own commodities; and those very men also dung their ground, plough, sow, harrow, reap and in [i.e. harvest];... and when there happens any shipwrecks as there do here now and then (for those shallows and shelves, so much dreaded by seamen lie over against it; namely the Godwin...the Brakes, the Fourfoot, the Whitdick etc.) they are extremely industrious to save the lading.

Visiting the site of Roman Rutupiae, later Saxon Richborough, he observed: "But, now, age has erased the very tracks of it: and to teach us that cities die as well as men, it is at this day a cornfield, wherein, when the corn is grown up, one may observe the draughts of streets crossing one another (for where they have gone the corn is thinner), and such crossings they commonly call there St Augustine's Cross."

[1] Contemporary opinion ascribed the loss of the harbour to the "inning" of the marshes in the neighbourhood.

[2] Another alien, Cornelius de Vos, was permitted to set up copperas works near Poole about 1564, and these works were taken over by a Londoner in 1587.

[3] Faversham, with Milton, shipped very large quantities of grain to London.

Having reviewed southern England in order, from Land's End to Thanet, Camden turned back to the west, and described Gloucestershire in terms of that triple regional division, Cotswolds, Severn Vale, and Forest of Dean, which has always been fundamental. The Forest, on which both Gloucester and Bristol drew for timber, was already much thinned by the local iron industry: in the orchard-rich vale, linked together by the waterway, lay the chief cities, Tewkesbury "famous for the making of woollen cloth and smart-biting mustard", and Gloucester, with its smiths, brewers and drapers and its Customs Quay of which Bristol was so jealous.[1] On the Cotswold "which takes its name from the hills and sheep-cots...[were] fed large flocks of sheep with the whitest wool: having long necks and square bodies, by reason, as is supposed, of their hilly and short pasture, whose fine wool is much valued in foreign parts". And then followed the enumeration of the wool-marketing and clothing towns, which lay either under the escarpment or down on the dip-slope, wherever the running water for the fulling mills was available. Oxfordshire lent itself less easily than Gloucestershire to simple characterisation, being a transition county. Northward, it reaches towards the midlands, to which Banbury, famous for its cheeses (Camden would not admit the cakes and ale, considering them unworthy of mention), served as a gate. Eastward, it reaches to the lower Thames, where at Henley "the greatest part of the inhabitants are barge men, and get their livelihood by carrying wood and corn to London by water". In neighbouring Buckinghamshire the contrast between the beech-clothed Chilterns and the Vale [of Aylesbury] has always been recognised. In both divisions "the soil is generally very fruitful, and the inhabitants thick-set and numerous, who generally follow grazing". "From the top of these hills [the Chilterns] we have a clear and full prospect of the vale," wrote Camden, "it is almost all champain; the soil is chalky, stiff and fruitful: the rich meadows feed an incredible number of sheep, whose soft and fine fleeces are sought after, even from Asia itself." The devotion of potential arable land to sheep grazing was a change accepted in Camden's day, and he quoted the protests of Thomas Ross and Hythloday as something belonging to a past age. Grazing, however, had reference not only to sheep. The butchers rented grazing lands, around the larger towns

---

[1] Judging by the number of able-bodied men enumerated in the Armada year (1588) Gloucestershire competed with Surrey for the honour of being the most densely populated English county, while next in order came Somersetshire and Kent. Such a distribution of population was to be expected; the west country cloth industry and the growing metropolis alike determined local concentration.

and in the home counties, in which to fatten oxen and condition them for killing; while the increase of road travel created a growing demand for both grazing land and hay for horses, contiguous to the posting towns which were strung along the highways.[1] On its north-western border Buckinghamshire reaches to the oolitic limestone, and when describing Stony Stratford Camden followed for once Leland's useful practice of mentioning the quarry (near Calverton) whence were dug the stones for the buildings and for the bridge that carried the queen's highway over the Ouse "which uses, in winter floods, to break out into the neighbouring fields with great violence".

The neighbouring county of Bedfordshire he found specially notable for barley ("plump, white and strong") rather than for wool, and at Biggleswade there was a famous horse fair. At this point of Camden's narrative is to be found one of the few passages in which he described part of his actual route, namely from Woburn to Dunstable, that is to say from the Lower Greensand, across the Gault and on to the Chalk. "We had not gone far from the place, by Hockley in the Hole, heretofore a dirty (but now a very good) road, extremely troublesome to travellers in winter-time: and through fields of excellent beans yielding a pleasant smell, but by their fragrancy spoiling the scent of dogs to the great regret of hunters; till we ascended a white hill into the Chiltern, and presently came to Dunstable, seated in a chalky ground, pretty well inhabited and full of inns."[2] It is clear from this passage that the Gault section of the highway (Watling Street) on which Hockley lies, had been improved, as had been the case with the corresponding section of the main road to Aylesbury described by Leland.

Similar in position to Dunstable on Watling Street was Royston on the old North Road. "Richard I granted it a fair as also a market, which is now very famous and much frequented, upon account of the malt trade: for it is almost incredible what a multitude of corn-merchants, maltsters

[1] The Purveyors for the Royal Household went as far afield as Herefordshire to requisition beef oxen, which were driven up to London at the rate of about eight miles a day. This involved putting them into fields each day to graze and rest. The elder Hakluyt reared stall-fed oxen for meat on his estate near Leominster, and his complaints to the Privy Council about the price they realised afford details as to current practice. *The Original Writings of the two Richard Hakluyts* (ed. E. G. R. Taylor, Hakluyt Society, 1935).

[2] The frequency with which Camden mentioned towns as "full of inns" is a commentary on the increase of travel. There was an influx of gentry and their servants into London when Term began, and an exodus to country estates when it closed.

and the like dealers in grain, weekly resort to this market; and what a vast number of horses, laden with corn, on those days fill all the roads about it."

Nearly half-way from Royston to London, at the inner edge of the chalk belt, stood Ware which had already eclipsed its neighbour Hertford, on account of its more favourable position on the highway and on the river Lea. In the year 1581 there rose to a head the perennial conflicts between those interested in water carriage and those interested in land carriage along what was a major route supplying London with the materials for bread and beer. Some mischievous persons burnt Waltham Lock, and a cut was made to let off the water at Enfield, while the miller owning the lock near Stratford gave orders to the "lockyer" to let boats pass only at still water, so that at neap tides they were kept waiting for as long as four days. It would seem that the millers and mealmen controlled the pack-horse and cart transport, and that the efforts made to increase the carriage of grain by water (which provoked these disorders) were aimed at breaking a monopoly and lowering London prices. Among the arguments put forward for using the river were that horses took up land that might be used for sheep or corn, that wherever they grazed weeds were disseminated, and that they consumed grain that might otherwise serve for human food. Moreover, the heavy land traffic "pestered" the highway, and rendered it less apt for her majesty's posts. Besides the easy movement of grain down-stream, water-borne iron and coals could be carried up-stream, more cheaply than by road, to the smithies within a range of eight or ten miles of the river. A contemporary enumeration gives twenty-two boat and barge owners, employing a hundred boatmen, available for the twenty or so miles of waterway (a number that seems rather small), and it was not possible to use the favourite argument of the day, that the change would set more people "on work", and help to cure unemployment. The grist-mill owners, as well as the land carriers, were opposed to the development of navigation, for the passing boatmen were a nuisance to them. The same type of controversy cropped up with regard to the lower Medway, where the carriage of timber and iron to Thames-side from Yalding in the Weald was in question. The boatmen declared that their passage was obstructed by mills and fish-weirs, and their supporters maintained that the oxen used to draw the great waggons of the land carriers were so beaten and overdriven as to become unfit for human food, and further that the 2-ton iron-shod carts damaged both roads and bridges. On the other side, it was urged that the boatmen were petty thieves and damaged the crops, and that to stop the land traffic would be to rob of their livelihood those farmers who reared draught oxen.[1] The question of obstructions on the Thames itself was raised by the inhabitants of market towns along its banks, who were interested in cheaper transport. The citizens of Abingdon, for example, in 1580, listed twenty-five weirs, locks and mills hampering movement between their own town and Maidenhead. The riparian owners, however, appear to have been successful in maintaining their right to erect such mills and weirs.[2] Staines, near the boundary of Middlesex, was the limit of the jurisdiction of the City of London over the Thames waterway, and here, as at Maidenhead and at Henley, stood a wooden bridge, that at Staines carrying the highroad from London to Exeter and the south-west. Uxbridge, where the highway to Banbury crossed the Colne, was described by Camden as "a modern town, full of inns", no doubt because it was just an easy day's journey out from the capital, the focus of road and water movement.

Camden rejoiced in what to many was the dismaying growth of London, "the epitome of all Britain, the seat of the British Empire", and he spoke proudly of the range of splendid buildings and stately houses which had bridged the mile between the City and Westminster. "On the west side of the city the suburbs run out in another row of beautiful buildings, namely Holbourn...where are some inns for the study of the common law.... The suburbs have grown likewise on the north side... where is now a stately circuit of houses", and they had grown equally towards the north-west. The borough of Southwark had been annexed to the city in 1549, and a continuous line of houses ran along the south bank of the river, where there were the places for bull-baiting and bear-baiting, and the kennels of fierce dogs. Camden declared that it was incredible how London had grown and was still growing in public and private buildings "while the rest of the cities in England are rather decaying". As Thomas Milles, Customer for Sandwich, wrote in 1604, "All our creeks seek to one river, all our Rivers run to one Port, all our Ports join to one Town, all our Towns make but one City, and all our Cities and Suburbs to one vast

---

[1] Persistent rains brought such heavy traffic to a standstill across the clay belts, and lack of water-ways made the timber reserves of north Essex and south-west Surrey, if not inaccessible to the Thames shipbuilders, at least subject to weeks' delay in transport. The detailed accounts for fitting out Frobisher's voyages of 1577 and 1578 (Add. MSS. 39,852) throw an incidental sidelight on the growing scarcity of big oak timber from the Weald.

[2] The rush of water through the locks, when the "flashes" were opened, made them very dangerous, and John Bishop, in an appeal to the queen in dolorous vein, written in 1586, gives the names of twenty men recently drowned in the middle reaches of the Thames in this way (Lansdowne MSS. 49, no. 39).

unweildy and disorderly Babel of buildings, which the world calls London."[1]

A contemporary writer remarked in explanation that the Court was much greater and more gallant than in former times, while

gentlemen of all shires do fly and flock to this city: the younger sort of them to see and show vanity, and the elder to save the cost and charge of hospitality and housekeeping. For hereby it cometh to pass, that the gentlemen being either for a good portion of the year out of the country, or playing the farmers, graziers, brewers or such like more than gentlemen were wont to do within the country, retailers and artificers, at least of such things as pertain to the back or belly, do leave the country towns where there is no vent, and do fly to London, where they be sure to find ready and quick markets.[2]

The change over from individualist to capitalist production, and the expansion of overseas enterprises with the related developments of banking and insurance were of course factors in this metropolitan growth, but Camden was no economist, and took his London as he found it. It was left to Norden to describe the way in which the metropolis influenced the life of the country folk of Middlesex.

Not medlinge with the higher sorte, I observe this in the meaner, and first of such as enhabyte near the Thames, they live either by the bardge, by the wherrye or ferrye, by the sculler or by fishinge, all which live well and plentifullye, and in decent and honest sort releve their famelyes. Such as live in the inn countrye, as in the body or hart of the Shire, as also in the borders of the same, for the most part are men of husbandrye, and they wholly dedicate themselves to the manuringe of their lande. And theis comonlye are so furnished with kyne that the wyfe or twice or thrice a weeke conveyethe to London mylke, butter, cheese, apples, peares, frumentye, hens, chickens, egges, baken, and a thousand other country drugges which good huswifes can frame and find to get a pennye. And this yeldeth them a large comfort and reliefe.... Another sorte of husbandman or yoman rather ther are, and that not a few in this Shire, who wade in the weedes of gentlemen...who havinge great feedinges for cattle, and good breede for younge, often use Smythfelde and other lyke places with fatt cattle, wher also they store themselves with leane. And thus they often exchaunge, not without

[1] E. G. R. Taylor, *Late Tudor and Early Stuart Geography*, 1933. London rapidly outgrew its natural supply of spring water, and the project for bringing water from the river Lea into the city was mooted in 1577. *S.P. Dom.* 118, no. 67. For its food supply see F. J. Fisher, "The Development of the London Food Market, 1540–1640", *Econ. Hist. Rev.* v, 2. Chapmen from all parts of the country resorted to London to lay in their store of goods from the wholesalers.

[2] Anonymous *Discourse* in Stow's *Survy of London*, 1598.

great gayne.... Ther are also that live by carriage for other men and to that ende they keepe cartes and carriages, carry meale, milke, and manie other thinges to London, and so furnish themselves in their returne with sundry men's carriages of the countrye, wherby they live very gainfully.[1]

Such pictures of diversified rural activity indicate that round London, as indeed round other large towns and also in areas where there were cottage industries, the old common-field agriculture, in so far as it had actually existed, had been replaced by the large or small individual holding. The enclosure by great landowners of commons and wastes had also begun: there were many disorders on this account, as for example at Enfield in 1589 where about 90 acres had been filched from the commoners, and in Westminster in 1592.

The marshy borders of neighbouring Essex were famous for their mutton-sheep and milking ewes, as were the grazing lands of the south-east for butter and for "those cheeses of an extraordinary bigness which are used, as well in foreign places as in England, to satisfy the coarse stomachs of husbandmen and labourers".[2] Apart from the good grazing, only the famous oysters and the saffron fields were noticed by Camden, but Norden mentioned the hop-growing of the northern (chalky) parts of the county, and the oats of the south-east "whence her Majesty hath great store of provision of avenage". Sheep bred for wool were comparatively few, but some of good quality were found on the heathy lands (for example on Tiptree Heath), and a woollen industry flourished at Colchester and in the small towns and villages behind it which went back to medieval days, although given fresh impetus in the later part of the sixteenth century by the importation of foreign technique.[3] In the west of the county,

[1] Harleian MSS. 570.
[2] Huge Essex and Dutch cheeses were in demand for provisioning ships going on long voyages.
[3] John Johnson the woolstapler wrote in 1571: "It is not 40 yeares paste that Anthony Bouvise an Italiane merchant brought in to England the practice of fyne clothinge at Coxshall (Coggeshall), etc. And sithence that tyme and evene presently thear is a greate increasement of clothing in this Realme of bayes, frisadoes and other thinges made of wolles by Straungers, whearof as there is groune great benefite to this Realme, and London, and divers other citties" (Harleian MSS. 253). A colony of Flemings came to Colchester in the troublous 1560's, and ten years later a request was made that some forty households might settle at Halstead, and make "bayes and muscadoes" as their countrymen did elsewhere. See also George Unwin, "The History of the Cloth Industry in Suffolk", *V.C.H. Suffolk*, II, 254–71.

nearest London, lay Epping Forest and Hatfield Chase, well stocked with red and fallow deer, the fattest (Camden says) in all England. Norden, with his keen surveyor's eye for man's common needs, remarked that these were "noe good neighbours to the forest inhabitants", and comments also on the absence of springs (as was natural in a clay belt) which in many areas made it necessary (as it continued to do right down to the last century) to rely on standing waters "bad in winter and in summer worse" for domestic use.

In the lowland margin of Suffolk the rich grazing lands produced a surplus of butter and cheese as in Essex, and the clothing industry also extended across the border, to Lavenham, Clare and Long Melford. The fertile boulder-clays ("compounded of clay and marle") called forth one of Camden's rare comments on the roads, which, because of this clay, were "deep and troublesome" all round Debenham. Of Norfolk, a clearer picture emerges than of Suffolk; its soils were "in some places fat, luscious and moist, as in Marshland and Fleg: in others, especially to the west [i.e. Breckland], lean and sandy, and in others clayey and chalkey". The city of Norwich, industrious, prosperous and populous, was reckoned among the most considerable in the kingdom, being "partly indebted for its prosperity to the people of the Netherlands, who when they could no longer endure the tyranny of the Duke of Alva, nor the bloody inquisition then setting up, flocked hither in great numbers, and first brought in the manufacture of slight stuffs".[1] Norwich profited also, of course, by the prosperity of its out-port Yarmouth, where "it is almost incredible what a great and thronged fair here is at Michaelmas, and what quantities of herring and other fish are vended". Lynn, at the edge of the marshland, where tens of thousands of sheep, besides young cattle, were grazed, was still, when Camden wrote, second only to Norwich, but it was presently to fall behind Yarmouth in activity and wealth.[2] The draining of the fens

[1] The Borough of Thetford "now almost desolate" petitioned the queen in 1580 for some staple trade to be located there, to induce strangers, i.e. foreigners, to settle among them (Lansdowne MSS. 31, no. 29). By that time some English clothiers had already undertaken the manufacture of the new stuffs, but others merely complained that their kersey- and broadcloth-making were decaying owing to the increasing use of the new light fabrics for garments and linings, and of the worsted yarn for knit hose. When the rates of subsidy and alnage on the "new draperies" were fixed, the following types of stuffs were mentioned: bays, frisadoes, naples fustians, grograynes, broad and narrow russets, serge, sayes, worsteds, tukes (or turks), Spanish and English rugs, Coxsall bayes (Lansdowne MSS. 26, no. 61).

[2] In 1572 one Thomas Fermor offered to farm the Customs of Lynn for £1000 annually (Lansdowne MSS. 14, no. 52). According to accounts submitted in March 1590 relative to Sir Francis Walsingham's farm of the Customs of a number of ports, this

was under constant discussion at this time, but the idea (and the tentative experiments), while attractive to the great landowners, was passionately resented and resisted by the fenmen, who saw their grazing and fowling and fishing, their withies for basket-making, their reeds for thatching, all threatened by the innovation.[1] Nor was it certain (in spite of comparisons with Holland) that drainage would be successful, and Camden himself was inclined to think that it was tempting Providence, or at best "lining private purses by promise of public good". The Newleame ditch was made early in the century by Bishop Morton "for the better convenience of water-carriage, and thereby increasing the trade and wealth of this his town [Wisbech]; though it has fallen out otherwise, for it is but of small use, and the neighbours complain, that this has quite stopped the course of the Avon or Nen into the sea by Clowcross".

Huntingdonshire, like Cambridgeshire, was noted for grain so far as the "upland" areas were concerned, and shared in the valued fens. Westwards, the landscape altered, and Northamptonshire, through which runs the broad outcrop of oolitic limestone, was "everywhere adorned with noblemen's and gentlemen's houses" including that of Sir Christopher Hatton, who had himself enjoyed it for so brief a time. Northamptonshire, like Leicestershire, showed little change since Leland had viewed it: grazing was all-important for sheep and horses and cattle. At Rothwell, midway between Pytchley and Market Harborough, was a frequented horse-fair; and at Harborough itself a great beast-fair. Even a Camden could not overlook the stone quarries in such a country of stone architecture as this was, and he described the famous Collyweston roofing slates, and the great quarries of Barnack, from the stones of which the abbeys of Peterborough and Ramsey had been built. In the same corner of the county was the fine stone bridge of Wansford, carrying the main north road over the Nene, which was damaged by a storm in 1586 and which was forthwith ordered to be repaired by royal command at the expense of the county. Only in Leicestershire was there a faint hint of change to come. Cole Overton

figure would have shown a handsome profit in normal years. The trade of Boston had shrunk to very trifling dimensions. "The trades before tyme used were principally by the merchants of the staple and of the styllyard, both which are now decayed ther, as is manifest. . . . And nowe the trade remaininge is with the Scottes for salt principally, and secondlie for fishe, herings, course drapes, lynnen clothes, course tapistry." So the citizens declared in 1586, when entering their protest against the grant of a patent for making white salt on the north-west coast to Mr Wilkes (Lansdowne MSS. 47, no. 68). There is a hint in Camden, however, that a new prosperous class of graziers was growing up in Boston, surrounded as it was by fens and water-meadows.

[1] See chap. XII.

(Orton) was so called said Camden "from pit-coals, being a bituminous earth hardened by nature, and here (to the great profit of the lord of the manor) dug up in such plenty as to supply the neighbouring country all about with firing". Rutland merely repeated the characteristics of east Leicestershire, but it is worth noting that Camden reported the recent founding of grammar schools at Uppingham and at Oakham, and that such schools were now normal even in the smaller market towns, where in Leland's day they had been exceptional.

In Lincolnshire Camden returned to a consideration of the problems of the fens, and remarked how in Holland the people were at pains to protect themselves on the one side from the sea floods, on the other from the land floods; the latter were the more constant menace, and in the year 1599 (he recorded) a new channel had been begun at Clows Cross to carry them off.[1] Of Crowland he wrote:

Beyond the [triangular] bridge formerly stood that famous monastery, though small in compass; about which, unless on the side where the town stands, the ground is so rotten and boggy that a pole may be thrust down thirty feet deep; and there is nothing round about but reeds, and next the church a grove of alders. However, the town is pretty well inhabited, but the cattle are kept at some distance from it, so that when the owners milk them, they go in boats (which will carry but two) called by them skerries. Their greatest gain is from the fish and wild ducks that they catch, which are so many that in August they can drive into a single net three thousand ducks at once: and they call these pools their corn fields, there being no corn growing within five miles of the place. For this liberty of catching fish and wild ducks they formerly paid yearly to the abbot, as they do now [1607] to the king, three hundred pounds sterling.

In north Lincolnshire, in the marshy vale of the "Ankham" (Ancholme), wild fowl were again a source of wealth. "This country is, at certain seasons, so stocked with fowl (to say nothing of fish) that their numbers are amazing: and those not the known ones, of greatest value in other countries, teal, quails, woodcocks, pheasant, partridge, etc., but such as no other language has names for, and are so delicate and agreeable that the nicest palates and richest purses greatly covet them, viz. puits, godwits, knots...and dotterels." But of the towns and ports, the bridges and ferries, and the crops of Lincolnshire, Camden had little new to say. At Barton-on-Humber was the famous ferry into Yorkshire: across the

---

[1] Letters and petitions from Lincolnshire landowners in 1598 indicate that a series of wet years had caused serious hardship, and they sought State aid in draining their "fenny tracts" (Lansdowne MSS. 87, no. 4).

Trent, a little below Gainsborough, came a route from Doncaster; "it is a great road for pack-horses, which travel from the west of Yorkshire to Lincoln, Lynn and Norwich." Lincoln's fifty churches had diminished in two generations to eighteen, but the lesser market towns appeared flourishing.

Moving westward into Nottinghamshire, Camden was once more in a country where the contrast of soils dictated to the country folk a simple regional division, namely the "sand" and the "clay". Over the sandy soil stretched Sherwood Forest "formerly one close continual shade" but now "it is much thinner, and feeds an infinite number of deer and stags: and it has some towns in it whereof Mansfield is the chief". Through the clay vale ran the Trent, with Newark and Nottingham on its banks. "In respect of situation the latter town is very pleasant; on one side to the river are very large meadows, and on the other side hills of an easy and gentle ascent: it is also plentifully provided with all necessaries. Sherwood supplies them with great store of wood for fire (though many burn pit-coal, the smell whereof is very offensive) and the Trent serves them plentifully with fish."

Derbyshire likewise had its two-fold division, in this case into lowland and upland: "The west and south are well cultivated and pretty fruitful and they have many parks. The west part, beyond the Derwent, called the Peak, is all rocky, rough and mountainous, and consequently barren: yet it is rich in lead, iron and coal, and convenient enough for the feeding of sheep." Derby itself, seated between the two regions, was an old wool-stapling town:

Its present reputation is from the assizes of the county that are held there, and from the excellent ale brewed in it.... The wealth of this town depends in a great measure upon a retail trade: which is to buy barley and make it into malt, and sell it to be sent into the high-land countries, for the town consists chiefly of this sort of merchants.... The west part of the county...though rough and craggy in some places has also grassy hills and vales which feed abundance of cattle and great stocks of sheep very securely.... It produces so much lead that the chymists (who condemn the planets to the mines, as if they were guilty of some great crime) tell us, ridiculously as well as falsely, that Saturn, whom they make to preside over lead, is very good to us.... Out of these mountains, lead stones (so the miners call them) are daily dug up in great abundance, which they melt down with large wood fires upon those hills exposed to the west-wind (about Creach and Worksworth, so called from the lead-works) at certain times when that wind begins to blow, which they find by experience to be the most constant and lasting of all winds: and then digging channels for it to run into they work it into sowes. And not only lead, but stibium also, called antimony in the shops, is found here in distinct veins.... Mill-stones also and grind-stones are dug here, and sometimes there is found in these mines a kind of white fluor that is in all respects like crystal.

As Camden briefly stated, the Peak district was becoming familiar to people outside Derbyshire, since both Buxton and Matlock Bath were in Queen Elizabeth's reign being increasingly frequented for their medicinal and thermal waters. At the former, the Earl of Shrewsbury had built comfortable lodgings and places of amusement so as to ensure the well-being of the gently and nobly born visitors. The arrangements at Bath were very squalid and unhygienic by comparison, and Buxton drew its fashionable crowd each season. In Protestant England a visit to a spa replaced the older pilgrimage to a holy well or sacred shrine, but it required all the eloquence of the English physicians to convince their patients that the home waters were equal to those of the more famous continental resorts.

The order in which Camden described the counties of England was dictated by his views as to the British tribes inhabiting them in Roman days, and hence he turned from Derbyshire to the west midlands, War-wickshire, Worcestershire, Staffordshire, and then to the northern borders of Wales, Shropshire and Cheshire. Warwickshire was traditionally divided into the Feldon and the Woodland, "the champain" and the wooded country, the one on the hither- the other on the far-side of the river Avon. The "champain" country included the famous sheep pastures of the limestone belt, as well as the richer meadows and cornfields of the vale beneath Edge Hill, from which point of vantage Camden viewed it. That Rugby was described as "abounding with butchers" bears witness to the importance of cattle-grazing in the vale of Avon; but of the geography of other parts of the valley Camden found nothing to say. In the woodland, which was "for the most part cloathed with woods, yet wants not pastures nor corn-fields", he visited Kenilworth, given by Queen Elizabeth to Lord Robert Dudley who spent money lavishly upon it.

Whether you regard the magnificence of the buildings, or the nobleness of the chase and parks, it may claim a second place among the stateliest castles of England.... From hence (that I may pursue the same course that I did in my journey) I saw Solyhill in which was nothing worth the sight, besides the church. Next, Birmingham, swarming with inhabitants, and echoing with the noise of anvils, for here are great numbers of smiths. The lower part of the town is very watery: the upper part rises with abundance of handsome buildings. . . . From hence going southward I came to Coleshill. . . . Lower, in the middle of this woody country is seated Coventry, so called (as I conjecture) from a convent. . . . Wheresoever the name was taken, the city, being some ages since inriched with the manufacture of clothing and caps,[1] was the only mart-town of this county and of greater resort than could be expected from its midland situation. . . . A little higher, upon Watling Street, where is a bridge of stone over the river Anker, *Manduessum* is seated; a town of very great antiquity. . . . This name (since a quarry of free-stone lies near it)[2] was probably given it from the stone there dug and hewed. . . . But, how great, or of what note soever it was in those times, it is now a poor little village, containing not above fourteen small houses. . . . Atherstone on one side, a well frequented market, and Nuneaton on the other side, have by their nearness reduced Mancester to what you see.

Worcestershire presented more variety of activity than Warwickshire, it had its famous salt wiches, its long-established pear orchards (for perry), its fat lampreys from the Severn, its rich clothiers of Worcester. But the salt manufacture was a serious tax on fuel, as Leland had already noticed, and the substitution of coal for wood had not yet taken place. "What a prodigious quantity of wood [says Camden] these salt-works consume, though men be silent, yet Feckenham-forest, once very thick with trees, and the neighbouring woods, by their thinness declare daily more and more." The Forest of Wyre, near Bewdley, had also decayed since Leland saw it, "it was lately remarkable for the wonderful height of the trees. . . which are now in a manner all gone".

In the absence of a vocabulary for forms of surface relief other than mountain, hill and vale, the contours of England largely escaped the six-teenth-century topographer, but Camden remarked upon the Malvern Hills seen across the Severn, "rising like stairs one higher than the other", and he also realised the unity of the Pennines.

The north part of the county [Staffordshire] rises gently into small hills; which begin here, and like the Apennine in Italy, run through the middle of England in one continued ridge, rising higher and higher, as far as Scotland, under several names: for here they are called Moorland, after that Peak, then Blackstone Edge,

---

[1] As was noted on p. 352, Coventry was at the mercy of fashion. One John Green complained in 1591 that his trade of cap-making had decayed through the now common wearing of hats. Legislation to make cap-wearing compulsory on holy days had failed.

[2] In the Keuper sandstone.

anon Craven, next Stanmore, and last of all, when they branch out into horns, Cheviot. This Moorland...is a tract so very rugged, foul and cold that the snows continue long undissolved.... It is observed by the inhabitants that the west wind always causes rain; but that the east and south winds, which are wont to bring rain in other places, make fair weather here, unless the wind shifts about from west to south; and this they ascribe to their nearness to the Irish Sea.

South of the Pennines, Staffordshire was remarkable for its grazing lands, its cattle- and horse-fairs, and its output of leather; in Needwood Forest there were many parks, "wherin the gentry hereabouts frequently exercise themselves, with great application, in the agreeable toil of hunting". But in the south there was to be found something new—"much pit-coal mines and iron: but whether to their loss or advantage, the natives themselves are the best judges, and to them I refer it". Camden seems to have had some prescience of the industrial revolution, and of the price which posterity was to pay for it in loss of simplicity and beauty. In Shropshire, as in Staffordshire, the future was already discernible: "Not far from the foot of this hill [the Wrekin] in a deep valley, and upon that Roman military high-way [Watling Street], is Okengate, a small village, noted for the plenty of pit-coal that it affords." But just as Lichfield then dominated Staffordshire, so Shrewsbury was the metropolis of the neighbour county.

At this day it is a fine city, well inhabited and of good commerce: and by the industry of the citizens, their cloth-manufacture, and their trade with the Welsh, is very rich; for hither the Welsh commodities are brought, as to the common mart of both nations. Its inhabitants are partly English and partly Welsh: they use both languages; and this among other things must be mentioned in their praise, that they have erected one of the largest schools in England for the education of youth.

Shrewsbury was the focus (as it is to-day) of a chain of secondary markets lying along the edge of the Welsh upland block. Such, for example, was Oswestry "a place of good traffic, for Welsh cottons especially, which are of a very fine, thin, or (if you will) of a slight texture: of which great quantities are weekly vended here".[1]

Chester, too, owed much to English intercourse (not always peaceful) with the Welsh, as also with the Irish, but although still the post-town for Ireland its doom was already sealed.

[1] Exactly when so-called "cottons" were in fact made partly of cotton yarn is not very clear, since the substitution of cotton for wool or flax was looked upon as producing "false" cloth and was not done openly. The change is no doubt to be associated with the reopening of the English Levant trade and was effected by 1600.

The city is of a square form, surrounded with a wall about two miles in compass, and it contains eleven parish churches. Upon a rising ground, near the river, stands the castle, built by the earl of this place, wherin the courts palatine are held, and the assizes twice a year: the buildings are neat, and there are piazzas on both sides, along the chief street. . . . Nor is there anything wanting to make it a flourishing city, except it be that the sea is not so favourable as it has been to a few mills that were formerly situated upon the river Dee, from which it has gradually withdrawn, and the town has lost the advantage of a harbour, which it enjoyed heretofore.

But if one town had decayed, another elsewhere had expanded: Nantwich (Camden declared) was the greatest and best built town of the county; and Congleton, without so much as a parish church of its own, "in consideration of its greatness, populousness, and commerce, has deserved a mayor and six aldermen to govern it". Congleton, it may be remarked, lay on the highway to Stockport and Manchester, but, before describing the changing face of Lancashire, Camden turned southwards to Herefordshire as a prelude to treating of the lands of the ancient Silures.

In Herefordshire lies the Golden Vale of the Dore, and in his penpicture of its landscape Camden epitomised all that was held desirable in the English country-side and yet was doomed, even in his day, to pass away. "The hills that incompass it on both sides, are cloathed with woods; under the woods lie cornfields on each hand; and under those fields lovely and fruitful meadows. In the middle, between them, glides a clear and crystal river." For the rest, Camden could pluck his description of Herefordshire from Leland's pages, for fortunately it had as yet no mines, and no manufactures (save of minor woollens) to give a first impulse to the accelerating change of which the onset elsewhere has been noted.

In Yorkshire the story was a different one. The Sheffield smiths, it is true, had already been familiar figures for centuries, and hence Camden passed over the iron-working with a bare remark: but the West Riding woollen industry had but lately expanded with a remarkable rapidity, while that of York itself, like that of Beverley, had decayed. Halifax, in the fifteenth century a mere village, had

about twelve thousand men in it, so that the parishioners are wont to say that they can reckon more men in their parish than any kind of animal whatever; whereas in the most fruitful places of England elsewhere, thousands of sheep shall be found, but so few men in proportion, that one would think they had given place to sheep and oxen, or were devoured by them. The industry of the inhabitants is also admirable, who notwithstanding an unprofitable, barren soil, not fit to live in, have so flourished by the clothing-trade (which within these

25

70 years[1] they first fell to) that they are very rich, and have gained a reputation for it above their neighbours; which confirms the truth of that old observation, that a barren country is a great whet to the industry of the natives; by which we find that Norimberg in Germany, Venice and Genoa in Italy, and lastly Limoges in France (all situated on barren soils), have ever been very flourishing cities.

An entirely new industry had been developed at Aberford, on the main North Road: it was "famous for its art of pin-making, the pins made here being in particular request among the ladies". Not far from Aberford, and situated likewise on the Magnesian Limestone belt, stood Hesselwood, under which town "is the remarkable quarry called Peter's post, because the stately church at York, dedicated to St. Peter, was built with the stones hewed out here, by the bounty of the Vavasours". This limestone formation was all-important from many points of view. Tadcaster stood where its outcrop approached most nearly to York, and from this town lime for mortar was carried ten miles to the city of York and to other places in the vale. In Elmet also (near Hesselwood) "limestone", said Camden, "is plentifully dug up; they burn it at Brotherton and Knotting-ley, and at certain seasons convey it in great quantities for sale to Wake-field, Sandall and Standbridge; from thence it is sold to the western parts of this country, which are naturally cold and mountainous; and herewith they manure and improve the soil.   But leaving these things to husband-men, let us return". He could not, however, quite leave the subject, but spoke more than once of a yellow marl (actually part of a glacial deposit) found in the Aire valley "of such virtue, that the fields once manured with it, prove fruitful many years after". It is used to-day, and still works wonders!

The impression Camden leaves of the West Riding is of a region whose diversity called forth an active circulation and promoted an active develop-ment: in the East Riding, on the other hand, all was much as in Leland's day, and, as regards the North Riding, his narrative (of a stamp which grew more familiar as the seventeenth century advanced) was of the Dutchman's seizure of the opportunities which the Englishman had neglected.

It is worth remarking [he wrote] that those of Holland and Zealand carry on a very great and gainful trade of fishing in the sea here for herrings, after they have, according to ancient custom, obtained licence for it from this

---

[1] So said in 1607. The West Riding woollen area was remarkable for the large number of cottager weavers forming a relatively dense population on the poor moorland country within six or seven miles of the towns.

castle [Scarborough]; for the English always granted leave for fishing, reserving the honour to themselves, but, out of a lazy humour, resigning the gains to others, it being almost incredible, what vast gains the Hollanders make by the fishery on our coast. These herrings (pardon me if I digress a little, to show the goodness of God towards us) which, in the times of our grandfathers, swarmed only about Norway, now in our times by the bounty of Providence, swim in great shoals round our coasts every year. About Mid-summer they draw from the Main Sea towards the coast of Scotland, at which time they are immediately sold off, as being then at their best; from whence they arrive on our coasts, and from the middle of August to November, there is excellent and most plentiful fishing for them, all along from Scarborough to Thames-mouth. Afterwards, by stormy weather, they are carried into the British Sea, and are there caught till Christmas; thence having ranged the coast of Ireland on both sides, and gone round Britain, they return into the northern ocean, where they remain till June; and after they have cast their spawn, return again in great shoals.

But in spite of the laxness of its fishermen,[1] the North Riding had a new industry to show, that of alum mining,[2] behind and near Whitby, one of the greatest importance because of the relation of the product (used as a mordant) to the woollen manufacture. Apart from this the North Riding, like the East Riding, was chiefly notable for its agriculture and live-stock rearing, and Camden remarked that in the Forest of Galtres a yearly horse-race was run for the prize of a little golden bell, while at Northallerton, on the main highway, "which is nothing but a long street", there was held "the throngest beast-fair, on St. Bartholomew's day" that he ever saw.

Richmondshire made the fourth division of Yorkshire, and lay almost wholly in the Pennines, where, on the Lancashire border "the prospect among the hills is so wild, solitary and unsightly, and all things are so still, that the inhabitants have called some brooks there Hell-becks.... There is a safe harbour in this tract for goats, deer and stags, which for their unusual bulk and branchy heads are very remarkable and extra-

---

[1] Robert Hitchcock in his *Pollitique Platt*, published 1581, but written several years earlier, advocated a levy for the building of a new fishing fleet to compete with the French and Dutch, "For that fishe thei now utter unto us, we should receive of them the comodities of the Lowe Countries, viz. Hollande clothes, Rape oile, Hoppes, Madder, all sorts of Wier, and diverse other Merchandizes, Or else their ready Gold and Money."

[2] Previous to Elizabeth's reign, the chief sources of alum had been from Spain and Civita Vecchia, and the Italian banker, Horatio Palavicini, offered in 1578 to procure supplies at cheap rates, acting no doubt in the interests of the foreign producers and shippers.

ordinary". Nevertheless, the road into Westmorland and Cumberland lay right across Stanemore, "entirely desolate and solitary, except one inn in the middle for the entertainment of travellers", who could not therefore have been many.

To the north, in the bishopric of Durham, the veins of iron, the increasingly worked coal-pits, the glass-houses on the Wear and Tyne, and the great salt-pans, all pointed forward to the industrial revolution; and, in the county of Lancashire, Manchester had already surpassed "all the Towns hereabouts in its building, populousness, woollen manufacture, market-place and church...", while Liverpool had become "the most convenient and usual place for setting sail into Ireland, but not so eminent for antiquity as for neatness and populousness".[1]

In Westmorland, the landscape changes, although Camden pointed out that the county was far from being (as many believed) entirely mountainous and barren: in the vales and bottoms was plenty of arable land, and there had been more, for the ridges crossing the ground indicated that what was now moorland (i.e. common pasture) had once been under the plough. "The gentlemen's seats in this county are large and strong, and generally built castlewise, for defence of themselves, their tenants, and their goods, whenever the Scots should make their inroads, which, before the time of King James I, were very common." Kendal, with its centuries-old cloth industry, had been incorporated by Queen Elizabeth and was further advanced by her successor; but it was in Cumberland, where the Keswick copper mines had been extended and worked by Daniel Hochstetter, that a more important development had taken place, for copper had hitherto been very deficient in England. Here, too, were the blacklead mines.

The Union of Crowns which put an end to border castles and border warfare (and so made no longer necessary the fortification of gentlemen's seats) had not yet taken place when Camden made his journeys to the north. Hence, there were parts of the "Picts' Wall" which he did not think it prudent to examine, for "we...durst not go and view them, for fear of the moss-troopers", but it was as far as possible along the line of this wall that he travelled from Cumberland into Northumberland. Here were bred the "bog-trotters", which could cross the mountain bogs impassable to ordinary horsemen, and here "all over the wastes...you would think you see the ancient Nomades,...a martial sort of people, that

---

[1] Figures for the coal export to Ireland, for example, show steady (though small) shipments from Liverpool, and only occasional shipments from Chester. J. U. Nef, *Rise of the Coal Industry*, 2 vols. (1932), App. D.

from April to August lie in little huts (which they call sheals and shealings) here and there, among their several flocks".

At length, he reached the point where the Wall meets the Tyne, near which meeting

stands Newcastle, the glory of all the towns in this country; it has a noble haven on the Tyne, which is of such a depth as to carry vessels of a very good burthen, and of that security, that they are in no hazard of either storms or shallows. . . . We have already treated of the suburbs called Gateshead, which is joined to Newcastle by a bridge and belongs to the bishop of Durham. This Town, for its situation and plenty of sea-coal (so useful in itself, and to which so great a part of England and the Low-countries are indebted for their good fires) is thus commended by Johnston in his poems on the cities of Britain:

Newcastle

From her high rock great Nature's works surveys,
And kindly spreads her goods through lands and seas.
Why seek you fire in some exalted sphere?
Earth's fruitful bosom will supply you here.
Not such whose horrid flashes scare the plain,
But gives inliv'ning warmth to earth and men.
Iron, brass, and gold its melting force obey;
(Ah! who's e'er free from gold's almighty sway?)
Nay, into gold 'twill change a baser ore;
Hence the vain chymist deifies its power:
If't be a god, as is believ'd by you,
This place and Scotland more than heaven can shew.

It is fitting thus to close this résumé of Camden's survey of his England with his quotation of the Aberdonian scholar's panegyric on coal, for the coal question was typical of the new, and indeed novel, economic problems which a changing England had now to face. London had long known sea-coal in smithies, bakeries, brew-houses, and even in the domestic fire-place; but by 1600 coal was being burned wherever it could be cheaply enough procured, not willingly indeed, but inevitably, because of the scarcity of wood. Chimneys and hearths constructed to burn logs, and stoves made for charcoal fires, were not well adapted to coal,[1] and the "sulphurous" fumes and smoke were a source of endless complaint, but

[1] The manufacture of cast-iron fire-places and fire-backs became an important Sussex industry.

it was recognised that coal had become a necessity, and that its price was a matter of national concern. In 1591 it had risen from 8s. or 9s. to 13s. or 14s. a chaldron,[1] while at the same time inferior coal was being mixed with the better types. The queen looked upon it chiefly as a taxable commodity; and in this particular year the mayor and aldermen of Newcastle found it necessary to protest to the Dean of York against the grant of an impost on every chaldron. This, they declared, would drive the Frenchmen and Dutchmen, who now came to Newcastle to buy coal, to seek their supplies in Scotland (from the Fifeshire field). New mines (they continued) were constantly being discovered, and these of the best coal, so that supplies were in sight for hundreds of years to come. Hence it would be bad policy to ruin the export trade, for this would mean the ruin of Newcastle. From the more general standpoint it was argued that the export of coal should be encouraged, as it paid for English imports from countries in which there was either little demand for other English staples, or one only for goods (especially provisions) which could be ill spared.

John Keymer, in the first of his economic tracts, written in 1601, drew attention to the disquieting fact that almost the entire export of coal was in foreign bottoms, and huge profits were made on its resale by the Dutch and French shippers. The Newcastle colliers were almost wholly engaged on the coastwise trade,[2] to which in build and tonnage they were better adapted. No doubt Keymer exaggerated the situation, for a few years later another publicist declared that, owing to foreign competition, as many as 30 or 40 boats were laid up at Aldeburgh which used to carry coal to France and fetch back bay-salt. Exact figures were not available to pamphleteers, but they were right in pointing to the coal trade as the greatest bulk trade of all England; and Northumberland and Durham came to be regarded as the breeding ground of ships and seamen rather than the fishermen's counties of Devon and Cornwall.

The coal question belonged to the end of Camden's thirty years of observation. From the beginning of the same period there had been the question of the marketing of English cloth, which continued to be an increasingly

---

[1] For coal prices see J. U. Nef, op. cit. App. E. As might be expected, the contemporary figures giving the rise of prices are exaggerated, for Londoners were paying 10s. 6d. a chaldron in 1562. The account books of King's College Cambridge suggest that the price of coal at inland towns reached by water (in this case via King's Lynn and the Ouse) was very little, if at all, higher than that at London. It was said in 1578 that brewers, dyers, hat-makers and others "have long sithens altered there furnesses and fierie places and turned the same to the use and burninge of Sea Coale".

[2] The volume of coastwise trade was still four or five times as great as the export trade at the end of the sixteenth century.

acute problem until Far Eastern and colonial markets were thrown open
to England. An excellent analysis of the situation, as it appeared to an
experienced business man of the time, is afforded by Roger Bodenham's
memorandum prepared for Lord Burghley in 1571.[1] After listing the
staple products of western Europe and Barbary, he proceeded to note
those which formed articles of trade with England, and the English goods
bought by foreigners in return. Flanders was then the greatest market for
English cloth, but the revolt of the Netherlands had filled Antwerp with
soldiers, and the reorganisation of their trade by the Merchant Adventurers
had not yet taken place. France bought west-country kerseys and coarse
cloths, besides Cheshire cottons, but not in sufficient quantities to balance
the English imports of wine, wood, bay-salt, linen and haberdashery.
Spain bought coarse cloths "of Suffolk marking", Devonshire kerseys,
northern dozens, western kerseys, Kentish cloths, both coarse and fine.
Portugal bought (largely for resale in Spain at a profit) sorting cloths of
Bristol and London, made, of course, in other parts of the country.
Danske (Danzig), Rye (Riga), Reval and the Narve (Narvik, the new
Baltic port of Muscovy) bought large quantities of coarse cloth of divers
colours, and Barbary bought both coarse and fine cloths. The new
Muscovy trade by way of the White Sea ports Bodenham left for descrip-
tion by those who knew it at first hand, and this completed the list of
English foreign markets, for, as the writer well knew (having captained
the last voyage himself), English merchant ships no longer ventured
through the Strait of Gibraltar for fear of Turkish pirates. The Battle of
Lepanto (1571) had not yet borne fruit.

Turning from details to general economic policy, Bodenham pointed
out the fallacy of the commonly held view that foreign imports should
be discouraged, especially those of "sundry trifles which it is said we have
no need of":[2] imports pay for exports, "if we do not buy from other
countries, they cannot buy from us", and, moreover, imports are a source
of customs revenue to the prince. Since our most widely marketed export
was cloth, he said, we ought to have special care to its quality, as either
establishing or destroying our credit throughout the world. In point of
fact, our cloth was not so good as the high quality of our wool would
allow, and we should, he advised, follow other nations in insisting that
(1) only standard lengths and breadths were made, (2) all cloths were

[1] Lansdowne MSS. 100, no. 25.
[2] "Nifles and trifles", as the writer of the *Libelle of Englyshe Polycye* termed them.
See also *A Speciall Direction for Divers Trades* (c. 1575–85), in R. H. Tawney and
E. Power, *Tudor Economic Documents*, 3 vols. (1924), III, 199.

thoroughly shrunk before retailing, (3) they fell even, like a sheet, without "tuckelling, banding or rowing", (4) they were dyed in the wool where possible, (5) the blue dye of all black cloths must be dyed in the wool, while (6) poor dyes like brasil and orchel were entirely forbidden. That English manufacturers were capable of making cloth of the highest quality was proved by the fact that, in his day, Winchcombe's kerseys were sold by their mark alone, as were "Springe of Lanames" Suffolk cloths, while of "cloths called ocho and quatro made by John and George Barnes, there cannot be so many made as can be sold".[1] The superiority of Flanders cloth, the chief rival of English cloth, lay in the good workmanship, dyeing and dressing; and, if the Flemings were not allowed to import English wool, they would be unable to make their high-class cloth from Spanish wool alone (although unfortunately English sheep had been exported to Spain, so that the breeders there were improving their wool). On this subject Bodenham concluded by saying:

If you will have your Townes which do decay in most places of Inglande to floryshe as in tymes paste they did, you must cause your clothinge to be made in the same towns as in tymes paste it was wonte to be, and so to sett order for all thinges aperteyninge to clothe making...whereas nowe by meanes everye man makes in his owne howse good or badd such as he lykes beste, there is so much false dealinge as indede it bringe our clothe out of creditt in all places and is thereby cause of the decayed townes in the which in tymes paste the sayde clothinge was made....

But laws such as Bodenham demanded were already in the Statute Book and had proved impossible to enforce.

Another aspect of the clothing-trade problem was the disorganisation of the Antwerp market where the Merchant Adventurers had their head-quarters. This led John Johnson, a merchant stapler, to put forward a proposal for establishing a new mart at Ipswich. The objections raised against this proposal were, however, generally sound. Ipswich was only seven miles from the open sea, and therefore much more vulnerable than Antwerp in spite of the flanking fort of Harwich. The Italian and High German (Hanse) merchants might well object to a longer journey by

---

[1] Winchcombe was the famous "Jack of Newbury", who flourished in the reign of Henry VIII, and of whose "factory" it was said:

> "Within one room being large and long
> There stood two hundred looms full strong."

Thomas Springe of Lavenham in Suffolk was almost equally famous. He died on the eve of the Reformation. See E. Lipson, *History of the English Woollen and Worsted Industries* (1921), p. 48.

sea, and the merchants in general to an island as opposed to a continental location. The cost of carriage of woollens to Ipswich might prove greater than to Flanders; shipping would decay; the queen would lose her custom of 14s. 6d. on each parcel of cloth going to Antwerp; English subjects might object to the great influx of foreigners: London might be prejudiced by the change. The author of the project was not convinced by these arguments, and pressed his proposal again in 1578, but in fact there was an insuperable geographical difficulty: goods could only be distributed to the European consumer from a European entrepôt conveniently situated for the main continental roads and inland waterways. Hamburg and Emden were the official substitutes for Antwerp, but in point of fact some merchants managed to maintain their position in the latter city, while trade also drifted naturally to such points as Amsterdam.[1]

Even in 1571 no economic memorandum could be complete which did not touch on the timber question, and Bodenham pointed out that in Biscay (northern Spain), where iron was smelted to supply "the most part of Christendom", and where there was yearly building of ships of burden from 200 to 600 tons, the forests were yet undiminished. This was because for every tree cut six had to be planted, and for this purpose oak seedlings were grown from acorns and planted out when mature enough. Furthermore, England was very deficient in ships of burden (i.e. of 200 tons and upwards), and to remedy this, Bodenham recommended that a subsidy and a loan should be granted[2] to the builders, as in Spain, on every ton above a certain figure. There, the smaller ships were forbidden to go on

---

[1] Johnson, like many of his fellow-countrymen, had been greatly impressed by the history of the rise of Antwerp as told in Guiccardini's *Description of the Low Countries*, first published (in Italian) in 1567. That English trade was the making of Antwerp, and English cloth-finishing of the Low Countries, was, however, deplored by Englishmen at a much earlier date, as for example by Wm. Cholmeley, a London grocer in his *Request and Suite of a True-hearted Englishman* written in 1553. (R. H. Tawney and E. Power, *op. cit.* III, 130–49.)

[2] In H.C.A. 25, no. 1 (Public Record Office), will be found documents concerning subsidies granted to ships built between 1571 and 1587 (the last six years incomplete). A grant of 5s. per ton was made for every ton above one hundred. The average burden was 180 tons only, although one or two ships of upwards of 200 tons were built each year, and from 1577 onwards the tendency was to build larger vessels. The builders had to enter into recognisances not to sell the ships to any foreigner. More than a quarter of the ships subsidised during the period named were built on the Thames, while the Orwell, Bristol and Newcastle together accounted for another quarter. Out of more than eighty ships only seven were of 300 tons and upwards, 600 tons being the maximum. These seven were built at Bristol (2), London (1), Dartmouth (1), Southampton (2), and Ipswich (1).

long voyages, and in any port or haven the greater ships were impowered to take away the lading of the less; this again was an example which England could follow.

While such discussions as these were going on, an increasing number of public men were being attracted by the prospect of making England independent of continental markets, either by establishing direct trade with the Far East, or by establishing English colonies in America, or by a combination of both projects. Economic nationalism is no new dream, and it was the aim of Richard Hakluyt the Lawyer (and later of his younger cousin and disciple) to free England from any dependence on the Continent for food-stuffs and raw materials, as well as for markets. The eastern margin of North America, between 60° N. and 30° N. (to which England had right by "first discovery" through the Cabot voyages) offered itself as ideal for effecting this purpose, for its northern section could produce the whole range of Baltic "naval stores", its southern section the whole range of staples of France, Portugal and Spain. It would be no longer necessary to wait on the pleasure (and the exactions) of the king of Denmark to pass in and out of the Baltic, or to send English merchants and seamen to imperil their souls in Catholic Spain!

But overseas expansion had a further function. Beggary, vagrancy and crime at home, piracy on the high seas abroad, had alike swollen to grave proportions in England in the 1570's, and the more thoughtful (such as the Hakluyts) ascribed this state of affairs not to human wickedness, but to unemployment and lack of opportunity for legitimate advancement. They observed that Spain and Portugal, which had colonies, had no unemployed, or rather no "surplus population" as it was termed, and the inference was obvious. If England followed their example many persons who were now a national burden would find work in English colonies, in trading settlements in the East, and on the great fleet of ships that would be brought into being. And, further, there would be an expansion of industry at home, to supply cloth, metal wares, and "toys" to the native peoples with whom commerce would be opened. While men like Anthony Parkhurst, who knew Newfoundland, considered that it was this region, with its illimitable supplies of fish, furs and timber, which should be the first objective, others (including the Hakluyts) argued for a warm temperate location in 40° N. or thereabouts (Norumbega). Here, vines and olives could be planted, the latter for the sake of the olive oil necessary to the woollen cloth industry in which the elder Hakluyt was so deeply interested.

He was particularly anxious to see the end of the export of raw wool and of partly manufactured cloth, which was sent to Flanders for dyeing, dressing and finishing; in several of his pamphlets, therefore, he directed attention to the advantage of introducing new dye-stuffs into England for cultivation at home.[1]

The whole question of industrial crops was, however, highly controversial and bristled with difficulties. As has already been pointed out, the common view taken was that there was only arable land sufficient to keep the nation in food and fodder crops, and that to put the land to any other use meant dear food, or even a danger of scarcity. Only far-seeing and travelled men like Dr John Dee were aware that the country was capable of supporting at least twice the existing population, were its resources fully exploited. The case of dye-stuffs is instructive. Madder was imported in quantity from Flanders, woad from France and the Azores: both could be grown in England. The cultivation of the latter was attempted, and with such success that the cry of "dear food" began to be raised, and a proclamation of 1585 forbade the sowing of this crop in the neighbourhood or towns. In the following year an official survey of the ground under woad was taken, when it was found that nearly 5000 acres had been planted. Of this total, about one third was in Hampshire, the counties next in importance being Dorset, Sussex, Berkshire and Wiltshire. The former method of utilisation of each acre was noted, and while, as was to be expected, the greatest proportion was arable, pasture or meadow land, there had also been conversion of lands classified as follows: marsh; woodland, rough and bushy ground; sheep and downland; park and cony ground; barren and ley ground.

The arguments put forward for fostering the industry were the obvious ones: the need for woad in England, the saving of treasure (money) otherwise sent abroad, the provision of employment, the high return to

[1] Hakluyt wished to see England industrialised, an importer of raw materials only, and an exporter of highly finished goods. He urged merchants to study the needs of the foreign customer, and to maintain their credit abroad by strict attention to the quality of their goods. (E. G. R. Taylor, *The Original Writings of the Two Richard Hakluyts*, 1935.) Elizabethan economists had to contend with those who believed that the country was benefited by exporting raw wool for reimport as cloth, since there were both inward and outward customs to be collected and shippers made their profits on the double movement. The writer of the lengthy *Discourse on the Relief of the Poor*, addressed to the queen towards the end of her reign, shared Hakluyt's view that the wages paid when a pound of wool was transformed into a pound of cloth should go into English, not Flemish, pockets; and that, were this the case, the unemployment problem would be on the way to a solution. (Lansdowne MSS. 95, no. 3.)

the cultivator. Most of the arguments for restriction were obvious also: the queen lost the customs on imported woad; there was a loss of trade and shipping affecting merchants and sailors; there was a loss of good arable ground to tillage and a prospect that the price of arable land would rise to an exorbitant level; there was a diversion of compost (manure) from arable land, while the price of compost had risen; victuals such as butter, milk, cheese, lamb, veal and grain were made scarce and dear. It was further stated that woad made a big demand for labour just at harvest time, and that this adversely affected the husbandman, while, if it was permitted to be grown near the clothing towns, the clothiers would be unable to get spinners of wool. More fantastic were the arguments that woad destroyed the fertility of the soil, and injured the bees that visited the flowers. As might be expected, the queen's advisers so feared the cries of "dear food" and "scarcity of labour" that they decided upon restriction. Exaggeration had won the day: "By turning up both pasture, arable and meadow grounds to woad a husbandman shall neither be able to maintain his plow cattle in good state to till withall, nor sheep to maintain his arable grounde, neither kine to make his provision for his household, which is his chiefest stay."[1]

The case of hemp and flax presented other problems. Acts laying down the acreage to be planted were passed in 24 Hy. 8 and 5 Elizabeth: yet they remained a dead letter, and it was necessary to supplement home supplies by imported linen articles, sailcloth and cordage. A demand that the dormant Acts should be enforced came in 1590 from the soap-makers, for next to rape-seed oil (also used largely for frying fish), oil derived from hempseed, and flax-seed (linseed) oil, constituted the best ingredients for soap. Tallow and fish oil were also used, but the result was an inferior and foul-smelling soap.[2] Here, it seemed, was an industry that would set

[1] There are various papers dealing with woad in Lansdowne MSS. 49.

[2] In 1598, however, the rape-seed oil-makers declared that a majority of the thirteen soap-makers of London were refusing to buy their product, so that they were forced to sell it to the clothiers. No doubt vested interests were at stake, and to meet the argument that to stop the import of train-oil and blubber from Biscay would diminish the customs receipts and injure shippers, the complainants suggested that this oil was good enough to be used in the manufacture of the (despised) coarse cloths of the north. There is a strong suggestion that Elizabethan trade policy was dominated by rich merchants and by the powerful men who farmed the customs, two groups inimical to home industry, and careless of the interests of the consumer. Incidentally, the controversy over vegetable oil throws light on the price question. Seville oil which during the first half of the century had been sold in England at £12 to £13 a ton was said in 1578 to fetch from £30 to £36 in a good olive season and from £40 to £46 in a poor one. The reason ascribed to this rise was the opening to Spain of the

large numbers of people, including women and children, "on work" during those months when they were not needed for hay and corn harvest. The weeding of the hemp and flax fields, the stripping and retting of the fibre, and in winter the picking of tow and the spinning of thread, would all serve to lessen unemployment. But unemployment was a problem for the unemployed, not for the husbandman; and farming practice changed at a rate so slow as to be almost imperceptible in half a century. Experiment and innovation came from the great landowners, who were responsible for large-scale sheep husbandry, for enclosed fields, and for the draining of fens and marshes. The yeomen and peasant farmers remained stubborn conservatives.

Sailors and travellers, however, were not averse from novelty, and a new habit of great import for the coming century was spreading rapidly—the use of tobacco. This is amusingly illustrated by a letter from the Deputy-Customer of the small port of Penryn in Cornwall, begging for the support of his superiors, since no one heeded him when he demanded payment of customs.[1] "At the same tyme came a French shippe and a Flemynge from the Indians and they have solde Tobaco to the valewe of two thowsand pounds in money, and I have not any peny customes." The offenders declared that there was no custom on tobacco, and on a subsequent day the Deputy when walking on the cliffs accidentally met a party bringing 120 lb. of it ashore, who drew their rapiers and held them to his breast when he protested. A Cornish sailor, Hannyball Nangillion, boasted that he had landed four hundredweight, and told the unhappy Customer that he could do what he liked about it! And this was in 1597, scarcely a dozen years after the return of the colonists from Ralegh's Virginia who, according to tradition, were the first English smokers.

Idealists and theorists in every age press for a planned economy: Governments are content to leave the development of the nation's resources to individual enterprise, and to the moulding pressure of external circumstances. While the Hakluyts were urging men and women to sail westwards to Ralegh's Virginia, the London and Bristol merchants were making handsome presents to the Grand Turk who ruled the Near

---

West Indian market, which affected other Spanish commodities in the same way. Actually, of course, the influx of bullion into Spain caused a serious fall in the purchasing power of gold and silver, and a rise in wages and cost of production. (See Lansdowne MSS. 26, no. 55.) For the price question, see E. J. Hamilton, *American Treasure and the Price Revolution in Spain* (1934).

[1] Lansdowne MSS. 84, no. 20.

East in order to establish the Levant Company on a firm foundation.[1] The opening of the seventeenth century saw Englishmen's feet set firmly on the twin paths of colonisation for permanent settlement in the West, and settlement for exploitation in the East. John Dee, a quarter of a century earlier, had coined the phrase "the British Empire", and had cast the queen of England for the rôle of Mistress of the Seas: when he died, in 1609, both Empire and Sea Power were appreciably nearer. In Europe the "industrious" Dutchman had superseded the "insolent" Spaniard as model and rival: England was now aware of five continents and a shadowy sixth, English ships now sailed on every sea and ocean. Dimly in Leland's day, with growing clarity in Camden's, Englishmen realised the significance of their country's new world position, and after a tedious period of trial and error they learned how to exploit it.

For Bibliographical Note see p. 353.

[1] Lord Burghley's notes in his own handwriting on this new venture are extant (Lansdowne MSS. 34, no. 65). At the chief entrepôt of Aleppo, besides drugs, spices and jewels, the goods to be obtained in return for English kerseys and broadcloths included indigo and white soap (alike useful in the cloth industry), raw silk, cotton wool and cotton yarn, the bases of new textiles, pot ashes for glass and soap manufacture, and cotton fabrics. From Alexandria, it was proposed to bring home currants, sweet oil, honey, raw silk, goat skins, buff hides and sponges. The French had already made a bid for Levantine trade, and Burghley notes that they had established coral fisheries off the Barbary coast, since coral was as good as money in the East, and were manufacturing at Marseilles imitation Venetian fabrics of the type the Levant market was accustomed to buy.

Chapter XI

ENGLAND IN THE SEVENTEENTH CENTURY

J. N. L. BAKER, B.Litt., M.A.[1]

Foreigners, who looked at this island narrowly but not unkindly, saw in the England of the seventeenth century a land of plenty and prosperity. Molin, for instance, reported that it was "comfortable, pleasant, and rich beyond all other islands in the world". To Lando it was "as fruitful in commerce as by the gifts of nature". And the Venetian Foscarini declared in 1613:

In England I have seen a country of great beauty and fertility and very populous. Four things in particular have struck me as being especially worthy of notice. First that in the 420 miles that I have traversed in my journey from London to the frontiers of this realm [of Scotland] I have not seen a single palm of unfertile land. Second that every 8 or 10 miles I have found a city or at least a town comparable to the good ones of Italy. Third a number of navigable rivers including the Thames, the Trent, and the Severn which in their long course to the sea widen to a mile or more. Fourthly, I might have said first, a quantity of most beautiful churches so numerous as to pass belief. The kingdom is most rich in the fertility of the soil and by its extensive commerce with all parts of the world.[2]

With this general enumeration may be compared the "7 things wherein England may be said to excell", which Peter Mundy noted.[3] First was "a peacable and quiett enjoying off Gods true Religion. Secondly a temperate ayre and healthy climate, taken one with another". In the third

---

[1] I wish to express my thanks to my colleague, Mr C. F. W. R. Gullick, for reading the manuscript and for much useful criticism; and to Mr J. R. Crompton of the Public Record Office for advice on the Hearth Tax Returns and for permission to use his figures for Surrey. I am especially indebted to the late Sir Charles Firth who placed some of his unpublished work at my disposal, read the manuscript, and gave me much valuable advice. Without his generous help and kindly criticism, extending over a long period of years, I would not have ventured to write this chapter.

[2] These descriptions are by Venetian Ambassadors in 1607, 1622, and 1613 respectively. They are printed in the *Calendars of State Papers, Venetian*.

[3] *Travels* (ed. Sir Richard Temple, Hakluyt Soc. 1925), IV, 47–51. This part of Mundy's diary was written in 1639.

category he placed "our aboundance and plenty off whatt most usefull for the liffe of Man, especially in these Northerne parts, as Corne, Woolle, Flesh, Fish, Tynne, Iron, Lead, Seacole, etts., with all which our owne land is not only sufficiently served, butt many Countries and Nationes Farre and Neare are supplied From us". Education and discipline came in the fourth category. Into the fifth went the advantage "for Trafficke and discoveries, viz. soe many encorporated companies off Merchantts For Forraigne trade who employ their study and Meanes for the Encreas thereof by adventuring their goodes and sending Fleetes and shippes into Most parts off the knowne world". And in the sixth and seventh categories were "excellencies off art" and "Naturall wonders" respectively. Among the latter were the medicinal waters of Bath and the salt deposits of the midlands.

But the physical features of England and Wales as a whole received no attention from contemporary writers, although many of them noted the salient facts of topography. By contrast with Scotland, "full of steep mountains and barbarous people",[1] or "all mountainous and barren", England appeared to be "flat and fertile". Another observer (Molin) remarked, more accurately, "the land is not really flat but is broken into little hills, so low, however, that seen from a distance they cannot be distinguished from the plain". Of the smaller topographical features there was more detailed, though not always accurate, appreciation. Celia Fiennes saw, from the Berkshire downs, the Vale of the White Horse extending "a vast way". She visited the Malvern Hills, called by some the English Alps, and by Camden "hills indeed or rather great and high mountains". Of these she wrote: "they are at least 2 or 3 miles up and are in a Pirramidy fashion on ye top.... On the one side of this high ridge of hills lies Worcester: Oxfordshire, Gloucestershire, etc. appears in plains, enclosures, woods and rivers and many great hills tho' to this they appeare low: on the other side is Herrifordshire which appears like a country of gardens".[2] Camden noted that the Chiltern Hills divided "the whole region acrosse from the South-west to the North-east", and that from the ridge was "a large prospect every way downe into the Vale beneath". He also recognised the unity of the Pennine mass.

The North part [of Staffordshire] riseth up and swelleth somewhat mountainous, with moores and hilles, but of no bignesse, which beginning here, runs

---

[1] The phrase is Contarini's.
[2] *Through England on a side saddle in the time of William and Mary, being the diary of Celia Fiennes*, ed. by E. W. Griffiths (1888), p. 33.

like as Apennine[1] doth in Italie, through the middest of England with a continued ridge, rising more and more with divers tops and cliffs one after another, even as far as to Scotland, although oftentimes they change their name. From heere they are called Mooreland, after a while the Peak, Blackstone edge, then Craven, anon as they goe further Stanmore, and at length being parted diversely, as it were into hornes, Cheviot.[2]

Outside the English plain, Devonshire and Cornwall were "mountainous" countries. And Cumberland, in another work, was dismissed as containing "several meers or small lakes". Windermere, "separating part of Lancashire from Westmorland, is about 10 miles long, and four broad, the biggest standing water in England".

It is necessary to go to later writers for a concise description of the physical features of England in the seventeenth century: and it may be said with certainty that, with few exceptions, these have changed but little with the passage of time. The greatest transformation has been that of the Fenland which is being dealt with separately.[3] Round the coasts, many local changes were taking place slowly. Chester was a port whose glory lay in the past; for the silting Dee estuary was rapidly being surpassed by the Mersey. "The Ocean being offended and angry (as it were) at certain Mills in the very chanell of the River Dee, hath by little withdrawne himselfe backe, and affordeth not into the City [of Chester] the commodity of an Haven, as heretofore."[4] Camden, too, recorded a similar silting up of Budley, Seaton and Sidmouth in Devonshire. Of Bridport, Gibson wrote "the tides [are] perpetually barring it with sand against which they could not find any remedy".[5] Near Braunton, many acres of land were "overflown by the sands". On the south coast, Dungeness was

---

[1] A later writer calls these hills the Alps. "On th'English Alps where Darbie's Peak doth lie" is the opening line of *De Mirabilibus Pecci: being the wonders of the Peak*, Latin verses by Thomas Hobbes, translated by a person of quality (1678). (The person of quality was Charles Cotton.)

[2] W. Camden, *Britannia*, ed. by P. Holland (1637), p. 586. Cited hereafter as Camden.

See pp. 371–2 above for a different rendering of this paragraph. The first (Latin) edition of this work appeared in 1586 and the sixth (Latin) edition in 1607. This last was translated and edited by P. Holland in 1610, but Holland did not see the translation in the press and disapproved of it when printed. He issued a revised edition in 1637. A new translation was made by E. Gibson in 1695. This edition included some additions to the original text, but these were distinguished from the rest. In many cases they referred to developments that had taken place since Camden wrote.

[3] See chap. XII.   [4] Camden, p. 606.

[5] W. Camden, *Britannia*, ed. by E. Gibson (1695), p. 51. (Cited hereafter as Gibson.)

growing at the rate of about seven feet a year, at least during the latter part of the century. The detail of the Yorkshire coast was also altering. The large sandbanks west of Paull in the Humber, which existed in 1660, have by now become joined to the mainland. Spurn Head was growing rapidly, at the rate of about 44 feet a year, and the light erected thereon in 1676 had become useless as soon as 1684. On the other hand, serious losses were being recorded to the northward. The village of Kilnsea, for example, then in existence, has now disappeared. But all these numerous alterations cannot be catalogued here.

The climate of England[1] was of greater interest to contemporary writers; the occupations of the country were still predominantly agricultural, and the fluctuations of the weather had direct economic effects. No records exist for the whole century, nor indeed for any large part of it; but towards its close several sets of observations were kept, and then it becomes less difficult to reconstruct the sequence of weather. Many observers remarked on the equable nature of the climate.

The ayre of England is temperate, but thicke, cloudy, and misty, and Caesar witnesseth, that the cold is not so piercing in England as in France. For the sunne draweth up the vapours of the sea which compasseth the Iland, and distills them upon the earth in frequent showers of rain, so that frosts are somewhat rare; and howsoever snow may often fall in Winter time, yet in the Southerne parts (especially) it seldom lies long on the ground. Also the coole blasts of the Sea winds mittigate the heat of Summer.[2]

"Although so far north and though the sun rarely shines in its full splendour for a whole day, yet the air is temperate."[3] "It is certain a more temperate climate could not be wished."[4]

While there is no comprehensive treatment of the variation of the weather from year to year, it is possible, by combining a large number of scattered references in manuscript diaries, in memoirs, in the *Philosophical Transactions of the Royal Society*, and in the Calendars of State Papers, to make a tentative graph (Fig. 65) of the general conditions of the century. This only shows very broadly whether a year was wet or dry: no indica-

---

[1] See J. N. L. Baker, "The Climate of England in the Seventeenth Century", *Q. J. R. Met. Soc.* LVIII, 421 (1932).

[2] Fynes Moryson, *Itinerary*, IV, p. 165. (Maclehose edition, 1907.)

[3] Lando, 1622, *C.S.P. Venetian*, p. 423.

[4] M. Misson's *Memoirs and Observations in his Travels over England* (1719), p. 318. (*Temp.* William III.)

tion of the degree of wetness or drought is possible, and the most that can be claimed for the graph is that it is not wholly wrong. For purposes of comparison another graph is imposed upon it showing the fluctuations of wheat prices as given by Thorold Rogers.[1] It must not be assumed that climatic variations were solely or always responsible for these changes in price and, in view of the fact that little is known of the degree of variation of the weather, it is imprudent to press any correspondences in the graphs too far. A complaint from Norwich was made in December 1630 "that the scarcity of wheat is due to brewers, who continue to make use of barley which is required to be used for bread corn" and "to starch makers who buy great quantities of wheat weekly". Similarly, in 1637, the bakers and maltsters were held responsible for dear corn in Cheshire. Yet, the weather

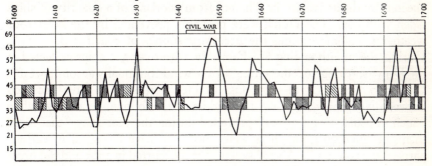

Fig. 65

This attempts a reconstruction of the weather of the seventeenth century. The shaded area above the line (39) indicates wet years; that below the line indicates dry years; the lighter shading denotes less certain evidence. The graph shows the variations of the price of wheat in shillings per quarter.

itself was frequently responsible for poor harvests and consequent dearth. "The weather has been so bad lately that it has been impossible to gather the crop", wrote the Venetian Ambassador in 1609, "the corn is suffering and rotting in the fields, just as the drought in the early season caused a poor hay crop which is of great importance in this country because of the numbers of animals fed on it."[2]

[1] *Agriculture and Prices*, v, 268–74.

[2] *C.S.P. Venetian*, 3rd Sept. 1609. Two years later it was reported that drought had caused scarcity. And the dry weather in the early part of 1620 reduced the hay crop "to a wretched condition, while many animals are perishing in the country from hunger". Sir William Brereton reported that, as a result of the drought of 1635, between Bath and Bristol "where they had twenty loads of hay last year they had not four load this year"—*Travels in Holland, The United Provinces, England, Scotland, and Ireland in 1634–5* (ed. E. Hawkins, Chetham Soc., 1844).

In a country which depended much upon external communications by water, storms naturally received mention. Particularly bad were the storms of December 1626, December 1627, and January and February 1642. In 1637 "furious storms" raged at sea for a week, and kept the Venetian Ambassador without news; he reported another four days' gale on 13th November, and was himself delayed for three weeks by storms in December. In January 1644, "an extraordinary snowfall which lasted for eight days without intermission" suspended all military operations. On 26th July 1652 a pious man wrote "God grant that the late terrific thunder and tremendous hailstorms and lightning that accompanied it, which are very unusual in this climate, do not prove the language and direct menace of Heaven". Later in the same year the city of London was "much cheered by the arrival in the Thames from Scotland of a quantity of ships laden with coal. The Dutch were probably prevented by the boisterous weather from intercepting the necessary supply of coal for this city". During the month ending 8th February 1617 over 300 ships collected in the Downs, owing to "the setting of the wind continually at south-west". Over 300 wrecks, with great loss of life, were said to have occurred between November 1612 and the middle of the following January, while "such unusual storms at sea" prevailed during the year 1627 "that more than one hundred good ships were known to have been lost in various parts of the country". A very severe storm occurred in 1703 and did great damage throughout the kingdom. The recently built lighthouse on Eddystone Rock was destroyed.[1]

With evidence of such a fragmentary character, it is perhaps rash to draw up a chart of the weather of the century but it is still more dangerous to base a scientific argument on the few facts known. Theories of the

---

[1] On this storm, and Defoe's account of it, see J. N. L. Baker, "The Geography of Daniel Defoe", *Scot. Geog. Mag.* XLVII, 258 (1931). Frequent references to storms will be found in the Calendars of State Papers, Domestic and Venetian, from which the details above have been taken.

For the later years of the century the general weather conditions are recorded in the diary of Elias Ashmole, the chief features of which have been discussed and need not be repeated here. J. N. L. Baker, *Q. J. R. Met. Soc.* LVIII, 421 (1932).

Temperature conditions are more difficult to determine because they are less frequently mentioned. Occasionally there are records of severe winters, among which were 1607–8, 1615–16, 1634–5, 1642–3 and the two succeeding winters, 1657–8, 1666–7, 1683–4, 1688–9, 1694–5: while among the summers reported to be "cold" were 1618, 1621, 1622, 1663, 1669 and 1692. Unseasonable warmth in winter was recorded in 1617, 1661–2, 1675–6 and 1692–3, while certain summers were abnormally warm, as in 1636, 1667, 1694 and 1699.

variation of the climate within the century must remain as yet unproved.[1] It was only when the Royal Society had become firmly established, and after the work of the older Oxford Philosophical Society had matured, that observations of the weather became at all scientific.

R. Bohun, the author of *A discourse concerning the origine and properties of wind*, which appeared in 1671, declared "I have here in England for some years past kept by me an exact table or Ephemeris both of the Vernall, and Summer Etesians; but found the winds no lesse variable in those months than at other seasons".[2] Bohun, too, recognised some of the fundamental features of the meteorological conditions of England. Thus he wrote "with us in England the easterly are at certain seasons of the yeare exceeding cold and very often the most freezing winds especially if they hang somewhat towards the north. I need assign no other cause for the frigidity of the easterly winds than that they have their first rise from the continent where the midland air is much colder than the maritime".[3] In contrast, the west winds bring great storms and often bring rain "and especially the south-west which are the most humid and pluvious because they travel by sea many thousand miles". This important recognition of the maritime and continental influences in the weather of England carries with it an appreciation of the significance of the length of the wind current, often, relatively, a more important factor than wind direction.

Bohun, however, is a small figure on the stage at this time, and is dwarfed by Edmund Halley whose important work on winds appeared in 1687. Halley communicated his findings to the Philosophical Society in

---

[1] Francis Bacon, in some ways, anticipated meteorologists in his appreciation of cycles of weather. These had been dimly recognised since classical times, but Bacon specifically referred to the thirty-five-year cycle, both in his essay "Of vicissitude of things" and in the *Historia Ventorum* (1622). "It has been observed", he wrote, "by the dilligence of some that the greater and more remarkable seasons of weather, as great heats, great snows, great frosts, warm winters, and cold summers, generally come round in a circuit of thirty-five years" (*The Works of Francis Bacon*, collected and ed. by J. Spedding, R. L. Ellis and D. D. Heath, v, 174). In the essay, he adds: "It is a thing I do rather mention, because, computing backwards, I have found some concurrence." See E. Huntington and S. S. Visher, *Climatic Changes* (1923), p. x; Sir Napier Shaw, *Manual of Meteorology* (1926), 1, 118; and C. E. P. Brooks and J. Glasspoole, *British Floods and Droughts* (1928), p. 182.

[2] These observations are, unfortunately, lost. His book was based on earlier and better known works, such as the *Geographia Generalis* of Varenius (1650), which was subsequently edited by Newton, and on the writings of that great Cambridge scholar himself. In turn, it came to supply the basis for the opening chapter of Daniel Defoe's book *The Storm* (1704) which dealt with the storm of 1703.

[3] Bohun, *op. cit.* p. 142.

Oxford and his results, which are well known, were of far greater practical importance than any obtained by either Bacon or Bohun, since they offered an explanation of the winds of the oceans. His conclusions were incorporated in the English edition of the standard *Geography* of Varenius.

AGRICULTURE

With these minor changes in the weather, from year to year, the Englishman was profoundly concerned, for they affected his daily life in a way which now can be appreciated only with difficulty. The England of the seventeenth century depended for its food upon the produce of its fields. It is not possible to say how much of the country was cultivated; here, as everywhere, the historian is baffled by a complete lack of accurate statistics. At the end of the century, Halley calculated from a map that the area of England and Wales was 39,938,500 acres, which is about 1,600,000 acres in excess of the truth. On this basis G. King (1698) estimated the land utilisation of the country; his figures, which were accepted by C. Davenant,[1] are here given (in millions of acres):

Arable	9
Pasture and Meadow	12
Woods and Coppices	3
Forests, Parks and Commons	3
Heaths, Moors, Mountains and Barren Land	10
Houses and Homesteads, Orchards, Churches and Churchyards	1
Rivers, Lakes, Meers, and Ponds	0·5
Roads, Ways and Waste Land	0·5

This estimate may be approximately true for the end of the century, but it is certainly incorrect for the beginning. Among the great improvements in the forty years following the middle of the century which Petty noted[2] were the draining of fens, the watering of dry grounds, the improving of forests and commons, the "making of heathy and barren grounds to bear saint-foyne and clover grass", and the improvement of fruit cultivation. Many of these improvements had been initiated earlier than 1650. Fen draining was perhaps the most spectacular of these changes. Sir William Dugdale's *History of Imbanking and Drayning*, published in 1662, gave an

---

[1] C. Davenant, *Works* (ed. Sir Charles Whitworth, 5 vols. 1771), II, 216.

[2] In his *Political Arithmetic*, printed in *The Economic Writings of Sir William Petty* (ed. C. H. Hull, 2 vols. 1899), I, 303.

account of much activity in different parts of the country; the largest of these schemes concerned the great Level of Norfolk, Lincoln and Cambridge.[1]

The woodlands of England were gradually diminishing during the century. In Warwickshire, for example, the ironworks of that and neighbouring counties consumed much timber "whereupon the inhabitants, partly by their own industry, and partly by the assistance of marle, and other useful contrivances, have turn'd so much of wood and heath land into Tillage and Pasture, that they produce corn, cattle, cheese, and butter enough, not only for their own use but also to furnish other Counties; whereas, within the memory of man, they were supplied with corn etc. from the Feldon".[2] The wood was used for shipbuilding, domestic purposes, fuel, implements and machinery, and for all kinds of industrial purposes. In some cases, forests suffered through the action of the Crown, which was able to raise money by disafforesting or by selling certain areas.

The first two Stuarts [wrote R. G. Albion] greatly hastened the danger of an oak shortage by extending the exploitation of forests commenced by the Tudors. That practice of deriving revenue at the expense of the future oak supply must stand as the real forest policy of England from 1535 to 1660. The Civil Wars and the Interregnum were to give the *coup de grâce* to England's forest plenty. The forces let loose during those twenty years destroyed whatever surplus was left after the previous royal exploitation and failure to check waste.[3]

The attempts of Charles I to stop encroachments of the forests went too far; and the Long Parliament, in 1641, passed an act declaring that the limits of the forest should be fixed as they were in the twentieth year of James I's reign, i.e. 1623. The policy of Charles I was directed primarily to secure money not at the expense of the future oak supplies, but at the expense of those who lived within the ancient limits of the forest. But, although the Forest Courts were particularly active between 1634 and 1637, only £23,000 were obtained from all the forests of England.[4] The Grand Remonstrance, however, which complained of unreasonable encroachments and of the enforcement of obsolete laws, protested at "the general destruction of the King's timber, especially that in the forest of

[1] See below, chap. XII.
[2] Gibson, p. 509. For the location of "Feldon" see p. 370 above.
[3] R. G. Albion, *Forests and Sea Power* (1926), p. 127.
[4] Thus Charles "allowed himself, for the sake of a few thousand pounds, to be regarded as a greedy and litigious landlord rather than as a just ruler or as a rational King". See S. R. Gardiner, *The Personal Government of Charles I* (1877), II, 183.

Deane, which was the best storehouse of this kingdom for the maintenance of shipping". In spite of this, during the next twenty years destruction went on, and its record fills many pages of John Evelyn's famous
Sylva; or a Discourse of Forest Trees, published in 1664. There was much
destruction of timber during the Civil War. The case of the Duke of
Newcastle, if perhaps an extreme example, illustrates this tendency. His
total losses were estimated at £45,500, made up as follows:

> Clipston Park and woods cut down to the value of £20,000.
> Kirkby Woods, for which my Lord was formally proffered £10,000.
> Woods cut down in Derbyshire, £8000.
> Red Lodge Wood, Rome Wood, and others near Welbeck, £4000.
> Woods cut down in Staffordshire, £1000.
> Woods cut down in Yorkshire, £1000.
> Woods cut down in Northumberland, £1500.[1]

Towards the end of the century, economists seemed to be resigned to the
situation; Sir William Petty thought that the matter was not serious since
timber could be imported,[2] while A. Yarranton, in refuting an argument put
forward by Evelyn that the iron industry should be discouraged, claimed
that "if the iron works were not in being these coppices would have been
stocked up and turned into pasture and tillage as is now daily done in
Sussex and Surrey where the iron works, or most of them, are laid down".[3]
It was this diminishing supply of timber which gave to English policy in
the Baltic much of its significance during the second half of the seventeenth
century.[4]

It is not possible to give an accurate account of the distribution of the
forests of England and Wales. In 1608 a survey of royal forests, parks and
chases was made.[5] But, as this survey applied only to Crown lands, it was
not complete for the whole country. It left out the forests of Dean and
Wychwood, nor was there included "timber on any estate where the
quantity did not exceed what was deemed necessary on any such estate".[6]
A much later survey, in or about 1783, showed the great decline in certain
areas.[7]

---

[1] *Life of William Cavendish, Duke of Newcastle* (ed. C. H. Firth, 1907), p. 79. Sir
Charles Firth quotes the similar case of the Marquess of Worcester, whose losses are
given in E. B. G. Warburton's *Memoirs of Prince Rupert* (1849), III, 515.

[2] Petty, *op. cit.* I, 294.

[3] A. Yarranton, *England's Improvement by Sea and Land* (1677), p. 60.

[4] R. G. Albion, *op. cit.* pp. 164–76.

[5] Printed in *House of Commons Journals*, XLVII, 284–6 (1792). Reprinted here on
pp. 398–9.   [6] *Ibid.* p. 266.   [7] *Ibid.* p. 351.

| | 1608 | | 1783 and since | |
	Timber trees	Decayed trees	Timber trees of 30 ft. and upwards	Scrubbed, dotard, decayed and defective trees of all sizes
New Forest	123,927	118,072 loads	32,611	2,067
Alice Holt	13,031	23,934 loads	9,136	21,353
East Beere	5,363	8,814 loads	256	530
Whittlewood	51,046	360 trees	5,211	10,768
Salcey	15,274	440 trees	2,918	9,565
Sherwood	23,370	34,900 trees	1,368 (in 1789)	8,749

Among the most important changes affecting the agriculture of the country were those usually designated under the term "Enclosures". The enclosure movement had been going on in the previous century, and continued to transform the country-side throughout the seventeenth century. Enclosure took one of several forms. It might result in the abolition of the intermixed strips in the common fields,[1] and in a re-allotment of the land to private owners who hedged and ditched their new and compact holdings. In many cases, this process led to a conversion of arable to pasture land; but there was also some addition to the arable acreage. Generally speaking, it implied improved agriculture and a greater yield of crops. Rather different in character was the enclosure of waste lands and commons, but the result was similar—improved land utilisation, and an increase both in arable land and in efficient pasture. Entirely different from this again were those enclosures made to provide suitable residential estates, and gentlemen's seats. Agriculturally, of course, this third implication of the term was not as important as the other two.

Though there is evidence that enclosures were widespread during the century, certain parts of England were already enclosed when the century opened.[2] Mr E. C. K. Gonner's enclosure map (Fig. 66) shows that large areas, especially in the south-east and in the west and north,[3] had less than half their area in common fields: here the proportion of land enclosed in the seventeenth century was perhaps small, but by 1700, as the map (Fig. 67)[4] makes clear, the solid block of enclosed land in these

[1] See Miss E. M. Leonard, "The Inclosure of Common Fields in the Seventeenth Century", *Trans. Roy. Hist. Soc.* (new series), XIX, 101 (1905).

[2] Some of the land may have never been unenclosed. See Fig. 23 above.

[3] E. C. K. Gonner, *Common Land and Inclosure* (1912), Map D. Of course it should be remembered that much of the country was still waste, e.g. in Northumberland.

[4] *Ibid.* Map C.

Survey of Royal Forests, Parks and Chases, 1608.

County	Forest, etc.	Timber trees (no.)	Decaying trees (no.)	Coppices in leaf (area)	Coppices out of leaf (area)	Coppices (area)	Miscellaneous
Bedfordshire	Ampthill Parks	25,112	1,018	—	123 ac. 2 r.	—	—
Berkshire	Windsor Forest	3,147	1,541	—	—	—	—
	East Harrell						
	Sunning Great } Parks	7,596	7,270	—	—	15 ac.	—
	Little Moat						
Buckinghamshire	Barnewood Whittlewood }	21,613	408	1562 ac.	238 ac.	—	—
	Olney Park	1,012	2,103	—	—	—	640 ashes
Carmarthenshire	Penrin Pennyheath Cardiffe }	6,542	17,296	—	—	—	—
	Riffigg Park	727	3,054	—	197 ac.	—	—
Denbigh	Parks	50	1,595	—	—	—	9500 young saplings
Derbyshire	Wardes	1,774	4,426	—	—	—	—
	Mansfield Ravensdale } Parks	—	4,217	—	—	—	—
Dorset	Gillingham Forest	4,651	19,816	—	—	—	1871 saplings
	Gillingham Park	408	2,427	—	—	—	14 saplings
Essex	Havering Park	78	860	—	—	—	8176 other decaying trees
Hertford	—	1,281	7,083				
Huntingdon	Somersam Park	3,485	20	—	230 ac.	—	—
	Somersam Close	14,425	—	—	541 ac.	—	—
Kent	Eltham Parks	4,927	15,918				
Leicester	Leicester Forest }	3,055	1,170				
	Tooly Park	2,102	1,086				
	Forests	163	9,570				
	Parks	2,753	2,102	—	—	—	270 small beeches
Montgomery	Park	640	3,290				
Nottingham	Sherwood Forests	23,370	34,900				
Northampton	Parks	93,942*	712	—	6342 ac.	—	—
		14,198*	119	—	348 ac.	—	50 ashes
Oxfordshire		58,936	3,722	1663 ac.	1981 ac.	—	2678 ashes
Pembrokeshire		2,666	22,884	—	786 ac.	—	21,032 saplings
Rutland	Lyefield	6,955	2,326	—	—	—	1152 coppices fellable
	Park	3,361	1,017	—	—	697 ac.	—

County	Forest, etc.	Timber trees (no.)	Decaying trees (no.)	Coppices in leaf (area)	Coppices out of leaf (area)	Coppices (area)	Miscellaneous
Shropshire	Forests	6	142	—	—	—	—
	Parks	154	699	—	—	—	1524 saplings
Somerset	—	1,362	3,902	—	—	—	Aldermores 96 loads
Southampton	Forests	151,753†	154,252§	—	—	1304 ac.	Lopps of trees 25,000
Stafford	Walds and Lands	36,372	41,830	—	—	—	—
	Parks	5,981	12,728	—	—	38 ac.	—
Surrey		6,157	5,083	603 ac.	—	—	—
Worcestershire	Wyer and Feckenam	12,520	25,900	—	447 ac.	—	—
	Bewdley Park	1,887	2,265	—	—	—	—
	Pensham ⎫						
Wiltshire	Blackmore ⎬ Forest	41,792	36,404	320 ac.	89 ac.	—	2234 great trees; no timber
	Braden ⎭						
	Alborn Chase	990	3,128	—	—	—	—
York	Galtres	2,797	14,360	—	—	—	Ashes 257 loads
	Warlas Park	—	2,494	—	—	204 ac. wasted	—
	Coverdale Chase	—	—	—	—	—	—
	Knaresboro' ⎫						
	Pickering ⎬ Forest	1,384	7,534	—	—	—	—
	De la Hay ⎪						
	Pontefract ⎭						
	Ackworth ⎫						
	Cirdling ⎬ Parks	1,776	2,781	—	—	—	—
	Altofts ⎪						
	Bilton ⎭						
	Hatfield Chase	—	—	—	—	—	Utterly wasted

			Timber trees	Decaying trees
TOTAL	Forests, Parks and Chases	Exchequer	566,179	455,259
		Duchy	14,014	36,032
	Manors and Tenements	Exchequer	175,352	146,195
		Duchy	29,203	44,572
			§ acres.	

† loads.

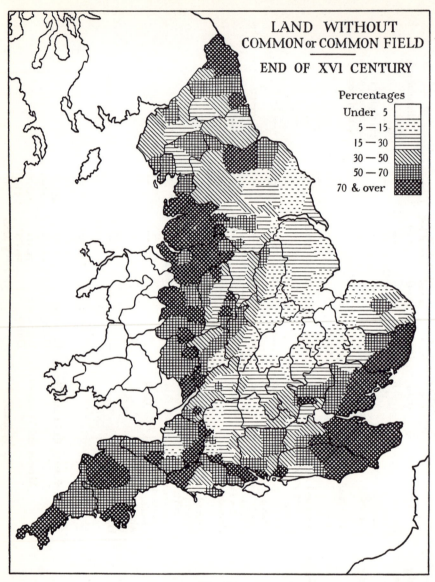

LAND WITHOUT
COMMON or COMMON FIELD
——
END OF XVl CENTURY

Percentages
Under 5
5 — 15
15 — 30
30 — 50
50 — 70
70 & over

Fig. 66

Redrawn from Map D at the end of E. C. K. Gonner's *Common Land and Inclosure*
(Macmillan & Co. Ltd., 1912). The percentages are those of the total area of each county
—or district within a county—without common or common field towards the end of
the sixteenth century. Wales and Scotland are not included.

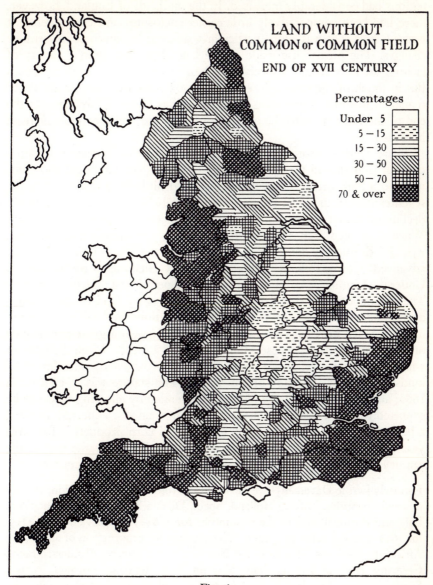

Fig. 67

Redrawn from Map C at the end of E. C. K. Gonner's *Common Land and Inclosure*
(Macmillan & Co. Ltd., 1912). The percentages are those of the total area of each
county—or district within a county—without common or common fields towards the
end of the seventeenth century. Wales and Scotland are not included.

regions had considerably increased. In the midlands of England, however, there had been comparatively little earlier enclosure; and consequently it was this area that was more particularly affected by seventeenth-century changes. The precise use made of the land depended upon a variety of factors. In East Anglia, especially in Norfolk, the arable land increased. But in much of the midlands, enclosure took place on heavy lands which remained in grass, supporting cattle as well as sheep. Only towards the end of the century, with the coming of new crops, did the arable increase here also.

There were obvious economic reasons why this extension of agricultural land should take place. In the sixteenth century wool had commanded a high price. Wool prices remained high for a few years in the seventeenth century, but thereafter were probably stationary or actually declining. Grain prices, on the contrary, increased during the century. Wheat averaged 35s. 3½d. a quarter during the decade 1603–12, and 43s. 2¾d. during the decade 1693–1702, while other commodities similarly advanced. Corn-growing, too, was indirectly encouraged by legislation. Under the Commonwealth an Act was passed permitting the export of corn. The preamble stated that "it hath pleased Almighty God to bless the industry and endeavours of the people of these nations, in the great improvement of Fens, Forests, Chases and other lands, with a great redundancy of corn, cattle, butter, cheese and other commodities". When the price of wheat did not exceed 40s. a quarter export was allowed; rates for other kinds of grains were made proportionate. The statute was re-enacted in 1660; and in 1663 the liberty given was extended by allowing wheat to be exported when the price at home was not more than 48s. a quarter. In 1672, a bounty of 5s. per quarter further stimulated corn production for three years, and this provision was revived in 1689. Thus, by the end of the century, England had become an important corn-exporting country.

Formerly [wrote Davenant] [1] we carried grain from the port of London, and but in small quantities, only to Holland, Spain, Denmark, Africa, the plantations, Italy and Portugal, and to these countries for 1662–3 to the value of £4315; for 1668–9 to the value of £2011. Whereas now we export grain of all sorts to Africa, the Canaries, Denmark and Norway, East Country, Flanders, France, Germany, Holland, Ireland, Italy, Madeira, Newfoundland, Portugal, Russia, Scotland, Spain, Sweden, Venice, Guernsey, and the English plantations, by a medium of years, from Christmas 1699 to Christmas 1710 to the value of £274,141...Holland in particular £151,934 in that period.

[1] *Works*, v, p. 424.

Other causes at home also stimulated arable farming. The landed gentry found land profitable and, forming a majority in both Houses of Parliament, they endeavoured to promote tillage and agriculture by their legislation. Towns were growing as the result of the development of industries, and their population required food. The population of the country as a whole was also growing. London, that "great mouth",[1] was fed upon the surplus of the surrounding counties, especially Cambridgeshire, Hampshire, Buckinghamshire, Berkshire, Middlesex and Hertfordshire. Other parts of the country, too, such as Somerset and east Yorkshire, frequently had more corn than they required.

At the end of the century, Davenant[2] estimated production as follows (in millions of bushels):

Wheat	14	Peas	7
Rye	10	Beans	4
Barley	27	Vetches	1
Oats	16		

It is not possible to give more than a very approximate account of the distribution of these crops. Writing of Cumberland and Westmorland Celia Fiennes said "In these northern countyes they have only the summer graine, as barley, oates, peas, beans, and lentils, noe wheat or rhye for they are so cold and late in their yeare they cannot venture at that sort of tillage." She further noted that "they have much rhye" in Lancashire, Yorkshire, Staffordshire, Shropshire, Herefordshire and Worcestershire, while she had met with it also in Suffolk and Norfolk. Her own dislike of rye bread enabled her to detect rye cultivation even when the inhabitants denied its existence! Rye was frequently grown with wheat, when the crop was known as "maslin", and there is reason to believe that rye growing was more extensive at the beginning of the century than at its close.[3] But the distribution of the crops did not closely follow any geographical feature: wheat gradually became the more profitable crop on the good soils, and rye was relegated to the poorer areas. Certain areas, however, had a reputation for fertility quite early in the century. Camden described Kent as "full of corne fields";[4] southern Cambridgeshire was a country of corn and barley; in Rutland, lay the "fruitful vale of Catmose" in which Oakham was situated; Feldon, the eastern division of Warwickshire, was the richest part of the county; the Vale of Evesham, sometimes called the

[1] Yarranton, op. cit. p. 49.   [2] Works, II, p. 216.
[3] See Sir William Ashley, The Bread of our Forefathers (1928), passim.
[4] Camden, p. 324.

Vale of the Red Horse, yielded corn abundantly and was famed as the granary of the neighbouring counties. A writer who described the agriculture of Herefordshire in detail claimed "the rye of Clehanger and of some parts of Irchinfield is as good as the muncorne or miscellane of many other countreys, and our wheat is upon the ground far richer than I saw in the fair Vale of Esome".[1]

Probably the nearest approach to a crop-distribution map that is possible is Sir William Ashley's map of markets returning rye and wheat prices during the period 1692–1703 (Fig. 68). In the south and south-west wheat appears more important than rye: in the midlands and eastern counties, which were the most important agricultural areas in the country, both crops were cultivated. Barley was widely cultivated and was of great importance to the brewing industry. Oats also had a wide distribution, but was naturally more adapted to the colder and wetter parts of the country. Both grains were frequently grown together as "dredge". Again, certain areas attracted the notice of contemporary writers. In Shropshire the Clee Hills grew very good barley; in north Lancashire the land was said to be good for oats "but not so apt to beare barley", while in the hills of Derbyshire oats again constituted the more important crop. Celia Fiennes's description of the crops of the northern counties has already emphasised the predominance of oats and barley in those areas.

A few special crops require a brief mention. In 1608 many thousands of mulberry trees were sent to Devonshire "for the relief of the silke wormes in this countie", to be divided among such landowners as chose to pay three-farthings apiece for them. This was part of the attempt of James I to introduce the manufacture of silk into the country, but it was unsuccessful. Later in the century Hartlib[2] reported that silkworms were bred near Charing Cross and at Dukenfield in Cheshire. Tobacco was successfully cultivated in many parts of the country (especially in Gloucestershire), despite the gibe of Bacon who said, "the English tobacco had small credit, as being too dull and earthy". It was suppressed in order to leave plantation tobacco without a rival. Tobacco planting was prohibited in 1652 and there were further prohibitions in 1654 and 1657. Writing of Winchcombe, Gloucestershire, Gibson remarked: "the inhabitants made planting of tobacco their chief business, which turned to good account, till restrained by the 12 Car. II they decayed little by little and are now generally poor".[3]

[1] "I. B." *Herefordshire Orchards, a pattern for all England* (1657), p. 37.
[2] S. Hartlib, *his legacy of husbandry* (3rd ed. 1665), p. 55.
[3] Gibson, *op. cit.* p. 249.

MARKETS RETURNING
RYE & WHEAT PRICES
to Houghton's *Collections*
1692 - 1703

- Chester Wheat and Rye Prices
- Bristol Wheat but no Rye Prices
- Lewes Rye quotations very irregular:
  supplies probably small

Fig. 68

Redrawn from Sir William Ashley's *The Bread of our Forefathers* (the Oxford University Press, 1928), p. 13. In the absence of agricultural statistics, the map shows, as well as any map can, the significant facts about the crop distributions.

27

Hops, said to have come into England in the year 1525,[1] were success-fully established particularly in the south-east. From Sittingbourne to Canterbury, Celia Fiennes passed by "great hop yards on both sides of the road", and Maidstone was the centre of another producing district. The crop was probably much more widely distributed than at present, and it only became restricted to a few localities by the process of trial and error. Many hops were grown in Essex at the end of the century. And Hereford-shire was famous by the middle of the century, for an "abundance of the fairest and largest sort of hops. All about Bromyard in a base soil there is great store".[2] "At first", wrote "I. B.", "we adventured only upon deep, low, rich, and moorish grounds; now we climbe up the hills with wonderful success. We find also that the bottoms are apt to gather heat as an oven and that begets hony-dews, when the more open air escapes." In contrast, no hops grew in Derbyshire; the requirements of its large brewing industry were met by purchase from Stourbridge Fair and from Shrewsbury.

Saffron, a crop of declining importance, was practically limited to the east and south-east. It was widely grown in Essex, where the town of Saffron Walden lay "among the fields looking merrily with most lovely saffron",[3] and in south Cambridgeshire and Norfolk. Camden noted of Essex that the merit of the crop was that "the ground which three yeares together hath borne saffron, will beare abundance of Barley eighteen yeares together without any dunging or manuring, and then againe beare saffron".[4] In Cornwall, Stratton was the centre of saffron cultivation.

Flax was another crop which was grown locally. Yarranton advocated a wide extension of planting in order to revive the linen industry. The soils of Oxford, Warwick, Leicester and Northamptonshire were "very good, being rich and dry" and, he believed, well suited to flax.[5] Celia Fiennes saw the crop growing on the borders of Berkshire and Wiltshire. Camden described it in the "Island of Axelholme" in Lincolnshire; while Hartlib, in the middle of the century, reported that "about London far greater quantities of flax is sown than formerly". Thread was made from it at Maidstone, "the only place in England" that he knew.[6] The flax grown in west Lancashire supplied some of the raw material for the linen industry of that county.

Market gardening and fruit growing increased in importance as the century progressed. The former, according to Hartlib,[7] "began to creep

---

[1] But see p. 329 above.
[2] I. B. *op. cit.* pp. 47–8.
[3] Camden, p. 452.
[4] *Ibid.*, p. 453.
[5] Yarranton, *op. cit.* p. 47.
[6] Hartlib, *op. cit.* pp. 31–2.
[7] *Ibid.* p. 9.

into England" from Holland about the beginning of the century, was established near Sandwich, in parts of Surrey, and on the borders of London, and soon produced a variety of vegetables including "Cabbages, Colleflowrs, Turneps, Carrets, Parsnips, Raith, Pease". The progress of this branch of agriculture kept pace with the growth of towns; though, in the middle of the century, Hartlib declared that "the name of Gardening is scarcely known" in many parts even of Kent, as well as "in the north and west" of England.

Fruit growing was particularly important in Worcestershire, Shropshire, Gloucestershire, Somersetshire, Kent and Essex: outside these areas, orchards were "very rare and thin".[1] In Worcestershire, pears and cherries claimed first place; while in Hereford apples were "more proper".[2] In Kent, and round London generally, there were many orchards, though the Kentish codling had the reputation of being "a flat, insipid apple". Here, too, the cherry was widely cultivated, for Camden records[3] that thirty parishes round Tenham grew this crop. In Somerset, apples and pears were important, although Celia Fiennes observed that the cider produced from them was inferior to that of Hereford because "they press all sorts of apples together". Herefordshire claimed to be the orchard of England, and as such it appeared to Celia Fiennes who saw, from the Malvern Hills, "the whole county...very full of fruit trees"[4] of which apples and plums were the most important. A local and enthusiastic gardener described the situation thus:

About Bromyard a cold air and a shallow barren soyl yet store of orchards of divers kinds of spicey and savoury apples. About Rosse and Webley, and towards the Hay, a shallow, hot, sandy or stony rye-land, and exposed to a changeable air from the disgusts of the black mountain; yet here, and all over the Irchinfield, and also about Lemster, both towards Keinton and towards Fayremile (which makes a third difference of shallow and starvy land) in all these barren provinces as good store of undeceiving orchards as in the richest vale of the county, even by Frome banks.[5]

Despite the widespread character of the crop, there was some relation between soil and produce, for where the former was hot, shallow and dry, "we must be content with a full and certain blessing every second year". It was a Herefordshire apple which produced the well-known "Redstreak" cider.

---

[1] Hartlib, *op. cit.* p. 15.   [2] I. B. *op. cit.* p. 10.
[3] Camden, p. 334.   [4] Fiennes, *op. cit.* p. 33.
[5] I. B. *op. cit.* p. 9.

The vine had little importance in England, but a number of writers found themselves constrained to observe that its disappearance was due to a lowering of the standard of fruit cultivation rather than to any change in climate. In Hereford, however, "our Gentry have lately (c. 1657) contended in a profitable ambition to excell each other: so that the white muscadell is vulgar, the purple and black grape frequent, the parsly grape and Frontiniack in many hands".[1] The same writer pointed to the weakness of the situation. Hereford hoped to gain the name of a vineyard; and "the late dry summers did swell us with hopes", but "the later fickle spring and moist autumn did blast or drown our expectation".[2] Elsewhere the vine was grown sporadically. Camden reported it in Hampshire, near Basingstoke, but observed that it was no longer grown in Gloucestershire. Commenting on this fact, Moryson said that "no doubt many parts would yeald at this day, but that the inhabitants forbeare to plant vines, as well because they are served plentifully, and at a good rate, with French wines, as for that the hilles most fit to beare Grapes, yeelde more commoditie by feeding of sheepe and cattell".[3]

Animals played an important part in English agricultural life. Davenant[4] estimated their number as follows:

Cattle	4,500,000	Deer and Fawns	100,000
Sheep and Lambs	12,000,000	Goats and Kids	50,000
Swine and Pigs	2,000,000	Rabbits	1,000,000

The climate of the country was such that "all beasts bring forth their young in the open fields, even in the time of winter".[5] Innumerable foreign observers remarked on the abundance and the excellence of pasture lands at the end of the century; Davenant declared that "no country but England and Ireland has a sward or turf that will rear sheep producing the wool of which most of our draperies are made";[6] while Houghton argued, on economic grounds, that "it seems more the national interest of England to employ its land to the breeding and feeding of cattle than to the produce of corn".[7]

Cattle were bred all over England, but the pasture lands of some areas were of special importance. Devonshire and Somerset in the west, and Kent in the east, were counties very rich in pasture. Cattle from "the

---

[1] I. B. *op. cit.* p. 33.   [2] *Ibid.* p. 49.
[3] *Itinerary*, IV, 165.   [4] Davenant, *Works*, II, 216.
[5] Moryson, *Itinerary*, IV, 165.   [6] Davenant, *Works*, II, 235.
[7] J. Houghton, *A Collection of Letters for the improvement of husbandry and trade* (1861–3), revised by R. Bradley, II, 228.

farthest parts of Wales and England" were sent to be fattened on the "ranke greene grasse" of Romney Marsh.[1] In Buckinghamshire, Camden noted that the "inhabitants chiefly raise cattle"; and "grazing and feeding cattle" was an important occupation in the Vale of Aylesbury. In Suffolk, the richness of the pasture was famous, while the same may be said of the Isle of Ely. The cattle fair at Market Harborough drew its supplies from the rich pastures of the Vale of "Beaver", as Camden called it, "lying stretched out in three shires, of Leicester, Nottingham, and Lincolne".[2] In Camden's day Rugby was a town of many butchers.

Cattle rearing gave rise to two important industries. Cheese was mainly produced in three areas—Cheshire, Somerset and Suffolk. Of the first, Celia Fiennes observed that "this shire is remarkable for a great deal of great cheeses and dairys", while a century earlier Camden, in praising this commodity declared that "all England again affordeth not the like". In Somerset, there were many pastures, some on the flooded lands of the coastal belt, others on the Mendip Hills: from these came the renowned Cheddar cheese. In Suffolk, the cheese produced was, in Camden's time, "vented into all parts of England: nay into Germany, France and Spaine also". The other industry dependent on cattle was that of leather. Its distribution was widespread and calls for no particular mention. It was with pride that E. Chamberlayne declared that the English went shod with leather, while some of their less fortunate neighbours were obliged to wear wooden shoes.

The rearing of sheep was more widespread than that of cattle. Four areas, Leicestershire, the Cotswolds, the Isle of Wight and Hereford, were particularly famous for their sheep walks. In Hereford, sheep did best on "the hot rye-land where the pastures have a coarse sea-green blade, or short and poor, and where the fields refuse wheat, pease and fitches".[3] In the Cotswolds, the "weally and hilly situation of their pasture" was thought to produce "flocks of sheep, long necked and square of bulke and bone".[4] On the rich pasture of Romney Marsh, three sheep could be fed to every acre.

INDUSTRIES

The sheep of England gave rise to what was rightly regarded as the most important industry in the country, partly because of its wide-spread character and partly because of its predominance in the foreign trade of the country. Many writers argued that enclosures were not disadvantageous

[1] Camden, p. 350.   [2] *Ibid.* p. 536.
[3] I. B. *op. cit.* p. 51.   [4] Camden, p. 364.

because, even if the rural population were displaced, work could always be found in the towns; and also the supply of raw wool to the woollen industry would be increased. This argument in itself is symptomatic of the industrial changes of the time. The essentials of the industry, apart from labour, which was usually available, were raw wool, water power, fuel, and fuller's earth, "a precious treasure whereof England hath better than all Christendom besides....Nature intended England for the staple of Drapery".[1]

There was an abundance of raw wool, both of the short-staple wool of Hereford, Wiltshire, Dorset, and the chalk country, and of the long-staple wool of Devonshire, the Cotswolds, Leicestershire and Lincoln. Fuller's earth, or some suitable equivalent, was widespread in all the important manufacturing areas of the country. Water power was available both in the areas which manufactured woollens or serges and in those which produced certain kinds of woollen articles for which scouring but not milling was necessary. Fuel, in the shape of wood, or coal later in the century, was required for making the machinery used in the industry, for "heating dyers' vats...and the kilns in which [the clothiers] scoured their wool",[2] and was supplied from England's diminishing forests or from her coal-fields whose exploitation was increasing.

By the seventeenth century, the industry was roughly fixed in certain localities. Fuller's well-known list[3] gave the main distribution in 1655 as follows:

East	Norfolk. Norwich. Fustians.
	Suffolk. Sudbury. Bayes.
	Essex. Colchester. Sayes and Serges.
	Kent. Kentish Broadcloths.
West	Devonshire. Kerseys.
	Gloucestershire and Worcestershire. Cloth.
	Wales. Friezes.
North	Westmorland. Kendal Cloth.
	Lancashire. Manchester Cotton.
	Yorkshire. Halifax Clothes.
South	Somersetshire. Taunton Serges.
	Hants., Berks., Sussex. Cloth.

More than twenty years later, Yarranton noted that woollen manufacture was well established in Gloucestershire, Worcestershire, Shropshire,

---

[1] T. Fuller, *The Church History of Britain* (1655), p. 112.
[2] J. U. Nef, *Rise of the British Coal Industry* (1932), I, 214.   [3] Fuller, *op. cit.* p. 112.

Staffordshire, part of Warwickshire, Derbyshire, Nottinghamshire, York-shire, Suffolk, Norfolk, Essex, and in some measure in Kent, Sussex and Surrey.

Within these counties, there was specialisation[1] depending upon the kind of wool available, the power resources and, in some cases, the intro-duction of new methods. Thus serges were made from a mixture of short and long wool. Exeter was very famous for this industry, and, by the end of the century, was said to "deal for £10,000 a week" in cloth.[2] The serges were sent by sea from Topsham to London. Taunton, which was reputed to employ 8,500 persons weekly in the making of cloth, belonged to the same group. Devizes, "a rich trading place for the clothing trade", was one of the west country woollen towns, using short fine wool, as was Stroud, well known for its dyeing of scarlet and its broadcloth. Salisbury, once "a fine town for trading in the woollen manufactures", was said by Yarranton to be "much decayed of late years";[3] and in Reading the "seven score clothiers" had been reduced to "a very small number" in 1695.[4] In contrast, Newbury was a flourishing centre of the cloth industry. Kidderminster, famed for its "stuffs for hangings", used yarn to the value of £100 a week, imported from Germany; but Yarranton, who made the statement, also referred to the "poor decayed clothiers of Worcester and Kidderminster", who were compared with the "cappers" of Bewdley whose trade was "wholly decayed".[5] Tewkesbury, noted for its mustard, also manufactured woollen cloth. Coventry, "growing wealthy by clothing and making of caps", was thriving in 1637, and "very rich" when Celia Fiennes visited it about sixty years later.

East Anglia was the home of the worsted industry which used long wool, but for which water power was not required. The importance of Norwich as the centre of the "Norwich Stuffs" industry was recognised by the establishment of a Corporation of Worsted Weavers of Norwich in 1650, whose duties included that of "keeping up the goodness of that valuable manufacture". Eleven years later, an Act of Parliament to regulate the industry further became necessary because "the said trade of weaving of stuffs hath of late times been very much increased, and great

---

[1] See the classification of fabrics in R. H. Kinvig's "Historical Geography of the West Country Woollen Industry", *Geographical Teacher*, VIII, 301 (1916).

[2] *Angliae Notitia* (1694). Cf. Celia Fiennes, who speaks of "an incredible quantity", *op. cit.* p. 207.

[3] Yarranton, *op. cit.* p. 207.

[4] Gibson, p. 152. The reason for the decline was given as "the convenience of the river giving great encouragement to the mault trade".

[5] *Ibid.* pp. 146, 162.

variety of new sorts of stuffs have been invented". Celia Fiennes noted that Norwich made yarn for crapes, callemancos and damasks, and that the surrounding country was full of spinners and knitters. The value of the stuffs produced was £100,000 a year at the end of the century, while another £60,000 came from the stocking manufacture. Not a little of the prosperity of the Norfolk industry was due to the migration of Flemings in the sixteenth century, who brought with them great skill and the "New Draperies". Suffolk was mainly engaged in the manufacture of yarn for the Norwich weavers,[1] and in other preliminary branches of the worsted industry. In Essex, Colchester was the most important centre, the whole town being "employed in the spinning, weaving, washing, drying, and dressing their bayes".[2]

A third important group of woollen towns lay on the flanks of the Pennines, particularly in Yorkshire,[3] where the two chief products were the "Northern Dozen", a broadcloth, and the kersey, also a cloth but longer and not so wide as the "Northern Dozen". Both required long wool and employed water power. The chief centres of the industry were at Leeds, Halifax and Wakefield. Here, as elsewhere, the industry was not confined to the towns but spread over the surrounding country. "There was a small clothier making one piece weekly, and living from hand to mouth; the yeoman, who combined agriculture and industry, either making cloth or finishing it, or both; the large clothier, with his flock of spinners and weavers, and with apprentices learning their trade under his care."[4] On the western side of the Pennines, the woollen industry flourished in Lancashire, and Manchester was well known for its clothing. But during the century the Lancashire woollen industry was threatened by the rising cotton industry which had become important by 1700.

In addition to these three main areas, there were many other centres of the woollen industry scattered throughout the country. Kendal, "a place for excellent clothing" in 1637, was still the centre of a flourishing industry at the end of the century. In Kent, Surrey and Sussex the woollen manufacture was declining and although broadcloth was still made in Kent, "the manufacture for cloth has been much greater than now" (1694).[5]

[1] E. Lipson, *The History of the Woollen and Worsted Industries* (1921), p. 230. Many details have been taken from this work.

[2] Celia Fiennes, *op. cit.* p. 115.

[3] For the Yorkshire Industry, see H. Heaton, *The Yorkshire Woollen and Worsted Industries* (1920).

[4] H. Heaton, *op. cit.* p. 203.

[5] *Angliae Notitia* (1694), p. 19.

The woollen industry suffered periodical setbacks during the century, for which numerous causes were responsible. Plague and famine took their toll of the workers; foreign competition, particularly in Holland and France, was increasing; bad quality or dishonest measure did harm in foreign markets; political conditions at home and abroad were disturbed; these and other reasons were believed to account for trade depressions. It was said that Irish wool caused "the greatest mischief". William Carter recommended that it should be imported only through "Liverpool, Chester, Bristol, Minehead, and the river of Barnstaple".[1] Yet, by the end of the century, there can be no doubt that the industry had progressed as a whole. The home market was increasing, while the exports, if not actually increasing, provided the largest and richest item in the list.

Another textile industry of importance was that of silk. Attempts to introduce the breeding of silkworms early in the century failed; but in 1620 a certain Mr Burlamach "brought from beyond the sea silk-throwsters, silk-dyers, and broad-weavers";[2] and the industry gained a firm foothold in London. Within nine years, the silk throwers of London had been incorporated, and in 1630 a proclamation, dealing with fraudulent practices, referred to the industry as "much increased within a few years past". The preamble of an Act of 1661 for regulating the trade referred to 40,000 men, women and children employed in London. The arrival of Huguenot refugees in 1685 still further stimulated the manufacture of silk, while aid also came from the prohibition of the import of Indian silk in 1699. The industry spread beyond London into Kent. Celia Fiennes reported it at Canterbury, where French people were "employed in the weaving and silk winding"; one house contained twenty looms. Silk weaving had long been established at Sandwich; and elsewhere there were local industries, as that of ribbon at Coventry, stockings at Nottingham, and buttons at Macclesfield.

The silk industry owed its origin to foreign workers, and depended on imported raw material: the cotton industry,[3] however, although also drawing its raw material from abroad, imposed itself upon the woollen industry and ultimately displaced it in Lancashire. The Manchester "cottons" of the early part of the century were made of wool, but "cotton-

[1] *Eighth Report of the Royal Commission on Historical Manuscripts* (1881), p. 138.
[2] A. Anderson, *An Historical and Chronological Deduction of the Origin of Commerce* (1764), II, 4.
[3] See especially A. P. Wadsworth and J. de L. Mann, *The Cotton Trade and Industrial Lancashire, 1600–1780* (1931); and for the geographical factors, see L. D. Stamp and S. H. Beaver, *The British Isles* (1933), pp. 482–7.

wool" was imported from Cyprus and the Levant, and cotton was used in Lancashire "in combination with flax yarn in the making of fustians. The beginnings of this manufacture are to be found about the turn of the century, but cannot be dated with certainty."[1] It spread rapidly and widely, and within twenty years had become well established. Mr A. P. Wadsworth and Miss J. de L. Mann[2] have summarised its growth:

By the beginning of the seventeenth century the three textile industries of Lancashire were becoming localised. The area which specialised in the woollen manufacture shrank until it lay mainly along the eastern fringe of the county, in the deep-cut Pennine valleys between the Calder and the Tame. The valleys of the Roch and Irwell were wholly occupied with it and, until ousted by fustians, it was fairly extensive round Bolton. Blackburn had been important enough to have (along with Rochdale, Manchester, and Bolton) a deputy ulnager in 1566, but by 1681 the woollen industry there had so far declined that the ulnager had ceased to act. What woollen lost, linen and fustians gained. Manchester as the commercial centre of south-east Lancashire, "the very London of those parts, the liver that sends the blood into all the countries thereabouts", retained an interest in the merchanting and distribution of woollen goods, but ceased to manufacture them, although the fulling, dyeing, and finishing branches remained, owing to their dependence on the merchant, in some strength in the close neighbourhood of the town. The place of woollens as a Manchester manufacture was taken by linen and cotton smallwares, an expansion of linen (later cotton-linen) weaving, and some small manufacture of fustians and silk. All the outlying villages of the great parish of Manchester were given over to linen weaving.

Fustians took hold first in the hilly country between Bolton and Blackburn; by 1630 they had also become firmly established in the neighbourhood of Middleton, Chadderton and Hollingwood (the present Oldham area). The linen industry predominated in the rest of Lancashire, except in the parishes of "Lonsdale north of the sands"—that part of the county north of Morecambe Bay—which fell into the Westmorland woollen clothmaking area. With the linen industry itself differentiation appeared in the late seventeenth century. In the districts nearest Manchester and the fustian area cotton came to be used in mixed goods. The more remote parts of the county, which are even now little industrialised—the western seaboard, the Fylde, the Wyre valley, and the northern banks of the Ribble—continued to make pure linen goods. The specialisation here roughly described was to last until, a century and a half after the new textile branches had appeared, machine industry overrode the local boundaries of custom.

---

[1] A. P. Wadsworth and J. de L. Mann, *op. cit.* p. 15.   [2] *Ibid.* pp. 23, 25.

As the century progressed, the Lancashire textile industry expanded, capturing some of the business of the other two centres in East Anglia and the west country, and coming to have connections all over the world. Its linen manufacture was drawing yarn from Ireland, Scotland, and Germany, and using it in combination with cotton in new goods. Its fustians, made from linen yarn and the cotton of the eastern Mediterranean and the West Indies, were branching out into new varieties. Its smallwares were now able to enter the competitive market with the recent adoption of the Dutch loom, a relatively complicated piece of machinery, whose advantage Holland had enjoyed for a century, but which gild hostility still kept out of many European textile centres.[1]

The linen industry obtained part of its raw material from home-grown flax, but this was insufficient to meet all requirements. The Lancashire industry has already been mentioned: it imported yarn from Ireland for its workers in Manchester and elsewhere, while some came from Russia and north Germany. By comparison with other textile industries, it seems to have been of small importance in England.

Lace making was chiefly carried on in two districts. The centre of the Devonshire industry was at Honiton, where Celia Fiennes saw the making of "the fine bone lace in imitation of the Antwerp and Flanders lace". In the east midlands, Stony Stratford, Newport Pagnell and Bedford were the leading towns in an area where this industry was the chief manufacture.

It is impossible to enumerate all the other industries of importance in England, but three call for special mention. Of the metal industries, iron was, perhaps, the chief. There were many small producing areas, but there were three main groups. In the south-east, Sussex and Surrey[2] had been the chief centres, supplying the market of London and having the practical monopoly of cast-iron guns in England. In 1607 Norden had counted 140 "hammers and furnaces" in Sussex, while in 1611 it was estimated that of the 800 "iron mills" in the country one-half were in Sussex and Surrey. Lists of Wealden furnaces and forges[3] in 1664 show a decline:

	Furnaces	Forges
Blowing in 1653	35	45
Discontinued before 1664 but repaired and stocked on account of the [Dutch] War	12	—
Ruined before 1664 and so remain ...	9	19
Laid aside and not used	—	5
In repair at the beginning of 1664 ...	14	21

[1] A. P. Wadsworth and J. de L. Mann, *op. cit.* p. 71.
[2] See E. Straker, *Wealden Iron, passim.*   [3] E. Straker, *op. cit.* p. 61.

Soon after this date Yarranton declared that "most of the works in Sussex and Surrey are laid down and many in the North of England and many in other parts must follow if not prevented by inclosing commons to supply them with wood".[1] There was still abundant iron in the Weald; and the rivers provided fair transport facilities and water for use in the hammer ponds which, in turn, supplied power to drive the machinery in forge or furnace. But fuel was getting increasingly dear; legislation prevented indiscriminate cutting; there were long droughts in this part of England so that water power was not always available and furnaces had frequently to be blown out; while many economists were opposed to the industry. Evelyn declared,[2] "'twere better to purchase all our iron out of America than thus to exhaust our woods at home". Perhaps the exhaustion was more apparent than real, for sixty years later Defoe could speak of this area as "one inexhaustible storehouse of timber".[3] Yarranton suggested that the reason for the decline was that "Iron from Sweadland, Flanders, and Spain comes in so cheap that it cannot be made to a profit here".[4]

The Forest of Dean was another centre where the iron industry had been long established. Although it suffered through the rise of the Wealden industry, it was flourishing and producing "infinite quantities of raw iron" with bar iron and wire when Yarranton wrote. "The greatest part of this sow-iron", he added, "is sent up Severn to the forges into Worcestershire, Shropshire, Staffordshire, Warwickshire, and Cheshire, and there it's made into bar iron. Because of its kind and gentle nature to work it is now at Stourbridge, Dudley, Wolverhampton, Sedgley, Walsall, and Birmingham, and thereabouts, wrought and manufactured into all small commodities and diffused all England over."[5] The basic factors upon which the industry of the Forest of Dean rested were the cheapness of coal, abundance of timber, and facilities for transport by water. Yarranton declared that it gave employment to "no less than sixty-thousand persons".

In Worcester, Shropshire, Staffordshire, Warwickshire and Derbyshire, iron produced was used for a different purpose. "This iron is short, soft iron, commonly called Coldshore iron, of which all the nails are made and infinite other commodities. In which work are employed many more persons if not double to what are employed in the Forest of Dean."[6] The location of the industry there was explained by Yarranton. "In all these counties now named there is an infinite quantity of Pit coals, and the Pit coals being near the iron, and the iron-stone growing with the coals, there

---

[1] Yarranton, *op. cit.* p. 63.　　　　[2] Quoted in E. Straker, *op. cit.* p. 124.
[3] This may have been an exaggeration; see p. 489 below.
[4] Yarranton, *op. cit.* p. 146.　　[5] *Ibid.* p. 57.　　[6] *Ibid.* p. 59.

it is manufactured very cheap and sent all England over and to most parts of the world." There are many other witnesses to the rising iron industry of the midlands. "Bremicham" was described, in the 1637 edition of Camden's *Britannia*, as "full of inhabitants and resounding with hammers and anvils, for most of them are smiths". Much later Houghton refers to the steel mills made in Birmingham and costing thirty or thirty-five shillings. This third centre, later to be called The Black Country, was indeed the most prosperous iron area. Yarranton believed that "within ten miles round Dudley there are more people inhabiting and more money returned in a year", than in the counties of Oxfordshire, Warwickshire, Leicestershire and Northamptonshire combined.[1] And he thought the iron industry as a whole was "of the same value as...and much more to the publick than the woollen manufacture". This is probably an exaggeration, for the real advance in the iron industry did not come until the early years of the next century.

Lead and tin were two other minerals of importance. Derbyshire was important for the former; and Houghton, who describes the resources of the county at length,[2] details the routes by which the lead was exported. Some was taken to the smelting mill at Derby, thence to the ferry near Wilm, five miles away, and so by water to Hull. It was also sent from the mills about Critch and Worksworth to "Nottingham bridges". Another route was "from the Chesterfield district and all that side to Bawtree". Some was collected at Sheffield and went by Doncaster "to the German Ocean". The industry flourished in Derbyshire throughout the century: in the Mendip Hills it reached a peak about 1670.[3] Numerous mines were worked in Durham and elsewhere. In all, England was producing about 12,000 tons of lead in 1636;[4] and it seems likely that this figure was not exceeded during the century.

The tin mining industry was of greatest importance in Devonshire and Cornwall. Tin had for so long been a recognised commodity of English commerce that most writers mentioned its production. Yet the total amount produced seems to have been small, and had declined seriously from that of the first half of the sixteenth century.

For this, several reasons may be assigned. The general European rise of prices which marks the period was for the English tinners much more pronounced

---

[1] Yarranton, *op. cit.* p. 52.   [2] Houghton, *op. cit.* 1, 124 *et seq.*

[3] Nef, *op. cit.* 1, 167. The Cardigan mines of Wales were producing large quantities of lead early in the century and much silver was coined at the mint in Aberystwyth Castle. L. B. Cundall and T. Landman, *Wales* (1925), p. 207.

[4] Nef, *op. cit.* p. 194, note 1.

in the material necessary for mining than in the products of the mines themselves; the price of tin rose absolutely, but was stationary and even declining if considered relatively to that of timber, rope, iron and corn. By a fatal coincidence, the mines at about the same period began to attain a depth at which water flowed in faster than the clumsy drainage devices then in use could get rid of it.[1]

Tables prepared by G. R. Lewis show that production fluctuated considerably from year to year, but was always very much greater in Cornwall than in Devon, where alluvial deposits had been of relatively small importance since the thirteenth century. In Cornwall itself there was a westward drift of industry which is illustrated by the establishment of a coinage in remote Penzance in 1663. All the tin produced was, at least in theory, taken to a coinage town to be stamped and taxed and from these towns was exported.[2] "Some found its way to local consumpton in the shape of small bars for use as solder; some was melted down for the casting of bells; but by far the greater part of the domestic consumption was in the manufacture of pewter."[3] The substitution during the century of china ware for pewter injured the domestic trade and it was not until the end of the period that some compensation was found in the tin-plate industry which was then making a somewhat uncertain beginning in south Wales.

Copper mining and smelting was in its infancy in England. In 1602 R. Carew[4] reported that "Copper is found in sundry places, but with what gain to the searchers I have not been curious to inquire, nor they hasty to reveal; for at one mine (of which I took a view) the ore was shipped to be refined in Wales, either to save cost in fuel or to conceal the profit". This is an early indication of the shipment of Cornish copper to the refinery which had been set up at Neath in south Wales during the closing years of the previous century. Nearly a hundred years later Celia Fiennes saw copper mines at Redruth from which some of the ore was sent to Bristol for "melting" while some was "melted" at St Ives. The same writer records the mining of copper at Ashbourne in Derbyshire; it was also mined at Caldbeck and near Newlands in Borrowdale, where Keswick was the chief centre of the industry.

Among the new industries introduced into England during Elizabethan

[1] G. R. Lewis, *op. cit.* p. 41.
[2] The towns were Plympton, Tavistock, Ashburton and Chagford, in Devonshire; and Helston, Bodmin, Truro, Lostwithiel and Liskeard, in Cornwall (with Penzance after 1663).
[3] G. R. Lewis, *op. cit.* p. 45. The pewter industry was carried on chiefly in London.
[4] R. Carew, *Survey of Cornwall* (1811 ed.), p. 21.

times was the making of glass. Progress was slow until the second half of the seventeenth century. J. Cary, in his *Essay on Trade*, written at the end of that century, said: "Glass is a manufacture lately fallen on here and in a short time brought to great perfection, which keeps many at work; and as the materials whereof it is made are generally our own, and in themselves of small value, it costs the nation little in comparison of what it formerly did to fetch it from Venice." Although not enough glass was produced in England to meet the home demand, the industry had undoubtedly spread widely, and Houghton[1] in 1696 gave a list of "glass houses", summarised in the table on p. 420.

It will be noticed that the most important centres, with the exception of London, were situated near to salt workings. The manufacture of salt from the brine pits of Cheshire and Worcestershire was an industry of long standing and increased in importance during the century. Evaporation of sea water was carried on at a number of places round the coast, but the most important centres were on the north-east coast and particularly at the mouth of the Tyne where the industry grew from small beginnings and by the end of the century was producing about the same quantity of salt as the older established plants of the midlands.[2] Rock salt was discovered in Cheshire in 1670[3] and the rock salt manufacture was added to the earlier methods of producing salt before the century had closed.

It is unnecessary to continue the catalogue of English industries with references to alum, copperas, saltpetre, gunpowder, soap making, and others of less importance. One interesting development was the making of tobacco pipes at Broseley on the Severn[4] and in London. Clay, obtained locally or near Poole and shipped to London, provided the raw material for this trade. Clay was also dug for brick-making in many parts of the

---

[1] Houghton, *op. cit.* II, 48. Houghton gave a summary table, the totals of which do not agree with his earlier figures, showing 100 glass houses occupied as follows:

Making bottles ...	... ...	42
„ window glass ...	...	18
„ crown glass	... ...	5
„ flint, green and ordinary		28
„ looking glass plates	...	7

Not all these were working.

[2] As early as 1635 Sir William Brereton declared that at Shields there were "more salt works and more salt made than in any part of England that I know". He was well acquainted with Cheshire. See P. Pilbin, "A Geographical Analysis of the Sea-Salt Industry of North-East England", *Scot. Geog. Mag.* LI, 22 (1935).

[3] *Philosophical Transactions*, V, 2015.

[4] Nef, *op. cit.* p. 186.

country, among others at Nottingham, Gestingthorpe in Essex, Walpet in Suffolk, and Ebbisham in Surrey. For building stones, Colly Weston in Northamptonshire was well known; but many other quarries up and down the country were being worked increasingly. In the Potteries local clay

Glass houses in 1696.

In and about London and Southwark 24	9 for bottles	
	2 „ looking glass plates	
	4 „ crown glass and plates	
	9 „ flint, green and ordinary	
Woolwich (Kent)	1 „ crown glass and plates	
	1 „ flint, green and ordinary	
Isle of Wight	1 „ do. do.	
Topsham near Exeter	1 „ bottles	
Oddham near Bath	1 „ bottles	
Chellwood (Somerset)	2 „ bottles and window glass	
In and about Bristol 9	5 „ bottles	
	1 „ bottles and window glass	
	3 „ flint, green and ordinary	
Gloucester...	3 „ bottles	
Newnham (Glos.)...	2 „ bottles	
Swansea	1 „ bottles	
Oakengates (Shropshire)	1 „ bottles and window glass	
Worcester	1 „ flint, green and ordinary	
Coventry	1 „ do. do.	
Stourbridge 17	7 „ window glass	
	5 „ bottles	
	5 „ flint, green and ordinary	
Near Liverpool	1 „ flint, green and ordinary	
Warrington	1 „ window glass	
Nottingham	1 „ flint, green and ordinary	
Answorth (Notts.)	1 „ bottles	
Custons-More near Answorth ...	1 „ bottles	
Near Silkstone	1 „ flint, green and ordinary	
Near Fennybridge	1 „ bottles and flint, green and ordinary	
King's Lynn	1 „ bottles	
Yarmouth	1 „ bottles	
	1 „ flint, green and ordinary	
Newcastle-on-Tyne 11	6 „ window glass	
	4 „ bottles	
	1 „ flint, green and ordinary	

was being used for the pottery industry, though in some parts it was already becoming exhausted.

For this industrial development, power was necessary. It was obtained sometimes from water, as had been the case for centuries. Wood was used, but it was becoming scarce and expensive. Wind power was not as much developed as might have been expected, and Celia Fiennes remarked with surprise upon the absence of windmills in Devonshire and Cornwall "though they have wind and hills enough". Her comment was "maybe its too bleake for them".[1]

Coal, the remaining source of power, was being rapidly developed during the century. Dr Nef has exhaustively studied the growth of the coal industry and his conclusions can only be briefly summarised in the space available. His table[2] showing the expansion of the industry furnishes an explanation of its importance and its distribution.

Estimated Annual Production of the principal mining districts (in tons).

Coal-field	1551–60	1681–90	1781–90	1901–10
Durham and Northumberland	65,000	1,225,000	3,000,000	50,000,000
Scotland	40,000	475,000	1,600,000	37,000,000
Wales	20,000	200,000	800,000	50,000,000
Midlands	65,000	850,000	4,000,000	100,180,000
Cumberland	6,000	100,000	500,000	2,120,000
Kingswood Chase	6,000⎫	100,000	140,000⎫	1,100,000
Somerset	4,000⎭		140,000⎭	
Forest of Dean	3,000	25,000	90,000	1,310,000
Devon and Ireland ...	1,000	7,000	25,000	200,000
Total ...	210,000	2,982,000	10,295,000	241,910,000
Approximate increase ...		14 fold	3 fold	23 fold

An examination of the map redrawn from his book (Fig. 69) shows that, with the Kentish exception, there was no important coal-field on which some working did not take place. Nearness to the sea, or to navigable rivers, was obviously a factor of great importance in the distribution of such a bulky commodity, and contributed much to the development of

---

[1] Fiennes, *op. cit.* p. 219. The Cornish tinners were, of course, experts in the use of water power.

[2] J. U. Nef, *op. cit.* pp. 19, 20. The coal worked in Devonshire was lignite.

28

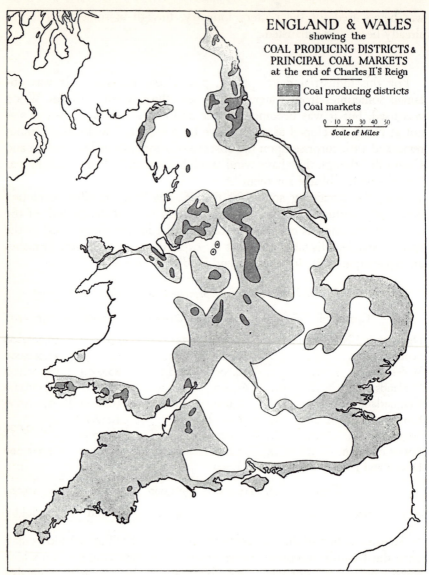

Fig. 69

Redrawn from J. U. Nef's *Rise of the British Coal Industry*, 2 vols. (George Routledge & Sons, Ltd., 1932), folding map, opposite p. 19.

All coalfields within reach of the sea exported coal, and the markets of southern and eastern England were supplied in this way. The Severn and the Trent rivers fed the markets of the Midlands.

some fields—Pembrokeshire, Whitehaven, and above all Newcastle.[1] But in many cases, collieries were located under roads. In Derbyshire, "the pits are so thick in ye roade that it is hazardous to travel for strangers" wrote Celia Fiennes. The same writer saw Warwickshire coal being unloaded from barges at Gloucester, and recorded that Bridgwater imported coal from Bristol and sent it on, by pack-horse, to Taunton: between the two places "the roads were full of these carryers going and returning".

In addition to stimulating industry in general, the increased use of coal contributed to the growth of shipbuilding, because many ships were required to carry it either to the numerous small ports with which the coasts and navigable rivers of England and Wales were liberally supplied, or to foreign countries. Dr Nef estimates that at the end of the century between three and four thousand ships of all kinds were engaged in transporting coal. This represents an enormous increase over the numbers so employed in Elizabethan times, and it probably accounted for about one-third of all English shipping; the tonnage, too, was greater. Both the Tyne and the Thames had numerous shipbuilding yards, while Bristol, Yarmouth, Whitby and Harwich were also important centres of the industry. By the end of the century, Celia Fiennes recorded that the dock at Ipswich was disused. On the other hand, she saw a ship of 200 tons on the stocks at Chester, and numerous ships being built at Plymouth, where two docks were completed in 1693. Yarranton advocated establishing a shipbuilding yard at Christchurch, which not only had a good harbour but was close to supplies of wood and iron. And it was nearness to such supplies that largely determined the position of the Royal dockyards at Portsmouth and Chatham.[2]

Not all English ships were built in England, nor were they built of English materials. The Baltic trade supplied many items of naval stores, and it was one criticism of the Navigation Acts[3] that they made these commodities more expensive and thus damaged the shipbuilding industry.

[1] Valuable details of the use of rivers for the transport of coal, and of river navigation generally, are given in T. S. Willan, *River Navigation in England, 1600–1750* (1936), which appeared after this chapter was written.

[2] Portsmouth was much developed by Charles II, and, at the end of the century, was "reckoned amongst the principal chambers of the Kingdom". At Chatham "large additions of new docks and storehouses" were made and Gibson thought "there may not be a more compleat arsenal than this in the world". Accommodation at Deptford was doubled; Sheerness, "an appendix to Chatham", was improved; and Woolwich, "the mother dock", maintained its earlier importance. Gibson, pp. 219, 229–30.

[3] See below p. 433.

INTERNAL COMMUNICATIONS

For a full development of the resources of the country it is essential to have good means of internal communication.[1] Of those provided by nature, rivers obviously claim first place and, in this respect, south-eastern England in particular was fortunate. And there were many people who appreciated the importance of river transport. The water poet, John Taylor, not only saw the advantages of cheap transport but advocated the construction of an artificial water channel to connect the Severn and the Thames. Fourteen years later, in 1655, Francis Mathew revived the project and discussed at great length the advantages of improved communication by water: in particular, he was anxious to see trade developed between Bristol and London, and he outlined a plan to link East Anglia with the Trent, for which purpose the Fosse Dyke, an old and long disused Roman canal or drain, was to be rebuilt. Somewhat similar, if less ambitious, proposals were made by Yarranton in 1677.

These men who proposed to use the rivers had to face strong, continuous, and not infrequently successful, opposition from other vested interests. Mills, weirs, and other impediments hindered the use of rivers for navigation purposes. A good illustration of the difficulties which faced the improvers can be instanced from the Wye. The objects of the improvements were to increase the export of grain and fruit from Herefordshire, to reduce the cost of wood and coal imported from the Forest of Dean, and to increase trade between the towns along the river and Bristol. The struggle with the owners of mills and weirs had been going on before the century opened: it was continued until 1662, when an Act of Parliament authorised improvements in navigation. The works on the river, carried out by Sir William Sandys, were not wholly successful, and it was not until the close of the century that another Act of Parliament was passed which allowed the removal of all weirs except one. The traffic on the river during the next century fully justified the improvements.

The case of the Wye was typical of many others. Macaulay declared that "there was very little communication by water" in seventeenth-century England. This was certainly an exaggeration. In 1605 an act was passed to complete the clearing of the Thames as far up as Oxford: hitherto the river was "very navigable and passable with and for boats and barges of

[1] On communications generally see Joan Parkes, *Travel in England in the Seventeenth Century* (1925), and W. T. Jackman, *The Development of Transportation in Modern England* (1916), vol. I. Many details of river improvement given above have been taken from Jackman's book. See above, p. 423, footnote 1.

great content and carriage" to "within a few miles" of that city and "for many miles" beyond it. The Act of 1605 was not, apparently, successful, and a further Act was passed in 1624 "for opening the Thames from Burcote by Abingdon to Oxford" to carry Oxford free-stone to London, and coal and other necessaries from London to Oxford. In August 1635 the improvements had been effected, with financial aid from the university, for in that year Laud, in his account of his Chancellorship, records that "the Thames was brought up to Oxford and made navigable for barges". This brought material advantages to the city, and cheer to the declining years of Robert Burton, of whom it was said that "nothing at last could make him laugh but going down to the footbridge in Oxford and hearing the bargemen scold and swear at one another, at which he would set his hands to his sides and laugh most profusedly". Later in the century, Celia Fiennes saw the Thames "full of barges" between Abingdon and Oxford. In 1677, according to Yarranton, the Thames was navigable to Lechlade; but an Act of 1737 says it was navigable to Cricklade. On other rivers similar improvements were being made.[1]

During the reign of Charles II many Acts relating to waterways were passed. Two such Acts in 1662 dealt with the Stour and Salwarp, and the Wye and Lugg respectively. In 1664–5 similar Acts were passed to deal with the Wiltshire Avon, the Medway, the Mole and the Itchen; while later Acts related to the Trent, Brandon, Waveney, Fale (in Norfolk), Witham and Wey. At the very end of the century, an Act was passed "for making and keeping navigable the Aire, the Calder, and the Trent". Even before the passage of this Act there was much traffic on these rivers. Houghton gave a list of navigable rivers in Yorkshire which included the Tees for 9 miles, the Hull for 8 miles, the Humber for 20 miles, the Ouse to York and, for smaller vessels, to Boroughbridge, the Wharfe to Wetherby, and the Aire to Otterton, or by Tunbridge Dyke to Thorne and thence by the Don to Doncaster. The Severn, which had received much attention in the sixteenth century, was said by Yarranton to be navigable as far as Welshpool in 1677. Yarranton himself supervised improvements on the

---

[1] During the century strenuous efforts were made to improve the navigation of the Lea, along which grain, malt, and provisions were carried to London, and of the Medway which was used for the conveyance of timber to Chatham: in both cases the efforts met with but partial success. In 1606 the Nene was surveyed with the object of extending navigation above Allerton. In 1634 the Soar was improved for about six miles from its junction with the Trent. Two years later the lower part of the Warwickshire Avon was improved, while about the same time a scheme was put forward to improve the Wey, which ultimately became navigable to Guildford in 1653. In 1657 an Act was passed for amending the navigation of the Ouse in the neighbourhood of York.

Dee, cutting a new channel, but the scheme does not seem to have been successful, for an Act of 1700 was obtained to enable Chester "to recover and preserve the navigation of the river". The traffic on the Derwent from near Derby has already been noted. It was only the opposition of Nottingham which prevented the river Derwent being made navigable to Derby itself. In the south-west, opposition to the improvements of the Exe had been overcome, and Gibson reported that as a result of the new works lighters could go up to the city quay in Exeter.

Of the road system little need be said as it is admirably shown on the map of seventeenth-century England published by the Ordnance Survey.[1] London had already established itself as the chief road centre in the country, while, in the west, Salisbury, Bristol and Exeter, though of less importance, each had five or more main roads converging upon them. In the midlands, Coventry claimed first place; while in the border country Gloucester, Worcester, Chester, and to a less extent Shrewsbury, were of importance. The large number of roads in the centre and south of England, and their comparatively poor development elsewhere, is an indication of the relative importance of the different parts of the country.

It has long been customary to describe the roads of the country as uniformly bad until the time of the great road-makers of the eighteenth century. Such a picture is exaggerated for the seventeenth century. As roads were unmetalled, their state naturally depended very largely upon the kind of rock over which they ran. Sussex was "very ill for travellers especially in the winter, the land lying low, and the wayes very deep".[2] Hockley-in-the-Hole was so named, according to Camden, because of the "miry way in winter time so troublesome to travellers".[3] The heavy Gault clay, over which Watling Street passed in this part of Bedfordshire, might give "fields smelling sweet in Sommer of the best Beanes" but made a poor road surface in the wet months. But much of this old road was in better condition, for Camden followed it "from Tamis to Wales" without comment. Celia Fiennes, who travelled extensively over England, has many interesting remarks to make on the roads. Thus the 22 miles from Dunmow to Colchester were "mostly clay deep way"; from Evesham to Weston in Gloucestershire, through the Vale of Evesham, was "all very heavy way"; Derbyshire was "full of steep hills and nothing but the peakes of hills as thick one by another is seen in most of ye country which

---

[1] *A Map of XVII Century England*, 1930. A map of London c. 1660 accompanies this.
[2] J. Speed, *The Theatre of the Empire of Great Britaine* (1611), p. 9.
[3] Camden, p. 402.

are very steep and which makes travelling tedious and ye miles long";
between Harfordbridge and Bagshot was "a heavy sandy way" while
from Dorking to Kingston was "a chalky hard road"; round Ely the
roads were under water; and in Yorkshire the 12 miles between Knaresborough and York were the "worst riding" in the county. From this it
is clear that the roads were not uniformly bad and, in view of the increasing
amount of traffic which passed over them, they were, perhaps, not so
unsatisfactory as might have been expected.

Many attempts were made to improve the highways. At first the
Government tried to regulate the kind of traffic which could use the
roads, by limiting the number of draught horses employed or the types
of vehicles used. One of the most effective of these measures was Cromwell's Ordinance of 1654. Its preamble began by stating that "the several
statutes now in force for mending highways are found by experience not
to have produced such good reformation as was thereby intended".
Henceforward surveyors of the highways were to be appointed yearly in
every parish. They were to have power to raise a rate in the parish for the
mending of the highways to the extent of one shilling in the pound, and,
in the case of important roads, the justices of the peace might levy a rate
in and upon neighbouring parishes. They had power also to hire labourers
and carts, and to take gravel and stones to mend the roads; and finally to
levy fines upon persons who damaged the roads by using too heavy carts
and teams. This last clause was important, for there were frequent complaints that the roads were ruined by overweighted waggons.[1]

[1] In 1638 the inhabitants of Milton, Lewknor and other villages complained to Laud,
the Chancellor of the University of Oxford, that carriers "do carry such unreasonable
carriages by which means they spoil the highways that notwithstanding the petitioners'
great and extraordinary charges in continual repairing of them, the ways are almost impassable" (Laud, *Works* (1847–60), v, 211). A proclamation of Charles I in 1630 had tried
to prevent this evil by prohibiting the use of four-wheeled carriages. No carrier was
permitted to use any but two-wheeled vehicles. But as a special favour to the University
the Oxford carrier was allowed to use four-wheeled carts, to the great damage of the
roads. Accordingly, Laud adopted a new plan, which was to limit the number of horses
the Oxford carrier might put to his cart. He was not to have more than five or six;
then, said Laud, he would not dare to put an excessive weight on to his cart for the sake
of the horses (Laud, *op. cit.* v, 211, 271). Cromwell's Ordinance made the system
general; no carrier was to put more than five horses or, as an alternative, six oxen and
one horse, to his cart or waggon. Of the roads in Cornwall, Carew wrote: "Intercourse
is obtained by highways and bridges: for highways, the Romans did not extend their's
so far: but these laid out of later times and in the eastern part of Cornwall uneasy,
by reason either of the mire or stones, besides many uphills and downhills. The western
are better travelable, as less subject to these discommodities; generally the statute
18 Eliz. for their amendment is reasonably well executed" (Carew, *op. cit.* p. 163).

In 1656 an Act was introduced which instituted a surveyor-general of the highways for the whole nation, with authority to supervise all roads; but it failed to pass. But Cromwell's Ordinance of 1654 was followed up

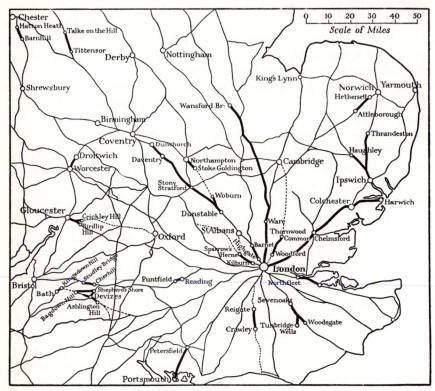

Fig. 70

Redrawn from Arthur Cossons' *The Turnpike Roads of Nottinghamshire* (issued jointly by the Historical and Geographical Associations, 1934), p. 7.

Continuous hair lines: Roads described in Ogilby's *Itinerary*, 1675. Thick lines: Roads for which Turnpike Acts had been passed up to the end of the reign of Queen Anne. Where roads not mentioned by Ogilby were turnpiked, they have been extended by broken hair lines.

Mr Cossons has recently found that small stretches of road near Bath were also turnpiked; but these are too small to be shown on this map.

by the Acts of 1662 and 1670. And besides these there were two special Acts, in 1663 and 1665.[1] The first provided for repairs to the Old North Road in the counties of Hertford, Huntingdon and Cambridge. Tolls

[1] On the earliest Turnpike Bill, 1622, see F. G. Emmison, *Bull. Inst. Historical Research*, XII, 108 (1934).

were to be levied at certain places. The second Act was similar to the first, but it applied to Hertfordshire only.[1] Other roads were subsequently dealt with on the turnpike principle, including the road from London to Harwich in 1695–6 and two smaller stretches of road in Sussex and Gloucester a year or two later (Fig. 70). This system of turnpikes was to be rapidly extended during the next century.

The character of the traffic using the roads is evidence of the efficacy of these measures. On all the great roads there were, during the latter part of the century, coaches running from London and accomplishing their journey in a fixed period. The first definite mention of such a service is in 1637 when two weekly coaches ran between London and St Albans,[2] but it is possible that some ran before that date. In 1657 coaches ran three times a week from London to Chester, taking four days; while a coach left London every Monday for Norwich and returned on the following Thursday. In 1658 coaches ran to Exeter and to York in four days. There was even a stage-coach which started once a fortnight for Edinburgh. These coaches gradually became more numerous, the service more frequent, the speed greater, and the fares lower. At the beginning of the reign of Charles II there were coaches running from Oxford to London, but they took two days over the journey, and passengers generally stopped a night at Wycombe or Beaconsfield. But in April 1669 new "flying coaches" were set up, which performed the whole journey in one day for a fare of 12s. But they undertook this only from 26th April to Michaelmas.

The obvious conclusion to be drawn from this evidence is that the main roads at least were not so bad as might be inferred from casual records of hardship and danger—records made, no doubt, because they were exceptional—and that communication between town and country was easier than a reader would gather from Macaulay's account of the state of England.

EXTERNAL COMMUNICATIONS

The external communications of England were as important as those within the country. There was a large number of ports scattered round the coasts and, in some cases, lying far up the rivers.[3] London surpassed all

[1] A. Cossons, *The Turnpike Roads of Northamptonshire* (Hist. Ass. Leaflet, 97, 1934), p. 6.
[2] J. Parkes, *op. cit.* p. 82.
[3] A list of 1578 enumerates 391 "landing places, portes, and creekes in England and Wales"; see D. and G. Mathew, "Iron Furnaces in South-eastern England and English Ports and Landing Places, 1578", *Eng. Hist. Rev.* XLVIII, 96 (1933). Another

other ports. Bristol was a large port where the Italian Foscarini in
1643 counted "43 vessels with tops besides ten of seven to eight hundred
tons and other smaller ones in the part nearest the sea"; at Newcastle, in
the same year, he counted 98 with tops. Twelve years later, another
Venetian, Gussoni, reported that "England can make use of from 940 to
1000 merchantmen.... Among these are included 400 ships which, from
Newcastle...transport combustible earths to all parts of the realm." About
the same time Peter Mundy wrote[1] of Bristol that "it is even a little London
for merchants, shipping, and great and well furnished marketts etc. and I
think second to it in the Kingdom of England". Among the east coast
ports mentioned in the 1637 edition of Camden's *Britannia* were Boston,
"well frequented"; King's Lynn, "the safe haven which yeeldeth most easie
access"; and Hull, "the most famous towne of merchandise in these
parts". In the south-west, Topsham has already been mentioned as the port
of Exeter. On the west coast, Peter Mundy referred to Barnstaple as "a
bigge towne, a sea port...[with] many vessells belonging to it"; and to
Bridgwater as "a great towne".[2] Sir William Brereton crossed from
Waterford to Minehead "whereinto there is great recourse of passengers
for Ireland".[3] Chester was declining in importance. In 1677 Yarranton
declared, "at present a vessel of twenty tuns cannot come loaded to that
old noble town"; and he attributed the silting of the river to the fact that
the wind was mainly on shore so that sand and gravel could not be removed
by tidal currents. But Chester strove to retain some of its trade by establishing out-ports along the Wirral side of the Dee estuary, and Celia
Fiennes saw "many ships riding along" the harbour of "high lake"
(Hoylake) at the mouth of the estuary. Liverpool, however, was gradually
taking the place of Chester; by the end of the century it had become
"a very rich trading town" and was "London in miniature".[4] Gibson
remarked upon its passenger trade with Ireland, its West Indian trade,
and the manufactures which were growing up near it, all of which were
responsible for "the vast growth of this town of late years.... Its buildings
and people are more than doubly augmented and the customs eight or
ten-fold encreas'd within these 28 years past."[5]

list for the reign of James I gives 194; R. G. Marsden, "English Ships in the Reign
of James I", *Trans. R. Hist. Soc.* (new series), XIX, 309 (1905). A great mass of
information in the Port Books in the Public Record Office awaits detailed study. Some
idea of the value of this can be gathered from *Royal Commission on Public Records*,
1st Report (1912), I, Part II, 45–51.

[1] Mundy, *op. cit.* p. 11.   [2] *Ibid.* pp. 2, 3, 4.
[3] Brereton, *op. cit.* p. 164.   [4] Fiennes, *op. cit.* p. 152.
[5] Gibson, p. 801.

The many ports of England and Wales had to deal with both coastal and foreign trade. The former was relatively greater than now because internal means of communication were poorer, and it covered many commodities. Besides bulky articles such as Newcastle coal, iron from the Forest of Dean, and fuller's earth from Arundel, there were cattle carried from Ireland to Liverpool; and provisions of all kinds were brought to London.

The staple export trades of the country were cloth, lead and tin, but many other articles figured in the lists of exports. Petty[1] estimated that the exports of woollen manufactures were worth £5,000,000 a year; and those of lead, tin and coal together £500,000, while "cloathes and household stuff carried into America" were valued at £200,000. The articles of commerce and the direction of trade changed but little during the first half of the century and a representative list recently compiled[2] will serve to show its general character about the year 1660:

COLONIAL.
Exports: Cloth (esp. serges), tin, leather goods, lead, glass, earthenware, hats, shoes, stockings, raisins (re-exported), slaves (from Guinea); woollen goods, cattle, utensils for sugar boiling, tobacco pipes, and provisions generally.
Imports: Tobacco (Virginia), sugar (West Indies), ginger, cotton, skins (a small quantity), indigo, fustic wood.

EAST INDIES.
Exports: Cloth, aqua-vitae, bullion.
Imports: Pepper, ginger, saltpetre, calicoes, silks, cotton yarn, nutmegs, various spices and drugs.

EASTLAND.
Exports: Cloths (esp. Spanish), woollen goods, men's woollen hose, leather goods. To Poland: shortcloth; to Norway: bullion.
Imports: Hemp, pitch, tar, potashes, tallow, masts, oars, deal.
NOTE. During the Protectorate, Whitelocke supported a proposal to secure a monopoly of Swedish copper.

FRANCE.
Exports: Cloth (esp. kerseys and serges), woollen goods, hose, refined sugar, wrought iron, butter, salt.
Imports: Wool, silk, wine (Bordeaux), beads, paper, olives, salt.

GERMANY.
Exports: Cloth (esp. bays and new draperies), lead, tin, corn, beer, gloves, stockings.
Imports: Wine (Rhenish), copper, hemp, linen, canvas, flax.

[1] *Op. cit.* I, p. 296.
[2] M. P. Ashley, *Financial and Commercial Policy under the Cromwellian Protectorate* (1934), pp. 184, 186. I am also indebted to Mr Ashley for further information by letter. For a full account see E. Lipson, *Economic History of England* (1931), vols. II and III.

GUINEA.
 Exports: Cloth, provisions.
 Imports: Slaves (re-exported to West Indies), ivory, dyeing woods.

LEVANT.
 Exports: Cloth (esp. light cloths) dyed and dressed, tin, bullion. To Italy: serges, perpetuanas, wax, tin.
 Imports: Cotton, galls, raw silk, currants (from Greece), hemp, mohair yarn, goat's hair, carpets. From Italy: wines, capers, beads.

NETHERLANDS.
 Exports: Cloth (mainly undressed), lace, red lead, skins, copperas (re-exported), hats, glass.
 Imports: Loaf sugar, flaxen goods, hops, tow, dressed hemp, indigo, cordage. From the Spanish Netherlands: Bruges thread, silks, tapestries.

PORTUGAL.
 Exports: Fish (esp. dried cod), cloth (esp. Spanish), shoes, stockings, silk and leather goods, gunpowder.
 Imports: Oil, sugar (from Brazil), fruit (esp. citrons), wine (port).
  NOTE. The proposal was put forward by English merchants at the Restoration that the English should acquire a monopoly of the Portuguese salt trade.

RUSSIA.
 Exports: Dyed and dressed cloth, aqua-vitae.
 Imports: Furs (esp. sables and squirrel), skins, red hides, hats, potashes, caviare.
  NOTE. There was very little trade between England and Russia during the Interregnum.

SPAIN.
 Exports: Fish, cloth (esp. bays, perpetuanas, Norwich stuffs), hats, stockings, gold and silver lace, bacon, hoops, glass, earthenware.
 Imports: Wine (Saragossa), raisins, wool (merino), silk, bullion (from Mexico), oil, indigo, iron, cochineal, tobacco.

During the second half of the century the foreign trade of the country expanded considerably. Figures[1] are unsatisfactory, but as all suffer from the same defects they can be used for comparative purposes. The following relate to England as a whole, except for 1663 when they are for London only: 27 per cent. should be added for the other ports in that year:

	Exports	Imports
1613	£2,487,435	£2,141,151
1622	£2,320,436	£2,619,315
1663	£2,022,812	£4,016,019
1700	£6,477,402	£5,970,175

The volume of trade during the century was affected by a variety of events. Wars between European powers, wars with the Dutch, general

[1] Quoted from E. Lipson, *Economic History of England* (1931), II, 189.

commercial rivalry, the increasing importance of America, the introduction of new commodities, and the legislative activity of the Government; all these played their part. In 1651 and 1660 came the Navigation Acts,[1] and these may have helped the increase of shipping. They also changed the direction of a section of English trade, for, while they strengthened relations with the plantations, the Levant, and Mediterranean Europe, "they caused loss of the Greenland, Eastland, Scottish, Irish and Guinea trades, the East India trade in spices, and the Norwegian trade in timber; the last to the Danes, most of the others to the Dutch".[2] Mr Ogg suggests that this may have been intentional policy, because "in the forty years preceding the Restoration...there was a serious diminution in the exports of woollen manufactures to the Hanse and Baltic towns; there was also a threat that the Dutch might drive us out of the Mediterranean trade, as they already, in effect, excluded us from the African trade and the spice trade of the East Indies". Faced with this situation the framers of the Act of 1660

surrendered the hemp, pitch, and tar of New England on which earlier legislators had set so much store; they ignored the pepper and cacao-nuts of Jamaica and the white flax of Virginia, for which a ready market might have been had at home, and they set themselves both to encourage and to monopolise commodities for the production of which it seemed that the empire had special advantages not shared by other colonial empires, namely tobacco, sugar, cotton, and dyeing woods. In return for these, English shippers would send out not cloth, but iron, tin, and leather manufactures, with provisions and re-exported wines—commodities for some of which we had practically no European market. By exchanging our manufactured goods (other than cloth) for sub-tropical raw material, much of which might be re-exported, a diversion would be created from the old continental trade conducted mainly from the ports on the eastern coast of England.[3]

This diversion "helped to develop such ports as Liverpool, Plymouth and Bristol and so relieved the congestion in London".

Government activity also affected trade in particular commodities or with particular countries. Exports of leather, which were important up to 1661, declined with the prohibition of the export of raw hides and tanned leather except as boots, shoes or slippers. Imports of French wines were prohibited from 1649 to 1656 in retaliation for the exclusion of English woollen goods from France. The embargo was then raised, but

[1] There were certain later supplementary Acts.
[2] D. Ogg, *England in the Reign of Charles II* (1934), I, 240.
[3] *Ibid.* pp. 241, 242.

duties were increased in 1664, and three years later imports were again prohibited. In this year the French tariffs virtually excluded manufactures from France and a long tariff war between France and England began. The French wine trade evaded the prohibition by sending its produce through Spain and Portugal, while the Portuguese took advantage of the situation to increase their production of red wine, much of which came to England. At the end of the century the average imports of wine were as follows:

From Spain and the Canaries	9039	tuns
„ Portugal	6897	„
„ Italy	1508	„
„ France	1245	„
„ Rhine lands	736	„

Among the new trades were those of tea and paper. Pepys mentioned the "new china drink" in 1651, but the tea trade did not develop until after 1667, when the East India Company ordered tea from Bantam. Thereafter, that Company had the virtual monopoly of the tea trade, and derived great financial profit from the increasing consumption of tea by the aristocracy of England. The paper industry dates from 1588 but was of small importance until after 1685. Even then little but brown paper was made, but after 1690, as a result of the tariff against France, white paper was manufactured, and its import declined enormously.

Many other items on the list of exports and imports were affected by restrictions or regulations. The lowering of the price of Barbados sugar deprived the Lisbon sugar dealers, who imported Brazilian sugar, of their market, and so increased imports of sugar from English plantations, while the tax of 1685 hit the refiners of sugar and trade in refined sugar went first to Spain and then to Holland. Prohibition of the export of Irish cloth and woollen manufactures to any place but England did much to crush a rival industry and to start a brisk smuggling business with France. The prohibition or high taxation of French silk goods after 1667, and the influx of foreigners after 1685, helped the silk industry, which received further protection in 1700 against imports from the Far East. This policy was not without its critics. N. B[arbon], in *A Discourse of Trade* published in 1690, remarked that "Flanders lace, French hats, gloves, silks, Westphalia bacon, etc. are prohibited because it is supposed they hinder the consumption of English lace, gloves, hats, silk, bacon, etc." But, he pointed out, the Englishman "may desire to eat Westphalia bacon when he will not English". Somewhat earlier (1669) Sir J. Child, in evidence before a Committee of the House of Lords appointed to consider the causes and grounds of the decay of trade, attributed losses to dishonest aulnage, dishonest packing of fish, the Statute of Bankruptcy, taxes on home manufactures, the export of coin, trade bye-laws, scarcity of labour, the Fire and Plague, and heavy land taxes.[1] Yet English trade prospered and was more important at the end of the century than it had been at the beginning.

It was universally recognised that the fishing industry[2] was one of the soundest foundations upon which the maritime expansion of England could be built, yet throughout the century efforts to stimulate it met with little success. James I and Charles I, both anxious to check the activities of the Dutch, failed to accomplish anything of value, and their successors did no more. Numerous Fishing Companies were proposed or promoted, but none survived long. Yarmouth, Scarborough and Bridlington were the only towns of importance engaged in the herring fishery; while Hull sent ships to the more distant Icelandic and Spitzbergen fisheries whence whale oil was obtained. But better methods of curing fish and better sales organisation enabled the Dutch to control the foreign market and even to sell their produce in this country. One matter of interest arising out of the claims of James I was the preparation of a map[3] in 1604 showing the "position and limits of the King's Chambers and ports and the sailing directions for the same".

THE POPULATION OF ENGLAND

The increase in the country's trade was in some measure a reflection of the increased population during the century. Similarly, the developments in agriculture and in industry which had taken place during the century were in part necessitated by, and in part the cause of, a growth in the population. Many estimates of the population of the country were made by contemporary writers, and to these must be added the calculations of later historians. It must be emphasised that all estimates are little more than guesses. For the population at the beginning of the century two foreign observers agree within 40,000 that it was about 3,500,000. It has been calculated, from the number of fighting men in the last quarter of the sixteenth century, that the population was 4,688,000. Thorold Rogers, from other sources, made an estimate of 2,500,000, but this is probably too low. Towards the end of the century, a number of estimates were made.

[1] *Eighth Report of the Royal Commission on Historical Manuscripts* (1881), p. 134.
[2] See T. W. Fulton, *The Sovereignty of the Seas* (1911).
[3] Reproduced in T. W. Fulton, *op. cit.* p. 121.

Petty, in 1682, estimated the population of London at 669,930, and believed that the population of England was eleven times that number. King believed that the population was 4,885,696 in 1600, and 5,500,520 at the Revolution; and Davenant, following up his calculations, made the total 6,596,073 in 1690. Petty made a second guess that the population in 1682 was only six millions. Davenant's figure was based on the details of houses paying tax in 1690, the number of which was 1,319,215. But whereas King allowed five persons per house Thorold Rogers believed four to be more nearly correct, and so made the total population five and a half millions. Macaulay quotes the calculation of an actuary named Finlaison who arrived at the total of "a little under five million two hundred thousand". The writer of *A Discourse on Trade*, in 1690, declared that the population was seven millions. Thus estimates vary from 5,200,000 to 8,039,160. Macaulay himself pronounced "with confidence" that "when James the Second reigned England contained between five million and five million five hundred thousand inhabitants".

Accepting either the lowest or the highest estimates, it is clear that a large increase in population took place. King calculated the excess of births over deaths at the rate of 20,000 a year, which certain exceptional causes of mortality reduced to a normal increase of 9000. According to other figures, the increase in the number of houses between 1685 and 1690 was about 20,000, or at the rate of about 4000 a year. This probably meant an increase in the population during that period of at least 8000 annually. Over and above the natural increase was that due to foreign immigration. A certain number of foreigners were brought to England in order to establish new industries or to settle on drained land. Others fled from religious persecution. Huguenots from France furnished the first large numbers under this latter category, but they were exceeded by the great migration which took place immediately before and after the Revocation of the Edict of Nantes in 1685. It is possible that about 80,000 Huguenots came to England during this great persecution, of whom at least 40,000 or 50,000 remained in the country.[1] In all it seems reasonable to assume that at least 80,000 foreigners migrated to England during the century, for, in addition to refugees, a number came whose chief business was trade.

---

[1] "The Protestants", wrote one of their pastors, "carried commerce with them into exile." Different forms of silk manufacture, linen manufacture, sailcloth making, glass making, paper making, and the production of felt and beaver hats, were all due to their industry. England gained considerably from their immigration. R. L. Poole, *The Huguenots of the Dispersion* (1880), p. 170.

On the other hand, there was a certain loss by migration, to Ireland, and to the colonies. Numerous small plantations, and the great plantation of Ulster in 1611, attracted a number of settlers. The later Cromwellian conquest and settlement of Ireland took at least 50,000 British-born settlers. The conquests of William III and the confiscations of land which followed introduced some additional population, although there was then no systematic attempt to colonise Ireland as had been made by Cromwell.

Meanwhile, the many persons who believed that England was over-populated turned with hope to the colonies in America. By 1650 it was estimated that there were 40,000 English colonists settled there; and by 1689 their number was approaching 250,000. Not all the settlers were of English birth, but, after making generous allowances for the exceptions, there can hardly have been less than 100,000 English settlers. To this number should be added those who had gone to the West Indies, probably quite as many as had settled on the mainland. Thus the total population lost to England was perhaps about 300,000, and the net loss, after allowing for immigration, cannot have been less than 200,000 during the century.

The distribution of the population followed closely the relative agricultural importance of the different parts of England. Thorold Rogers drew up a list of counties on the basis of seven different assessments made during the century.[1] A consideration of the first of these (1636) gives an indication of the distribution of the population in the early years of the century. From the map (Fig. 71) it will be seen that a belt of highly taxed country runs northward from London, and is enclosed within a wider belt of country extending from the Bristol Channel to the Wash. The most striking feature of the map is the comparative poverty of the five northern counties; even Yorkshire, which had much valuable agricultural land, was only a little richer than these areas.

It is impossible to get comparable details for the end of the century. The assessment for 1693 shows a number of fluctuations within the area of great density, but the northern counties still remain at the bottom in the list. There is reason to believe, however, that the population of northern England had increased more than Thorold Rogers admitted. He printed a list of houses as recorded in the Hearth Books for 1690[2] and from the calculations made by him a map (Fig. 72) has been drawn which probably represents, as nearly as any map can, the distribution of the population at that date. The predominance of the south remains, but Yorkshire now

[1] *Agriculture and Prices*, v, 118–19.
[2] *Agriculture and Prices*, v, 120. The figures were first printed by J. Houghton in *A Collection for the Improvement of Husbandry and Trade* (1861–3), I, 71–84.

29

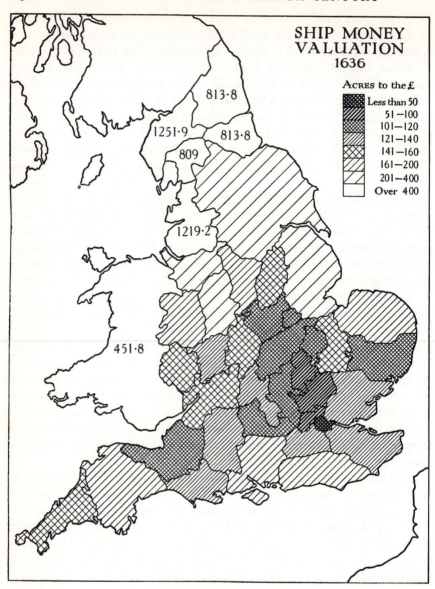

SHIP MONEY
VALUATION
1636

ACRES to the £

	Less than 50
	51—100
	101—120
	121—140
	141—160
	161—200
	201—400
	Over 400

813·8

1251·9   813·8

809

1219·2

451·8

Fig. 71

An approximate distribution of the population in 1636 may be inferred from this
map, which is based on the table in J. E. T. Rogers, *A History of Agriculture and
Prices in England, 1259–1793* (7 vols., 1866–1902), v, 104–5. Scotland is not included.

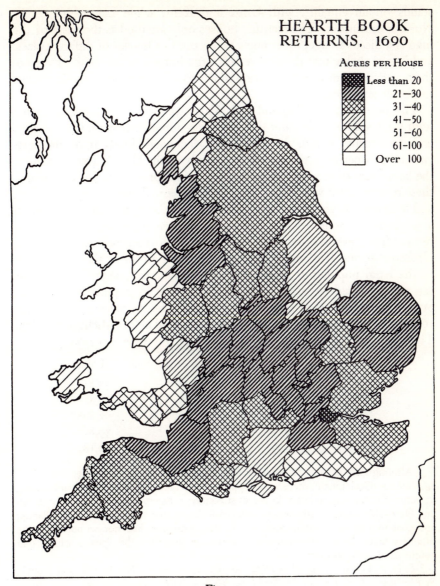

HEARTH BOOK
RETURNS, 1690

ACRES PER HOUSE

Less than 20
21—30
31—40
41—50
51—60
61—100
Over 100

Fig. 72

This map probably represents, more accurately than Fig. 71, the distribution of
population, although allowance must be made for hearths which paid no tax. The
information is based on the table in J. E. T. Rogers, *A History of Agriculture and
Prices in England, 1259–1793* (7 vols., 1866–1902), v, 120–21. Scotland is not
included. Compare with Fig. 83 on p. 524 below.

stands on an equality with Essex and Kent, while Lancashire equals Norfolk or Suffolk. The fact that houses only are used as the basis of the map is a disadvantage, for it takes no account of houses of different sizes, and it is probable that there were less large houses in the north. Even so, the map does no more than indicate what was a fact, the rise of the industrial midlands and north.

The population of England and Wales was still so closely associated with the produce of the land that it was predominantly rural in character; and the density, within the rural areas, depended closely upon the agricultural prosperity of those areas.[1] A study of the Hearth Tax Returns in the later years of the century makes this point quite clear. Unfortunately, these returns, even where they have been preserved intact, vary in quality, and it has not as yet been found possible to embody their information on a single map. Some returns have already been published.[2] A study of returns from several other counties, including Northumberland, Berkshire, Oxfordshire and Rutland, reveals very little difference in the distribution of the rural population about 1660 and that of, say, 1851. The returns can also be used to show the general increase in population. Thus in Surrey the return of 1670 recorded 990 newly built houses while that for 1674 recorded 356 hearths in houses not inhabited before Michaelmas 1673. The following table shows the returns for Surrey over a period of years:

	Hearths paid	Hearths exonerated	Total
1662	55,983	1,827	57,810
1670	53,904	13,948	67,852
1671	59,219	3,679	62,898
1673	56,857	2,993	59,850
1674	57,439	4,744	62,183

[1] Although the density of population in rural areas was primarily determined by agricultural development, there were supplementary means of employment influencing density—the woollen industry in the west country; mineral exploitation in Cornwall and Derbyshire; the worsted industry in Norfolk; the new industries of the midlands; and coal mining in Northumberland. As yet, industries had not withdrawn large numbers from the countryside: they had rather increased the density which that countryside could support.

[2] S. H. A. H[ervey], *Suffolk in 1674*, Suffolk Green Books, No. XI, vol. XIII (1905); Staffordshire, *Hist. Coll. of Staffordshire, William Salt Arch. Soc.*, 1921, 1923, 1927; Somerset, in E. Dwelly's *National Records* (ed. R. Holsworthy), vol. I (1916); Bedfordshire, *Publications of the Bedfordshire Historical Record Society* (ed. Lydia M. Marshall), XVI (1934). The last contains a valuable introduction covering far more ground than might be supposed from the title of the work, and should be consulted by all who use the Hearth Tax Returns.

From these figures it can be seen that returns for any year may be misleading, and the whole must be used with caution in plotting the distribution of population.

Macaulay wrote that "great as has been the change in the rural life of England since the Revolution, the change which has come to pass in the cities is still more amazing.... In the reign of Charles II no provincial town in the Kingdom contained thirty thousand inhabitants, and only four provincial towns contained as many as ten thousand inhabitants." The first of these was Bristol, which had gained from the growth of the Atlantic trade and which appears to have prospered as the result of the Navigation Acts. Macaulay put its numbers at 29,000: Petty had believed them to be very much larger. Norwich was about equal in size to Bristol: it derived its numbers from the industries of which it was the centre, and, according to Petty, held 1/300th of the people of England. In the three lists compiled by Rogers[1] for 1641 and 1649 the town of Norwich is always second after London, and this was probably its place until Bristol surpassed it towards the end of the century. York was considered to be the second town in the country in 1635, when Gussoni wrote his Report to Venice, and it appears third on the list of Rogers for 1641. But it failed to keep pace with other growing towns, and it is doubtful if it had as many as 10,000 persons in the reign of Charles II. Exeter, the fourth of the cities referred to by Macaulay, was, like Norwich, an important manufacturing centre. The other towns he named were Gloucester (4000–5000), Worcester (8000), Nottingham (8000), Derby (not quite 4000), Shrewsbury[2] (7000); Manchester (less than 6000), Leeds (less than 7000), Sheffield (less than 4000), Birmingham (less than 4000), and Liverpool (less than 4000). Most of Macaulay's figures were guesses. It is now known that the population of Birmingham rose from 5472 in 1650 to 15,032 in 1700; that Manchester had about 10,000 persons in 1700; while Liverpool had a population of about 6000: and the two last towns were growing rapidly. For the rest, it cannot be said that any estimates are very useful. Petty noticed that certain towns had increased in size—London, Newcastle, Yarmouth, Norwich, Exeter, Portsmouth and Cowes; in each case the growth of these is explained by industrial or commercial developments. Of Newcastle, Brereton wrote as early as 1635, "this is beyond all compare the fairest and richest town in England, inferior for wealth and building to no city save London and Bristow and whether it may be accounted as

[1] *Agriculture and Prices*, v, 120.
[2] Brereton referred to this as "a very great town", *op. cit.* p. 186.

wealthy as Bristow I make some doubt".[1] At the same time three Norfolk visitors found "the people and streets much alive, neither sweet nor cleane".[2] But few people seem to have noted the rise of the new industrial centres which were transferring the larger part of the urban population from the old towns to the growing cities of the midlands and the north. Cornwall had twenty-one towns which sent members to Parliament; Devonshire twelve; and Wiltshire sixteen. In contrast, Lancashire had only six (Lancaster, Clitheroe, Liverpool, Preston, Wigan and Newton), Durham one, Northumberland three (Newcastle, Berwick and Morpeth), and although Yorkshire had fifteen, such towns as Leeds, Halifax and Sheffield were unrepresented. In the midlands, no members were returned by Burton, Wolverhampton, Dudley and Birmingham. The country had to wait until 1832 before the balance of urban representation was even partially redressed. In 1696 King calculated that all the cities and market towns together had a population of 870,000 and, with his estimate for London of 530,000, this would make the urban population about one-quarter of the total.

London stood in a class by itself. Petty believed that one-eleventh of the population of England lived within the capital, and that it contained 669,930 persons in 1682; Graunt, in 1676, thought the number was 384,000; Gibson, in 1695, gave the figure as 700,000; while King, in 1696, put it at 530,000. Accepting the last as a reasonable figure, it would appear that the capital doubled in numbers during the century. As early as 1603, London and Westminster were united by houses; and, two years later, an Act of Parliament was required to augment the water supply of the city. Shortly after this, attempts were made to regulate the number and the height of new buildings. Hackney coaches increased so rapidly that an Order in Council was issued in 1635 to check their number: it seems to have failed in its object, and, in 1654, an Act of Parliament limited the number of such coaches to 300. This Act was followed by another, in 1661, which provided for the paving and widening of streets and the licensing of 400 coaches. The fire of 1666 gave an opportunity to rebuild much of London, and the City soon recovered from the disaster and continued to increase in size. It was the centre of many industries, the leading port of the country, and incomparably the greatest city.[3]

The seventeenth century ended with England engaged in war with the greatest European power of the time. That she was to emerge successfully

---

[1] Brereton, *op. cit.* p. 85.
[2] *A relation of a short survey of 26 Counties...*, Newcastle Tracts (1904), VII, 19.
[3] For further details of the growth of London see below, chap. XIV.

was no doubt due, in part, to factors quite unconnected with geography, and in particular to the development and organisation of finance. But her resources were considerable, and she was making increasing use of them. It may safely be said that the growing recognition of the possibilities of England's geographical advantages, stressed by all observers, and outlined in the foregoing pages, materially helped Englishmen to make England what she soon became, "the acknowledged head of European civilisation on its political side".[1] The century had been one of gradual change; the kingdom was, to use Defoe's phrase, "daily altering its countenance"; and its people were, in more senses than one, looking out upon a new world.

BIBLIOGRAPHICAL NOTE

A full bibliography covering this period will be found in the *Bibliography of British History, Stuart Period, 1603–1714*, edited by Godfrey Davies (1928). This can be supplemented by *Agricultural Writers from Sir Walter of Henley to Arthur Young, 1200–1800*, by D. MacDonald (1908). I have consulted both these books freely. In addition to the sources mentioned in the footnotes the following have also been used:

G. N. CLARK, *The Later Stuarts, 1660–1714* (1934).

M. COATE, *Cornwall in the Great Civil War and Interregnum, 1642–1660* (1933).

E. G. R. TAYLOR, *Late Tudor and Early Stuart Geography, 1583–1650* (1934).

The most convenient summary of the cartographical work of the century will be found in Sir George H. Fordham's *Some Notable Surveyors and Map Makers of the Sixteenth, Seventeenth, and Eighteenth Centuries and their Work* (1929).

[1] G. M. Trevelyan, *The England of Queen Anne* (1932), p. 177. In writing this chapter use has also been made of the same writer's *England under the Stuarts* (8th ed. 1919).

Chapter XII

THE DRAINING OF THE FENS, A.D. 1600–1800

H. C. DARBY, Litt.D.[1]

During the Middle Ages the drainage of the 1300 square miles of the fenland had remained largely a matter for local concern. When necessity
arose, owing to the ravages of the sea or to the overflowing of the watercourses, the Crown granted a commission to remedy the evil. A succession
of Commissions of Sewers combined with the *consuetudo loci* to maintain
a local habit of life quite different from anything else on the English plain.
The upkeep of any single channel involved many interlocking interests,
and the dissolution of the monasteries in 1539 served only to increase the
confusion of divided responsibilities. But, as Samuel Hartlib wrote, "in
Queen Elizabeth's dayes, Ingenuities, curiosities and Good Husbandry
began to take place". The time was becoming ripe for a "greate designe"
in Fenland. During the later years of the sixteenth century various
schemes and experiments prepared the way.[2] At last, in 1600, there was
passed "An Act for the recovering of many hundred thousand Acres of
Marshes...". Of the many stretches of marsh in the kingdom, that of the
great Fenland itself promised the most spectacular transformation.

But there were difficulties. The fenmen, living largely upon the perquisites of the commons, resented any suggestion of interference with
their traditional livelihood. Capital was needed for the extensive operations of draining. Many people were totally opposed to any improvement
which necessitated the expenditure of money or of energy. Consequently,
one of the most typical figures of the time became the "undertaker". By
the terms of the Act of 1600, lords of manors, and a majority of the
commoners in any fen, together with the owners of any "several fen
ground" lying near, might bargain and contract part of such fens to any
person or persons who would undertake their "draining and keeping dry
perpetually". In other words, it was recognised that outside financial
assistance, upon the security of part of the reclaimed lands, had to be
procured before the Fens could be improved.

[1] I am indebted to Major Gordon Fowler and to Mr C. F. Tebbutt for reading
through the proof-sheets of this chapter and for making suggestions.
[2] See pp. 366–7 above.

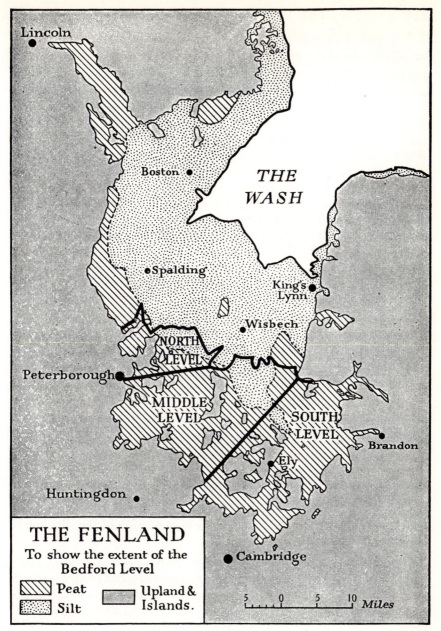

Fig. 73

S. B. J. Skertchly, in *The Geology of the Fenland* (1877), p. 129, noted that the precise boundaries of the peat and silt were "very obscure, for the peat thins out insensibly along its borders,..." The limits of the Bedford Level are taken from Samuel Wells' map of 1829 on a scale of $1\frac{1}{2}$ miles = 1 inch.

A picture of the region is given in Camden's *Britannia* which describes the inhabitants of the peat fens of Cambridgeshire as

a kind of people according to the nature of the place where they dwell rude, uncivill, and envious to all others whom they call *Upland-men*: who stalking on high upon stilts, apply their mindes, to grasing, fishing and fowling. The whole Region it selfe, which in winter season and sometimes most part of the yeere is overflowed by the spreading waters of the rivers *Ouse, Grant, Nen, Welland, Glene,* and *Witham,* having not loades and sewers large enough to voide away: But againe when their Streames are retired within their owne Channels, it is so plenteous and ranke of a certaine fatte grosse and full hey, (which they call *Lid*) that when they have mowen downe as much with the bettes as will serve their turnes, they set fire on the rest and burne it in November, that it may come up again in great abundance. At which time a man may see this Fennish and moist Tract on a light flaming fire all over every way, and wonder thereat. Great plenty it hath besides of Turfe and Sedge for the maintenance of fire: of reed also to thatch their Houses, yea and of Alders, besides other watery Shrubbes.[1]

Huntingdonshire likewise was not free from the "unwholesome aire of the Fennes"; and, amongst other stretches of water, it contained that "fishfull Mere named Whittlesmere".[2] To the north, in the silt zone of Lincolnshire, Camden described wide open flats left uncovered by the ebbing tides; the soil itself was soft, completely without stone, yet splendid churches testified to the wealth of the country, not in corn, but in grass and fish and fowl. Here, people lived in perpetual threat of

a mighty confluence of waters from out of the higher countries, in such sort that all the Winter quarter the people of the country are faine to keepe watch and ward continually, and hardly with all the bankes and dammes that they make against the waters, are able to defend themselves from the great violence and outrage thereof.[3]

Like conditions were to be found in the continuation of the silt zone in Norfolk marshland:

Over against *Linne,* on the farther side of the River lieth *Mershland,* a little moist mersh country, as the name implieth, divided and parted every where with ditches, trenches and furrowes to draine and draw the waters away: a soile standing upon a very rich and fertile mould, and breeding abundance of cattell:

---

[1] *Britannia,* 1637 edition, p. 491.   [2] *Ibid.* p. 500.
[3] *Ibid.* p. 529. In Lincolnshire, too, there were "foule and flabby quauemires, yea and most troublesome Fennes, which the very Inhabitants themselves for all their stilts cannot stalke through". (Pp. 530–1.)

in so much as that in a place commonly called *Tilney Smeth* there feed much about 30000 sheepe: but so subject to the beating, and overflowing of the roaring maine Sea, which very often meaketh, teareth, and troubleth it so grievously, that hardly it can be holden off with chargeable wals and workes.[1]

During the early years of the seventeenth century there were many schemes afoot. At the accession of James I, in 1603, the Crown held several estates in the region, and the king himself showed a very lively interest in the possibilities of draining. Many proposals had been discussed in the previous reign, but now surveys were made and projects examined. In 1604, on July 11th, the king directed letters[2] to the commissioners of sewers for the Isle of Ely and the adjoining counties, encouraging their proceedings and asking them to certify the names of those who would be willing to give up part of their surrounded lands to such as would drain them. Events promised to move quickly.[3] During the years immediately following, there were numerous petitions to the king, and many letters from the Privy Council dealing with the condition of various channels and drains. Much debate about ways and means took place, but nothing effective was done; general dissatisfaction was felt everywhere.

At length, King James "for the honour of his kingdom would not longer suffer these countries to be abandoned to the will of the waters, nor let them. lie waste and unprofitable"; in 1620–1 he declared that he himself would undertake the work in return for 120,000 acres of the marsh. The Dutch engineer Vermuyden, who reported this,[4] was invited to consider the matter, but "by reason of their other great and more important concerns", and owing to the king's death in 1625, nothing was done.

Charles I was inclined at first to leave the reclamation to his subjects. There were various proposals in 1626;[5] and, in 1628, a Session of Sewers at Huntingdon suggested a levy of six shillings per acre on all the marsh as a contribution to a general drainage. Nothing came of it, and, in the following year, the king instructed the commissioners "to receive of the Undertakers fair propositions for draining the fens".[6] The entries in the State Papers relating to the "fen business" grew more and more frequent. The commissioners, meeting at King's Lynn, proposed a contract with Vermuyden;[7] but this had to be abandoned owing to local opposition.

---

[1] *Britannia*, 1673 edition, p. 481.

[2] State Papers, Domestic, Jas. I, VIII, 99 (referred to as S.P.).

[3] *Ibid.* XVIII, 100–4, which contain numerous arguments and petitions. See also XXI, 11, and XXVI, 37.

[4] *A Discourse touching the drayning the great Fennes*...(1642). See S.P., Jas. I, XVIII, 101.   [5] S.P., Chas. I, XXXII, 45.   [6] *Ibid.* CXLIV, 83.   [7] *Ibid.* CLXXIII, 29.

From the reports and surveys of the time, however, it is evident that the main rivers had deteriorated to such an extent that adequate repairs by individuals were out of the question. Many people began to look upon the draining as being really feasible, and, accordingly, they approached Francis, 4th Earl of Bedford, the owner of 20,000 acres near Thorney and Whittlesey, who contracted within six years to make "good summer land"[1] all that expanse of peat situated in the southern Fenland, later known as the Bedford Level. An agreement was drawn up in 1630. In the following year, thirteen Co-Adventurers[2] associated themselves with the earl; and in 1634 they were granted a charter of incorporation by which

it was hoped in those places which lately presented nothing to the eyes of the beholder but great waters and a few reeds scattered here and there, under Divine mercy might be seen pleasant pastures of cattle and kyne and many houses belonging to the inhabitants.

To assist in the enterprise the earl secured the services of the despised Vermuyden who had already been at work upon the reclamation of the Axholme marshes. Although the fenmen had preferred the Earl of Bedford to a foreigner, they had been by no means unanimous in approaching him. Consequently, the engagement of Vermuyden by the earl produced complications. But, still, the work proceeded. The discussions about the methods of draining ranged around two main opinions: one, held by Vermuyden, advocated that the courses of the main rivers be shortened to increase their gradient; the other urged that existing waterways should be deepened and embanked and kept clean. The triumph of Vermuyden meant that, wherever was expedient, new straight cuts replaced the older channels. Cuts, drains and sluices were consequently made. The chief work was the Old Bedford River extending from Earith to Salter's Lode, 70 feet wide and 21 miles in length.[3]

In 1637, at a Session of Sewers held at St Ives, the Level was adjudged to have been drained according to the true intent of the agreement of 1630. But complaints and petitions were showered upon the Privy Council, and royal feeling turned against the Corporation. In the following year, the award was set aside; it was declared that, as the Level still remained subject to inundation, the undertakers had not fulfilled their contract. A General

[1] I.e. land free from floods in summer.
[2] So called because they "adventured" their capital. Sometimes the "adventurer" was also the "undertaker".
[3] See W. Dugdale, *History of Imbanking and Drayning* (1662), p. 410, for list of cuts made at this time.

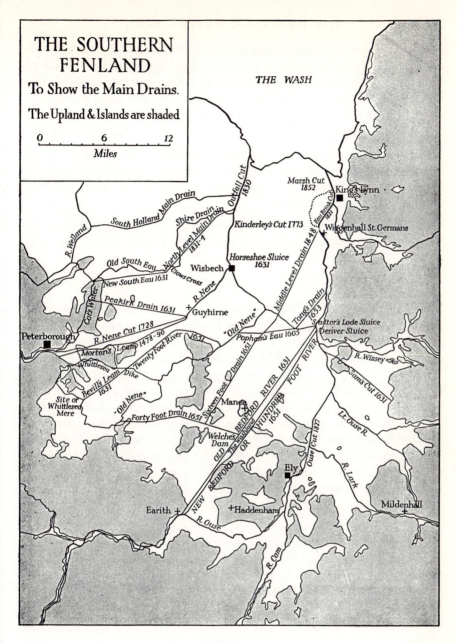

THE SOUTHERN
FENLAND

To Show the Main Drains.

The Upland & Islands are shaded

0   6   12
Miles

THE WASH

Marsh Cut 1852

King's Lynn

Outfall Cut 1830

South Holland Main Drain

Shire Drain

North Level Main Drain 1831-4

Kinderley's Cut 1773

Eaubrink Cut 1821

Wiggenhall St. Germans

R. Welland

Old South Eau

New South Eau 1631

Clows Cross

Horseshoe Sluice 1631

Wisbech

Middle Level Drain 1848

Peakirk Drain 1631

Cat's Water

R. Nene

Tong's Drain 1653

Guyhirne

"Old Nene"

Salter's Lode Sluice

Denver Sluice

Peterborough

R. Nene Cut 1728

Morton's Leam 1478-90

1631

Pophams Eau 1605

R. Wissey

Whittlesea

Twenty Foot River

HUNDRED FOOT RIVER

Sam's Cut 1631

Bevill's Leam 1631

Dike

Sixteen Foot Drain 1651

BEDFORD RIVER 1651

Lt. Ouse R.

Site of Whittlesea Mere

"Old Nene"

Forty Foot Drain 1651

Manea

OLD the Washland

Ouse Cut 1827

R. Lark

Welches Dam

OR

Ely

Mildenhall

Earith

NEW BEDFORD

+Haddenham

R. Ouse

R. Cam

Fig. 74

The approximate dates of the main drains are given; in some cases there was an appreciable interval between the start of a project and its completion.

Session of Sewers at Huntingdon, on April 12th, judged the earl's under-
taking to be defective; and, on July 18th, the king himself was declared
undertaker, to make the Fens "winter as well as summer lands".[1] The
services of Vermuyden as engineer were retained, and some works were
started. In 1638 he presented a scheme for the guidance of the king. This,
which was not published until 1642, declared that His Majesty had seen

that these lands being a continent of about 400,000 acres, which being made
winter ground, would be an unexpected benefit to the Commonwealth of six
hundred thousand pounds per annum and upwards, and a great and certain
revenue to all the parties interested.

Vermuyden's suggestions did not go unchallenged. Soon, however, the
fen difficulties were overshadowed by greater difficulties. The country was
at war within itself.

During the Civil War, the draining of the Fenland was in abeyance. The
entries relating to drainage in the State Papers grow less and less frequent,
until at last they disappear. Those works of the drainers that were not
destroyed were neglected and became decayed. As the Fens lay along the
frontier of the Associated Counties, there was much passage of troops
through them; and many rumours spread in and around the district. But,
during these years of alarms and excursions, the project of draining had
not been forgotten. After many committees and sub-committees, an "Act
for the draining the Great Level of the Fens..." was passed on 29th
May 1649; and the 5th Earl of Bedford and his associates were "declared
to be the undertakers of the said work". The geographical consequences
and wider economic bearings of the proposed changes were stated at once
in the preamble to the Act. The fen was to be drained and made

fit to bear coleseed and rapeseed in great abundance, which is of singular use to
make soap and oils within this nation, to the advancement of the trade of clothing
and spinning of wool, and much of it will be improved into good pasture for
feeding and breeding of cattle, and of tillage to be sown with corn and grain,
and for hemp and flax in great quantity, for making all sorts of linen cloth and
cordage for shipping within this nation; which will increase manufactures,
commerce, and trading at home and abroad, will relieve the poor by setting

[1] I.e. land free from floods in winter. The motives behind these proceedings are
obscure and probably not unmixed. Sir John Maynard, in *The Picklock of the Old Fenne
Project* (1650), declared that the king had been informed privately of "all the cheats of
the Undertakers; how, in particular, they had claimed land which was not drowned".

them on work, and will many other ways redound to the great advantage and strengthening of the nation.

Even so, there were some who were not convinced. Many petitions had been delivered during the discussions preceding the Bill, and it was among doubts and fears that the second drainage Act was launched.

In his *Discourse* of 1642, Vermuyden had divided the Great Level into three areas:

1. The one from Glean to Morton's Leame.
2. From Morton's Leame to Bedford River.
3. From Bedford River southwards, being the remainder of the level.

These became the North, the Middle and the South Levels respectively.[1] Work upon the North and Middle Levels was started first. The works of the earlier drainers were restored; banks were made, sluices built, and channels secured. In particular, the New Bedford River was cut, running parallel to the Old Bedford River.[2] On 26th March 1651[3] the first warrant of adjudication was given at Peterborough.

Work upon the South Level was then started. The original hostility to a general draining still continued, and complaints again became frequent in the State Papers. In some cases, the works of the drainers were thrown down, and had to be guarded with armed force. Nor was it easy to discover the offenders; "the business is so much in the dark, and so subtly and cunningly carried out by the country, that no considerable discovery"[4] could be made. The "actors and abettors" remained at liberty to plot further destruction. As a result of their commotions, there were difficulties in collecting the taxes imposed by the Adventurers upon the drained land; and the wages of the labourers frequently remained unpaid. Petition after petition was presented by the disappointed parties to the Commissioners of Adjudication, and to Protector Cromwell;[5] and the Adventurers became involved in great legal expenses. The labour situation was saved by the use of Scotch prisoners after Dunbar;[6] and, later, Blake's victory over

---

[1] Later, however, the North Level did not extend beyond the Welland.

[2] The New Bedford River was alternatively known as the Hundred Foot River, and, for the greater part of its course, ran half a mile to the east of the older cut.

[3] Confusion is caused in the dates by the fact that the year commenced on 25th March.

[4] S.P. XL, 19 (5th Sept., 1653).

[5] S.P. 1654, LXXI, 30. Col. Simon Rugeley wrote to the Protector: "As I have sold 500*l* a year to pay my debts, and as a last refuge mortgaged my debentures, I beg an order to put me in possession of the lands."

[6] S.P. 1651, XVI, contains many references to the arrangements for these.

Van Tromp in 1652 supplied the Government with a large number of Dutch prisoners,[1] some of whom were at once sent to the Fens to work as labourers.

At length, an application for an adjudication was made, and the lords commissioners arrived in the Fens in March 1652. Having viewed some of the works, they assembled at Ely where Vermuyden explained his "designe", and reported the progress that had been made in the two levels already adjudged to be winter grounds.

They are so far improved, that there are about 40,000 acres at this time sowne with cole seede, wheate, and other winter graine, besides innumerable quantityes of sheepe, cattle, and other stocke, where never had byn any before, and not surrounded since the adjudication, being now 2 yeares past.[2]

Successive changes in the administration of the realm witnessed the completion of the machinery for preserving the works of the drainers. In 1657, came "An Ordinance for the Preservation of the Works of the Great Level of the Fens"; and, after temporary Maintenance Acts in 1660 and 1661, there was passed ultimately the General Drainage Act of 1663.

In the northern Fenland no such transformation was effected. There, in Lincolnshire, the commissioners of sewers continued their vigilance. As early in the century as 1602 there were discussions "directing the cleansing and repairing drains in the fen ground lying along the river Witham".[3] The Domestic State Papers[4] record various grants and arrangements that were made for the draining of individual stretches of fen. But, taken together,[5] this effort in the north was sporadic in character, small in achievement, and temporary in result. Nor did the years following upon the Civil War produce any change. For the most part, the northern Fens "were wasted in the late times, and lie overflowed".[6] There were various negotiations concerning the fens east of the Witham (East, West and

---

[1] S.P. 1653, XXXIV, 57; and XXXVII, 78.

[2] Printed in S. Wells, *History of the...Bedford Level* (1830), I, 269–70. The term "surrounded" was used as equivalent to "drowned" or flooded.

[3] S.P., Jas. I, CCLXXXIII, 58 (1602).

[4] Among those of interest are: S.P., Chas. I, CCLXXXV, 55. Notes of the acreage and other particulars of the fens from Kyme to Bourn (1635); CCCVII, 28. Survey of the West Fens commons (1635); CCCCVII, 41. Map of Peterborough Little Fen alias Fleg Fen (1638); CCCCVIII, 74. Plan of lands in or near Stickney, Revesby and Hagnaby (1638).

[5] Dugdale (1662), pp. 208, 418, 423. See S.P., Chas. I, CCCCXVI (1639), "Papers relating to the Drainage of the Fens in Lincolnshire".

[6] S.P., Chas. II, XXIV, 116.

Wildmore Fens), and efforts were made by some of the original participants to renew a decree of adjudication that had been passed before the
Civil War.[1] But these and other agitations reaped but little success.
Generally speaking, the fen-commoners of Lincolnshire remained in
possession of their ancient privileges.[2]

THE CONSEQUENCES OF THE DRAINING

At first, great success followed upon the works of the drainers in the
Bedford Level. Cultivation was introduced on lands which, as far as
record went, had never before known a plough; and the faith of the
speculators in the economic possibilities of the late fens seemed to be
justified in all ways. Thomas Fuller, writing in 1655, was loud in his
praises. As he said, "the best argument to prove that a thing may be
done, is actually to do it".

The chiefest complaint I hear of is this, that the country thereabout is now
subject to a new drowning, even to a deluge and inundation of plenty, all
commodities being grown so cheap therein.

Not only was the soil improved but the atmosphere too; and, through this
country of plenty,

the river Ouse, formerly lazily loitering in its idle intercourse with other rivers,
is now sent the nearest way (through a passage cut by admirable art) to do its
errand to the German Ocean.[3]

---

[1] See W. H. Wheeler, *History of the Fens of South Lincolnshire* (2nd ed. 1896),
passim.

[2] Conditions in the northern Fenland during the closing years of the century were
described by Christopher Merret, Surveyor of the Port of Boston, in "An Account of
Several Observables in Lincolnshire, not taken notice of in Camden, or any other
Author", *Philosophical Transactions*, XIX, 343 (1696).
In the south of Lincolnshire, particularly between Spalding and Crowland, there
grew large crops of oats, rape-seed, cole-seed, flax and hemp. To the north, in the fen
itself he noted the "great plenty and variety of fowl and fish, particularly duck, mallard
and teal, which are usually taken in decoys and sent to London". On the pastures
between fen and sea, were cows and a "great number of fat oxen and sheep, which
are weekly sent to London in droves". Finally, along the coast there were salt marshes
continuous with those of Norfolk.

[3] *History of the University of Cambridge*, Section V. Fuller commented at length upon
the benefits of the new regime. For further eulogy, in 1685, see also Sir Jonas Moore,
*History or narrative of the great level of the fens called Bedford Level: with a large map
of the said level as drained, surveyed and described.*

30

The condition of the region after these changes was described in some detail by Sir William Dugdale, whose diary, "Things observable in one itinerie beginning from London 19 May 1657", is a record of channels and sluices and banks;[1] but here and there are references to the economic consequences of the new order. On the western side of the island of Ely, not far from Willingham, onions, peas and hemp were being grown in the fen. On the other side, between Ely and Southery, Dugdale found flax, hemp, oats, wheat, cole-seed and woad. At Whittlesey,[2] there were rich meadows and corn land, together with plantations of fruit trees and willows and vegetables. Thorney, he wrote, "was bordered with Fenns, but the Fens now environing it are by the Adventurers draynings, made so drye, that there are of all sorts of corne and grasse, now growing thereon, the greatest plenty imaginable". To the north, Deeping Fen was under water; and, as Dugdale travelled through the Lincolnshire fens, his descriptions of improvements ceased.

The margin of these pages bears a note "Will Dugdale's Diary in Order for his hystory of the Fenne drayning"; and, in 1662, there appeared *The History of Imbanking and Draining of divers Fens and Marshes both in Foreign Parts and in the Kingdom, and of the Improvements thereby. Extracted from Records, Manuscripts, and other Authentic Testimonies.* Dugdale's maps of the level undrained and drained stand in marked contrast. In the Preface, he gave a summary of the benefits of the recent changes;[3] and the point he emphasised was that, in addition to the new advantages of the drained fen in cattle and crops, there still remained the former benefits of the undrained fen in fish and fowl. So rich was the land, wrote John Evelyn, "that weeds grew on the banks, almost as high as a man and horse".[4] But, alas, despite all these improvements, the level still remained pestered with "swarms of gnats". And the impressions of Samuel Pepys a few years earlier had not been very cheerful. Visiting his poor relations near Wisbech he passed, with the eye of a visitor, "along dikes, where sometimes we were ready to have our horses

---

[1] Lansdowne MSS. 722.

[2] Hartlib, too, recorded the abundance of willows around Whittlesey "which in those vast and vacant grounds being alwaies very moist doth soon produce an incredible profit and increase of fire-wood and timber for many Country uses". Ashmole likewise mentioned a large plantation at Whittlesey (Ashmolean MSS. 784, Bodleian, Oxford).

[3] The valuation given in S.P., Chas. II, XXIII, 92, 1, is interesting. The 10,000 acres allotted to the king in the Great Level "would let for £3,781. 11s. 10d. a year; but the draining expenses would be £1,000 a year".

[4] John Evelyn, *Diary*, 22nd July, 1670.

sink to the belly", and "over most sad Fenns, all the way observing the sad life which the people of the place (which if they be born there, they do call the Breedlings of the place) do live, sometimes rowing from one spot to another, and then wadeing".[1] Although the enthusiasm of the drainers was, to a large degree, justified, we must remember that much of the Fenland was still covered in autumn by miry pools and not golden harvest.

The crops grown on the reclaimed ground were numerous and varied—wheat, oats, barley, turnips, beans, peas, cole-seed, chicory, mustard and woad. In the parishes along the coastal silt belt, there were two kinds of land and two kinds of management. The newer marshlands defended by walls along the edge of the sea were chiefly arable with some pasture; on the higher ground inland, where the villages stood, were luxuriant grazing lands. Behind the coastal silt belt lay the peat fen. Here, the process of paring and burning, introduced in the middle of the seventeenth century, remained the most important feature of cultivation for 150 years. In 1791 Arthur Young reported that in the fen district

they could not cultivate without this capital assistant. It is scarcely possible, profitably, to bring boggy, mossy, peat soils, from a state of nature into cultivation, without the assistance of fire. . . . In these fens, the original surface is rough and unequal, from great tufts of rushes, etc. called hassocks. . . . After this they go over it again with a very complete and very effective tool called a fen-paring plough, the furrow of which is burnt.[2]

When burning was in progress, clouds of white smoke, with an acrid bituminous smell, rose above the expanse of peat from small heaps of turf some three or four feet in diameter. The rotation that followed this treatment was: 1st, cole-seed; 2nd, oats; 3rd, oats; 4th, wheat; followed by several years of seeds laid down for grass. There were many objections to the practice of paring and burning, the chief being that it reduced the soil greatly and helped to lower the level of the ground. But this view was very much disputed, and improper management and bad rotations were held to be responsible for most of the other objections. Arthur Young, in his survey of Lincolnshire agriculture in 1799, considered that the

---

[1] Samuel Pepys, *Diary*, 18th September, 1663. He arrived at Parson's Drove to find his "uncle and aunt Perkins, and their daughters, poor wretches! in a sad, poor thatched cottage, like a poor barn, or stable, peeling of hemp, in which I did give myself good content to see their manner of preparing of hemp"—(Sept. 17th).

[2] A. Young, *General View of the Agriculture of the County of Suffolk* (1794), p. 30.

practice had resulted in "a great and permanent improvement".[1] Not until early in the nineteenth century were the practices of deep ploughing and claying the land undertaken.[2]

But, despite many praises, it was soon evident that all was not well in the Bedford Level. The immediate consequences of the draining promised more than was fulfilled for many years. Some of the complaints of the time were summarised in the General Drainage Act of 1663: that the draining of one place had made worse the lands in other places; that there were many discrepancies in the allotments of land between owners, commoners and townships; and that the future maintenance of the Level involved a clash with the interests responsible for navigation. But, as the century drew to a close, these legal and administrative difficulties were being overshadowed by natural difficulties which brought the very success of the drainers near to disaster. Without some idea of the nature and magnitude of these physical difficulties, it is impossible to understand the rapid deterioration of every scheme of drainage, and to appreciate the controversy that has raged round the Bedford Level up to the present day.

The principal feature of any river is its natural gradient or "fall"; together with the volume of the water, this controls the speed of the current and its power of scouring out a river bed. Normally, the natural gradient is sufficient to ensure a satisfactory outfall; the force of the current can grind a bed for the waters, and so keep the channel clear. But conditions were far from normal in the Fenland. For forty miles from the mouths of the rivers the gradient was very slight, and the current and its resulting "scour" were sluggish. It was able to maintain a channel of only limited capacity. In winter, therefore, very high water-levels were inevitable in the Fenland rivers. Moreover, it may not be irrelevant to note the postulated subsidence of the land in historical times.[3] Long before the project of general reclamation had been mooted, the decay of the river outfalls

---

[1] A. Young, *Agriculture of the County of Lincoln* (1799), p. 247. Of "the supposed mischief of paring and burning", he noted: "Many objections have been made to this practice in the fens, particularly that it reduces the soil greatly, visible in the sinking of drained lands that have been pared".

[2] That is to say, not until the surface layer of soil had thinned so much that the underlying clay was easily accessible. See what follows below on the shrinkage and wastage of peat.

[3] On the basis of evidence from the Thames estuary, the rate of subsidence seems to be 9 inches per 100 years. See T. E. Longfield, *The Subsidence of London*, Ordnance Survey Professional Paper, new series, no. 14 (1933). See also (1) the discussion in *Abstracts of the Proceedings of the Geological Society of London*, no. 1260,

was known to be the chief reason for the flooding of the Fens. A serious criticism of Vermuyden's scheme, made by many people, was that it stopped short at Salter's Lode and Guyhirne, and did not make adequate provision for the river outfalls.[1] While the drains were in their first state of perfection, the increased speed of the waters passing down to Lynn and Wisbech did improve the estuaries of the Ouse and the Nene. But, in a disturbingly short time, a rapid deterioration was noticed; and during the eighteenth century there were continual disputes about these outfalls.[2]

Underlying these problems, there were yet other difficulties of a more fundamental nature. Neither Vermuyden nor any of the other drainage engineers foresaw, when they planned and executed their schemes, that, as soon as they started to drain the Fens, the surface of the peat would become rapidly lower in level.[3] This lowering was due, in part, to the shrinkage of the peat as it dried, but also to the wasting away of the drying surface owing to bacterial action. Indeed, the better the drainage, the more rapid the wastage.[4] The effect of this disconcerting characteristic was that, soon after the completion of the reclamation, many of the main

p. 73 (March 3rd, 1933); (2) H. H. Swinnerton, "The Post-Glacial Deposits of the Lincolnshire Coast", Quart. Journ. Geol. Soc. LXXXVII, 360 (1931). Many other papers, too, discuss the evidence for different parts of the coast. See above pp. 61–2 and 94.

[1] But see footnote 3 below.

[2] See H. C. Darby, "Windmill Drainage in the Bedford Level", Official Circular No. 125, Brit. Waterworks Assoc. (1935); also in The Engineer, CLX, 75 (1935).

[3] The earlier critics of Vermuyden's schemes, apparently, neglected this point in their arguments. S. B. J. Skertchly, however, in The Geology of the Fenland (1877), showed a clear appreciation of the problem: "The neglect of the outfalls was doubtless a vera causa in respect to the rivers, but there is direct evidence to show that the drains were lowered at the same time that the river waters maintained or increased their level. In fact, from the moment the drainage works commenced the land began to subside, owing to the abstraction of water from the porous soil. Here, then, we have a natural and competent explanation of the phenomenon in question, yet it is a remarkable fact that until within the last few years no authority mentions the possibility of the subsidence of the land" (pp. 50–1). See ibid. pp. 42, 49 and 154–7. See also S. H. Miller and S. B. J. Skertchly, The Fenland Past and Present (1878), p. 160; "From the time the desiccation of the fens commenced, the drainage began to deteriorate in consequence primarily of the shrinkage of the peat which reduced the slope of the ground; and, secondarily, of the silting up of the outfalls of the rivers...." In this connection it is interesting to note what people said of the evil effects of paring and burning. See footnote 1, p. 456 above.

[4] S. B. J. Skertchly, op. cit. pp. 155–6, provides some interesting statistics to show the "Compression of Peat by Drainage". See also Gordon Fowler, "Shrinkage of the Peat-covered Fenlands", Geog. Journ. LXXXI, 149 (1933).

channels flowed at a level higher than that of the adjoining surface. This extraordinary paradox can still be seen along most of the fen rivers, where the heights of the barrier banks are far greater from the outside than from the beds of the rivers. Even the minor drains soon flowed at a lower level than the channels into which they tried to discharge! Colonel Dodson recorded in 1664 that the silt region of Norfolk Marshland, north Wisbech, and Lincolnshire Holland was about "five or six foot" lower than the peat behind.[1] It is difficult to interpret this statement because other evidence indicates that the undrained peat surface must have lain at about 10 or 12 feet above Ordnance Datum, i.e. at about the same height as much of the silt lands.[2] Whatever be the answer, the fact remains that today much of the peat surface lies at a foot above to a foot below Ordnance Datum. The heights of the river banks have had, therefore, to be increased.

Conditions were aggravated during years of great rain. In 1673 upland waters caused heavy floods

betwixt Spalding, Wisbech, Crowland, and the Isle of Ely. Many cattle are drowned, stacks of hay and grain are swimming or standing a yard deep in water, all the cattle are driven to small banks, the poor people's houses are full of water, and they are forced to save themselves in boats. All their cole-seed lost, and all they have besides, in and about Thorney fen, where were many farmhouses.[3]

These troubles are reflected in the Proceedings and the Order Book of the Conservators of the Level. The record of their London meetings began in 1663, that of their Ely meetings in 1665; and soon both were full of complaints from the owners of drowned lands, and full, too, of references to the rebuilding of broken banks, to the opening of blocked sewers, to

[1] Printed in S. Wells, *Bedford Level*, II, 454. Dugdale, *op. cit.* (1662) p. 179, also referred to the Great Level (i.e. the peat fens) as being "four foot higher than the Levell of [Norfolk] Marshland". When criticising Vermuyden's scheme, these facts have to be borne in mind. How the problem appeared in pre-drainage times can be seen from Humphrey Bradley's very interesting treatise on the draining of the Fens (Lansdowne MSS. 60, no. 34). It is dated 1589 and is written in Italian. Bradley stated the facts clearly enough: "The greater part of South Holland and [Norfolk] Marshland and many other places, lying lower than the inundated land [i.e. the peat fens] by six or seven feet and separated from the Marsh [i.e. on the seaward side] only by embankments, remain dry. And if the outlets suffice to drain the water of the low flats, should we be doubtful of that from the high ground?"
[2] H. Godwin, *The New Phytologist*, XXXIX, 238 (1940). See also H. and M. E. Godwin and M. H. Clifford, "Controlling factors in the formation of fen deposits, as shown by peat investigations at Wood Fen, near Ely", *Journ. Ecology.* XXIII, 513 (1935).
[3] S.P., Chas. II, cccxxxii, 3.

the difficulties of navigation, and to arrears of taxes. These minutes of successive meetings form an eloquent narrative of the struggle to maintain the drained level.[1] By the end of the century, petitions had become frequent. In 1694 the inhabitants of Ely, on behalf of themselves and of adjacent towns, were "setting forth what great losses they have sustayned by their Lands being so much drowned these last two years for want of better Outfalls". Their tale of woe was indeed a common one.

WINDMILL DRAINAGE

The only remedy for such conditions was to pump the water from dyke to drain, and from drain to river. In fact, the introduction of pumping engines was the critical factor that prevented most of the Fens from being re-inundated. Some such engine may have been in use before 1600,[2] and, at any rate during the early part of the seventeenth century, there were notices[3] of patents for engines "for raising water and draining surrounded grounds". These engines were now to save the situation. At the Board Meetings of the Corporation, discussion about the setting up of mills became more and more frequent. The Corporation frequently resented the employment of pumps because, in throwing up water, they endangered the banks of the drains or caused an overflow; and for some years they were frequently condemned and ordered to be pulled down. On 28th May 1696, at the Fen Office in the Inner Temple,

this Court being informed by Mr Wakelyn that a Mill is intended to be sett up by Toyes Foster to drayne the Lady Portland's Lands, doe hereby declare that they will assist Madam Coventry against his setting up such a Mill for that the same will be very prejudiciall to ye Adventurers Lands thereabouts, And also to Eldernell Farme.[4]

Yet, despite such complaints, the fact was that mills provided the only means of clearing the water; and, as the seventeenth century passed into the eighteenth, mill drainage became more and more prevalent.[5] By

---

[1] Preserved in the Fen Office at Ely.

[2] S.P., Eliz. cxxvii, 57; and S.P., Eliz. ccxli, 14.

[3] E.g. S.P., Chas. I, xc, 103 (1617), and cxxii, 57 (1622).

[4] Conservators' Proceedings, London Series, vol. 151, f. 8 (Fen Office, Ely).

[5] Defoe could note, in Letter vii of his *Tour*, "that here are some wonderful engines for throwing up water, and such as are not to be seen anywhere else, whereof one in particular threw up, (as they assur'd us) twelve hundred ton of water in half an hour, and goes by wind-sails, 12 wings of sails to a mill: This I saw the model of but I must own I did not see it perform" (A.D. 1724–6).

1726 the state of the rivers and interior drains had become so bad that the Corporation found it impolitic to resist local efforts at improvement any longer and, in the following year, the first sub-district was established at Haddenham in the South Level. A private Act of Parliament authorised this district to set up a pump, and to raise funds by rates within its boundaries for securing and maintaining immunity from floods. This precedent was followed at intervals through the next 150 years by other groups of landowners, supplementing the work of the Bedford Level Corporation.

Outside the Bedford Level, in the northern Fens,[1] the work of draining was taken up again during the eighteenth century. There, effort was more divided than in the Bedford Level, for the arrangement of upland and coast disposed the area into a number of separate tracts. Conditions varied from place to place; several districts had been brought into cultivation, but only temporarily. In the silt area of Holland there were some very wet low-lying lands in almost every parish. These were improved under various local acts for "dividing and inclosing the open common fields, common meadows and other commonable lands". Towards the end of the eighteenth century, these acts came in increasing numbers, and they consolidated the advantages resulting from the works that followed the Act of 1765 "for draining and improving certain low marsh and fen grounds lying between Boston Haven and Bourn, in the parts of Kesteven and Holland, in the County of Lincoln".

To the north of the Holland Fens lay those of the Witham. James Scribo's report of 1733 showed how bad conditions were here.[2] The fall of water from Lincoln to Boston was only 16 feet; for over 20 miles the river of the haven was very crooked, very shallow and, in some places, not more than 18 feet in breadth. Several of the large rivers and brooks that brought down water from the uplands were actually bigger than the winding haven that formed their outfall. Consequently, in winter, the banks of the Witham were generally overflowed, and several thousand acres of rich pasture lay drowned to the depth of three feet. This water remained on the land for three or four months to the prejudice of farming and navigation alike. There were discussions and memoranda, meetings of landowners and meetings of engineers, until, at last, in 1761, there was passed an Act for draining the fens "lying on both sides of the

---

[1] See W. H. Wheeler, *Hist. Fens, S. Lincs, passim.*

[2] This is printed *in extenso* in J. S. Padley, *Fens and Floods in Mid-Lincolnshire, with a Description of the River Witham* (1882).

River Witham ".[1] The improvements that followed, however, were not sufficient, of themselves, to render these lands fit for cultivation. Embankments and mechanical pumping were yet necessary, and so, during the rest of the century, there were many local enclosure and embanking acts dealing with individual parishes. Arthur Young, in describing the Witham lands at the close of the century, stated that the produce before enclosure was little, the land

letting for not more than 1s. 6d. an acre; now, from 11s. to 17s. an acre. ...This vast work is effected by a moderate embankment and the erection of windmills for throwing out the superfluous water.[2]

The district that benefited least by the activity following the act of 1761 was that to the north of Boston. Here, Wildmore Fen, West Fen and East Fen formed a tract of 40,000 acres which was for many years to remain a swampy forage ground. Thus Thomas Pennant, writing in 1774, noted that

the East Fen is quite in state of nature and gives a specimen of the country before the introduction of drainage: it is a vast tract of morass, intermixed with numbers of lakes from half a mile to two or three miles in circuit, communicating with each other by narrow reedy straits; they are very shallow, none above four or five feet in depth.[3]

In many other parts of the northern Fenland stretches of fen lingered on, nursing the older economy based on fishing, fowling, reed cutting and the like.

But despite the differences between the northern and southern Fenland, the problems were the same in essence, and the whole of the Fenland came to consist almost entirely of small sub-districts, each dependent for its internal drainage upon small cuts leading to a central drain, which, in turn, discharged its water by pumping into one of the arterial cuts. Each of these statutory districts had its own authority, with full control over the internal drainage of the area, with wide powers of local taxation, and with

---

[1] The Preamble of the Act summed up the situation at once. Owing to the obstruction of the outfall by sand and silt, "great part of the low lands and fens lying on both sides of the said river (and which contain together about one hundred thousand acres) are frequently overflowed, and rendered useless and unprofitable, to the great loss of the respective owners thereof, the decay of trade and commerce, and the depopulation of the country".

[2] A. Young, *Agriculture of...Lincoln*, pp. 239–40.

[3] *Tour in Scotland*, p. 10. Pennant's description referred to the year 1769. In 1799 Arthur Young also noted the undrained condition of these fens (*Agriculture of ...Lincoln*, cap. XI.). The East and West Fens were drained in 1803.

limited powers of negotiation with adjacent bodies over matters of mutual concern. The Bedford Level and northern Fenland alike came to exhibit on the map a network of drains, dykes, leams, eaus and lodes, in many places crossing one another in apparent confusion.

On the ground itself, windmills imparted a peculiarity to the landscape. The extent to which these remedies were being applied may be seen from a pamphlet written by Thomas Neale in 1748 dealing with *The ruinous state of the parish of Manea, in the Isle of Ely*, and the causes and remedy of it. A great number of mills had recently been erected:

there are now no less than two hundred and fifty in the Middle Level. In Whittlesey parish alone, I was told by some of the principal inhabitants there are more than fifty mills, and there are, I believe, as many in Donnington with its members. I myself, riding very lately from Ramsey to Holme, about six miles across the Fens, counted forty in my view.

But, at best, the windmill was a wayward co-operator, at the mercy of wind and gale and frost and calm. It was never powerful, and it soon ceased to provide a satisfactory solution to the problem of clearing water from the drains. And, as the peat surface continued to shrink and waste away, the windmill became less and less effective. It is true that there were improvements, but, even so, wind-driven pumps were becoming quite inadequate.

Hovering spectre-like over all agricultural activity was the threat of inundation. The story of Burnt Fen in the South Level is dramatic enough as told by Arthur Young:[1]

Forty years ago 500 acres were let for a guinea a year; but in 1772 an act was obtained for a separate drainage, and 1s. 6d. an acre levied for the expence of embankment, mills, etc. In 1777, the bank broke; and most of the proprietors ruined. In 1782, on the success of the machine called *the Bear*, in cleansing the bottoms of the rivers, and other reasons, occasioned some persons to purchase in this neglected tract. The banks were better made, mills erected, and the success great. Servants of the former proprietors bought lots for 200*l.* with almost newly erected buildings on them, that cost 3, 4, and 500*l.* Such lots now let at 100*l.* per annum. An estate of Mr Jones, bought of —— Chitham, Esq. for 200*l.* would now sell at 2000*l.*

Throughout the region, the price of land was most susceptible to local drainage rumours and devices, and it fluctuated in an extraordinary manner. Nor was the problem of draining usually settled as quickly as that of Burnt Fen. In 1777 Mr John Golburne was declaring: "Look

[1] A. Young, *Agriculture of…Suffolk*, pp. 31–2.

every way you will, you will see nothing but misery and desolation"; and other pamphlets of the time echoed the same feeling. That dealing with Grunty Fen, published in 1778 and entitled *Considerations and Reflections on the present state of the Fens*, was full of disappointment that the promises of the drainers had not been fulfilled. In 1800 Mr Chapman, with like complaint, described the condition of the northern Fens in a pamphlet entitled *Observations on the Improvements of Boston Haven*.

Of the last six seasons four have been so wet that most of the new enclosed fens bordering on the Witham were inundated and the crops either lost or materially injured. Many hundred acres of the harvest of 1799 were reaped by men in boats. Of the oats fished up in this way some sold in Boston market at 25/- per last, when good oats were selling at ten pounds.

Another pamphlet of the same year, written by "A Holland Watchman", describes the reaping as having been done by men standing up to their waists in water and clipping off the ears of corn wherever they peeped above the surface.

It is easier to enumerate pamphlets and to put down the statistics relating to such drowning than to imagine the bankruptcy and distress that followed when crops not merely failed but completely disappeared. During the latter half of the century according to Arthur Young there were many fens "all waste and water" where twenty years previously there had been "buildings, farmers and cultivation".[1] Near Ramsey, was "a tract of water, sedge, and frogs, which Mr Pooley remembers thirty-six years ago in a state of cultivation, and producing ample crops over hundreds of acres". It was with dismay that Arthur Young viewed the scene spread before him:

It was a melancholy examination I took of the country between Whittlesea and March, the middle of July, in all which tract of ten miles, usually under great crops of cole, oats and wheat, there was nothing to be seen but desolation, with here and there a crop of oats or barley, sown so late that they can come to nothing.

Most mournful were his general comments: "I was shocked at the sight of this desolation"; "The total ruin of the whole flat district must ensue";

[1] A. Young, *Annals of Agriculture*, XLIII, 539 *et seq.* (1805). Young's general remarks on the situation were indeed vigorous: "For the last thirty years many inundations have taken place, and lately with increasing power and immense mischief. The remedies that have been applied by numberless acts of parliament, obtained with merely local and distinct views, have been in vain and nugatory, but the burden by taxes immense" (*ibid.* p. 543).

"The fens are now in a moment of balancing their fate; should a great flood come within two or three years, for want of an improved outfall, the whole country, fertile as it naturally is, will be abandoned."

But salvation was at hand. As early as 1800, the question of the application of steam engines to draining seems to have been entertained. Arthur Young noted in 1805 that

The application of steam-engines to the drainage of the fens, instead of windmills, is a desideratum that has often been mentioned, but none yet executed:...it must be evident that the power of steam could nowhere be employed with greater efficacy or profit[1].

Some years later W. Gooch[2] was complaining that windmills froze easily in winter and, owing to calm weather, had been known to "stand still for two months together". Too often they were

useless when most wanted, and the proprietors consequently sustain material injury; to remedy this, steam engines have been recommended; and I have found many persons in the country entertaining an opinion that they would answer.

Until power could be commanded at will, the drainage of the Fens could never be absolute. The fenmen were slow to apply the new ideas. After some hesitation a 30 H.P. engine was put up along the Ten Mile Bank during 1819–20. Soon afterwards another was erected near Bottisham in 1820–21. It was the beginning of a new age in the Fenland.

BIBLIOGRAPHICAL NOTE

Among the works cited in the footnotes those of most general interest are by S. Wells (1830), S. B. J. Skertchly (1877), S. H. Miller and S. B. J. Skertchly (1878), and W. H. Wheeler (2nd ed. 1896). Samuel Smiles' *Lives of the Engineers* (1st ed. 1861) has much interesting material. Sir William Dugdale's *History* of 1662 is of course a classic account. A bibliography is given in H. C. Darby, *The Draining of the Fens* (1940).

[1] A. Young, *Annals of Agriculture*, XLIII, 569 (1805).
[2] W. Gooch, *Agriculture of the County of Cambridge* (1817), pp. 239 *et seq.*

Chapter XIII

ENGLAND IN THE EIGHTEENTH CENTURY

W. G. EAST, M.A.[1]

The human geography of southern Britain in the eighteenth century had more in common with that of earlier centuries than with that of the years to come. In fact, the differences between the England of 1800 and that of to-day are infinitely greater than the short interval of time might suggest. Industrial capitalism, when it matured—and it matured after 1800—produced changes on the map which were, perhaps, unequalled in magnitude by any others subsequent to the Anglo-Saxon settlement. But although many of the features of the eighteenth century were survivals, that is not to say there were no new elements in the scene. United as a single state for the first time in 1707, the island of Britain became increasingly powerful in the course of the century as an imperial and maritime state. The development of British agriculture, industry, and commerce both within and outside the Empire produced many changes on the map and, as Daniel Defoe wrote in 1724,[2]

the Face of Things so often alters, and the Situation of Affairs in this Great British Empire gives such new Turns, even to Nature itself, that there is Matter of new Observation every Day presented to the Traveller's Eye.

The Fate of Things gives a new Face to Things, produces Changes in low Life, and innumerable Incidents; plants and supplants Families, raises and sinks Towns, removes Manufactures, and Trade; Great Towns decay and small Towns rise; new Towns, new Palaces, new Seats are Built every Day; great Rivers and good Harbours dry up and grow useless; again, new Ports are open'd, Brooks are made Rivers, small Rivers, navigable Ports and Harbours are made where none were before, and the like.

Several Towns, which Antiquity speaks of as considerable, are now lost and swallow'd up by the Sea, as Dunwich in Suffolk for one; and others, which Antiquity knew nothing of, are now grown considerable: In a Word, New Matter offers to new Observation, and they who write next, may perhaps find as much room for enlarging upon us, as we do upon those that have gone before.

[1] I am grateful to Professor Rodwell Jones and Mr T. H. Marshall, M.A. for reading the proof of this chapter and for making suggestions.
[2] *Tour thro' the Whole Island of Great Britain, 1724, 1725 and 1727* (ed. G. D. H. Cole, 1927), I, 2.

In another word, the "period cross-section" of English geography in the eighteenth century is that of *un paysage humanisé* with a personality of its own.

THE COUNTRYSIDE

Despite the many changes of eight hundred years, the distribution of settlement sites upon the English plain was very similar in 1800 to that in 1086. The eighteenth-century traveller, as he left the built-up area of the towns (and, except in the case of London, this was not large) passed out into a countryside studded with nucleated villages, hamlets and homesteads, no less evident then than before. On the one hand, nucleated villages formed compact groupings of dwellings, usually around the parish church, and spaced out as a rule several miles from each other. On the other hand, were found dispersed habitations, either homesteads or single mansions, or hamlets. In some cases, these scattered settlements prevailed widely to the virtual exclusion of nucleated villages; in other cases they were dispersed between adjoining villages. The distributional problems connected with the sizes and shapes of parishes and settlement patterns remain a matter for local investigation. By the eighteenth century many complications had accumulated over much of the English plain.

One particular feature of these years was the distribution throughout the countryside of what topographers had long traditionally called "gentlemen's seats". These castles, halls or mansions, which increasingly affected the classical style, together with their gardens, parks and farm lands, afforded a visible symbol of gentility, and emphasised the presence of the landed aristocracy which then ruled Britain. "The whole country", wrote the Prussian traveller Archenholtz,[1] "is adorned with parks remarkable for their situation or picturesque views: almost at every step we meet with alleys of fruit trees, leading to elegant villages." And again he adds:

The grass in England is of incomparable beauty.... Hence comes the taste for those beautiful grass walks which are smoothed and made resplendent by means of a large stone rolled over the ground: these walks are often so level, that they can play at bowls upon them as on a billiard-table.

Among the topographical works and travel books of the century, many of which were published thanks to the subscription of the landed gentry, were some, like those of William Tunnicliff,[2] which carefully enumerated

[1] W. de Archenholtz, *A Picture of England* (a new translation, 1797), p. 91.

[2] *A Topographical Survey of the Counties of Somerset, Gloucester, Worcester, Stafford, Chester and Lancaster* (1789).

and mapped, alongside with cities, market towns and villages, these numerous country seats. It is clear that they were thickly grouped around London, about the regional capitals, such as Bristol and Chester, and about growing cities, like Manchester and Birmingham, being situated in regions which as yet afforded sufficient natural amenities for country life. Further, the practice of seeking political power, pleasure and gentility, through the building of country houses had long become common among successful merchants and industrialists, including the new rich from India and North America. Thus even Defoe, who considered evidences of active industry as "objects that rank in a class abundantly superior to brilliant palaces, and gew gaw gardens",[1] testified to the presence of country houses of city men at the villages of Stratford, "Leyton-stone", Walthamstow, Woodford, West Ham, Plaistow, Upton and elsewhere on similar sites to the south of London, whence by coach their owners made their way to the city.[2] The Swedish traveller, Kalm[3] noted that Portland stone was commonly used in the big houses around London. Occasionally, as Tunnicliff showed later in the century, industry and country houses stood side by side, since country gentlemen often took an active part in coal mining and iron working. At Pendleton Hall in Lancashire, for instance, stood "the Seat and very extensive Manufactory of William Douglas, Esq." In fact, it would seem that in some parts of the country, gentlemen's seats were as plentiful as villages themselves! Further, the contemporary fashion of placing country houses on hilly sites suggests the increasing control which the wealthy exercised over water supply. This practice attracted the attention of travellers; nor was it always approved. "The modern gazeabouts on the hill top", wrote the Hon. John Byng,[4] (are) "expos'd to every tempest, and distant from every comfort."

The varying densities of rural population from one part of the country to another were no less clearly marked than the varieties of settlement

---

[1] Op. cit. 1, 64.

[2] Ibid. 1, 6. "In all which Places, or near them, (as the Inhabitants say) above a Thousand new Foundations have been erected, besides old Houses repaired, all since the Revolution: And this is not to be forgotten too, that this Increase is, generally speaking, of handsom large Houses, from 20£ a year to 60£ very few under 20£ a Year; being chiefly for the Habitation of the richest Citizens, such as either are able to keep two Houses, one in the Country, and one in the City; or for such Citizens as being rich, and having left off Trade, live altogether in these Neighbouring Villages, for the Pleasure and Health of the latter Part of their Days." See also 1, 168.

[3] Kalm's Visit to England (trans. P. Lucas, 1892), p. 22.

[4] The Torrington Diaries (ed. C. Bruyn Andrews and J. Beresford, 1934–5), 1, 147.

types. It is certainly not safe to infer that those regions which had the best soils for agriculture were necessarily or solely the areas of greatest density of rural population. Geographical position was important, still more so were the facilities for communication with markets. Thus east Kent and east Suffolk, athwart navigable rivers, and close to the London market, were both able to exploit profitably their tracts of rich loam soils and their relatively favoured climate. Likewise, the country immediately around London profited from London manures and other refuse brought up by

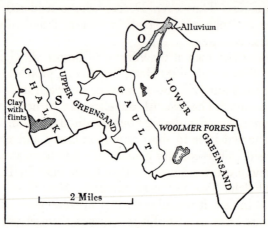

Fig. 75

The Parish of Selborne, showing the relation of the parish boundaries to the geological outcrops. Note the varieties of terrain. *S* marks the site of the nucleated settlement of Selborne situated on the Greensand plain below the Chalk plateau whence issued a spring. *O* marks the "large hamlet of Oakhanger". There were also "single farms, and many scattered houses along the verge of the forest".

river, whilst its foodstuffs and dairy produce found a ready market in the capital. Again, the economy of many rural regions depended not only on some branch of agriculture but on some processes of the textile industry. Thus, as Defoe so vividly described, in many parts of England, not only in good agricultural regions such as those of Norfolk, Suffolk and Essex, but also in remote Pennine valleys, industry supported a dense rural population. But the more ordinary village was based on a semi-self-sufficient economy such as that immortalised by Gilbert White. *The Natural History of Selborne* describes with brief yet scientific precision the varied geographical content of a parish in Hampshire—hill pasture, culti-vated plain and sandy forest tract (Fig. 75).

It is probable that about one-half of the arable land of England in 1700 was cultivated on the traditional open-field system with intermixed strip holdings, with common rights in pasture and waste, and with many time-honoured customs.[1] But, the enclosure movement, so-called, had in many districts replaced this obsolete system by compact and enclosed farms.[2] "No generalisation", wrote Lord Ernle,[3] "will explain why these districts should have been enclosed sooner or more easily than elsewhere", and he added that either the open-field had never characterised them, or it had existed in a form which made its transformation easy. Geographical research presents a challenge to this view, and argues that, at least in part, the enclosure movement was governed by the nature of the terrain.[4] Fuller understanding of the geographical factor in the enclosure movement clearly awaits regional studies.

Conditions in the south-eastern lowlands—comprising the counties of Suffolk, Essex, Middlesex and Kent—had been initially different from those of the midland plain. These counties too had reaped the benefit of their soils, their climate and their nearness to the great London market; and, by the end of the sixteenth century, they were certainly full of hedged fields, devoted largely to pastures. In the western counties of Hereford, Monmouth, Worcester, Shropshire and Stafford, the open-field system had been, or was being, extinguished with little or no legislative help. It had also disappeared from most of Somerset and Devonshire, where open fields had been confined to lowlands and valleys. Likewise it was not to be found in the highland country of the north and west—in Cornwall, Cumberland, Westmorland, Lancashire and Northumberland—where a pastoral economy was dominant, and where other customs, the "infield-outfield" or "run-rig" system, had been native. Even as late as 1800, there were many survivals. In some Cornish villages, not only the nearer fields were cropped with corn, but land was taken in from the waste, cropped for a few years, and then allowed to revert to its former condition.[5] Villages of the East Riding wolds had their "infield" "por-

---

[1] Lord Ernle, *English Farming Past and Present* (4th ed. 1932), p. 154.

[2] It must not be inferred that open-field agriculture was always thoroughly bad and that enclosed farms were always efficiently run. Cp. Ernle, *op. cit.* pp. 199–200, and T. H. Marshall, "Jethro Tull and the 'New Husbandry' of the Eighteenth Century", *Econ. Hist. Rev.* II, 41 (1929).

[3] *Op. cit.* p. 164.

[4] See M. Aurousseau, "Neglected Aspects of the Enclosure Movements", *Economic History*, I, 280 (1927).

[5] G. B. Worgan, *General View of the Agriculture...of Cornwall* (1811), pp. 46 and 53.

tioned into several falls, annually cultivated on a fixed rotation", usually three-course, whilst "beyond this was an outfield cultivated only occasionally". Beyond that again was sheepwalk.[1] Something akin to the run-rig system was also to be found in localised areas within the English plain itself, where peculiarities of geography were favourable. In the forest region of western Nottinghamshire,[2] and upon the Breckland of southwestern Norfolk[3] examples of infield-outfield exploitation have been cited.

Standing between these two regions of north-west and south-east, and forming a great contrast with them, was a broad midland zone extending diagonally across England from the coasts of Dorset and Hampshire north-eastward into Northumberland. This consisted essentially, in the eighteenth century, as in Leland's day, of open-field country. No completely satisfactory explanation of the survival can be given.[4] But to the men of the time the important consideration was not the question of origins, but the fact that parliamentary enclosure of the open field became usual before the third decade of the century was over. From that time onward, Enclosure Acts provide a reliable, if not a complete, indication of the pace and progress of the movement. A great deal remained to be done, but by the end of the century, the greater part had been enclosed.[5] Dr Slater's[6] figures show how actively the work went on:

[1] H. E. Strickland, *General View ... of the East Riding of Yorkshire* (1812), p. 91. See also A. Young, *A Six Months' Tour through the North of England* (1770), II, 9.

[2] J. D. Chambers, *Nottinghamshire in the Eighteenth Century* (1932), pp. 155 et seq.

[3] J. Saltmarsh and H. C. Darby, "The Infield-Outfield System on a Norfolk Manor", *Economic History*, III, 30 (1935).

[4] See Ernle, *op. cit.* pp. 167–8.

[5] The enclosure of the open fields did not take place without much dispute. On the one hand were the obstacles to farming improvement which were presented by the open fields. On the other hand was urged the injury which the break-up of open-field forms and the partition of commons inflicted on small owners and occupiers of land. Thus, John Laurence, in *A New System of Agriculture* (1726), was loud in his praises of enclosure and separate occupation as the best means of increasing produce. An example from the other side was the vigorous pamphlet by John Cowper, entitled *Essay proving that Inclosing Commons and Common-Field-Lands is Contrary to the Interest of the Nation* (1732). "I myself", he wrote, "have seen within these 30 years, above 20 lordships or parishes enclosed, and everyone of them has thereby been in a manner depopulated." But in this literary struggle the advocates were gaining the upper hand and, in any case, enclosures continued increasingly to be made.

[6] Gilbert Slater, "The Inclosure of Common Fields considered Geographically," *Geog. Journ.* XXIX, 35 (1907).

Percentage of the Total Area affected by Enclosure

The Area covered by Acts for inclosing common fields expressed as a
percentage of the total area of each county

Northampton	51·5	Wiltshire	24·1	Stafford	2·8
Huntingdon	46·5	Gloucester	22·5	Essex	2·2
Rutland	46·5	Middlesex	19·7	Sussex	1·9
Bedford	46·0	Worcester	16·5	Northumberland	1·7
Oxford	45·6	Derby	15·9	Cumberland	1·7
Yorks, East Riding	40·1	Herts	13·1	Durham	0·7
Leicester	38·2	Yorks, West Riding	11·6	Westmorland	0·6
Cambridge	36·3	Dorset	8·7	Cheshire	0·5
Bucks	34·2	Suffolk	7·5	Monmouth	0·4
Notts	32·5	Hampshire	6·4	Shropshire	0·3
Norfolk	32·3	Surrey	6·4	Kent	
Lincoln	29·3	Yorks, North Riding	6·3	Lancashire	
Berkshire	26·0	Hereford	3·6	Devon	0·0
Warwick	25·0	Somerset	3·5	Cornwall	

But his map (Fig. 76) shows that here and there all through the belt, common-field parishes lingered on. Along the eastern margins, these surviving areas were particularly abundant. Their end was soon to be. By 1820 there were only six counties of whose area more than 3 per cent. remained in open field; and in these the great part of the work was done by 1830.[1] "Taken as a whole," writes Professor Clapham, "British agriculture was undoubtedly the best in Europe, and as a land of enclosure England was unique."[2]

But "enclosure" was a term applied not only to the rearrangement of the common-field patchwork. It included the reclamation of stretches of heath, fen, moor, down, woodland, hitherto of little value. From the standpoint of production the large amount of waste land in England was a standing reproach to the century. In 1696 Gregory King had estimated the barren lands of England and Wales at ten million acres, or more than a quarter of the total area. The estimate is probably inadequate, but in any case it can be compared with conditions at the close of the century. In 1795 the *Report of the Committee of the Board of Agriculture* stated that 7,888,777 acres in England and Wales, that is, about one-fifth of their area, were uncultivated. This of course included much rough upland grazing in Wales, but many English counties too were wild enough. In 1773 Arthur Young made his *Observations on the Present State of Waste Lands of Great Britain*, in which he called attention to the many uncultivated English areas.

[1] The counties were Bedford, Buckingham, Cambridge, Huntingdon, Northampton and Oxford. E. C. K. Gonner, *Common Land and Enclosure* (1912), p. 279.
[2] *An Economic History of Modern Britain, 1820–50* (2nd ed. 1930), p. 19.

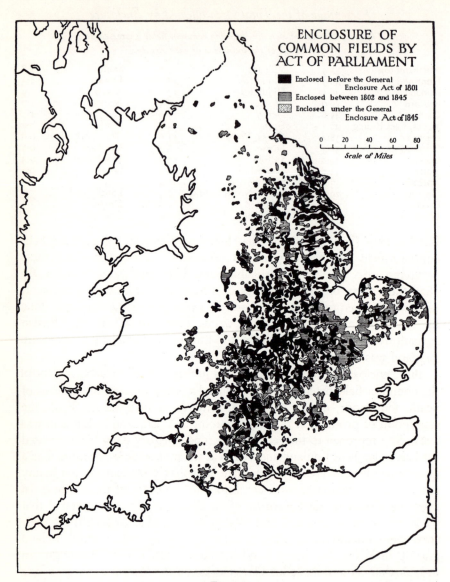

Fig. 76

From Dr G. Slater, *The English Peasantry and the Enclosure of Common Fields* (Constable & Co., 1907). This map, which should be compared with Fig. 67 (p. 401), shows where arable land was enclosed in England during the last phase of the enclosure movement.

There are at least 600,000 acres waste in the single county of Northumberland. In those of Cumberland and Westmorland, there are as many more. In the North and part of the West Riding of Yorkshire, and the contiguous ones of Lancashire, and in the west part of Durham, there are yet greater tracts: you may draw a line from the north point of Derbyshire to the extremity of Northumberland, of 150 miles as the crow flies, which shall be entirely across waste lands; the exception of small cultivated spots, very trifling.

Young found also that there stretched over Devonshire and Cornwall "immense" tracts of waste. Below the 800 foot contour line—to take a rough division—conditions were better, but Young found that the Mendip Hills were uncultivated, that 18,000 acres on the Quantocks lay desolate and that Sedgemoor was still one vast fen. Despite much cultivation, like conditions were to be found in other parts of the plain. Of the great Fenland itself there is no need to say anything here,[1] and there is no space to mention the neglected state of much sandy country. Even in the neighbourhood of London similar conditions were to be found. Nathaniel Kent, in his *Hints to Gentlemen of Landed Property*, in 1775, declared "that within thirty miles of the *capital*, there is not less than 200,000 acres of waste land". In 1791 the Weald of Surrey was still bare; as late as 1793 Hounslow Heath and Finchley Common were described as wastes fitted only for "Cherokees and savages"; and in the year 1794 the forests of Epping and Hainault in Essex were still good cover for "the more finished and hardened robber" retiring from justice.[2] But despite these facts, the eighteenth century has seen much progress. If the estimates of Gregory King (1696) and of the Board of Agriculture (1795) are approximately correct, upwards of two million acres were added to the cultivated area of England and Wales between 1700 and 1800.

The break-up of open-field farms and the reclamation of waste during the eighteenth century, especially in its latter half, reflected a wider movement of agricultural reform. On its economic side this was associated with a rise in the price of corn and with the pressure of a growing population. If the country was to be fed, more scientific methods of farming were necessary. The new spirit of enterprise in farming showed itself in many ways—in cattle shows, wool fairs, ploughing matches, experimental farms, the formation of local agricultural societies; numerous patents for new implements; improved stock breeding, new crops, new methods of cultivation, new schemes of rotation. This diffusion of knowledge was identified with theorists and experimentalists like Jethro Tull, Lord Townshend, Bakewell of Dishley, Coke of Norfolk and Arthur Young.

[1] See above, chap. XII.　　　　　　　　　[2] Ernle, *op. cit.* p. 154.

In an age when communications were only slowly improving it is not surprising that English farming revealed striking regional contrasts—contrasts which were not merely the reflection of differences in natural endowment. Certainly south-east England, near to London (already a great market) and to the Low Countries (whence new technical methods were introduced), boasted much of the best English farming of the century. But not every part of the south-east could claim this priority. Even in Middlesex farming had made little progress by the end of the century. Although it seemed to Defoe "Rich, Pleasant, and Populous",[1] Middlesex merited this description rather as a residential area for wealthy gentlemen from London than for any local excellence in cultivation or stock-raising. It was for the greater part a pastoral county, enjoying a high reputation for its hay which fed the horses of the capital.[2] From Teddington to Chelsea, along a belt of "excellent dry loam" which was well manured with London refuse, were found many orchards and market gardens, as also between Islington and Stepney[3] (Fig. 77). In 1798 barely one-tenth of Middlesex stood in commons, but there was a considerable area of unenclosed and ill-drained meadows, whilst the bulk of its arable land lay still in open fields.[4] The latter were ploughed by a team of no less than six horses and three men—a wasteful practice, despite the stiffness of its clayey soils. Nor again was Surrey "over fertile, especially in the middle Parts". It was compared with a coarse garment having a fine green border.[5] A broad area of open downlands, which supported sheep, and sandy heath, devoted to deer parks and rabbit warrens, separated the best arable parts of the country in the north and south. In the northern lowland towards the Thames, there was still some open field, but close to the capital, from Dulwich onwards, the country was chequered by ploughed fields, meadows, orchards and market gardens, interspersed with well-built villages and gentlemen's houses.[6] Similarly, in the south of Surrey, the clay vale of Holmesdale was cut up into hedged meadows, fields and parks. In the west of the county, around Farnham, which was a great corn market, hops were grown with much success, whilst the enclosed land of the upper Wey, from Farnham as far as Alton in Hampshire, was rich in

[1] *Op. cit.* I, 393.
[2] On haymaking in Middlesex in 1748, see *Kalm's Visit to England*, pp. 80–4.
[3] J. Middleton, *General View of the Agriculture of Middlesex* (1798), pp. 254–72. See also *Kalm's Visit to England*, pp. 8–9 and 33–5.
[4] Middleton, *op. cit.* p. 138.
[5] H. Moll, *A New Description of England* (2nd ed. 1724), p. 63.
[6] *Kalm's Visit to England*, pp. 53–4.

soil and well cultivated.[1] Even so, many inefficient practices in ploughing and in crop-rotation survived in Surrey at the end of the century.

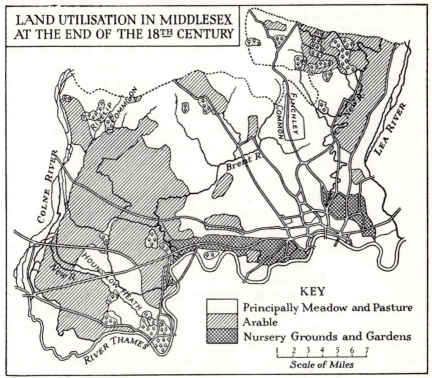

Fig. 77

This map is based on John Middleton's coloured map (1798). The tracts, stretching between Islington and Stepney and between Westminster and Brentford, which were devoted to vegetable and fruit cultivation, occupy loam soils covering river gravels. Foot's map (1794) continues the second tract to Teddington. The arable area included a variety of loam soils, and also heavy London Clay in north-west Middlesex. The meadow and pasture lands were largely London Clay. At the present day the greater part of the county is built over, but small areas of pasture are to be found in the west.

Kentish agriculture was certainly more flourishing than that of Surrey. The country between London and Blackheath, with its meadows and fields, was "almost one continued garden"; and around Dartford also the

[1] A. Young, *A Six Weeks' Tour through the Southern Counties of England and Wales* (1768), pp 180-2.

husbandry was excellent.[1] The loam tract of north Kent was covered with numerous enclosed farms, famed, above all, for their hops and fruit. East Kent, especially around Maidstone and Canterbury, was "the Mother of Hop Grounds in England".[2] So also it had numerous orchards; from the port of Milton, on the East Swale, London received "very great quantities of Fruit; that is to say, Apples and Cherries; which are produced in this County, more than in any County in England, especially Cherries".[3] The marshes of east Kent were dyked and afforded rich meadow grass.[4] Cattle and sheep were reared respectively in the Kentish Weald and on the pastures of Romney Marsh, and destined in part for the London market. But it was a reproach to Kentish agriculture that there were large areas of sandy heaths in the west of the county.[5]

The counties which were distinguished by the excellence of their farming were Hertfordshire, Essex, Suffolk, Norfolk and Leicester. The first of these, for centuries clad largely with woodland, had long become a land of enclosed, arable fields. In 1804 very few open-field villages survived; and only a small area, on the chalk plateau in the north of the county, remained as commons.[6] The best corn county in England, Hertfordshire had been among the first to introduce turnips and clover into its crop rotation. The river Lea, navigable up to Hertford and Ware, was used to bring from London all manner of refuse for fertilising,[7] whilst it served no less in the milling of wheat and barley, and as a means of transporting grain to the capital.[8] The roads of Hertfordshire, too, were described as relatively good. To improve their land Hertfordshire farmers used, besides much refuse, chalk on clay soils and clay on the lighter sands and gravels. Moreover, they grew tares on the turnip fallows to provide fodder for horses brought from Leicester. Already in the third decade of the century the loam soils of the Vale of St Albans, the chief town and corn market of the county, were excellently cultivated. The view of this vale, seen from Bushey Heath, made a striking impression in the time of Defoe.[9]

---

[1] *England Displayed...by a Society of Gentlemen* (1769) (revised by P. Russell and O. Price), I, 129.   [2] Defoe, *op. cit.* I, 113.

[3] *Idem.* On the orchards of Kent see also *Kalm's Visit to England*, pp. 375–76.

[4] *Ibid.* pp. 377–80, and *England Displayed*, I, 129.

[5] J. Boys, *General View of the Agriculture of Kent* (1794), p. 88.

[6] A. Young, *General View of the Agriculture of Hertfordshire* (1804), p. 55.

[7] On the utilisation of London refuse, see *Kalm's Visit to England*, pp. 54–5.

[8] The navigability of the river Lea was being improved in the 'sixties: see *England Displayed*, I, 284.

[9] Defoe, *op. cit.* I, 388–9.

I cannot but remember, with some satisfaction, that having two Foreign Gentlemen in my Company, in our passing over this Heath,...how they were surprized at the Beauty of this Prospect, and how they look'd at one another, and then again turning their Eyes every way in a kind of Wonder, one of them said to the other, That England was not like other Country's, but it was all a planted Garden.

They had there on the right Hand, the Town of St Albans in their View; and all the Spaces between, and further beyond it, look'd indeed like a Garden. The inclos'd Corn-Fields made one grand Parterre, the thick planted Hedge Rows, like a Wilderness or Labyrinth, divided in Espaliers; the Villages interspers'd, looked like so many several Noble Seats of Gentlemen at a Distance. In a Word, it was all Nature, and yet look'd all Art.

And in 1748 Pehr Kalm wrote[1] "it is commonly held that in Hertfordshire on this side [i.e. the western] there are the best Agriculturalists in England". Even so, the writers of *England Displayed* (1769) were critical of the county's farming, its ploughs and ditches being "very bad".

"The richness of the Lands, and the application of the People to all kinds of Improvement, is scarce credible."[2] Thus, already in the 'twenties Defoe emphasised the prosperity of Suffolk farming. High Suffolk, he added, was "full of rich feeding-Grounds and large farms, mostly employ'd in Dayries for making the Suffolk Butter and Cheese". It was there, too, according to Defoe, that turnips were first employed for the feeding and fattening of cattle and sheep, a practice, he noted, which did not adversely affect the flavour of the meat. From Suffolk, and south Norfolk also, large droves of turkeys and geese were reared for the London market, whither, in late summer, they were either driven or carried in specially made carts.[3] Nor was the coastal part of Suffolk unproductive: "The Country round Ipswich, as are all the Counties so near the Coast, is applied chiefly to Corn, of which a very great Quantity is continually shipped off for London; and sometimes they load Corn here for Holland, especially if the Market abroad is encouraging."[4] So also Defoe's picture of Essex farming was favourable, if less glowing than that of Suffolk. He distinguished between the interior upland of Essex and the low-lying tracts of the south-east. The first was primarily corn land; the second, the marshlands, provided pasture on which were fed calves "the best and fattest, and the largest Veal in England".[5] The accounts of

---

[1] *Kalm's Visit to England*, pp. 186–7. See also (p. 180) his glowing account of the prosperous husbandry from Woodford (Essex) across Hertfordshire to Little Gaddesden.

[2] Defoe, *op. cit.* I, 60.   [3] *Ibid.* I, 59.

[4] *Ibid.* I, 145.   [5] *Ibid.* I, 15.

Arthur Young, who toured East Anglia in 1768, and the agricultural reports at the end of the century alike testify to the rising prosperity of Suffolk and Essex farming. Manuring was practised on a generous scale and the refuse of the towns was used together with that brought from London by sea and river. In Essex much chalk was used to improve its heavy clays, notwithstanding that this had to be brought along bad roads from the lower Thames.[1] Again, Essex and Suffolk were famed for their "hollow draining",[2] which, little resorted to elsewhere, proved rewarding on waterlogged tracts. In ploughing and in crop rotation, Suffolk and Essex farmers adopted the best practices of the day. A few specialised crops, of use in the textile industry, were grown: teazels in Essex, and hemp and flax in Suffolk. Towards the end of the century, only small areas in these two counties stood unproductive; but part of Breckland, south of Thetford, though capable of improvement, still consisted of sheep-runs or wild and luxuriant heath.

Towards the end of the century, Norfolk became the premier farming county of England, but not many decades earlier much of its area had been waste and unproductive. "Norfolk, it is probable (speaking generally for the county), has not borne grain, in abundance, much above a century", wrote William Marshall,[3] "during the passed century [*sc.* the eighteenth] a principal part of it was *fresh land,* a newly discovered country in regard to grain crops." Even in 1794 it contained 80,000 acres of unimproved commons, but two-thirds of its area was arable land; of this, the bulk was enclosed, although some common field remained.[4] Much of east Norfolk was enclosed by the beginning of the eighteenth century, and was described as "the Woodlands", because of its high whitethorn hedges. Vast areas of heath were enclosed in the latter half of the century, thanks to the enterprise and capital of enlightened gentlemen farmers, who experimented in new technical methods and new crops and also encouraged their tenants to do likewise. The part played by Coke in developing the poor, sandy country between Holkham and Lynn has been exaggerated.[5] He began to

[1] Arthur Young refers to this in his *Six Weeks' Tour,* pp. 69 and 72, where he comments strongly on the badness of the road from Tilbury to Billericay, along which the chalk waggons passed.

[2] See C. Vancouver's *General View of the Agriculture of Essex* (1795), pp. 139–40.

[3] *Review of the Reports to the Board of Agriculture for the Eastern Department* (1810), p. 314.

[4] N. Kent, *General View of the Agriculture of Norfolk* (1794), pp. 5 and 22.

[5] Thus Ernle, *op. cit.* pp. 217–18, attributes to Coke the initial improvement of the country between Holkham and Lynn, which had clearly begun earlier, as Arthur Young records.

farm there in 1778, but already ten years earlier part of this tract was being transformed into small arable farms.

All the country from Holkham to Houghton [wrote Arthur Young[1] in 1768] was a wild sheep-walk before the spirit of improvement seized the inhabitants; and this glorious spirit has wrought amazing effects; for instead of boundless wilds, and uncultivated wastes, inhabited by scarce anything but sheep, the country is all cut into inclosures, cultivated in a most husbandlike manner, richly manured, well peopled, and yielding an hundred times the produce that it did in its former state.

The foundation of Norfolk agriculture was marling: marls, dug from underlying beds, were mixed with light sands and loams, whilst manure was freely used and sheep were folded on the fields after harvest. A four course rotation of wheat, turnips, barley and clover—the famous Norfolk "course"—served the double purpose of preserving the fertility of the soil and of providing fodder. Scottish cattle, bought at St Faith's Fair near Norwich, together with sheep, were fattened for the London market, oil cakes as well as turnips being used as winter feed. Moreover, cereals were produced on such a scale that much corn was shipped abroad and coastwise from the ports of Yarmouth, Lynn, Wells and Blakeney.[2]

Throughout the long midland belt of England, where considerable areas of open field and commons survived throughout the century, farming practices changed but slowly, and long retained features deplorable to progressive agriculturalists. Of Sussex, it was recorded in 1769:[3] "The husbandmen...have not adapted the modern improvements.... The turnip husbandry is little known, and their peas and beans are never hoed." In the downlands of Wiltshire, Hampshire and Dorset, Defoe noted[4] the decrease of sheep, "because of the many Thousand Acres of the Carpet Ground, being, of late years, turn'd into Arable Land, and sow'd with Wheat". Sheep were folded on these fields to improve the soil; in fact, the dung of the sheep-fold was said to be the chief reason for keeping sheep in Wiltshire.[5] The Wiltshire downs thus produced both corn and wool, whilst the valley lands were famous for their cheese and bacon. In 1794 it is recorded that very little land had been enclosed during the preceding fifty years, and that there was a considerable area of open field in the south of the county. On the north-west side of Wiltshire, in what was the best land of the county, there were many poor

---

[1] *A Six Weeks' Tour* (1768), pp. 21–2.   [2] N. Kent, *op. cit.* pp. 49–51.
[3] *England Displayed*, I, 116.   [4] *Op. cit.* I, 187 and 283.
[5] T. Davis, *General View of the Agriculture of Wiltshire* (1794), p. 20. Increased attention was being given to the rearing of sheep for meat.

commons, whilst the clay tract of pasture land between Westbury and Cricklade suffered from bad drainage, although in the irrigation of its meadows Wiltshire enjoyed some reputation. At the end of the century, sheep were still the mainstay of Wiltshire farming, but they continued to give way to cultivation.

The wet clay Vale of Aylesbury, long stripped of its woods, suffered from lack of drainage and was devoted rather to pasture than to corn: "All the Gentlemen hereabouts are Graziers," wrote Defoe,[1] "Tho' all the Graziers are not Gentlemen." Kalm noted[2] in 1748 the abrupt change from the excellent farming around Little Gaddesden in Hertfordshire on the Chiltern plateau to the inferior open-field cultivation in the Vale near Ivinghoe. When Young crossed the Vale between Aylesbury and Buckingham in 1768, the arable land lay open in striking contrast with the hedged fields on the Chiltern plateau in southern Bucks. A five-horse plough was still in use and inefficient rotations, such as fallow, wheat, barley and beans were still commonly followed.[3]

Farming in Oxfordshire left much to be desired. "Enclosing by no means flourishes," wrote Arthur Young[4] in 1768, "for from Tetford to Oxford enclosures are scarce; and from thence to North Leach, few or none." Woods were lacking, and inefficient rotations were largely followed. Between Witney and North Leach, "all the country is open, dull and very disagreeable...the crops were generally very poor, and mostly full of weeds". Bedfordshire farming was still more backward. More than two-thirds of the country stood in 1794 in open fields, common meadows, pastures and waste.[5] Drainage was badly neglected, and sheep were very subject to "rot". Cambridgeshire was reckoned at the end of the century the worst farmed county in the whole of England. It shared with Lincolnshire and Huntingdon wide areas of unimproved fen, outside of which—but in rather less than one-third of its area—it was almost wholly a corn county, barley being the chief crop.[6] It is a remarkable fact that in Cambridge as in Lincolnshire not only did no improvement in agriculture take place in the course of the eighteenth century, but the agriculture of the open-field villages was held to have deteriorated since the seventeenth century.[7]

[1] Op. cit. I, 394.   [2] Kalm's Visit to England, pp. 255–73.
[3] A Six Weeks' Tour (1768), p. 84.   [4] Ibid. pp. 102–3.
[5] T. Stone, General View of the Agriculture of Bedfordshire (1794), p. 11.
[6] See C. Vancouver, General View of the Agriculture of Cambridge (1794), p. 193.
[7] T. Stone, General View of the Agriculture of Lincolnshire (1794), p. 56, and C. Vancouver, op. cit. p. 97.

The midland plain of England was, generally speaking, a grassland country where "breeding, grazing and the dairy" prevailed. It contained one region of outstanding farming—the country between the Charnwood Hills, the southern bank of the Trent, the Tame, and the Ankar, i.e. Leicester, together with part of Staffordshire, Derby and Warwick. In these lands, the richest part of the midland plain, arable farming was combined with, though subordinate to, stock rearing, many farms being run by gentlemen of property and abilities.[1] Towards the end of the century, much land was enclosed—for pasture chiefly—but some of the open-field villages survived. In the absence of well-grown hedgerows, the prospect seemed to Marshall[2] "as naked to the distant eye, as the downs of Surrey or the Wolds of Yorkshire"; but the country generally was still moderately well wooded. In Leicester, as in some of the neighbouring counties, drainage was well understood and applied; the technique of cattle breeding for meat production had been largely revolutionised, whilst its arable farming, based on a scientific rotation and generous manuring, was widely famed.

The Vale of Pickering, when Marshall visited it in 1769, was completely enclosed. Much of this was recent, but the woody hedges in part of the vale indicated older enclosures.[3] The region was given over rather to grass than to corn, whilst on its northern side, the moorland valleys, into which cultivation extended, were well wooded. The roads at this time showed marked improvement but low-lying parts of the vale still awaited reclamation.

The East Marshes (and some smaller portions of the Vale) still remain a disgrace to the county; lying, chiefly in a state of fenn—provincially "Carr"; overrun with sedges and other palustrian plants; which afford, during a few months of summer, a kind of ordinary pasturage to young stock. In winter months they are generally buried under water, and in summer months, are subject to be overflowed.[4]

The north Yorkshire moors, bearing a "moory soil of fens", appeared to Marshall as "bleak mountains, covered with heath, and intersected by cultivated dales",[5] which supported some sheep on the heights and cattle on their borders. Cleveland, covered with much heavy clay, and given over more to corn than was the Vale of Pickering, was a great wheat-

---

[1] W. Marshall, *The Rural Economy of the Midland Counties* (1790), I, 4.
[2] *Ibid.* I, 63.
[3] *The Rural Economy of Yorkshire* (2nd ed. 1796), I, 47, and II, 215.
[4] *Ibid.* II, 183.   [5] *Ibid.* II, 265.

producing area;[1] but cattle and horses were reared there, whilst butter and bacon were staple products. The plain of Holderness still contained some woodlands and stretches of desolate marsh in addition to its fields. On the Wolds, long-fleeced sheep and rabbits were the chief resource, but excellent crops of oats and barley were raised on those parts which had been enclosed and broken up.[2] The Vale of York, except for "some fens at its base" and the heathy plain of Galtres beyond York, was very productive of wheat and barley. To the west, the vast moorlands of the West Riding, except for the dales, which yielded oats, rye and meadow grass, remained largely bog and open heath, on which were pastured horses, cattle and goats. Finally, it is worth noting the contrasts between the three Ridings in respect of waste land. The following figures[3] show the position in 1794.

	Acreage of waste land	Acreage of waste capable of cultivation or conversion to pasture	Area of Riding in acres
East Riding	254,588	254,588	819,200
North Riding	442,000	228,435	1,311,187
West Riding	405,272	265,000	1,568,000

Thus, in 1794, nearly a third of the West Riding and a full third of the North Riding lay waste, although a fair proportion of this was thought capable of reclamation for cultivation.

The four most northerly counties of England, characterised by their relatively scanty populations and by large areas of highland, were devoted, above all, to pastoral farming. In the west of Durham rose high moorlands, the eastern part of which provided good sheep grazing, whilst the western, higher part was almost entirely moorish waste. More than a fifth of the county was described as waste in 1794.[4] In contrast, eastern and southern Durham, particularly in the lowlands watered by the Tees and Wear and covered by dry loams and clays, was cut up into woods, fields and meadows. On the whole Durham was largely lacking in woodlands: "the face of the country is for the most part naked", Granger recorded[5] in 1794. Common-field farming, except for small traces in the

---

[1] J. Tuke, *General View of the Agriculture of the North Riding* (1794), p. 35.

[2] W. Marshall, *The Rural Economy of Yorkshire* (2nd ed. 1796), II, 243.

[3] Rennie, Brown and Shireff, *General View of the Agriculture of the West Riding* (1794), p. 140. It was claimed in 1769 that the West Riding still contained some native trees, seldom found in any other part of England, "particularly the fir, the yew, and chestnut" (*England Displayed*, II, 110).

[4] J. Granger, *General View of the Agriculture of Durham* (1794), p. 43.

[5] *Ibid.* p. 31, and *England Displayed* (1769), p. 162.

west, had long disappeared. Cattle and sheep rearing dominated in Durham farming, but oats, barley, wheat and turnips were grown. In the rotations of crops and the use of fertilisers the local farmers had much to learn, and much waste land was capable of reclamation. Even so, it is reported[1] at the end of the century that "the spirit of improvement was awake" in the county of Durham.

Northumberland presented within itself a striking contrast between the high plateaux of the north and west, which yielded rough pasture for sheep, and the coastlands and valleys—especially that of the Tyne—which provided rich meadows and ploughed fields. The Cheviots were "fine green hills, thrown...into a numberless variety of forms ... [enclosing] many deep narrow glens", whilst other mountainous districts, chiefly in the west, were "extensive, open, solitary wastes, growing little else but heath", on which fed vast flocks of sheep.[2] Even below and eastwards of the Cheviots, between Wooler, Rothbury and Alnwick, stretched wide areas of moorland waste, which experts considered well worth improvement for agriculture.[3] Northumberland, in contrast to Durham and Yorkshire, was cut up into large farms. In the moorland farms of the north and west, the shepherds lived in small huts called "sheals" or "shealings", keeping their sheep on the moor all through the year: only in very hard weather were they given hay during the winter.[4] In the lowland farms, some of the best of which stood in the country traversed by the Great North Road between Newcastle and Berwick, the husbandry was considered better than in Durham and Yorkshire.[5] Rye, barley, oatmeal, wheat and maslin—a mixture of rye and wheat—were grown, and these, together with peas, formed the chief breadstuffs in Northumberland.[6] In short, the husbandry of Northumberland showed both good and bad features. Some 200,000 acres of commons had been converted to tillage in the thirty years prior to 1794, although one-third of the county, much of which was private property, was reckoned as mountainous waste unsuitable for cultivation.[7] Again, it was reported[8] (in 1794) that "plantations [sc. of trees], on an extensive scale, are rising

[1] J. Granger, op. cit. p. 47.   [2] Ibid. p. 61.
[3] J. Bailey and G. Culley, General View of the Agriculture of Northumberland (1794), p. 8.
[4] Cp. A. Young, Six Months' Tour through the North of England (1770), III, 89.
[5] Ibid. III, 43, 78 and 89.
[6] Ibid. III, Letter XVI. It is of interest that tea was very little drunk in Northumberland, although this habit was common in Cumberland.
[7] J. Bailey and G. Culley, op. cit. pp. 7 and 51.
[8] Ibid. p. 15.

in every part of the county". On the other hand, Young styled the husbandry of the moorland farms as "vile" and "slovenly"; nor did the farming of the lowlands reach the best standards of the day.

In Cumberland the mountains were "perpetually covered" with great flocks of sheep, which were described as "ill-formed unprofitable animals".[1] "The face of the country", wrote the authors of *England Displayed*,[2] "is delightfully varied by lofty hills, vallies and water; but the prospect would be even more agreeable if it were not deficient in wood." Generally, arable farming was subordinate to sheep rearing and dairying. The plains around Penrith were divided into both enclosed and open fields. In Cumberland, as in Westmorland, much arable land stood open in the first half of the century, but enclosure took place gradually after the Union with Scotland in 1707 and the consequent pacification of the border.[3] Oats, barley and rye were the chief cereal crops. The county presented some striking contrasts of landscape, such as that, for example, between the wild and dreary moors about Shap and the lowland around Kendal to the south, with its "noble range of fertile inclosures, richly enameled with most beautiful verdure".[4] The northern part of Westmorland was described (in 1769) as "open Champaign country", devoted to mixed farming.[5] The southern part of the county consisted of mountains linked up by "roads, or rather paths...often frightful beyond description", but the valleys were fruitful, and many of the mountains yielded pasture for sheep and cattle. Elsewhere in the south stretched wide areas of moors: "their spongy surface serves the inhabitants for firing where they have no coals, and the turfs are called peat". Like Cumberland, Westmorland was thinly peopled, and in both counties arable farming, which played a minor part, was capable of much improvement. It was estimated in 1794 that more than one-half of Westmorland was uncultivated land.[6]

The western counties of England, although somewhat backward in farming practice, produced their several specialities, some of which reached the London market. Nearly one-half of Lancashire, consisting of moors, marshes, commons, mosses and fens, was recorded as "waste" in 1794.[7] The highlands of eastern Lancashire supported sheep, cultivation

[1] J. Bailey and G. Culley, *General View of the Agriculture of Cumberland* (1794), p. 17.   [2] I, 217.
[3] See G. Slater, *The English Peasantry and the Enclosure of Common Fields* (1907), pp. 257–60.
[4] A. Young, *Six Months' Tour through the North of England* (1770), III, 169.
[5] *England Displayed*, I, 237.
[6] A. Pringle, *General View of the Agriculture of Westmorland* (1794), pp. 6–7.
[7] J. Holt, *General View of the Agriculture of Lancaster* (1794), p. 52.

being restricted there to the valleys. Generally, grass farming was predominant in the county, and was growing at the expense of cultivation even, as on the loamy plain of the Fylde,[1] in the best arable lands. In the vicinity of the big towns dairying and market gardening became important: "there is not a town in the kingdom...London excepted, better provided with vegetables, roots, etc. than the town of Liverpool."[2] For the cultivation of potatoes Lancashire had a high reputation; it was, in fact, the first in the kingdom to introduce their cultivation.[3] It is clear that the growth of population in Lancashire, owing to its industrial activities, gave some stimulus to farming, but in arable farming bad rotations were adopted, although marling was practised with much success.[4] In the lowlands barley, wheat and, above all, oats were the chief cereal crops.

Oats, oats, oats, universally sown towards the north-east and south-east of Preston for years together, except this chain be broken occasionally by a crop of potatoes, and afterwards wheat, or wheat on a summer fallow. In the Filde, which, from its fertility, has been called the Granary of the county, the soil has never been worse abused.[5]

But something was being done to drain and improve some of the "moist and unwholesome" mosses of western and southern Lancashire. Finally, it is not without interest that in the Lancashire of 1769 the air was "serene", and the rivers abounded with fish.[6]

Of the Isle of Man it was recorded in 1794[7] that five-twelfths were "heathy mountains, and moorish ground"; the remainder, lying chiefly in the north and south of the island, was devoted to arable, pasture and meadow. The wide expanse of high country provided pasture for sheep, colts and young cattle, as well as peat fuel. In the broad, low plain in the north of the island an extensive moss had been reclaimed for grass and arable. Oats and barley were the chief cereals grown in the island, but scientific rotations were little understood or practised.

Cheshire, predominantly a grassland country, had its own peculiar commodities, salt and cheese, "both in request all over England", but it may be added that "Cheshire cheese" was made in many country districts, as, for example, in Warwickshire and in Wales.[8] In 1770 it was recorded that

---

[1] The Fylde was the plain between the Ribble and the Lune rivers.
[2] Holt, *op. cit.* p. 17.   [3] *Ibid.* p. 28.
[4] A. Young, *op. cit.* III, 198.   [5] Holt, *op. cit.* p. 26.
[6] *England Displayed*, II, 87–8.
[7] B. Quayle, *General View of the Isle of Man* (1794), p. 7.
[8] G. E. Fussell, "Agriculture and Economic Geography in the Eighteenth Century", *Geog. Journ.* LXXIV, 170–8 (1929).

Cheshire farmers were turning more of their abundant pasture into cultivation and that they were sowing clover and marling their land.[1] Even so,
the level of farming was low: some of the small farmers were described as
"more wretched than even day labourers".[2] Almost the whole of Shropshire, with the exception of some highland tracts in the south, was cut up
into hedged fields, as is clearly shown on Roque's large-scale map,
published in 1752. The authors of *England Displayed*[3] characterised
Shropshire in 1769 as follows:

The northern and eastern parts of the county yield great plenty of wheat and
barley; but the southern and western parts are hilly, are less fertile, though they
afford pasture for sheep and cattle, and along the banks of the Severn there
are large meadows, which produce abundance of grass.

Worcester was famed for "great plenty of all sorts of fruit, particularly
pears".[4] Sheep were pastured on the hilly ground of the county, whilst
the lowlands were devoted to corn, meadow and hops. So also was
Hereford noted for its orchards, above all for its apples: even the hedgerows
were said to be full of apple trees, and great quantities of cider were made.[5]
Its well-endowed plains, for example around Leominster, and its wealth
in woodland, pasture, corn, fruit and hops, impressed contemporary
writers. "Blessed is the eye, between Severn and Wye": so ran the proverb
which extolled the fertility of Herefordshire;[6] but some arable land lay in
open field there, as in Worcestershire, at the end of the century.[7] Finally,
Monmouthshire, the most southerly of the Border counties, was devoted
mainly to pastoral farming. In the west of the county the uplands were
thinly covered with peaty soil, upon which cattle, sheep and goats were
pastured, whilst the valleys were patterned with woods, pastures and even
a little cultivation. The south of Monmouthshire included a great deal of
moor and marsh, given over mainly to meadow, and in part to cultivation.
But the best part of the county was the eastern plain extending beyond the
Usk, since, rich in soils, it was "a treasure to the husbandman and the

[1] *England Displayed*, II, 72, and A. Young, *A Six Months' Tour through the North
of England*, III, 293.
[2] A. Young, *op. cit.* III, 298.
[3] II, 50–1.
[4] *England Displayed*, II, 43.
[5] It is of interest that in 1769 rough cider was preferred and made, whereas formerly
the taste was for sweet cider. Large quantities of cider were included in the daily wage
rates for farm labourers in the west country.
[6] *England Displayed*, II, 38.
[7] W. T. Pomeroy, *General View of the Agriculture of Worcester* (1794), p. 15.

grazier".[1] The county was well wooded; small enclosures were general; but there were still (in 1794) large tracts of waste "in a state of nature".

The Cotswold plateau stood almost entirely open in 1759; much of it, covered by furze and poor grass, afforded cow pasture and sheep grazing. By 1789, however, a great improvement was noted,[2] since the whole plateau had been cut up into enclosed fields. In the Vale of Gloucester, about one-half of the arable land was enclosed by 1789; and the grazing of hogs formed an important branch of farming. Towards the end of the century farming in the vale remained unprogressive; no carrots or potatoes were grown, fields were badly weeded and heavy ploughs, drawn by six or eight oxen, were still in use.

Defoe found Somerset a populous and flourishing county, conditions attributable in large measure to its textile industry and its sea-port towns. It contained sharply contrasted types of land. Among its highlands, Exmoor was "a vast Tract of barren and desolate Lands"; in the east of the county, and beyond it, stretching between Bristol, Cirencester, Sherborne and Devizes, the country was low, rich and enclosed; elsewhere, in the middle parts of Somerset, some land was "Marshy or Moorish, for feeding, and breeding, of black Cattle, and Horses".[3] The county was primarily engaged in dairy farming and Cheddar cheese had won for itself a high reputation.[4] The large areas of fen were described by Moll[5] as "bad in Winter by its being wet and moorish...which is troublesome to travellers". By 1797, however, 20,000 acres, forming the bulk of Brent Marsh, had been reclaimed and enclosed, and most of it had become fine grazing land.[6] Moreover, King's Sedgemore, a part of the extensive South Marsh, was then being drained.[7] Little had been done to reclaim the highland areas of Somerset, but half of the Mendips had been enclosed for sheep and arable farming. Some 60 per cent. of the county consisted, in 1797, of meadow and pasture land, and Somerset produced insufficient grain for its own consumption.

Despite its populousness and despite its industrial activity, Devonshire stood amongst the most backward agricultural regions of England.

Much inclined to Hills [wrote Moll in 1724], in some Places well cloathed with Wood, and the Ground generally ungrateful to the Husbandman, without

[1] J. Fox, *General View of the Agriculture of Monmouth* (1794), pp. 11–12.

[2] W. Marshall, *Rural Economy of Gloucestershire* (1789), II, 9.

[3] Defoe, *op. cit.* I, 267–80.

[4] *Ibid.* I, 278, and J. Billingsley, *General View of the Agriculture of Somersetshire*, (2nd ed. 1797), pp. 247–8.

[5] *Op. cit.* p. 38.   [6] Billingsley, *op. cit.* pp. 167–8.   [7] *Ibid.* p. 189.

great Pains and Charge in manuring, which they do by using Lime and Rags, and by paring the Surface of the Ground and burning it. . . . They have likewise a certain Sand from the Sea-Shore, which causes great Fertility, though several other Places are naturally fruitful.[1]

The reports on Devonshire agriculture at the end of this century show how little the spirit of improvement had penetrated there, despite the fact that its arable lands had long been enclosed. The practice of draining was little known; insufficient fodder for cattle was grown by the farmers of north Devonshire, largely "owing to the wet state of their lands". Although the roads around the towns were good, elsewhere the narrow, winding lanes were almost useless except for pack animals, wheeled vehicles being still few in 1808.[2] The paring and burning of land in order to improve it was widely practised, even on unsuitable soils.

Farming in Cornwall, too, remained very unenlightened, the land proving very unproductive, especially away from the coasts; in fact, in the interior it was mining, not agriculture, which supported the scanty population.[3] Pastoral farming predominated over arable, and the county abounded in game, fowl and fish. Moll[4] depicted the county in 1724 as:

generally very hilly, consisting ordinarily of Rocks and Shelves, but crusted over with shallow Earth, and more inclined to Barrenness than Fertility; but the Part to the Seaward, by reason of the industrious Husbandman's manuring the Ground with Sand and Ore-Weed taken from the Sea-Shore, bears good Corn, and feeds Store of Sheep and other Cattle; the inland Part, except the Inclosures about some Towns and Villages, lies pretty waste and open, bearing Heath and spiry Grass, and serves chiefly for Summer Cattle; and the Country is generally bare of Wood and Timber-Trees.

As late as 1794 there were "immense tracts of uncultivated wastes and undivided commons, entirely in a state of nature".[3] In fact, the fishing and coasting trade of the coastlands, together with the tin and copper mines and the woollen industry of the inland regions, formed the mainstays of Cornish economy. Even so, some improvement in agriculture was noted already in the 'sixties, since some turnips and potatoes were being grown and, in normal years, the county was no longer dependent on grain from outside.[5]

[1] *Op. cit.* p. 217.
[2] C. Vancouver, *General View of the Agriculture of Devon* (1808), p. 371.
[3] R. Fraser, *General View of the Agriculture of Cornwall* (1794), p. 8.
[4] *Op. cit.* p. 17.
[5] *England Displayed*, I, 4. Two crops of potatoes a year were grown around Penzance, thanks to the climate and to supplies of sea-sand and sea-weed for fertilisers.

The continued diminution of the woodlands of England, notwithstanding that successive governments had been aware of this for two centuries, was leaving its mark on the countryside and attracting the attention of eighteenth-century travellers and publicists. That this country had been plentifully clad with deciduous trees of many species, especially the oak, from time immemorial, is well enough known, although the studies of archaeology, of plant ecology and of climatology demonstrate clearly that there were always, from prehistoric times onwards, appreciable areas normally free of woods and undergrowth. It is true that even in the eighteenth century the country was not stripped bare of trees; but the supply of timber, it has already been shown,[1] was fast dwindling and was inadequate in relation to the increasing consumption. Defoe, asserting in 1724[2] that there was no lack of timber, tried to dispel the popular bogy of timber deficiency, but the weight of contemporary evidence discredits his assertion, and leaves no doubt as to the extent to which "destructive exploitation" of the English woodlands had gone. The demand for timber for many purposes had been met for centuries, but no effective measures had been taken, by means of a consistent forest policy, to make good the losses by new plantations; whilst the natural renewal of the woodlands was checked by browsing animals. Nor did the private owners, to whom the destruction and marketing of trees was *inter alia* a convenient means of paying debts, show that altruistic longsightedness required for forest plantation. As a rule they did no more than plant a few rows of trees for ornamental purposes.[3]

The growth of English maritime power and commerce, especially the latter, took heavy toll of the forests. In the eighteenth century the demands of the navy for timber for the hulls and planking of ships fell very far short of those of the merchant marine, but on the other hand the navy laid claim to the pick of the forest—stout, mature oaks of about a hundred years old.[4] Similariy, the industrial consumption of wood steadily increased. Charcoal was the essential fuel in smelting ores and, although coal was used for many industrial purposes, such as brewing, distilling, salt boiling, forging and glass making, its use was very restricted, partly owing to the difficulty and the high cost of inland transport. Moreover, for domestic purposes, both for building and for heating, the consumption

---

[1] See pp. 395 ff. above.   [2] *Op. cit.* I, 125.

[3] On the London Clay around London, as Kalm recorded, the elm was the favourite tree. *Op. cit.* p. 86.

[4] R. G. Albion, *Forests and Sea Power* (1926), chap. 1. This work contains a full bibliography.

of wood increased with the increase of population and the growing demands of the towns. Coal for domestic fuel was, Kalm[1] tells us, exclusively used in London by 1748, as in all the other towns.

In the villages which lay nearest around London, coal was also the principal fuel, although there they spin it out with sticks, cut in the hedges. But...about 14 English miles from London, and in places to which they had no flowing water to carry up boats loaded with coals, for the most part bare wood was used.

At the end of the century Archenholtz commented on the common view that one of the greatest inconveniences attributed to Britain was "the indispensable use of pit coal".[2] Nevertheless, in many parts of the countryside, firewood was the chief fuel, especially in the first half of the century, being largely obtained from quick growing coppice, from live hedges and from aged trees. Hedgerows and trees often suffered from excessive plundering: "I never saw", wrote John Middleton[3] in 1797, "hedgerows in any district so barbarously used by the tenants...as those in Norfolk and Suffolk." And Arthur Young[4] noted in 1768 that throughout southern England, except in East Anglia, trees were "stripped up like May-poles... with only a little tuft of leaves on the tops", so acute had become the dearth of even small wood. In some districts, too, as for example in Northamptonshire and Nottinghamshire, the extravagant practice long lasted of using dried cow dung as fuel. Finally, the enclosure movement, in the interests of arable or pastoral farming, made considerable inroads into the wooded wastes, since very often the best clay and loam soils, which were sought for cultivation or pasture, were those bearing the remnants of former woods.

Except perhaps in a few inaccessible districts, virgin woodlands had long ago disappeared in England. There is a sameness about the contemporary accounts of the decrease of woodland in almost every part of the country. "The timber in this county [sc. Devon]", it was noted[5] in 1808, "is wasting in a most alarming manner." In a similar vein a traveller in the 'eighties noted[6] how, even in the recesses of Merionethshire, the woods were "wasting every hour", enclosures being made for pasture. On the

---

[1] *Visit to England*, pp. 137–8.
[2] *Op. cit.* p. 88. Coal was thought "inconvenient" by some people because of the smoke.
[3] *Op. cit.* p. 275.
[4] *A Six Weeks' Tour*, p. 251. See also p. 77.
[5] C. Vancouver, *General View of the Agriculture of Devonshire* (1808), p. 457.
[6] *The Torrington Diaries*, I, 149.

other hand, there certainly existed a few extensive tracts of woods, such as the thick beech woods on the Chilterns in south Buckinghamshire;[1] plantations on a small scale were observed in different parts of the country, such as those of oak and ash which Marshall saw in the midlands;[2] whilst the planting of trees for ornamental purposes had a certain vogue and included species brought from America. Thus, even if at first glance much of England appeared wooded, its scenery was rather of the *bocage* or parkland type. The chief tree species were the oak, elm, beech and ash, but the Scots fir, the commonest tree in Scotland, was being introduced in many plantations throughout England. Part of the English woodlands still lay within the limits of royal forests,[3] but it was privately owned woodland which supplied the bulk of the timber consumed. The south-east of England contained its chief area of woodland: the oaks of Kent, Surrey, Hampshire, and above all those of Sussex, owed something of their high repute to the optimum conditions of soil and climate which obtained in the Weald. The best timber within the Weald came from the privately owned Ashdowne Forest, whilst of the royal forests that of Alice Holt in Hampshire was the most valuable. Farther to the south, the New Forest offered only moderate supplies of timber. A traveller[4] in the 'eighties wrote of the trees of the New Forest thus: "Neglected and wasted, as they are now, perhaps an enclosure were better than their continuing a wilderness of waste." In contrast, in south-west Gloucestershire, the Forest of Dean was still an important source of oak for the navy, and the most productive of all the royal forests. Finally, it is worth noting that the chief naval dockyards, at Woolwich, Deptford, Chatham, Sheerness, Portsmouth and Plymouth, lay around the south coast, close as the crow flies to the Weald, the New Forest and the Forest of Dean. Plymouth received its timber by sea, mainly from the Forest of Dean; the Portsmouth dockyards, which had exhausted the woods of the Isle of Wight,[5] were supplied to a small extent from the New Forest, timber being shipped from the many creeks which open out into Southampton Water. Timber from the Weald forests reached the Thames and Medway yards, either by way of the Medway, which had been made navigable below

[1] A. Young, *op. cit.* p. 86. These woods conditioned an active furniture industry: see Defoe, *op. cit.* I, 301.

[2] *Rural Economy of the Midland Counties* (2nd ed. 1796), I, 68.

[3] A forest was a tract of uncultivated land, not necessarily wooded, which belonged to the king and was subjected to a special law. See above pp. 173 *et seq.* and p. 349.

[4] *The Torrington Diaries*, I, 82.

[5] C. Vancouver, *General View of the Agriculture of Hampshire* (1794), p. 89.

Tonbridge in 1740, or from Godalming on the Wey, and thence into the Thames.[1]

Through the neglect of its woodlands, Britain had to rely increasingly on imported timber from Norway, the south Baltic ports, and New England. Stettin oaks, oaks and firs from Danzig and Riga, and firs from Norway and St Petersburg, were familiar cargoes brought into English ports. For masts and spars, as well as for naval stores generally, and at times even for ship's planking, the naval shipbuilding yards were essentially dependent on Baltic supplies, a dependence which had remarkable effects on British foreign policy.[2]

TOWNS, INDUSTRY AND COMMERCE

It is a truism to observe that urban development indicates economic specialisation—the grouping within towns of people who are engaged primarily in trade and industry and are dependent on the countryside or on imports for their supplies of food. Although this specialisation is sufficiently apparent in the eighteenth century, it must be emphasised that, to a large extent, industry was rural rather than urban. In some cases, notably in that of the woollen textiles, production had moved from the old corporate towns in search of freer habitats elsewhere; in other cases, for example in those of iron and coal, industry had been traditionally located in the country; finally, in the case of the cotton industry, of relatively recent development, the various processes of manufacture were conducted for the most part outside the towns. But even if towns had only a share in industrial production, they were all-important in the activities of exchange. The chief towns of the early eighteenth century were essentially "metropolitan" cities, the economic relations of which extended over a wide region. It is significant, however, that there was no suggestion on the map of 1700 of the great industrial conurbations of to-day; in fact these are only foreshadowed in 1800. Large-scale maps, which are available towards the end of the eighteenth century, emphasise the low degree of urbanisation in regions of expanding industry. Thus William Yates's map of Lancashire (2nd edition, 1800) does not show urban development around

[1] The river, navigable up to Guildford, was canalised in 1780 up to Godalming. Timber from the Alice Holt Forest was, as Gilbert White noted, shipped from Godalming.

[2] Thus on nearly twenty occasions between 1658 and 1814 British fleets were sent to maintain the so-called freedom of the Sound. C. E. Hill, *The Danish Sound Dues and the Command of the Baltic* (1926), p. 165.

Manchester-Salford, but indicates a general scattering of rural settlement throughout Lancashire, except in the many areas of peat bog or "moss".

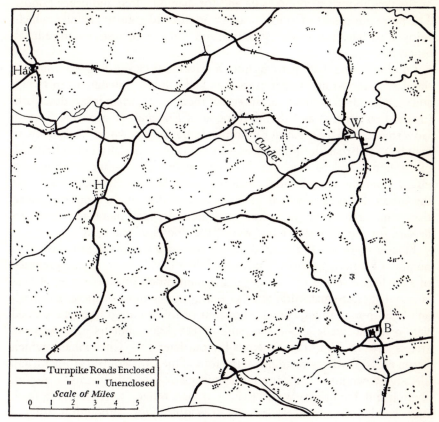

Fig. 78

The building pattern in a portion of the West Riding as drawn from John Tuke's map of Yorkshire in 1787 (2 miles = 1 inch). *W, B, H*, and *Ha* mark the towns of Wakefield, Barnsley, Huddersfield and Halifax. The map emphasises the low degree of urbanisation in the woollen area. This diffusion of rural settlement stands in contrast to the nucleated villages of the Vale of York. As water was available everywhere, the high lands, as well as the valleys, were settled. The many turnpike roads passed through country which was mainly enclosed and they tended to run obliquely to the valleys.

Similarly John Tuke's map of Yorkshire (1787) shows how the West Riding textile region was essentially rural, and not urbanised to any extent (see Fig. 78).

Apart from the metropolitan cities, towns were small and widely spaced : they were centres of industry, as well as markets for the produce of the neighbouring countryside and for manufactured and other goods. The more important of these towns were the old county capitals, which enjoyed parliamentary representation and had administrative and judicial functions.

London alone had grown into a great city according to modern standards, since it had rather more than half a million inhabitants in 1700,[1] and nine hundred thousand at the end of the century, when it was described (in the first census) as "the Metropolis of England, at once the Seat of Government and the greatest Emporium in the known world".[2] Finally, it may be noted that the population of the towns, especially in the first half of the eighteenth century, was able to grow only by reason of the constant influx from the countryside, since it is now fairly well established that the urban death-rate, which much exceeded that in the country, more than counterbalanced the urban birth-rate. In fact, there was much to be said for the view that cities were "the graves of mankind".

Finally there already existed at the end of the seventeenth century a number of watering-places which owed their development to the presence of a medicinal spring or to other natural amenities, but these were at best large villages or small rural towns. Bath, the foremost of these, although an old corporate town, consisted of four or five hundred houses crowded within its old walls above the bank of the Avon, but its narrow mean streets gave no hint of its spacious lay-out later on.[3] Cheltenham, which lay on the banks of a stream and at the foot of the Cotswold escarpment, had already grown into "a quiet country village of 1000 inhabitants". Again, Buxton and Tunbridge Wells, thanks to their waters, were changing from mere hamlets into small resorts. Even Brighton, too, although it suffered badly in the seventeenth century more than once from inundation and erosion by the sea, had its few visitors to supplement the fishing which was then its chief activity.

The development of English towns during the eighteenth century,

---

[1] This was the population within the somewhat comprehensive Bills of Mortality area. On this, see N. G. Brett-James, *The Growth of Stuart London* (1935), chap. xx and map, p. 248. See also chap. xiv below.

[2] But to Arthur Young, as an exponent of scientific agriculture, London appeared rather as "a market and as a dunghill".

[3] Thus Archenholtz wrote in his *A Picture of England* (1797), p. 325: "Bath is a very beautiful town and the public edifices that adorn it are superb. It is the rendezvous not only of the sick, but of persons in health, whom the variety of pleasures to be found there attracts from all the corners of the kingdom."

although it throws a forward light on the functions and grading of present-day towns, in some measure reflects conditions which had long endured. Among the most populous towns at the end of the century stood new industrial and mercantile centres, such as Birmingham and Liverpool, but the second and third grade towns were mainly old corporate cities which had important economic interests. It may be worth tabling below the populations of the largest English towns as these were for the first time accurately revealed in the first census. (The towns are divided into three categories, and the figures are given to the nearest thousand.)

The Chief Towns of England in 1801

Towns with between 50,000 and 85,000 inhabitants

Birmingham... ...	74,000
Bristol	64,000*
Leeds	53,000†
Liverpool	78,000
Manchester-Salford ...	84,000
Southwark	67,000

Towns with between 25,000 and 50,000 inhabitants

Bath...	32,000
Hull	30,000
Newcastle	28,000
Norwich	37,000
Nottingham	29,000
Plymouth	43,000
Portsmouth-Portsea...	32,000
Sheffield	31,000

Towns with between 10,000 and 20,000 inhabitants

Cambridge	10,000
Carlisle	10,000
Chatham	11,000
Chester	15,000
Colchester	12,000
Coventry	16,000
Derby	11,000
Dover	15,000
Exeter	17,000
Greenwich	14,000
Huddersfield ...	11,000
Ipswich	11,000
Leicester	17,000
Lynn	10,000
Oxford	12,000
Reading	10,000
Shrewsbury	15,000
Stockport	15,000
Sunderland	12,000
Wigan	11,000
Worcester	11,000
Yarmouth	15,000
York	16,000

* This figure includes the population of Barton Regis Hundred.

† This figure includes the population of seventeen townships within the Liberty of Leeds.

Further, the first census throws many other interesting sidelights on the development of English towns. The old corporate city of Lincoln, formerly so important, had only 7000 inhabitants; Southampton had only 8000. In contrast, Brighton, growing fast, was the most populous of the Sussex towns.

It remains to look in turn at some of the old established and some of the "mushroom" towns, which figured prominently in the eighteenth century. In the first class, Bristol and Norwich and in the second class, Manchester and Liverpool may be briefly examined.

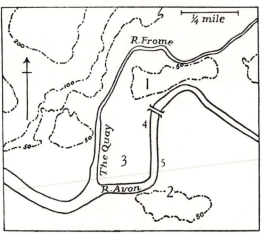

Fig. 79

The Site of Bristol. 1. The nucleus of the town, originally girdled on the west by the river Frome. 2. The analogous site of Redcliffe. 3. Canons' Marsh, reclaimed as Queen Square. 4. The Welsh Back. 5. Redcliffe Back.

Ships came up to 4 and 5 on the Avon and to "The Quay" on the Frome. Contours are shown in feet.

In its ground-plan eighteenth-century Bristol had grown well beyond its walls, within which its four principal streets intersected. On the western side of the town the river Frome had long been canalised to serve more effectively the needs of harbourage and warehousing. The harbours of the town consisted of "The Quay", on the town side of the Frome or the "Trench", which was used by river craft; the Avon both above and below the bridge, which was reached by sea-going ships; and finally Redcliffe Back to the south (Fig. 79).

The port is good [wrote Moll][1] and the principal Key stands on the Frome, which at Spring Tides flows about Forty foot, bringing in Ships of very great

[1] H. Moll, *op. cit.* p. 43.

Burthen;...it is well frequented and inhabited by merchants and tradesmen, having Plenty of Provisions of all Sorts.

Contemporary plans mark the growth of the town in the eighteenth century. The walls for the most part and the castle had been demolished, whilst "The Marsh", an area originally liable to continual floods, was reclaimed and built on, to be renamed, after Queen Anne, Queen Square. Towards the end of the century there was a rapid suburban development at Clifton.[1] On account both of the narrowness of the streets within the walls and of the many vaults or cellars, dog-drawn sleds instead of horse-drawn waggons were employed for carrying wares; nor could private carriages be used. On the other hand, Bristol seems to have been relatively well favoured in respect of sewerage and water supply, amenities which were greatly deficient in English towns, especially in the first half of the century. Water was supplied by several public conduits, and there was a spring at the foot of Brandon Hill on the northern flank of the town, and this hill, on the summit of which Cabot Tower now stands, remained open, although buildings were beginning to surround it.[2]

Bristol owed its success in the eighteenth century to the exploitation of its commercial opportunities. It was related landwards with the zone of populous agricultural, and in part industrial, counties of southern England: carriers continually passed and repassed between Bristol and the other important towns of the west country.[3] Seawards its relations were both local and extended. On the one hand, Bristol traded coastwise along the western coast of Britain and with Ireland; it had long established relations with the ports of western France, the Mediterranean, Holland and the Baltic, whilst on the other hand, it had been one of the first to forge links with West Africa and the Americas, above all, with the highly valued West Indian islands: "for domestick and foreign Trade, especially to the West Indies, [it] is justly reputed the second City in England".[4] Its commercial activities, which included the sale of West African natives in Central America, were reflected in some of the industries of the town (such as sugar refining) and in the commodities for which its fairs were renowned throughout Western Europe. Tobacco from Virginia, sugar and rum from the West Indies, wines and salt from France, Spain and Portugal, raw wool, woollen yarn, flax and cattle from Ireland, copper from Cornwall, coal and copper ware from south Wales, together with

[1] J. Latimer, *Annals of Bristol in the Eighteenth Century* (1893), pp. 493–4.
[2] On Millerd's plan (1673), Brandon Hill is described as a "publick convenience to ye Cittie for ye use in drying Cloaths" (cloth?).
[3] *England Displayed*, I, 52.   [4] Moll, *op. cit.* p. 43.

wool and wheat derived from Hereford and south Wales generally—these were among the staple commodities brought to the town. Bristol, again, had a local importance in relation to the West Country woollen industry: it imported wool and exported finished cloth, some of which was made in the town. Finally, as the greatest town of the west, Bristol played the part of a provincial capital, its influence extending throughout the West Country and south Wales.

Alike on the basis of population and industrial activity Norwich was reckoned one of the chief cities of eighteenth-century England. In the 1720's, as Defoe relates,[1] Norwich was still contained within its medieval walls, despite its growth in population and industry.

The Walls of this City are reckon'd three Miles in Circumference, taking in more Ground than the City of London; but much of that Ground lying open in Pasture-Fields and Gardens; . . . the Walls seem to be placed, as if they expected that the City would in time encrease sufficiently to fill them up with Buildings.

Like Bristol, it reached its greatest relative importance in this century. There were many facets to its economic activity.. As a busy centre of woollen, worsted and even silk manufacture, it provided employment for many workers both in the city and in the surrounding country. Again, as its cattle and corn markets showed, it was a market for an increasingly productive and populous agricultural region. Similarly, it was a bishop's see and a regional capital. Accessible by boats directly from the sea, as Defoe put it, "without Locks or Stops", it was placed at a point where bridging was relatively easy. Further, its geographical position was one of geometric centrality in relation to the county of Norfolk, and contemporary maps show how nearly a dozen routes converged on the town. It is noteworthy, too, that owing to the broad belt of water meadows to the east of the town, which formed an obstacle to north-south movement, some roads were deflected to the river crossings at Norwich.

The relative fall of cities, no less than their sudden rise, is illustrated in the urban history of eighteenth-century England. It must suffice here to note the continued decline of Chester (although, as Defoe and others testified, numerous instances of urban decline could be cited). The decline of Chester was essentially relative and not absolute, since this historic city had established for itself a regional position which could scarcely be effaced. Its port had long suffered from silting and in 1674 small ships of twenty tons burden could approach the town no nearer

---

[1] Op. cit. I, 63.

than Neston.[1] After many efforts to improve its river, Chester succeeded by 1737 in cutting a new channel from the town and at the same time in reclaiming an area of "white sands", formerly inundated by tidal water. In result, ships of 350 tons burden were for a time enabled to reach the town. Other canal projects were attempted, but without complete success: thus the Weaver navigation, designed to connect the town with the salt-producing parts of Cheshire, was carried only to Nantwich, instead of to Middlewich, and proved largely a failure. In fact, the task of maintaining the port proved uneconomic, and as a maritime city Chester decayed. Liverpool came easily to dominate as the north-western outlet of England, whilst Holyhead, in the first decades of the nineteenth century, was linked up by a genuine national road with London, and thence were conveyed the Irish mails. Even so, as a provincial metropolis for its own county and also for north Wales, no less than as an important market town for horses, cattle and dairy produce, and as a small industrial centre, Chester preserved its urban importance.

Even if much remained constant in the urban geography of nineteenth-century England, signs were not lacking to indicate the rise of new towns, the fortunes of which measured some of the changes in the concentration of industry and population. Many of the great cities of the last century, it is true, had not made spectacular advances by the end of the eighteenth century. Thus, for example, the West Riding towns of Halifax, Huddersfield, Wakefield and Bradford numbered between six and nine thousand souls.

In Lancashire, on the other hand, where industrialisation had already made much headway, Manchester and Liverpool, especially the latter, had made the most striking progress among the new towns. To Defoe,[2] however, Manchester was only "one of the greatest, if not the greatest mere village in England". It grew in the course of the century, side by side with the industrial development of east Lancashire generally, both as a market and as a centre for certain processes in the textile industry, such as printing and dyeing. The available figures suggest its rapid growth in the latter half of the century, at the end of which it had reached some 84,000 souls. The provisioning of the town, which stood amidst barren country, involved considerable claims both on the more distant countryside and the coasts.[3] Vegetables, including

---

[1] W. T. Jackman, *The Development of Transportation in Modern England* (1916), I, 197.   [2] *Op. cit.* II, 69.

[3] J. Aikin, *A Description of the Country from Thirty to Forty Miles round Manchester* (1795), pp. 204-5.

potatoes, came from the country around Warrington, Runcorn and Frodsham; fish from the Isle of Man and even, by pack-horse, from the Yorkshire ports; oats were brought from the Fylde, the low plain of north-western Lancashire; and coal, by canal, from the Worsley collieries near-by. The needs of the textile trade were served at first by pack-horses and waggons along the roads and later also by canals: by canal Manchester was linked with the Mersey estuary at Runcorn, with Huddersfield and Halifax, and with Hull, by way of the Grand Trunk canal. When Aikin wrote in 1795, he claimed that Manchester had assumed "the style and manner of one of the commercial capitals of Europe".[1]

Liverpool was originally a small agricultural settlement occupying a small sandstone plateau, below which, at the widening mouth of the Mersey, stood a small tidal creek, called "the Pool". In the Middle Ages merely a "limb" or creek in the port of Chester, it grew suddenly and rapidly in the eighteenth century in direct ratio with the growth of British imperial and commercial interests. What is more, it grew faster than, and ahead of, industrial east Lancashire. Defoe, who paid three visits to the town, was amazed at its remarkable advance. "Liverpool", he wrote,[2] "is one of the wonders of Britain, and that more, in my opinion, than any of the wonders of the Peak;...I am told that it still visibly increases in wealth, people, business, and buildings. What it may grow to in time, I know not." And Arthur Young, who—unlike Defoe—had little interest in towns, thought the town in 1770 "too famous in the trading world to allow me to pass it by".[3] In some respects the site and position of Liverpool were distinctly disadvantageous. Like Hull, the town lacked adequate local supplies of water, and at the end of the century water was actually sold in the town from carts at a half-penny a bucket.[4] Again, Liverpool suffered at first by marked isolation, if only landwards: the Wirral across the Mersey estuary lacked towns and effective thoroughfares until the latter half of the century, whilst there were no carriage roads from Liverpool, and only pack-horse routes were available between the town and east Lancashire. Not until 1765 did the stage coach from London, which formerly stopped at Warrington, reach Liverpool for the first time, and from this time onwards, thanks to turnpikes and canals, the town began to enjoy relatively good communications by land. In the last decades of the century Liverpool was drawing food supplies from the Wirral by ferry and from the plain to the north of the town, whilst it was

[1] Aikin, op. cit. p. 184.   [2] Op. cit. III, 200.
[3] Six Months' Tour through the North of England, III, 168.
[4] Aikin, op. cit. p. 363.

supplied by canal and river with coal from St Helens and Wigan, and with salt from Cheshire. Its hinterland relations were extended by canals, especially the Bridgewater canal to Manchester and the Grand Trunk to the Trent and Severn (Fig. 82). Towards the end of the century the commerce of Liverpool expanded rapidly; it included not only local trade with Wales[1] and Ireland,[2] but an extensive overseas trade with West Africa, the West Indies and North America. Like Bristol, it owed some of its wealth to the infamous "triangular trade";[3] it was a great importer of tropical produce—sugar, tobacco, coffee and cotton, and, further, it functioned as an entrepôt port to an ever-extending hinterland. Favoured by the facility of water-borne materials, the town developed successfully a number of industries: glass-making, salt and sugar refining, copper and iron smelting, shipbuilding, brewing, and the milling of corn. Liverpool led the way in dock construction, and had several docks and tidal basins along Merseyside by 1795.[4] To Defoe it was "already the next Town to Bristol", but in the 'sixties more ships were frequenting Liverpool than Bristol.[5] The former had become in fact the premier seaport of western England. On the basis of population, too, it had (by 1801) supplanted Bristol as queen of the west.[6]

The growth of English overseas commerce, especially rapid after 1780, and the mercantile conceptions which governed it, need not be examined here. It is true that, even in 1800, British commerce, although its expansion preceded and stimulated industrial development, was very small in scale measured by later standards. Britain began in this century to reap the fruits of its new location.[7] By the year 1763 it stood at the centre of widely spaced imperial possessions, in India, in the West Indies, in North

---

[1] Some young men at Liverpool found it worth while to learn Welsh, *The Torrington Diaries*, I, 149.

[2] On the basis of shipping tonnage leaving the port the Irish trade was easily the most important (Aikin, *op. cit.* pp. 367–8). It should be remembered that Ireland was relatively well populated in the eighteenth century.

[3] This consisted in collecting slaves in West Africa, selling them in the West Indies, and bringing thence rum and sugar. Defoe, *op. cit.* II, 202–3.

[4] See the plan of Liverpool in 1795 in Aikin, *op. cit.* p. 331.

[5] A. Anderson, *Origin of Commerce* (2nd ed. 1789), IV, 97.

[6] The population of Liverpool is estimated at about 4000 in 1700, about 34,000 in 1773. On the development of Liverpool in the eighteenth century see T. Baines, *History of Liverpool* (1852), and W. Enfield, *Essay towards the History of Leverpool* (1773).

[7] See above, pp. 353 and 386.

America, in West Africa and in the Mediterranean, and, further, it had achieved world ascendancy in naval power and maritime commerce. From India, China and the East Indian islands, ships brought to the port of London cargoes of spices, silks, dye-woods, porcelain, tea, and finely woven cotton goods.[1] From the West Indian islands in the New World, islands highly prized because their southern latitudes permitted the production of tropical commodities, were brought sugar, cotton and rum. Canada, in contrast, had little value: even in 1804 it was called "the wilderness across the Atlantic",[2] and it is well known that an English statesman considered exchanging it in 1763 for the West Indian island of Guadeloupe. The Thirteen American Colonies, on the other hand, were valuable for their supplies of tobacco, timber, iron, and, at the end of the century, cotton. The Guinea trade, too, was profitable enough: it centred on the collection of slaves in West Africa for trading in British and Spanish territories[3] in tropical America. To these more distant fields of commerce were added those nearer to hand, which had long been traditionally sought: the North Sea and Baltic lands, fundamentally important for their supplies of iron and naval stores; the Mediterranean lands for their cotton; Ireland for its flax and wool; France for its wine and salt, and the Iberian peninsula for its iron, wine and wool.

The growth in the volume and extent of overseas trade had its reactions on the seaports of England, and many changes took place in their relative values. The prosperity of Bristol and the rise of Liverpool, noted above, were in some part conditioned by their trade with the New World and with West Africa. On the east coast, London, by far the most active port, monopolised the East Indian trade, and received ships from almost every other part of the world. Hull and Newcastle specialised, above all, in the North Sea, Baltic and Muscovy trades. The repercussions of overseas commerce extended, too, beyond the seaports into the interior, which formed their hinterlands. Improved communications were relating the ports more closely with extending hinterlands, from which they derived a large part of their return cargoes. Thus the port of Hull, which built its first tidal basin in 1778 to meet the growing needs of shipping, was connected by river, canal and road with many regions of expanding industry:

---

[1] In the interests of the woollen industry, the import of cotton goods was more than once prohibited by Act of Parliament.
[2] G. Chalmers, *An Estimate of the Comparative Strength of Great Britain* (1804), p. 144.
[3] The Asiento clause in the Treaty of Utrecht gave England in 1713 the monopoly of the slave trade with Spanish America.

not only with the West Riding, but also with east Lancashire, Nottingham-shire and the Potteries of south Staffordshire.

But trade was domestic as well as colonial and foreign. It will be noted later how the difficulties of inland transport set obstacles to internal mobility and trade, but it would be rash to infer that English regions were self-sufficient. Many instances can be given to suggest the remarkable fluidity of internal trade. The dominance of London as the chief market and seaport of England was, it is true, no new thing, and its large and varied trade with the whole countryside points to the mobility of internal traffic. Defoe's *Tour* refers often to the flow of goods to the capital from every part of the country:[1] only Cornwall was an exception, or rather a partial exception, to this rule.[2] Coal from Newcastle and Sunderland by sea, cheese from the counties of the Welsh border, cattle from many parts of the country including even Anglesey, cloth from the West Country and elsewhere, salmon from the Tweed,[3] wool and grain from many parts of the country—these are but a few instances of the way in which the great demands of the capital were met. Travellers noted the effects of this trade: only in London, it was said, was the best of everything to be had, and at Cheltenham "even salmon is 7 or 8 pence the pound, owing to London's powerful indraught".[4] Similarly, outside London, commodities were moved over long distances, especially to the chief fairs. Scots cattle were brought south for sale in Norfolk; Welsh cattle were brought down from the hills for sale to midland graziers; wool was widely interchanged, both coastwise and by road; and finally, in the last decades of the century, coal, lime and stone were widely distributed by water.[5]

Even so, the contacts established by trade and travel between different parts of the country were not strong enough to erode the many local peculiarities of English life. In language and dialect differences were still

[1] Defoe, *op. cit.* I, 265. Defoe decided, on setting out on his travels, to "take Notice how every County in England furnish'd something of its Produce towards the Supply of the City of London". See pp. 541–2 below.

[2] *Ibid.* II, 660.

[3] The salmon was brought by horse from the Tweed to Newcastle, there "curl'd" and "pickl'd", and sent thence to London, where it was known as Newcastle salmon.

[4] *The Torrington Diaries*, I, 9.

[5] The movement of stone was having its effects on the domestic building in English villages. Thus, in the midlands, thanks to the Grand Trunk canal, "the cottage, instead of being half covered with miserable thatch, is now [*sc.* 1782] secured with a substantial covering of tiles or slates, brought from the distant hills of Wales or Cumberland" (T. Pennant, *The Journey from Chester to London* (1811 ed.), pp. 75–6).

well marked. If Cornish was dying out, Welsh was yet spoken with vigour at Shrewsbury on market day;[1] whilst in Somerset Defoe[2] found that "the Dialect of the English Tongue...is not easily understood, it is so strangely altered". So also in food and in drink, regional differences survived. The traveller John Byng drank good perry at Worcester, and Defoe found in Herefordshire that "he could get no Beer or Ale in their Publick Houses".[3] If, as Kalm[4] tells us in 1748, "wheat flour bread" was the everyday food around London, and "other bread is next to never eaten", nevertheless other cereals, well adapted to the local conditions of climate, were grown and used as breadstuffs in Wales and the north (see Fig. 80).

The commodities which England shipped overseas were in part collected from many sources for redistribution, but in part also derived from home industries. Chief of these was the woollen industry in its many forms; and, since it was so important, it must be noted here in its geographical aspect.

At the outset it should be emphasised that the marked localisation of industry which exists to-day, based on new organisation and technique of production, was still at the end of the eighteenth century in a rudimentary stage. Those familiar features of modern industry—the use of steam power derived from coal, the grouping of workers in factories and the concentration of factories in towns, and further, the growth of industrial conurbations on the coal-fields—those features occurred only exceptionally, if at all, at the close of the eighteenth century. The woollen industry, far from being confined as it is to-day within a few specialised regions, was spread very widely though not uniformly throughout England. There were, it is true, many parts of the country, for example, in the midlands, in Kent, in Surrey, in Sussex, and in Hampshire, where it was scarcely to be found; but, elsewhere, within towns and in rural settlements some branch of this many-sided industry formed a staple occupation. It is true, again, that changes in its localisation were taking place during the century: the West Country and East Anglia, which held the lead in the woollen industry, were losing some of their importance towards the close of the century, whilst the West Riding was rapidly expanding its production. In general, despite the changes which occurred, the geographical distribution

---

[1] Defoe, *op. cit.* II, 75: "On Market-Day you would think you were in Wales."
[2] *Ibid.* I, 219.   [3] *Ibid.* I, 448.
[4] *Op. cit.* pp. 55 and 88.

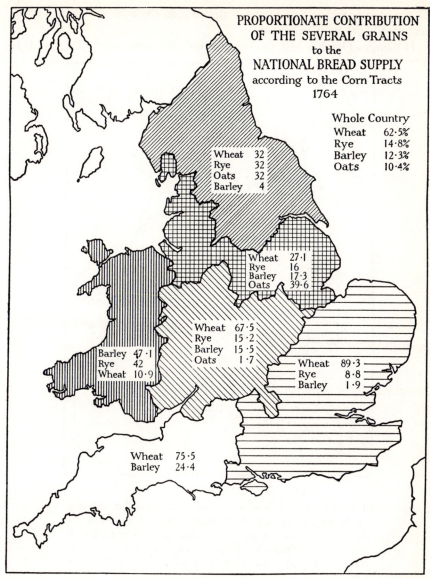

PROPORTIONATE CONTRIBUTION
OF THE SEVERAL GRAINS
to the
NATIONAL BREAD SUPPLY
according to the Corn Tracts
1764

Whole Country
Wheat	62·5%
Rye	14·8%
Barley	12·3%
Oats	10·4%

Wheat	32
Rye	32
Oats	32
Barley	4

Wheat	27·1
Rye	16
Barley	17·3
Oats	39·6

Wheat	67·5
Rye	15·2
Barley	15·5
Oats	1·7

Barley	47·1
Rye	42
Wheat	10·9

Wheat	89·3
Rye	8·8
Barley	1·9

Wheat	75·5
Barley	24·4

Fig. 80

Drawn from Sir W. Ashley, *The Bread of our Forefathers*, p. 7 (Oxford: Clarendon Press, 1928). The wheaten loaf was characteristic only in the southern half of England, and the map emphasises the reliance of different areas on locally grown cereals, adapted largely to local conditions of climate and soil.

of the textile industry, as Defoe sketched it in the 1720's, affords a good picture for the whole century.[1]

If then, as Defoe noted, the woollen industry was widely diffused, it is

Fig. 81

The distribution of towns engaged in the woollen industry on the basis of Defoe's *Tour* (1724–7). Many branches of the industry were carried on outside the towns. The rural character of the industry was most marked in east Lancashire and the West Riding where there were relatively few "woollen towns".

equally clear that there were three outstanding regions within which it flourished (Fig. 81). These were the West Country, East Anglia and the

[1] See also for a fuller treatment of the changes in localisation, E. Lipson, *The History of the Woollen and Worsted Industries* (1921), pp. 220–55.

West Riding of Yorkshire, to which should be added east Lancashire, which came to be identified principally with the production of cotton goods only in the last decades of the century.

In the West Country the localisation of the various branches of the manufacture was related above all to the availability of water supplies and of water power.[1] Certain cloths, such as broadcloths, serges and kerseymeres, required fulling, a process which demanded not only abundant water for cleansing the wool and for dyeing, but also water power to drive the mills. In contrast, other fabrics, for example blankets, flannels and stockings, needed little or no fulling, so that, although water was still indispensable in the preparation of wool, water power as such was not required. Thus the broadcloth manufacture of Gloucestershire was located along the base of the Cotswold escarpment and along the Stroud valley where good heads of water were provided. Similarly, in Devonshire, which was famous chiefly for its serges, many clothing towns and mills stood at the foot of the Dartmoor plateau and along the banks of rivers entrenched within it. Again, even in those districts where the manufacture required water only for cleansing and dyeing, the abundance of, or deficiency in, this natural resource was often a contributory factor in the growth or decline of their industrial activity. The wool used in the West Country, although in part derived from local sources, was largely supplemented by supplies brought from other parts of England, Wales and Ireland, and also from Spain. The Cotswolds, the downlands of Hampshire, Wiltshire and Dorset, and the highlands of Devon, were great producing areas of the coarser kinds of wool, but the woollen industry became so increasingly specialised in response to market conditions, that its various branches became largely dependent on particular types of wool. For the finest quality cloths the soft, short-stapled merino wools from Spain were essential, and among the English wools only Hereford fleeces approached these in fineness and softness. Cirencester was a great wool market for the West Country. Irish wool, a long "combing" wool used chiefly in the manufacture of serges, was sent over in the form of yarn to Bristol and Barnstaple. The cloths of the West Country, in common with those of the other major regions of the industry, served not only a home market but distant markets in continental Europe, North America and the West

[1] See R. H. Kinvig, "The Historical Geography of the West Country Woollen Industry", *The Geographical Teacher*, VIII, 243 (1916), and J. Morris, *The West of England Woollen Industry, 1750–1840* (unpublished M.Sc. (Econ.) thesis of the University of London).

Indies. The ports of Bristol and Exeter,[1] situated respectively on the northern and southern confines of the region, served it not only as outlets of finished goods but as inlets for raw materials, but more goods were sent to London and shipped therefrom.

The East Anglian textile industry, like that of the West Country, was carried on widely throughout the countryside, although it was associated with one dominant business centre, the city of Norwich. In Norfolk, wrote Defoe,[2] "we see a face of diligence spread over the whole country": the industrial and populous area in this county lay east of a line drawn between Brandon and Walsingham. Norwich itself was surrounded by a dozen market towns and "a throng of villages", but the industry extended also into Suffolk and Essex. The organisation of the industry was largely in the hands of rich clothiers or merchants who lived in the big towns, especially in Norwich. They bought the raw wool—especially the long stapled fleeces from Leicestershire and Lincolnshire—and distributed it throughout the countryside, where it was carded and spun and subsequently woven into worsteds. The specialisation on worsted production in Norfolk reflected the lack, in this level country, of water power, for, since worsteds did not need fulling, water was not needed as a source of power.

Finally, there were two other regions of thriving industry: first, the West Riding, which was destined, on the decline of the Norfolk[3] and West Country industries, to become the chief centre of the woollen and worsted production, and second, eastern Lancashire, which shared in the woollen industry, but was becoming also the specialised and exclusive home of the cotton manufacture in England. The presence of coal in both regions was useful for firing; and the scattering of the industry in cottages both in the valleys and the higher ground, was most characteristic (Fig. 78). Defoe[4] vividly describes the geographical conditions of the in-

[1] The port of Exeter was improved by the cutting of a channel, with locks, from the town to the sea, which enabled ships to avoid the difficult channel up the Exe estuary. Exeter was a famous market for serges. See W. G. Hoskins, *Industry, Trade and People in Exeter, 1688–1800* (1935).

[2] Defoe, *op. cit.* II, 268.

[3] The eventual decline of the Norfolk worsted industry, which began only towards the close of the century, was due to the competition of the West Riding. The success of the latter owed something to its geographical advantages of local coal, iron and water power, but owed more, perhaps, to the relative lack of enterprise shown by Norfolk capitalists. See J. H. Clapham, "The Transference of the Worsted Industry from Norfolk to the West Riding", *Econ. Journ.* XX, 195 (1910).

[4] *Op. cit.* I, 61.

dustry in the upper Calder valley around Halifax. The "Clothiers" kept a horse and a cow or two, and grew in enclosed plots a little corn, "scarce enough for their cocks and hens".

These Hills are so furnished by Nature with Springs and Mines, that not only on the Sides, but even to the very Tops, there is scarce a Hill but you find, on the highest Part of it, a Spring of Water, and a Coal-Pit....Having thus Fire and Water at every dwelling, there is no need to enquire why they dwell thus dispers'd upon the highest Hills, the Convenience of the Manufacture requiring it.

The working of wool, much more than agriculture, was the mainstay of the West Riding population between Leeds, Wakefield, Huddersfield, Halifax and Keighley. There, as in east Lancashire, a high rainfall and, in result, copious supplies of running water, much of which was naturally softened, served the needs of the industry. The banks of the swift Pennine streams were sought as sites for fulling mills. The development of the industry stimulated the growth of many small non-corporate towns, notably Leeds, which stood second only to London as a cloth market by 1750,[1] as well as Halifax, Huddersfield and Wakefield. Even so, it should be emphasised that "the towns were not the centres of manufacture, but were chiefly engaged in the finishing processes, in the marketing of raw material, cloth and foodstuffs, and in providing accommodation for merchants, clothiers and travellers".[2] In the marketing of wool and woollens the West Riding was related essentially with London and Hull. To both of these ports wool was carried coastwise from the eastern counties of England.[3]

In eastern Lancashire the manufacture of cotton goods became the chief branch of its textile industry in the second half of the century. In this instance an industry was developing on the basis of raw materials brought by sea from many distant sources, and was making a product in imitation of fabrics traditionally manufactured in the Far East. The localisation of the industry in the valleys and foothills of east Lancashire, owed something to the non-exclusiveness of its towns, to the local craftsmanship in working wool and flax, and also to the suitability of its climate for cotton working, an advantage shared by many other parts of western Britain. Supplies of cotton were derived from Smyrna and Cyprus, by way of London, as well

[1] R. B. Westerfield, *Middlemen in English Business* (1915), p. 304.
[2] H. Heaton, *The Yorkshire Woollen and Worsted Industries* (1920), p. 289.
[3] See on Hull, W. G. East, "The Port of Kingston-on-Hull during the Industrial Revolution", *Economica*, No. 32, p. 190 (1931).

as from the West Indies and South America, by way of the ports of Liverpool, Lancaster and Warrington. It was only at the end of the century, when large supplies of cotton were available in the United States, that Liverpool became the chief importer of cotton for Lancashire. Certainly the industry prospered, and it is well known that it was the first to introduce machinery, driven at first by water power, and later—mainly after 1800—by steam. In the 1780's, Arkwright's machines for carding and spinning cotton required motive power: in consequence mills were erected along the steeply graded Pennine valleys of Lancashire, where water power could be employed.[1] Although, both in Lancashire and in the West Riding the steam engine had made its appearance as a source of power in spinning by the end of the eighteenth century, its use did not become typical until several decades later.[2] It was only then that eastern Lancashire and the West Riding began to enjoy fully the advantage of their local coalfields, and it was only then that the use of steam proved a stimulus to industrial concentration in the towns.

Certainly in the eighteenth century the cotton industry of Lancashire was already well established, and Manchester merchants were exporting cotton goods all over the world,[3] by way of the ports of London, Liverpool and Hull. The domestic market was served, at first by packhorse and river transport, and later by means of waggons, using turnpike roads, and by canal barges.

The coal industry, although it made strides, played only a small part in the eighteenth century: its rapid growth came only later in the age of iron and steel. Even so, the transport of coal required a greater tonnage of shipping than any other commodity. Some estimates have already been given[4] which indicate the scale and distribution of coal production. Very little coal was exported, and perhaps twice as much went to Ireland as to other countries outside Britain.[5] Ireland received supplies mainly from Cumberland, but also from the Tyne and Wear, by the sea route around the north of Scotland. Within England coal supplies were distributed mainly by water: Newcastle and Sunderland practically monopolised the London market throughout the century, and supplied not only all the east coast, but also the south coast as far as Cornwall.

[1] For a contemporary account of this industry and its localisation see J. Aikin, *op. cit.*

[2] See J. H. Clapham, *An Economic History of Modern Britain, 1820–50* (2nd ed. 1930), pp. 441–3.

[3] See A. P. Wadsworth and J. de L. Mann, *The Cotton Trade and Industrial Lancashire* (1931), Book III.   [4] See above, pp. 421–2.

[5] T. S. Ashton and J. Sykes, *The Coal Industry of the Eighteenth Century* (1929), p. 228.

Newcastle, already in the 'twenties, boasted "the longest Key for land-ing and lading goods that is to be seen in England, except that at Yarmouth...and much longer than that at Bristol"; and, partly because of the "Smoke of the Coals", it was "not the Pleasantest Place in the World to live in".[1] And in 1770 Arthur Young[2] was impressed by the coal waggon roads from the pits to the water: "great works, carried over all sorts of ground, so far as...9 or 10 miles. The track of the wheels are (*sic*) marked with pieces of timber let into the road, for the wheels to run on, by which means one horse is enabled to draw, and that with ease, 50 or 60 bushels of coals". The western coastlands of England and Wales received supplies from many fields: from Cumberland, Denbigh and Flint, south Wales and Monmouth, each of which served its neighbouring stretch of coast. The Cumberland field, which had some of the deepest pits, was one of the most active in England, and led to the rise of Whitehaven.

The Town of Whitehaven [wrote Defoe] has grown up from a small Place to be very considerable by the Coal Trade, which is increased so considerably of late, that it is now the most eminent Port in England for shipping off Coals, except Newcastle and Sunderland, and even beyond the last, for they wholly supply the City of Dublin, and all the Towns of Ireland on that Coast.[3]

Inland districts were supplied either from the ports by waterways or from local fields. Thus Shropshire coal was moved along the Severn between Bristol and Bridgnorth, and London redistributed coal up the Thames. On the other hand, the midlands, the West Riding, Lancashire and south Staffordshire, were supplied from their own collieries. Coal mined around Leeds and Wakefield was shipped to York and Hull by water. Where, in the absence of convenient waterways, coal was carried by road, transport was largely limited to the summer months, when the roads were better fitted for waggons, and when, too, as in the midlands, the harvest was over and the waggons and horses of the farmers were available.[4] Certainly, the presence of convenient sources of coal near to hand was already stimulating industries in many parts of the midlands and north. The iron and cutlery trades around Sheffield and Barnsley were using coal from near-by pits; so in Staffordshire local coal was being used in the manufacture of tiles, pottery and hardware.

---

[1] Defoe, *op. cit.* II, 659–60. The coal trade of Newcastle fostered a great deal of shipping, and was looked upon as the nursery of the English navy. See J. Brand, *The History...of Newcastle* (1789), II, 273.
[2] *Six Months' Tour through the North of England*, III, 12.
[3] *Op. cit.* II, 683.
[4] Ashton and Sykes, *op. cit.* p. 122.

As with coal, so with iron, large-scale developments which were to leave their mark on the map awaited the nineteenth century. It is a remarkable fact that, despite its enormous resources of ironstone, England suffered from a scarcity of iron, which had to be imported, mainly from Sweden, Spain, Russia and the North American colonies. The explanation is that iron smelting depended on abundant supplies of wood for charcoal—supplies which the English woodlands could no longer provide. It was only by *c.* 1709 that coke was successfully used to smelt iron ore, and not until after 1750 that its use became at all the common practice. In consequence, it is not surprising that in 1720 there were only about fifty-four small blast furnaces in the whole of England, and these were distributed over fifteen counties, the distribution being related to supplies of ore, to the presence of running water to work the bellows for the blast, and, most important of all, to local supplies of wood for charcoal.[1] Fifteen of these blast furnaces lay in the south-east, in the Weald; eleven in the Forest of Dean district; twelve in the midlands; eleven in the region around Sheffield; and five in Cheshire and Cumberland. Even the furnaces of the Sussex Weald and of the Forest of Dean, which were formerly the most productive centres of the industry, were definitely declining. The total output of pig iron in 1720 was only 17,000 tons.

Iron working, as distinct from the smelting of the ore, was still small in scale and widely diffused. In some degree, iron was worked in almost every village and market town; but certain towns and regions were noted for specialised, or miscellaneous, iron goods. Bristol and Gloucester made pins, and Newcastle, knives. In the district around Sheffield and Rotherham, where swift streams and millstone grit were available, steel cutlery and other goods had long been made, albeit with Swedish ore or bar iron, imported *via* Hull. Along the banks of the Don, above and below Sheffield, and along many tributary streams to the west of the town, forges and grinding mills were, as John Tuke's map shows in 1787, strung out in close proximity. So, too, the industrial development of the region west of Birmingham, to become the Black Country, was very marked by 1800.[2] The discovery that coke could be used instead of charcoal in the smelting of ore gave much stimulus to the production of both coal and iron in the south Staffordshire coalfield. Birmingham,

[1] P. Mantoux, *The Industrial Revolution in the Eighteenth Century* (revised ed. 1927), p. 278.

[2] See G. C. Allen, *The Industrial Development of Birmingham and the Black Country*, (1929), chaps. I and II.

standing near this field, specialised already in small finished metal products; since it lacked river navigation, many canals were cut to link it with the Severn, the Thames, Trent and Mersey (see Fig. 82); but it had good local water resources, a favourable factor to a town which numbered 74,000 in 1801.

COMMUNICATIONS

The public rivers of England which, as the king's highways, were traditionally free to the passage of all his subjects, played, notwithstanding their many shortcomings, an important part in internal communications both before and after the period of canal construction. The Thames, the Trent, and the Severn were the chief arteries of river traffic. On the two last rivers, at points near their heads of navigation[1] converged pack-horse routes; up- and downstream boats plied, carrying commodities by way of the Severn to and from Bristol, and by way of the Trent to and from Hull. The Thames, too, was navigable between London and Oxford, whilst boats could ply even as high up as Lechlade.[2] But closer examination shows the many hindrances and difficulties of river traffic. Low water in summer was a standing difficulty, whilst mills or fish weirs were liable to obstruct the channel. In the Thames it was necessary during low water, at points where water was held up by mill dams, to resort to the practice of "flashing" or "flushing" in order to allow of the passage of boats.[3] Writers were not lacking to extol the virtues of English rivers,[4] as also to press continually for their thorough improvement by sundry means.[5] Certainly very little was done to the three major rivers, but in other instances works were carried out early in the century which proved effective for some time, and served in fact until better facilities were provided by

[1] Welshpool and Willingdon stood at the heads of navigation of the Severn and Trent respectively.

[2] See the *Report on the Navigation and Trade upon the Thames and Isis Rivers*, 1793. (*Reports from Committees of the House of Commons*, XIV, 230 (1803)).

[3] See Jackman, *op. cit.* I, 162. At certain points of the river boats could continue in times of low water only by the opening of the gates of mill dams, a practice very wasteful of water. This practice was known as "flashing".

[4] See J. Phillips, *A General History of Inland Navigation* (1795). On pp. 152–3 an exponent of waterway transport wrote: "There is no river that has such a length of navigation (in England) as the Severn: you may navigate a vessel of fifty tons, and not a lock the whole way up to Welsh Pool, except in excessive drought, which does not happen every year, and, when it does, not above a month, seldom two." There were a number of stretches, however, which gave trouble owing to shallows.

[5] Jackman, *op. cit.* I, 377 *et seq.*

canals. Thus the rivers Aire and Calder were improved up to Leeds and Wakefield respectively; the Irwell was made navigable up to Manchester, the Don up to a point three miles below Sheffield, and the Weaver as far up as the salt pans of Nantwich. In these instances the improvements were attempts to meet the needs of industrial towns and regions. Similarly, in other instances, rivers were improved in order to facilitate the movement of coal; thus the deepening of the river Douglas and the canal cut alongside the Sankey Brook served respectively the collieries around Wigan and St Helens.

The construction of canals effected the greatest change in the internal communications of England during the eighteenth century (Fig. 82). The increase in overseas trade and the consequent expansion of production in the latter half of the century provided the economic basis for, and stimulus to, an improvement in the means of distributing commodities. So also the need for effective inland transport for coal was an important motive for canal construction.[1] Many roads were of little use except for pack-horses, whilst rivers suffered from many physical disadvantages and were not equally serviceable in all parts of the country. It is not surprising that the first canals radiated outwards from coalfields, from regions of developing industry, and also from regions where rivers afforded little or no utility as waterways. It is important not to underestimate the part played by canals in the commercial, agricultural and industrial development of the last three decades of the century, for in many parts of the country, despite the impediment of numerous locks, they were much used, above all for bulky commodities, but also for passengers. Their construction bore witness to the advances in engineering technique and involved the piercing of tunnels (for example, through the Pennines) and the carriage of aqueducts across the valleys of rivers. It is not surprising that these works struck the imagination of the time: "Our aspiring genius", wrote Thomas Pennant in 1782, "scoffs at obstruction, and difficulties serve but to whet our ardour: our aqueducts pass over our once admired rivers, now de-spised for the purposes of navigation: we fill vallies, we penetrate moun-tains."[2] In some cases canals were cut to provide waterways superior to existing rivers, which for various reasons—floods, seasonal changes of level and the obstruction of mills—set obstacles to navigation. In other cases, in contrast, the aim of the canal builders was to provide water links

---

[1] Thus, between 1758 and 1801, 165 Acts were passed for the construction of canals, and of these 90 had chiefly in view the carriage of coal. T. S. Ashton and J. Sykes, *op. cit.* p. 235.

[2] *Op. cit.* pp. 68–9.

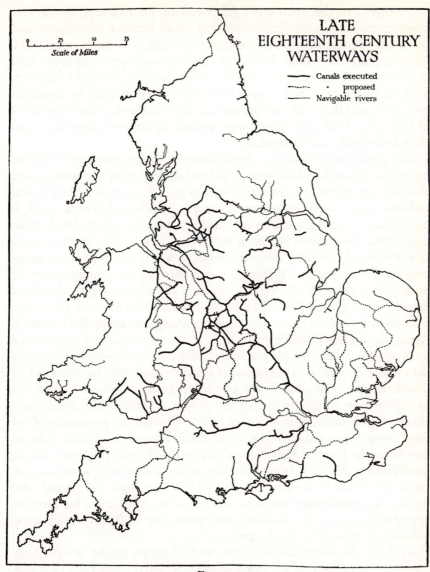

Fig. 82

This map is based on J. Phillips, *A General History of Inland Navigation* (new edition, 1795). Note the close network of canals in the industrial areas of the Midlands, Lancashire, and the West Riding, where, crossing watersheds, they linked up the navigable Severn, Thames, Mersey and Humber. The Liverpool-Leeds canal shown above was not fully completed until 1816.

or "feeders" to the great navigable rivers of the country. In the first category may be noted the Bridgewater canal, between Manchester and Runcorn, which replaced the inadequate navigation of the Irwell. In the second, stands the Grand Trunk canal, which was opened in 1777. This canal connected the Mersey with the Trent and thus with the Humber, whilst a branch canal linked up the Mersey and Trent with the Severn at Stourport, and thus with the Bristol Channel. In this way the three ports of Liverpool, Hull and Bristol were related by internal waterways, and their hinterlands were extended.

The preparatory works for this ambitious scheme reveal, not only the many considerations which were urged on behalf of canal construction, but also the care with which the canals were aligned in relation to the physical and economic conditions of the country.[1] The Grand Trunk canal, it was claimed, would serve alike the interests of landowners and industrialists, such as the masters of "great furnaces and forges" in Shropshire, all of whom would be particularly interested in a new means of cheap transport for heavy goods. Similarly, it would offer the pottery towns of Staffordshire cheap freight for clay, flint and hardware, and provide a short inland route between Bristol, Hull and Liverpool, in place of a long and sometimes hazardous coastwise sailing or a long and costly land journey. Again, the farmer would gain by easy access to supplies of manure and lime, and to markets for his produce, whilst the consumer would gain by the avoidance of dearths in regions of deficient foodstuffs badly served by inland transport. Mineral wealth of various kinds, which stood available to exploitation along the line of the projected canal, could be utilised: coal, building stone, lime and road metal. Finally, the promoters of the canal were not above claiming a strategical advantage for their scheme, and, since the Jacobite Rebellion of 1745 was still recent in men's minds, this claim was not entirely invalid: men, stores and cannon, it was argued, could be moved between the ports of Hull, Liverpool and Bristol; and further, the canals might serve as nurseries of seamen! Even the aesthetic appeal of the canal—its meadows, tow-paths, and its facilities for boating —were not ignored in the argument.

In short, the construction of this, and of other, canals, involved the co-operation of those to whom the scheme seemed economically desirable. Its alignment in detail was agreed on only after careful surveys had been made and compared. Thus it was decided that the canal should join the Trent, not at Burton, but at Wilden Ferry, nearly twenty miles below: this plan had the advantage of avoiding a difficult stretch of the river

[1] See Phillips, *op. cit. passim.*

between Burton and Wilden Ferry, where, owing to some twenty shallows, the river was navigable only during river freshes, and even then was liable to obstruction at the hands of mill owners who could control the water level. Similarly, the scheme for joining the western section of the canal to the river Weaver at Nantwich was rejected on the ground that this river, although it had been improved, provided an unreliable channel, liable to shoals. It was thought more convenient that the canal should make its eastern terminus at Preston Brook and thus merge with the Bridgewater canal, whence barges could continue to Runcorn at the head of the Mersey estuary. A question arose, too, as to where the branch canal, known as the Staffordshire and Worcestershire canal, should join the Severn. Bewdley, a flourishing river port, was originally marked out for this purpose, but since it objected, Stourport, consisting then of a single cottage, was chosen instead. In result, Stourport grew rapidly, whilst Bewdley declined.[1]

The inadequacy of the roads of England in the eighteenth century has been exaggerated, and the violent denunciations of Arthur Young, in particular, have often been cited to suggest that, even in the last decades of this century, long-distance travel by road was not only a difficult, but even a perilous, venture. It is well known, too, that the shortcomings of the English roads in the past are to be attributed largely to an inefficient system of road administration:[2] whereas in France the main arterial highways were constructed by the state, in England the legal responsibility and charge of maintaining the roads were fixed on the parishes for those roads, arterial or otherwise, which fell within their areas. Consequently, although road maps, which were becoming increasingly available for travellers, showed boldly a seemingly adequate network of main highways, there was a glaring discrepancy between the routes there shown and the roads as means of passage. "The hostler...derided my looking into my maps," wrote Byng,[3] "saying that such sorts of things, could give no information [sc. about the roads]." That hostler was wise.

Nevertheless, although it is true that the roads of southern Britain were generally neglected and certainly innocent of scientific construction, it is no less clear that distinct improvements were effected in the course of the century,[4] and that, despite their deficiencies, measured by Roman or

[1] Jackman, op. cit. I, 367.
[2] S. and B. Webb, The Story of the King's Highway (1920), passim.
[3] The Torrington Diaries, I, 72.
[4] See the chapter on Communications by H. L. Beales in Johnson's England (ed. A. S. Turberville) (1933), I. See also Lecky's History of England in the Eighteenth Century (3rd ed. 1883), VI, 167–80.

34

nineteenth-century standards, they played an essential part, together with the waterways, in linking up not only town and country but also hinterlands and ports. In order to discover the true conditions of the roads, it is necessary critically to examine the more colourful and vigorous accounts of contemporary travellers, notwithstanding that these had their springs in genuine and sometimes painful experience. It is important, in the first place, to note in relation to what means of transport the road was in each case criticised, and in the second, to consider the regional and seasonal conditions to which accounts relate. Thus to Defoe, who travelled on horseback in the 1720's, the roads seemed on the whole satisfactory, and his complaints referred usually to those parts of the country where, owing to the physique of the land or to the nature of the subsoil, passage was particularly difficult.[1] To Arthur Young, on the other hand, who travelled as a rule by chaise, the badness of the roads, as he often found them, derived partly from the fact that he judged them in relation to wheeled vehicles. At any rate, on two points contemporary experience was agreed. The worst roads were those which passed through clay-covered, and therefore often waterlogged, country; the best roads, in contrast, were usually found where sandstone or gravels provided a firm and porous subsoil. In fact, the modern geological "drift" sheets of the Ordnance Survey, which show the disposition of surface materials, throw a sidelight on the condition of eighteenth-century roads, since at this time these were still, in large measure, only roughly defined and unmade tracks. Moreover, there is no doubt as to the effects of seasonal weather conditions. In some parts of the country standing water, as in the Wealden claylands, or snow drifts, as in the Yorkshire moors, virtually suspended transport and travel during the winter months. In a few cases, too, as in Cumberland, as late as 1792, certain tracks over commons were re-opened to wheeled traffic each spring by the village surveyor.

The creation of turnpike roads, which had their beginnings in the seventeenth century, certainly did something to improve communications by land. Under Acts of Parliament from 1663 onwards turnpike trusts, set up in various counties or groups of counties, were authorised, in return for the right to collect tolls on traffic, to maintain specified lengths of road. The imposition of tolls, it is true, provoked outbreaks of rioting in many districts, and on this and other grounds the turnpikes were often denounced. Thus to the Hon. John Byng, writing in 1781, turnpikes were a "baneful luxury": "the country is only improved in vice and insolence", he noted

[1] *Op. cit.*

in his diary, "by the establishment of turnpikes".[1] Even so, it is significant that the ground of his objection—the results, to him unpleasing, of easier communication—implied the effectiveness of the new roads which he condemned. In the same way, despite his vehemence in the abuse of turnpikes, Arthur Young records how, in certain regions of superfluity, prices rose after the creation of turnpikes, another clear indication of their value.

It should be emphasised that the early phase of turnpike roads preceded the age of scientific road construction which awaited the efforts of John Metcalfe, Telford and Macadam. Of these Metcalfe was the first to apply new principles to road construction; his first works were done in the 1760's, when he was responsible for building roads from Harrogate to Boroughbridge and Knaresborough. In other words, from a technical standpoint, the turnpikes marked no advance. Roads were still roughly made by piling up loose material so that the section of the road was semi-circular; in steeply graded stretches the hollow road, in contrast, was often made. In any case, no effort was made satisfactorily to prepare a firm, well-drained basis for the road surface, so that complaints of the roads form the continual refrain of the travel books of the time. Thus the short stretch of road between Birdlip Hill and Gloucester was characterised by one traveller[2] as the worst turnpike road: "narrow, wet and stoney, and only mended with black iron ore, dangerous to man and horse". And again, the turnpike road between Wantage and Faringdon in Berkshire, which passed through "a filthy grazing country" of deep clays, was described in 1781 "as dirty 'as old Brentford at Xmas', though probably better now than when very rough".[3] Even so, if only by providing additional maintenance to certain well-frequented roads, the turnpike system made distinct improvements. Already by the time that Defoe was touring England a number of roads had been turnpiked. It is significant that most of these passed through clay-covered districts, above all through those to the north and north-east of London.[4]

Foremost among these turnpikes were the roads from London to Harwich, to Cambridge, and to Wansford Bridge near Peterborough, the

[1] *The Torrington Diaries*, I, 8 and 72. So also (on p. 6) Byng wrote: "I wish with all my heart that half the turnpike roads of the kingdom were ploughed up, which have imported London manners, and depopulated the country.—I meet milkmaids on the roads, with the dress and looks of strand misses; ...."
[2] *Ibid.* I, 261.   [3] *Ibid.* I, 252.
[4] See A. Cossons, *The Turnpike Roads of Nottingham* (1935), which contains a map of the roads turnpiked under Acts of Parliament between 1663 and 1715, and a diagram showing the times at which Acts were secured for turnpiking section by section of the Great North Road. The map is redrawn as Fig. 70, p. 428 above.

latter forming part of the Great North Road.[1] Defoe himself paid eloquent tribute to the ways in which the turnpikes had, already by 1724, greatly facilitated transport. Not only had the freights for the carriage of goods to the capital been reduced, but also London was being supplied in winter with sheep and cattle "from the remoter Counties of Leicester and Lincoln".[2] And there were other writers in the 'twenties who spoke in favour of English roads: "It is not only on the Thames that you travel with enjoyment...the high roads of England...are magnificent, being wide, smooth and well kept, rounded in the shape of an ass's back."[3] Nevertheless, the construction of turnpike roads was a slow and piecemeal process, as was well illustrated in the case of the northern arterial roads from London to Berwick and Carlisle, which were turnpiked, under the authority of more than twenty Acts of Parliament, between the years 1663 and 1775–6.[4] It was for the carriage of heavy bulk goods, such as coal and timber, that the roads, even when turnpiked, showed their inadequacy.[5] Where, however, towns were dependent on turnpike roads for the carriage of coal, freights remained high, especially in winter.[6] So also in many clay districts roads were still far from adequate. "The highways of the Weald", wrote Boys in 1794, "were the worst turnpike roads in the kingdom."[7]

Finally, since they indicate so vividly some of the background features of eighteenth-century life, it is worth noting the times taken by coach journeys between important cities. A traveller on horseback, and the post,[8] conducted by horse relays, could achieve faster rates of travel, but for the ordinary passenger the stage coach marked the pace. The times taken by coach services varied seasonally, being shortest in summer, and longest in winter, but it is clear that a very remarkable acceleration was made in the course of the century.[9] A few figures—which must be

---

[1] See Fig. 70, p. 428.   [2] *Op. cit.* II, 368.

[3] *A Foreign View of England* (the letters of de Saussure), trans. Van Muyden, 1902, pp. 146–7.

[4] See A. Cossons, *op. cit.*

[5] Thus Kendal, which stood near Morecambe Bay and not far from the Whitehaven collieries, since it had no river, suffered from a shortage of fuel in the earlier part of the century, and had to make do with turf cut from local mosses. Ashton and Sykes, *op. cit.* p. 229.

[6] Thus at Nottingham the price of coal in 1764 was 5*d.*–5½*d.* per hundred in the summer and 9*d.*–10*d.* in the winter (A. Cossons, *op. cit.* p. 12).

[7] *Op. cit.* I, 98.

[8] On the development of postal services, see L. W. Moffit, *England on the Eve of the Industrial Revolution* (1925), pp. 243–6.

[9] See W. T. Jackman, *op. cit.* I, 134–7 and 335–9.

regarded as rough averages—may be given to illustrate this increase in the rate of coach travel.

Journey from London	Year	Days	Year	Days
To Edinburgh	1754	10–12	1776	4
To Newcastle	1712	6	1776	3
To York	1754	4	1774	2
To Liverpool	1766	2–3	1781	2
To Shrewsbury	1753	4	1764	2
To Bristol	1750	3	1779	1–2

No account of the route-ways of eighteenth-century England would be complete without reference to the coastwise routes for which, as a small island with sufficiently accessible coasts, Britain was well adapted. It is suggestive of the importance of this means of transport, that Defoe, who had always a keen eye for evidences of economic activity, often devoted space in his *Tour*[1] to describe the sea lanes along which small coasters were continually plying. Although it is true that the whole subject of the coasting trade in this country awaits close study, it is possible from such accounts as those of Defoe, together with the information supplied by pilot books and maritime charts, to outline the contemporary conditions of coastwise navigation. The eastern waters of Britain, owing above all to two facts, the position of London and the coal production of the Tyne and Wear, were the most frequented. On the northern part of this route, situated on an inhospitable stretch of coast between Scarborough and the Tees, the small but busy port of Whitby played a useful part as a port of refuge.[2] Yarmouth with its river harbour and its roadstead, which was limited seawards by somewhat shifting sandbanks, occupied a strategical position on this route. Between Winterton Ness, a few miles north of Yarmouth, and Flamborough Head, in Yorkshire, the coast recedes westwards towards the Wash, making a broad and shallow bay. More-over, except for the Humber which offered considerable shelter to ships, either behind the Spurn, at Grimsby or higher up at Hull, the coast between Winterton Ness and Flamborough had only shallow water and no safe havens. Vessels sailing between London and the northern ports sought, therefore, to strike a direct course between Yarmouth and Flamborough Head, but, under adverse wind conditions, they ran grave

[1] *E.g.* I, 69–72.
[2] See W. G. East, "The Historical Geography of the Town, Port, and Roads of Whitby", *Geog. Journ.* LXXX, 484 (1932).

risk of foundering on the rocks or grounding on the shoals, along this dangerous length of coast. Thus, under heavy winds from the east, it was very difficult for ships sailing south to weather Winterton Ness, and if unable to do so, they had small chance to find refuge, since only Boston or Lynn, which were difficult to reach and accessible only at high tide, stood between them and disaster. In other words, the shallow waters of the Wash largely failed, as the deep waters of the Humber succeeded, in serving the needs of small sailing ships. Some efforts, it is true, had already been made by the 1720's to help ships around Winterton Ness by "Light-Houses kept flaming every Night", which were set above the shore to the north of Yarmouth, and also near by at Caister and Gorleston. Nevertheless, Cromer Bay, owing to its dangerous rocks, was known to sailors as "the Devil's Throat", and the coast about Yarmouth was reckoned among the most dangerous in the whole of Britain. The mariner's ill-luck, however, was the landsman's opportunity, and Defoe noted[1] that along the shore between Winterton and Cromer "the Farmers, and Country People had scarce a Barn, or a Shed, or a Stable; nay a Necessary-house, but what was built of old Planks, Beams, Wales and Timbers etc."

The southern half of the east coast, with its many small harbours and the wide estuaries of the Thames and Medway, afforded better facilities to shipping than the northern half. Off Deal, the "Downs" formed an excellent roadstead, except when strong winds were blowing from an easterly direction, since it was sheltered both by the South Foreland and a line of sandbanks. At the Downs, vessels usually anchored in order to despatch or receive letters and news, or to take on or drop passengers. Between the Downs and Portsmouth, along a stretch of coast which had already then much changed since the Middle Ages, there was no safe place of refuge for ships, except perhaps Rye, which, however, in the time of Defoe's travels, was no longer accessible to ships of over 200 tons.[2] Southampton had very little importance, and Poole was described by Defoe as the most considerable port in the middle of the south coast. The sea off, and to the west of, Portland, owing to the meeting of currents, proved a very dangerous part of the Channel, and lighthouses were set up to help shipping round Portland island. Beyond Exeter, the coasts of Devon and Cornwall abounded in small natural harbours, and coasting trade and pilchard fishing formed staple occupations of the coast towns.

[1] *Op. cit.* I, 71.
[2] Defoe speaks of "smuggling and roguing" as the reigning commerce along the coast from the mouth of the Thames to Land's End.

Entering into the fine harbour of Plymouth, owing to the Eddystone rock, was a danger to ships, and some four attempts were made from 1697 onwards to build there a lighthouse which would resist the onset of the sea. Finally, it is clear that much use was made of the sea-ways in the communications of the coastal regions of Wales, but it is significant that, especially in the first half of the century, in many cases, as at Conway in the north, towns well equipped as ports in respect of natural endowment, had in fact little traffic.[1]

POPULATION

The population of England and Wales in the eighteenth century cannot be assessed with any statistical accuracy. Although an attempt was made in 1753 to introduce the practice of an official census, which Sweden had already instituted in 1749, the project did not become law until 1800. Thus in any review of the population changes in southern Britain prior to the first census of 1801 the best available data consist of estimates. Among these, two sets of figures[2] derived from official sources may be regarded as fairly reliable. These are given below for the years 1700 and 1750, together with the census figure for 1801.

Population of England and Wales (in millions)

	A	B
1700	6·0	5·5
1750	6·5	6·5
1801	9·2	9·2

The extent to which the population increased between 1700 and 1750 is not therefore well established, but it may have increased by as much as 20 per cent. It is, however, remarkable, in view of the tendency to almost stable conditions in preceding centuries, that the population of England and Wales increased by about 40 per cent. in the latter half of the eighteenth century.[3] The explanation of this rapid growth need not be discussed

[1] Cp. Defoe, op. cit. II, 65.

[2] The figures in table A are taken from the preface of volume II of the 1841 Census, those in table B from the 1801 Census. They were reached, on rather different statistical assumptions, from a study of parish records of baptisms, burials and marriages. The figure given above for 1801 is Rickman's correction of the Census figure.

[3] On the population problems of this period see G. T. Griffith, Population Problems of the Age of Malthus (1926); M. C. Buer, Health, Wealth and Population in the Early Days of the Industrial Revolution (1926); E. C. K. Gonner, "The Population of England in the Eighteenth Century", Journ. Roy. Statistical Soc. LXXVI, 261 (1913); and T. H. Marshall, "Population and the Industrial Revolution", Economic History, I, 429 (1929).

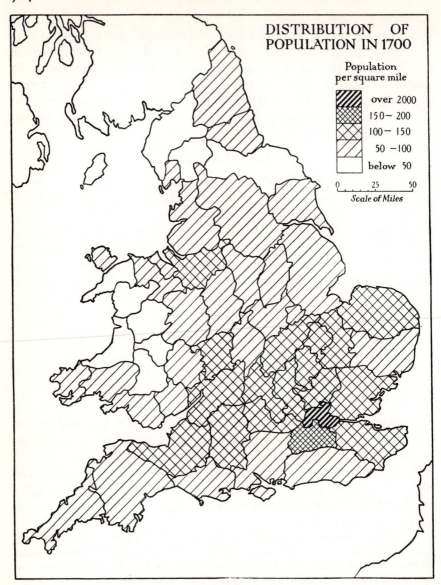

Fig. 83

This map is based on estimates contained in the 1811 *Census Report*, pp. xxviii–xxx. For a somewhat different distribution, see E. C. K. Gonner, *art. cit.* In detail the distributions shown above are certainly unreliable, but this map agrees with Gonner's in showing a broad well-settled belt astride a line joining London and Bristol. Note that the coalfield areas were relatively less populous. The average densities per square mile, on the basis of the Census estimates, were: England 102, Wales 45 and Scotland 36. No densities are shown for Scotland. Compare with Fig. 72, on p. 439 above.

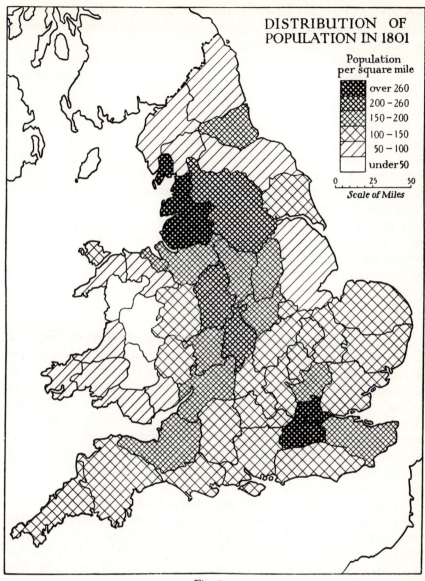

DISTRIBUTION OF
POPULATION IN 1801

Population
per square mile

over 260
200 – 260
150 – 200
100 – 150
50 – 100
under 50

0   25   50
Scale of Miles

Fig. 84

Note that the distributions given are averages and do not show the varying densities within county areas. Thus the high densities in Middlesex, Surrey and Kent are due to urban groupings at and around London. Note the concentration of population on the coalfields in England, but not in South Wales. That reflects not so much the importance of coal as the localisation of the major industries. No densities are shown for Scotland. The highest county densities are Middlesex (3200), Lancashire (380), and Surrey (370).

here. It may be noted, however, that the fundamental causes were both a high birth-rate and a declining death-rate, the latter being due above all to advances in medical knowledge.

The geographical significance of these figures can be appreciated only by answering a number of questions. How did Britain compare with the other European states of the time? How was the population distributed? And how did the distribution change in the course of the century?

In the first place, Britain was out-distanced by several continental states both at the beginning and the end of the century.[1] The territorial increases effected in 1707 and 1801 did something to redress the balance. The Union with Scotland and the effective incorporation of Ireland brought the total figure for the United Kingdom up to fourteen millions in 1801. On the Continent, in contrast, both Russia and France had populations well exceeding those of Britain at the beginning and the end of the century; and this despite the fact that during these years Britain became the most powerful imperial and maritime state. Even so, it is suggestive that, prior to its industrialisation in the nineteenth century, Britain supported a population little in excess of that of Italy and considerably less than that of France.

The distribution of this population revealed distinctive patterns which contrasted markedly with those of subsequent periods. At the beginning of the century the most densely populated counties formed an east-west belt on either side of a line between London and Gloucester (Fig. 83). At the end of this belt stood the two chief seaports of the time, albeit of very contrasted scale and importance, namely London and Bristol. Towards the north, south and west of this belt, population tended to decrease in density. Thus Wales, Durham, Northumberland, except along the lower Tyneside coal-field, Cumberland, Westmorland and the North Riding of Yorkshire, in all of which pastoral farming formed the characteristic economy, were alike scantily peopled.

By the end of the century the beginnings of a drastic transformation of the population map were already well apparent in the growth of certain industrial regions of the midlands and the north. In 1801, except for Middlesex and Surrey, which remained well populated thanks to London, the most densely populated counties were Lancashire, Warwick, the West Riding and Staffordshire (see Fig. 84). Within these four counties the areas of thriving industry roughly coincided with coal-fields, but it must be emphasised that coal did not then explain the redistribution of population.

[1] See Woytinsky, *Die Welt in Zahlen* (1925), I, p. 24, for estimates of the populations of European states.

That was due to the marked localisation of certain industries. Thus in Warwick and Staffordshire the iron industry, together with pottery in the latter, were the most striking instances of industrial localisation; whilst in eastern Lancashire and the West Riding of Yorkshire, textile industries of various kinds had become remarkably concentrated by the end of the eighteenth century. In fact, population had redistributed itself to such an extent that parliamentary representation bore very much less relation to the population conditions of 1800 than to those of a century earlier.[1]

One further point about the distribution of population is the significant fact that as late as 1801, nearly 78 per cent. of the population of England and Wales lived in the country. Even so, comparatively considered, southern Britain took the lead in the urban development of eighteenth-century Europe. In the Scandinavian countries and in European Russia the urban population scarcely exceeded 5 per cent. of the total at the end of the century; in France the figure was approximately 11 per cent., but in Italy, where civic traditions remained strong, the figure stood rather higher.

In short, England in 1800 was essentially a country of rural settlement; but its towns, though small, were growing throughout the century by immigration from the countryside. But the picture of England, as it is revealed in the writings of contemporaries such as Daniel Defoe, Arthur Young and William Marshall, is one in which the farm rather than the town, and the field rather than the factory, were the characteristic features. Even so, a foreign traveller wrote of it: "there was more motion on its surface, and in its ports, than in the ocean that surrounds it."[2] No longer situated, as in the Roman and medieval periods, at the western confines of the civilised world, Britain stood, in the eighteenth century, at the focus of maritime routes from Europe, America and the Far East. These routes, secured by sea-power, opened up pathways to trade, colonisation and imperialism which enabled Britain to play a part in European politics and commerce incommensurate with its small area and population. If, in retrospect, the beginnings of the Industrial Revolution loom large in its eighteenth-century history, Britain was still far from becoming the workshop of the world.

[1] See, on this distribution, L. B. Namier, *The Structure of Politics at the Accession of George III* (1929), I, 79.
[2] Archenholtz, quoting Bossuet, *op. cit.* p. 346.

BIBLIOGRAPHICAL NOTE

Some idea of the available literature on different aspects of the human geography of England in the eighteenth century has already been given in the footnote references. The following bibliographies contain particulars of many works of geographical interest: Eileen Power, *The Industrial Revolution, 1750–1850* (published by the Economic History Society, 1927. Blackwell's, Oxford); J. P. Anderson, *The Book of British Topography* (1881); Dorothy Ballen, *Bibliography of Roads and Roadmaking in the United Kingdom* (1914).

Reference should also be made to the current journals of geography and economic history, in particular to *Geography, The Geographical Journal,* and *The Economic History Review.* The latter contains bibliographical lists of article literature on English economic and topographical history.

No Ordnance Survey maps exist for the eighteenth century; the first sheets, covering Kent, a part of London and a part of Essex (on a scale of 1 inch to the mile), were published on January 1st, 1801. There are, however, a large number of county maps and town plans, of varying scale and value, which must be sought in libraries, and especially in the British Museum. For these, as for earlier maps, an invaluable guide is Thomas Chubb's *The Printed Maps in the Atlases of Great Britain and Ireland, A Bibliography, 1579–1870* (1927).

Chapter XIV

THE GROWTH OF LONDON, A.D. 1660–1800

O. H. K. SPATE, B.A.[1]

"The City of London is built upon a sweet and most agreeable Eminency of the Ground, at the North-side of a goodly and well-conditioned river, towards which it hath an aspect by a gentle and easie declivitie...the soil is universally Gravell, not only where the City itself is placed, but for several Miles about."[2] Evelyn's sketch of 1661 was redrawn by traveller after traveller in the early eighteenth century. The high ground of the City yet shared significance with the navigable estuary as a site-factor; but the growth which has left it a mere pedestal for St Paul's had already begun. Westminster itself, hemmed in by marsh and park, expanded but little (Tothill Fields were not built upon until about 1830),[3] but the hinterland of the Strand developed into a sort of neutral ground between Court and City, a district of coffee-houses, theatres and less reputable places of resort. The lawyers had always occupied a similar strategic position between the two disputant sections of society. The West End was beginning to take form; indeed, Evelyn himself reluctantly assisted in the process.[4] John Graunt noted in 1662 that the increase of population had been least in the parishes within the Walls; here, there was no room for new dwellings except in the great houses abandoned by the nobility and minutely subdivided. The trades of the City were gradually following the Court westward—at least those which were not dependent on the vicinity of the Royal Exchange, or of the Pool. Moreover, there was a greater demand for light, air and elbow-room, which made itself felt despite the short-sighted Tudor and Stuart ordinances against building. These restrictions, and the cramping occupational regulations of the City, had been responsible for surrounding London proper with a belt of warren-like tenements and cellars in the buildings and gardens of old mansions and religious foundations, so that Defoe called the Savoy "not a

[1] I am indebted to Professor J. H. Clapham for much general guidance; and I must also thank Dr Hilda Ormsby for help on specific points.

[2] Evelyn, *Fumifugium* (1661), pp. 4–5.

[3] J. Richardson, *Reminiscences of the Last Half-Century* (1858), chap. 1; W. Bardwell, *Westminster Improvements* (1839), pp. 6, 58.

[4] *Diary*, June 12, 1684 (Berkeley Square).

House, but a little Town, being parted into innumerable tenements". In other cases, new buildings of the most inexpensive and impermanent kind were run up, nicely calculated to last a 31 years' lease and no more.[1] Graunt and Evelyn, with the mass of less informed opinion, thought that London was a head too big for the nation, "to such a mad intemperance was the age come in building about a city, by far too disproportionate already".[2] The opposite case was argued by Petty and Davenant, and very ably in Barbon's anonymous *Apology for the Builder* (1685). In any case, the tendencies to expansion and to differential growth of population were much accentuated by the Fire.

But the consequences of the Fire were not so revolutionary as some have wished. The hurriedly drafted schemes of Wren and Evelyn would have taken years to complete, and would have been much too expensive.[3] Actually, the task of surveying was admirably done; and while many of the provisions of the justly famous Act for Rebuilding (1667) were no doubt counsels of perfection, and streets were built in Pepys's words "by fits and not entire", the comfort and health of the city must have been greatly improved, if only by the substitution of brick for wood. Evelyn's complaint of six years before could still be urged:

that the Streets should be so narrow and incommodious, in the very center, that there should be so ill and uneasie a form of paving under foot, so troublesome and malicious a disposure of the Spouts and Gutters overhead...the Deformity of so frequent Wharfes...imploying the Places of noblest aspect for the situation of Palaces towards the goodly River.[4]

But, on the whole, the streets were wider, the houses more solid, more convenient, and much more widely spaced.

Broader changes than these were involved. Many of the rich, driven out by the Plague and the Fire, remained in more spacious quarters; some City magnates, like Sir Josiah Child, began to live, for part at least of the

---

[1] It was in these areas that the mortality of the Plague was fiercest; the worst areas were localities such as St Giles, Cripplegate, a mass of slums near the stagnant ditches of Moorfields, and Stepney, where the deaths totalled three-fifths of those within the Walls (W. G. Bell, *Great Plague in London* (1920), pp. 145, 274).

[2] *Diary, loc. cit.* See also J. Graunt, *Observations on the Bills of Mortality* (1662); *Particulars of the New Buildings* from 1656 (Cal. S. P. Dom., Ch. II, 1678, p. 32); *The Grand Concern of England Explain'd* (1678); Sir W. Petty, *Essay... concerning the Growth of London* (1682); Sir W. Davenant, *Essay on Ways and Means* (1695), pp. 114–16; the useful discussion in C. H. Hull's introduction to *The Economic Writings of Sir W. Petty* (1899).

[3] Sidney Perks, *Essays on Old London* (1927), pp. 33–74.

[4] *Fumifugium*, "to the Reader".

year, outside the City in the surrounding villages.[1] At the same time, the distinctively working-class areas were steadily growing to the east, especially Spitalfields where the Huguenot refugees formed, after 1685, a large section of the weavers. The Fire also dealt a blow at the old industrial organisation of the City; the demand for labour in the building trade was so great that the Act of 1667 had to remove restrictions on "Carpenters, Bricklayers, Masons, Plaisterers, Joyners, and other Artificers, Workmen, and Labourers to be employed in the said Buildings", thus specifically admitting non-freemen. All this meant increasing complexity and differentiation in the occupational topography of the town, and brought new administrative problems which were hardly faced till 1888. At the same time, the City itself was much impoverished by the strain of the Plague, the Fire and the Dutch wars, coming so soon after the huge expenditure of the civil wars. "In England it was inevitable that recovery should be slow"; and the advance of 1668–71 was halted by Charles II's Stop of the Exchequer which precipitated a new commercial depression.[2] The small resistance which the City was able to make against such encroachments as this, and against the proceedings directed towards its Charter in 1683, was at once an index and a factor of its weakness. In some ways, indeed, the power of the "City" waxed greater than ever; the Bank of England, for instance, may be regarded as the master-stroke of the "moneyed men", which took the power of the purse away from the City as a corporation.[3] This is not without geographical significance in relation to the contemporary development of world trade and of a British empire, together with the growth of a huge population dependent on manufactures and on the distributive services.

It is "a labour of little less difficulty to attempt to describe the varying form of a summer cloud, than to trace from year to year the outline of

[1] See for instance Defoe, *Tour through the Whole Island of Great Britain, 1724–7* (ed. by G. D. H. Cole, 1927), 1, 6, 161, 382.

[2] W. R. Scott, *Joint Stock Companies to 1720* (1910), 1, 232 *sqq*.

[3] See e.g. G. K. Stirling Taylor, *Robert Walpole* (1931), pp. 87, 195; also "G. Stonestreet," *Domestic Union, or London as it should be* (1800), p. 49: "... The Bank of England by creating a sort of *Imperium in Imperio* within the City, has somewhat diminished the lustre of its ancient Municipality. In most of the antecedent periods, the Crown and its Ministers were in the habit of applying to the City, under every exigency, for aids; the principal bankers and money-factors were then usually members of the Court of Aldermen, and, by their services on such occasions, added to the consideration of the whole Court...."

London".[1] The expansion of the West End was largely a self-contained and formal process: estates were laid out on long leases and on a more or less definite plan. The class of occupant demanded a certain spaciousness; not that Westminster was without its slums, the worse, as contemporaries observed, because there was no regular industry in this enlarged royal village for the support of the poor.[2] From the topographical point of view, Westminster itself, centred on the gravel patch of Thorney, was unable to expand very far; the low marshy ground of Tothill Fields to the

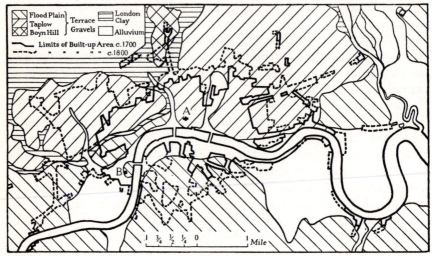

Fig. 85

London: Geology, and Growth in the 18th Century. A, St Paul's Cathedral. B, Westminster Abbey. Bridges shown as in 1800; East to West, London Bridge; Blackfriars' Bridge (opened 1769); Westminster Bridge (opened 1750).

south was liable to flooding. To the west and south-west were the marshy alluvial areas of Pimlico and the Five Fields. But to the north and west of Charing Cross stretched the great Taplow gravel ledge of central London, admirable ground for building, broken here and there by the marshy bottoms of streams such as the Tybourne and the Westbourne,

[1] J. P. Malcolm, *Londinium Redivivum* (1803), I, 5.

[2] J. Norden, *Speculum Britanniae* (1593), p. 47. By the end of the eighteenth century the problem had to some extent changed its incidence, and Patrick Colquhoun complained of a sort of "Poplarism" problem: "in the eastern part of the Town, the Poor may actually be said to be assessed to support the indigent. In Mile-East New Town ... the rates are treble the assessment in Mary-le-bone, where opulence abounds" (*State of the Poor in the Metropolis* (1799), pp. 29–30). Stonestreet (*op. cit.* pp. 24–5) advocated equalisation.

which only occasionally became nuisances when the natural drainage was interfered with. These gravel areas provided good brickearth in places— e.g. the long strip south of Piccadilly from Knightsbridge to Leicester Square.[1]

Defoe gave a vivid account of this "Prodigy of Buildings, that nothing in the World does, or ever did, equal except old *Rome*". In 1658 Faithorne and Newcourt's plan shows only a few houses near Piccadilly Circus, to the north and west of Long Acre. About 1668 St James's Square was built; Golden and Soho Squares followed by 1690. Hanover and Cavendish Squares were begun about 1717; the latter's progress was retarded by the disasters of South Sea year. When Defoe wrote, Seven Dials, the St Martin's Lane area, and part of Soho had been built up for some decades; but the buildings on the north side of Piccadilly "and that new City on the North side of *Tyburn*-Road called *Cavendish Square*" had been built very recently.

Building was but little hampered by the sharp slopes of the Fleet valley, even where they were formed of London clay; and only locally was the valley floor itself too swampy for mills or crooked slum alleys. North of Clerkenwell Road, however, the valley widened out somewhat, especially to the west of the Barnsbury spur where the Taplow ledge was broken by a wide expanse of London clay in the King's Cross area. The 1800 building line shows the inhibiting effect of this tract, intersected as it was by several marshy streams. Between the Fleet and the Lea, however, the Taplow terrace was very wide, and on these great gravel flats "all those numberless Ranges of Buildings called *Spittle Fields*" had been built within Defoe's (somewhat elastic) memory. "That part now called *Spittlefields-Market* was a Field of Grass with Cows feeding on it, since the year 1670...*Brick-Lane* which is now a long well-pav'd Street, was a deep dirty Road, frequented by Carts, fetching Bricks that way into White Chapel from Brick-Kilns in those Fields."[2] To the south-east, the restrictive legislation of Elizabeth's day had made exception in favour of the water-side areas, where new building was essential to keep pace with the growing population and trades of the district. Houses were built on the river walls to strengthen these works: "about 1571, when the Water had again broken in, a View was made by Commissioners of Sewers, who thought it necessary, that the walls should be builded upon by any who would",

[1] See, for example, R. Dobie, *History of St Giles and Bloomsbury* (1829), pp. 172–3. See also R. Mylne's *Geological Map of London* (1856).

[2] Quotations in this and the preceding paragraph from Defoe, *op. cit.* Letter v, *passim.*

35

though prohibition was reimposed for a time in 1583.[1] Houses also lined Ratcliff Highway which ran along the Taplow gravel above the alluvium inside the river-bend at Wapping, until it met the river's bank at Shadwell and Ratcliff, where the gravels came down to the water's edge.

By the middle of the eighteenth century, London east of the Fleet already showed a good deal of differentiation. The City sheltered a host of small tradesmen, and their wholesale establishments, banks, markets, and exchanges; the administrative offices of the City companies; and the buildings of the great joint-stock monopolies—the East India, South Sea, Hudson's Bay, and other companies. Around the City were grouped three industrial sectors, very different not only from this nucleus but from one another. Clerkenwell, sprawling up the eastern slopes of the Fleet valley, formed, together with Shoreditch and Bishopsgate, a district of innumerable twisting lanes and alleys, the homes of artisans of many trades, but especially those employed—or employing themselves—in the making of clocks and jewellery. To the north-east Spitalfields extended its monotonously regular streets of mean cottages, inhabited by a more homogeneous industrial population of weavers. Finally, in the waterside parishes there was the labyrinth of rope-walks, breweries, small foundries, anchor forges, docks, sugar boileries, oil, colour and soap works, coopers' and boat-builders' yards. Here, too, were the irregularly built homes and taverns for artisans, sailors, coal-heavers, and other waterside workers. South of the river, there was a similar area in Rotherhithe, joining up with Deptford along the river-wall. The chief industry on the Surrey side was tanning, connected, in its earlier history at least, with the tidal ditches which intersected the district round St Saviour's Dock. With tanning were associated kindred trades and industries such as fell-mongering, wool-stapling and glue-making. A little farther up the river were the timber-yards of Lambeth, well placed for the convenience of water carriage.[2]

The great achievements of the mid-century were the construction in 1756–7 of the "New Road", and the blocking out of south London by the opening of approaches and links between the two new bridges—Westminster (opened in 1750) and Blackfriars (opened in 1769). In 1811 Brayley remarked that "with very few exceptions the whole line of each road is now skirted on both sides with houses";[3] and streets were being built to

[1] J. Strype, *Stow's Survey of London* (1720), Book IV, p. 39.
[2] S. Lysons, *Environs of London* (1792), I, 317. See *ibid.* p. 547 for Bermondsey trades; also *V.C.H. Surrey*, II, 336 and T. Pennant, *Some Account of London* (1813 ed.), pp. 46, 74 *sqq.*
[3] E. B. Brayley, *London and Middlesex* (1811), Part II, p. 86.

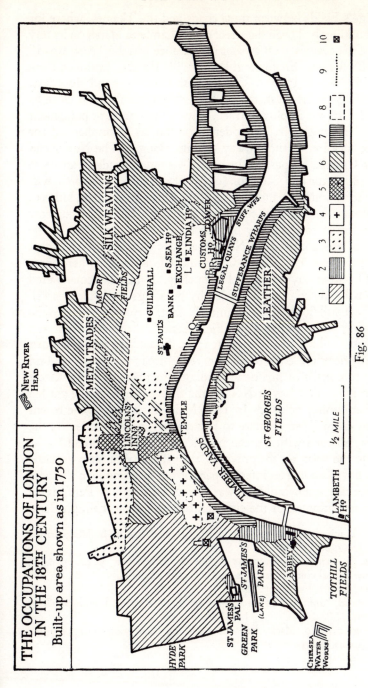

Fig. 86

Reference: 1, aristocratic residential quarter; 2, Government Offices, etc.; 3, middle class and professional residential quarter; 4, amusements and vice area; 5, legal quarter; 6, industrial areas and artisans' dwellings; 7, wharfs, warehouses, waterside trades, including labourers' dwellings, seamen's taverns, etc.; 8, "the City" —commerce and finance; 9, boundary of the City Liberties; 10, West End shopping and hotel centres round Haymarket and Charing Cross. *Principal Markets:* S, Smithfield (meat, hay); L, Leadenhall (meat, provisions, leather); G, Covent Garden; Q, Queenhithe (corn, meal, malt); B, Billingsgate; R, Roomland (coal). The boundaries shown are, of course, only approximate.

fill up St George's Fields between these roads and the Thames, where for long the only building had been the street called Lambeth Marsh on its tiny rise, the edge of the gravels.[1] Of greater importance was the "New Road" (now named Marylebone, Euston and Pentonville Roads) continued from the Angel by City Road (1760) to link up with Old Street and Moorgate. This line, especially in the west, ran on the gravels, parallel and close to the northern edge of the Taplow terrace, a line of the greatest permanent importance in the topography of modern London, as is emphasised for instance by the location of the railway termini along it. The New Road formed a North Circular by-pass for its period—a much-needed transverse link between Paddington and Islington (only villages then, but soon to wax exceedingly), and a boon to the inhabitants of the northern suburbs, hitherto compelled to plunge deep into the congested town by radial roads or narrow devious lanes in order to reach any but the most contiguous parts of the city.[2] Most of the area south of the New Road was soon built over, except in the central part (between King's Cross and Clerkenwell) where there was a salient of London clay along the Fleet valley. It must be remembered that though mostly arched over or hidden by buildings from Clerkenwell onwards, the Fleet still flowed past Mount Pleasant, "giving motion to the Flour and Flatting mills at the back of Field Lane".[3] It then disappeared under Fleet Market, which was opened in 1737. Some way to the west of the stream itself, near the edge of the gravels, "in 1708 Great Ormond Street was said to consist of fine new buildings; in 1784, when the Great Seal was stolen from Thurlow at No. 45, the thieves got over the garden wall from the adjoining fields".[4] In 1803 Malcolm lamented the recent loss of the view from Queen Square to Hampstead, now "hidden by the majestic houses, adorned with Tuscan pillars" of Guildford Street.[5]

Still farther to the west, on the gravels, "the village of St Mary-le-bone may be said to have become an integral part of the metropolis in the year 1770".[6] To the east again building was going on apace at Pentonville in the 'eighties and 'nineties of the eighteenth century, on the high Boyn

[1] On Fig. 86 "Lambeth Marsh" appears as the isolated line of buildings to the south-west of St George's Fields.

[2] M. E. Cooke, *St Pancras* (1932), pp. 45–6. See also Stonestreet, *op. cit.* pp. 64–5, for the difficult peripheral communications and bad exits of the City.

[3] E. B. Brayley, *op. cit.* Part I, pp. 68–9.

[4] C. N. Bromehead, "Influence of Geography on the Growth of London", *Geog. Journ.* LX, 125 (1922).

[5] J. Malcolm, *op. cit.* I, 6. Macky had eulogised the same view in 1714 (*Tour through England* (1714), I, 243).

[6] Brayley, *op. cit.* Part II, p. 98.

Hill gravels of the Barnsbury spur: the clay slopes south of Pentonville Road, however, long remained open fields cutting off Islington from Clerkenwell. Two centres on the clay were exceptions to the general rule—Camden Town (c. 1790), a ribbon development along the Hampstead Road, and Somers Town, about the turn of the century a respectable suburb but soon to deteriorate. Contemporaries rightly looked with the most pride on the splendid planned development of the Bloomsbury Squares on the Bedford and Foundling Estates, the work of the great speculative builder James Burton. Bedford House, on the north side of Bloomsbury Square, was pulled down in 1800, and from then until 1810 the building of the fine series of squares and places between the British Museum and the Foundling Hospital was practically continuous, despite the war which played havoc with other schemes. The last great planning in this part of London was slightly later in date, but in taste and temper it cannot be isolated from the eighteenth century. Nash carried through the great sweep of Regent's Street and the grandiose surroundings of Regent's Park between 1813 and 1823. It is worthy of note that we owe the very preservation of the Park not only to Nash's genius but also to its "want of water and clayey soil, ill-adapted for building on".[1] Alternative schemes provided for laying out the Crown's Marylebone property on the model of the Portman and other West End estates.

Expansion farther eastward was less spectacular. Finsbury Square was built about 1790, but on the whole the district at this time merited Sharpe's description: City Road was "the Rotten Row of the East End...in a respectable if not aristocratic suburb of the city". Farther east still the unchronicled growth of Spitalfields, of Bethnal Green and of Whitechapel went on unceasingly, checked by bad years but always ready to continue. One great event, at least, signalised this workaday end of the town, the beginning of the long process by which the riverside suburbs such as Limehouse ceased to justify Malcolm's pleasant account: "several descriptions of naval artificers reside within it, whose houses are frequently decorated with masts and flags...waving through the trees, and interspersed among large mansions of opulent merchants".[2] The turn of the century saw the inauguration of changes which were almost to revolutionise the port, the real life-source of the metropolis. The greatest artery of the East End, the Commercial Road, was opened in 1803.

---

[1] Lysons, *op. cit.* v, 225–6 (1811); *Some Account of Proposed Improvements* on this estate, 1814.
[2] *Op. cit.* II, 85.

It must not be assumed that the population always kept pace with the expansion of bricks and mortar. The complicated problems of London demography in the eighteenth century have caused much discussion; all

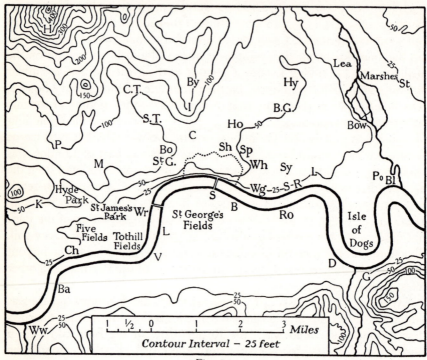

Fig. 87

London: Contours and Sites of Surrounding Villages. Dotted line—City boundary (approx.). London and Westminster Bridges shown.

Key. North of River: **B.G.**—Bethnal Green; **Bl**—Blackwall; **Bo**—Bloomsbury; **Bow**—Bow; **By**—Barnsbury; **C**—Clerkenwell; **Ch**—Chelsea; **C.T.**—Camden Town; **H**—Hampstead; **Ho**—Hoxton; **Hy**—Hackney; **I**—Islington; **K**—Kensington; **L**—Limehouse; **M**—Marylebone; **P**—Paddington; **Po**—Poplar; **St.G.**—St Giles; **Sh**—Shoreditch; **Sp**—Spitalfields; **S-R**—Shadwell-Ratcliff; **S.T.**—Somers Town; **St**—Stratford; **Sy**—Stepney; **Wg**—Wapping; **Wh**—Whitechapel; **Wr**—Westminster. *South of River:* **B**—Bermondsey; **Ba**—Battersea; **D**—Deptford; **G**—Greenwich; **L**—Lambeth; **Ro**—Rotherhithe; **S**—Southwark; **V**—Vauxhall; **Ww**—Wandsworth.

that can be done here is to mention the few undisputed facts. The pioneer statistical work of Graunt and Petty, based on the Bills of Mortality, indicates a very rapid rate of increase during the last forty years of the seventeenth century, from say 450,000 in 1660 to some 675,000 in 1700. This rate was not maintained. The population seems to have remained practically stationary till about 1760. Infantile mortality was appalling: Hanway boasts pathetically that of the 1384 children taken in during the first fifteen years of the Foundling Hospital's existence (to 1756) it was "very remarkable that *only* 724...died".[1] There is some evidence, too, of a decline in the birth-rate in the worst gin decade, 1740–50. It has been suggested that a large proportion of the big immigration after the Fire consisted of people between the ages of 15 and 30, past the dangerous years of infancy: if that were so, the high death-rate from 1730 to 1750 would be due to their natural mortality about the age of 60.[2] Making all allowances for inaccuracies, for changing administrative areas, and for defective parish registers, the fact remains that for more than half the century the number of burials exceeded the number of baptisms; and yet the city continued visibly to increase. Heberden, after pointing out that despite many woeful inaccuracies the large scale of the Bills of Mortality precluded really great errors, observed that the annual mortality increased between 1700 and 1720, and decreased after 1750; while christenings, after increasing, declined from 1727 to 1740, and were again on a gradual up-grade after 1760.[3] The 1801 census set the depopulation controversy at rest, giving a population of some 900,000, of whom 123,000, as compared with an estimated 9150 in 1700, lived in the five parishes "without the Bills" —Kensington, Chelsea, St Marylebone, Paddington and St Pancras.

The supply of this mass of people with food, water, fuel and other necessities of life, early evolved a most complex economic organisation whose ramifications, by the eighteenth century, extended throughout the country. The development of the water supply throws some light on the topography of expansion and on sociological differentiation within the city. The main sources of supply before the seventeenth century were conduits from various wells and springs in the gravels, or in the London clay, or at their junction. The London Bridge Water Works had been built in 1582, and the New River was made between 1609–13. Only the New River Company was at first able to take advantage of the demand for a better supply after the Fire, but about 1690–4, and again in the 1720's, some new companies were formed. The increase in the built-up and paved area did away with the old sources in the gravel; and the eighteenth-century companies took water direct from the Thames, from the Lea, or from the

[1] Jonas Hanway, *A Candid Historical Account of the [Foundling] Hospital* (1760), p. 23.

[2] J. Brownlee, "Health of London in Eighteenth Century", *Proc. Roy. Soc. Med.* XVIII, Part 2, p. 73 (1925).

[3] W. Heberden (the Younger), *Observations on the Increase and Decrease of Different Diseases* (1801), p. 30.

springs issuing from the Bagshot sands at Hampstead.[1] But in the middle
of the century, the conduits of Richard Soames still supplied houses from
Bond Street to Whitechapel, while his London Bridge engines pumped
water to the low-lying areas between Blackfriars and Goodman's Fields,
a territory narrowly limited by the rise of the ground.

The huge and formless expansion of the East End had to wait nearly
until the nineteenth century to secure anything like an adequate supply
of water. Even then, many of the poorer streets and courts were only
supplied by stand-pipes with an intermittent flow.[2] The Shadwell service,
pumped from the Thames, was extended by the introduction of steam
engines in 1750 and 1774, and the West Ham company set up an engine at
Old Ford (drawing from the Lea) in 1745,[3] but generally in this area
only the basements were supplied—even after the change to iron pipes.
In 1827 east London (including Spitalfields, Whitechapel, and all the
waterside below St Katharine's—with many of the largest industrial
consumers) received a daily average of 143 gallons per house.[4] The New
River area (north and central London, including the territory served
before 1822 by London Bridge) received 182 gallons. In strong contrast
to the East End, the lines of metropolitan expansion in the west were
firmly blocked out by 1720, and the Chelsea Water Company, established
under an Act of 1723 specifically to supply the West End, was able to
expand *pari passu* with the buildings, and to supersede various small
wells and water-wheels such as the one at Bayswater which had supplied
Kensington Palace. It is significant that this company, supplying the
richest quarter of the town, was the first to experiment with iron pipes (in
1756, but they were not generally adopted until 1810–20), and to construct
filter beds (about 1827–9). The daily supply per house in the West End in
1827 was 218 gallons, a high-level supply of relatively pure water from the

[1] Small wells continued to be used for local purposes—e.g. in Piccadilly in 1776 for
watering the streets (Malcolm, *op. cit.* IV, 323); and the poorest quarters continued to be
poisoned by contaminated well-water past the middle of last century. Authority to
use the Hampstead springs (including powers of compulsory entry) had been given
to the City as early as 1546, but they were not developed till after the Fire, and then by
private enterprise. They supplied some of the northern suburbs till 1833, and are still
used for non-domestic purposes (W. Garnett, *A Little Book on Water Supply* (1922),
p. 124; M. E. Cooke, *op. cit.* p. 34).
[2] A. S. Foord, *Springs, Streams and Spas of London* (1910), pp. 317–18—in 1850
not less than 80,000 houses with 640,000 inhabitants had no internal supply. See also
J. Sutherland's *Reports on Epidemic Districts* (1852), *passim*.
[3] W. Matthews, *Hydraulia* (1835), pp. 111–18.
[4] J. R. McCulloch, *Dictionary of Commerce* (1844), p. 1334.

Colne and Brent (*via* the Grand Junction canal), and from the Thames at Hammersmith and Chelsea. But south London was served, and ill-served, by three companies taking their water direct from the Thames between Vauxhall and Southwark, in at least one case without even a settling reservoir or any filtration except a wire grate.[1] The daily supply per house was only 93 gallons, good witness to the general insalubrity and squalor of the area, and of ill omen in view of cholera epidemics to come.

The organisation of London's food supply is one of the major elements in any study of the economic geography of eighteenth-century England. Already in 1727 Defoe remarked, at the very beginning of his *Tour*, that "it will be seen how this whole Kingdom, as well the People, as the Land, and even the Sea, in every Part of it, are employ'd to furnish something, and I may add, the best of everything, to supply the City of *London* with Provisions... Fewel, Timber, and Clothes also".[2] If this generous promise is not quite borne out by the actual picture he gives, there is ample evidence for the wide influence of the metropolitan market, and for its geographical significance. The internal economy of London also showed relics of topographical influences in the siting of its markets. We can only glance at two of the most essential commodities of the country, hay and corn, saying nothing for instance of the meat and fish supply, nor of the extension of market gardening and dairying late in the century.[3]

The hay and corn markets were very significantly placed. Hay, a commodity of great bulk in proportion to its value and of most essential importance to a great population entirely dependent on horse transport, was sold chiefly at four markets, originally "stones' end" sites: the Haymarket, Smithfield, Whitechapel and the Borough. These places lay at the four corners within which the mass of seventeenth and eighteenth-century London was contained; they developed as road terminals with inns and stabling for travellers. Later, with the completion of its branch from the Grand Junction canal about 1801, Paddington market served similarly the great growth round Oxford Street. In contrast

[1] McCulloch, *op. cit.* p. 1334.

[2] *Op. cit.* i, 12; cf. i, 3. See also p. 503 above.

[3] See the Board of Agriculture Reports on the Home Counties, 1794–1811, especially Middleton's *Agriculture of Middlesex* (1798). Lysons has an excellent "General View of... Market Gardens" in the *Environs*, IV, 573–6, with additions to 1811 in V, 446–8. These contain many references to the belt of "open development"—dairies, market gardens, brickfields—which surrounded London. See the map of land utilisation in Middlesex, Fig. 77, p. 475 above.

to the hay supply which came mainly from Middlesex, the corn, flour and malt trade was at once far more widely diffused as to its areas of supply, and more centrally organised within London itself. By 1700 the London corn trade, with its associated mealing and malting, had become one of the greater economic activities of the country. From all the upper Thames basin; from the large area centering on Farnham, perhaps the most important provincial corn market, and thence down the Wey; from Chichester by Farnham till about 1725, and after that date by "long sea" round to the Thames; from Kent and East Anglia up the estuary of the Thames; from the east midland counties by land to the great mills, still to be seen, of Hertford and Ware, and thence down the Lea—from all these areas, corn and corn products converged on the riverside markets of Bear Quay and Queenhithe: "Monsters for Magnitude, and not to be matched in the world...here Corn may be said, not to be sold by Cartloads, nor Horse Loads, but by Shiploads."[1] There was also a corn market "on the pavement" at Newgate, where the road from the famous wheat district of west Middlesex (Heston, noted by Norden) entered the City. To the east again was a subsidiary corn and flour centre at Stratford, probably dependent in the first place on local grain supply, local water power, from the Lea, and perhaps local manorial rights. Its importance as a baking centre, already noted in the fourteenth century, was doubtless influenced by the water power of the river Lea.[2]

By 1760 the leading merchants felt that the open market at Bear Quay, however accessible to shipping, was far too exposed to the elements (and probably to the competition of the smaller men). The main dealings in most grain crops were transferred to an exchange built and owned by the greater dealers. This move merely sets the seal on the domination of the corn trade in southern England by the metropolis: "the price was practically determined by the London price", and the Corn Exchange served a wider area than its narrow accommodation would suggest.[3] A last relic of the older conditions may be seen in the eight stands (out of 72) still reserved for the Kentish hoymen in 1800.

The corn which came from Farnham and the Thames valley, the metal goods of the midlands, used a river navigation, which, although essential to the economic life both of London and of the south midlands, was very neglected.[4] Even the Thames below bridge, vital as it was, was

[1] Defoe, op. cit. p. 347.
[2] H. T. Riley, Memorials of London (1868), pp. 71, 80, 121.
[3] Report on High Price of Provisions (1801) (Reports 1715–1802, IX, 149).
[4] 1793 Committee on Amendment of Thames and Isis, pp. 26–32.

hardly regulated at all in some respects, absurdly over-regulated in others, until the very end of the century, and beyond.

During the first half of the eighteenth century the river of London was physically adequate for its predominant share of the country's shipping. Few ships drew more than 11 or 13 feet of water, and they generally came up with the flood and went down with the ebb, except in the broader reaches below Gravesend where there was room for small ships to tack. There were occasional complaints against the Trinity House, which was supposed to clear away shelves and shoals, using the dredged material as ballast: but these protests were rather against exorbitant charges and an inadequate supply of ballast (the Trinity Brethren had a monopoly) than against obstructions in the channel. The only serious impediment to navigation during the earlier part of the eighteenth century was caused by the material scoured out from the famous breach at Dagenham. In 1713 the bank so formed was increasing daily, "though many Thousands of Tons have been taken up yearly by the Ballast-Men"; ships of any considerable size could get round Erith Point only at "Pitch of Flood", and the three-decker moorings at Woolwich had to be moved some 20 fathoms into the channel.[1]

Adequate as the river itself might be, the same could not be said of the port accommodation for the continually increasing trade:

1700	1750	1790
*Imports £4,875,000 (80 per cent.)	5,540,000 (71 per cent.)	12,275,000 (70·5 per cent.)
Exports £5,388,000 (74 per cent.)	8,415,000 (66 per cent.)	10,716,000 (56 per cent.)

* Figures in brackets equal percentage of total trade of country.

Between 1700 and 1790, the total tonnage entering the port (foreign and coastal) rose from 435,000 to 509,000.[2] The import of coal alone rose from 335,000 London chaldrons in 1700, to 862,088 in 1799.[3] Yet throughout the century there had been no increase or improvement of real moment in the accommodation of the port. The Legal Quays, where alone most dutiable goods could be landed, had not been increased since the Commission of 1665 which finally established them. The complaints of the merchants in 1705, and the appointment of a new commission in 1765 produced no result. The quays were less than 500 yards long, and between

[1] *Reports* of Water Bailiff and of Woolwich Pilots to the Lord Mayor, 1713; J. G. Broodbank, *History of the Port of London* (1921), p. 71.

[2] Table based on App. D of Committee on Improvement of Port of London, 1796. Tonnage figures, *ibid.* App. G and H.

[3] *1800 Committee on Coal Trade*, App. 19 (Reports 1715–1802, x, p. 591). T. S. Ashton and J. Sykes, *The Coal Industry of the Eighteenth Century* (1929), App. E.

a third and a quarter in bulk of the port's trade should have been landed there; by the end of the century this was about three times as much as they could handle, and it was calculated that the West India trade alone could give them full employment for three or four months of the year.[1] The Legal Quays stretched from London Bridge to the Tower, with one or two gaps, and during the century some relief had been given by the provision of Sufferance Wharfs on both sides of the Upper Pool, where a limited range of dutiable goods could be landed—if and when a Customs officer was available. By 1796 there were about 1200 yards of these. But the total quay accommodation was not much more than that of Bristol.

Notwithstanding a distinct set-back about 1780, due to the American War, the commerce of the port advanced rapidly until, by the 1790's, congestion became intolerable. The great Parliamentary enquiries of 1796–1800 provide a detailed survey of the position in the years since the close of the war.

The statement given above of the trade of London in the eighteenth century contains two very significant facts. The first is that during the century the hegemony of London in the country's trade lessened, and this relative diminution (for the absolute values doubled and trebled themselves) was much more marked in exports than in imports. The second is that in the 1790 figures the imports for the first time exceeded the exports.

Not only the volume of shipping but the size of the ships increased a good deal during the century. In 1732 there were 1212 ships belonging to London under 200 tons, and only 205 of tonnage over that figure. In 1792 the figures were 1109 and 751.[2] It must be remembered that the biggest ships on the river (the East Indiamen, one or two of which attained 1400 or 1500 tons and drew up to 22 feet of water) did not come up above Blackwall, the marine depot of the East India Company since 1612. Despite this limitation, there were sometimes 1200 or 1400 ships in the pools above Limehouse. This was in a strip of water full of small shoals, subject to a strong tidal current, and losing at low spring tides not less than a third of its average high-water breadth of 900 to 1200 feet. During the ten years preceding the enquiry of 1796 there were also complaints of a general loss of depth, by as much as four or five feet, from Deptford upwards. This was attributed to a variety of factors—spilling of coal and ballast, scouring of mud from small docks, discharge of ballast, and, most

[1] *Report*, 1796, pp. 51, 56. C. Capper, *Port and Trade of London*, 1862, pp. 144–6.
[2] *Report*, 1796, App. M. In 1702 London had a total tonnage of 84,560; Bristol, the next on the list, only 17,325. See also W. Maitland, *Survey of London* (1739), p. 618.

important of all, the mud and silt from the sewers of a continually increasing paved area. One witness had had a lighter, lying aground off a storm sewer, filled with twenty tons of soil in one very hard rain. Besides removing only material fit for selling as ballast, and so leaving mud shoals untouched, the Trinity House lighters made no continuous channel but only a series of holes.[1] In any case, the number of ballast lighters was insufficient; when a large coal fleet came in a collier might be delayed even longer in waiting for ballast than in discharging its coals.

The traffic of the port was largely seasonal, and in war-time the convoy system accentuated the crowding of the river. The biggest branch in bulk was the import of coals from the north-east coast. During its long history, this trade had become very highly organised at both the shipping and receiving ends. A spell of easterly winds meant that a fleet of loaded colliers was held up in the Tyne for days, to arrive at London all at once, mooring in the river and unloading into lighters. The distributive organisation was excessively complicated, hampered by trade customs and by a whole series of duties which involved a cumbrous system of metage; it almost seemed to have been devised to secure the maximum of delay and fraud. In 1796 there were about 1200 barges employed in the coal trade, and 500 in the timber trade, out of a total of some 2500 lighters and 1000 miscellaneous small craft on the river.[2] The coal lighters were frequently used as warehouses, and their working was usually confined to the short stretch they could cover on one tide from the ship to the coal wharves (many above bridge, at Scotland Yard, etc.) so that the colliers had generally to come up well into Limehouse Reach, which was thus the most crowded part of the river.

In summer, a few of the colliers were occasionally employed in the Baltic timber trade. This trade also was at its height during the summer months when the Baltic was open and, in later years, when the logs had come down to the St Lawrence on the spring floods. It occupied a disproportionate amount of space: the timber was lightered or rafted up to Lambeth, and a cargo occupied ten or twelve times the water surface of the ships, which might have to wait some days with the timber floated round them for a meter to arrive and measure it for the duty—"the greatest nuisance on the River" said a timber lighterman.[3]

The great West Indian fleets also arrived in summer, and practically all of their cargoes (the bulk of which consisted of sugar, rum and dye-woods) had to be lightered to the Legal Quays. The Legal Quays and Sufferance

[1] *Report*, 1796, pp. 101, 143.   [2] *Ibid.* pp. x, xi and App. Ss.
[3] *Ibid.* p. 184.

Wharves together could accommodate some 92,000 hogsheads of sugar if all their storage space were devoted to that article alone;[1] but 122,000 hogsheads arrived in the river in four or five months in 1793. Even when sufferances were granted to the West India trade (which happened rarely in peace time) there were not enough Customs officers on the Sufferance Wharves, so that, on the testimony of a lighterman who thought his trade would be ruined by the docks, there could be "upwards of a Thousand Tons of Goods lying in Lighters at a Sufferance Wharf for some Days without that Tonnage being decreased; they had only One Landing Waiter and One Scale employed".[2] But there were not wanting suggestions that merchants themselves were sometimes responsible for delays by omitting to make the required Customs entries. Thirty days were allowed to unload rum, and it might be sold several times over on the ship before being landed. But the net result of the alternate excess and lack of system was that losses by plunderage, waste, and exposure to weather, not to mention losses to the revenue by smuggling, imposed a great and increasing toll which fell heaviest on the West India trade. The East India Company did not suffer to anything like the same extent, thanks to its semi-military organisation and to the system of carrying goods in locked wagons from Blackwall to the company's city warehouses. While Colquhoun's estimate in his *Police of the Thames* (1801), that the West India trade lost over 2 per cent. of its annual value by plunderage may not be too high, the evidence of the 1796 series of reports suggests that such losses were regarded—rightly—as no more than a collateral factor, of great importance no doubt, in an urgently unsatisfactory general situation. The prime causes were (1) the congestion of the river itself, now barely adequate physically for the increased and fast increasing volume of trade, (2) the total inadequacy of the Legal Quays, (3) cumbrous business and revenue arrangements.

The full weight of the well-organised and immensely powerful West India trade, backed by a crowd of lesser mercantile interests, was thrown into the movement for reforms. Briefly their proposal, as put forward by William Vaughan, its most active exponent, was to remove three or four hundred of the most valuable ships and cargoes out of the river into docks. The 1796 Committee agreed with them, and its recommendations were acted upon with relative promptness, considering the opposing interests—wharfingers, lightermen, warehouse owners, riverside parishes, some merchants and revenue officials. There was a good deal of debate as to the best situation for these proposed docks. The great difficulty was that

---

[1] Capper, *op. cit.* pp. 145–6.　　　　[2] *Report*, 1796, p. 149.

of entry; it was suggested that only a few ships (say seven or nine out of sixty) would be able to get in upon one tide, and that the rest would impede the passage of other ships and themselves be in danger of running aground.

The sites considered were at Wapping and across the neck of the Isle of Dogs. It was objected that the ground at Wapping was too costly, being partly built upon, and that it was too far up the river, and while the Parliamentary discussions were going on, the majority of the West India merchants went over to the Isle of Dogs plan. This alternative had certain topographical advantages, since the land already lay below high-tide level, and the sometimes dangerous passage from Blackwall to Limehouse could be avoided by an entrance at Blackwall. The committee approved of both schemes but thought the Isle of Dogs scheme the more urgent. By the turn of the century work had begun, and the laying of the first stone of the West India Dock on 12th July 1800, in some sense laid the foundation of the great port as we know it to-day.

During the earlier part of the century, besides this grip on foreign trade, the City of London "controlled the wool and woollen trades, the coal trade, the livestock trade; and it affected considerably the trade in butter, cheese, fish, poultry, wine, horses, tin and linen".[1] Survivals of these controlling influences still exist; as late as 1929 45 per cent. of the raw wool brought to the country came through London.[2] But there is the significant absence from this list of iron and steel and of cotton.[3] The rise of the industrial north created a new economy less dependent on London. But the metropolitan market remained. The political centre of an ever increasing empire continued to exert its magnetism for whole classes of men, of arts, and of trades. These forces not only remained but grew; and behind them was the vast generative force of the accumulated capital of the City, with all its inter-reciprocating commercial advantages. The momentum of London, the attractive and reproductive power of that vast aggregate of human needs and wills, carried it on to a new domination. While the north had an industrial, London had a financial revolution. Losing its petty and detailed administrative control over the rude trades of this island, the City took into its hands the less direct but infinitely more powerful financial empire of the world.

[1] R. Westerfield, *Middlemen in English Business* (1915), p. 420.
[2] L. Rodwell Jones, *The Geography of London River* (1931), p. 135.
[3] The iron and steel trades of the port (anchor-smiths, etc.) and the import of cotton from the Levant (see p. 509 above) were on too small a scale to affect this argument.

BIBLIOGRAPHICAL NOTE

The literature of London is so vast that any essay will owe much of its general background to works which may not be drawn on for any specific point.

A very stimulating survey of the sources will be found in Miss E. Jeffries Davies's *London and its Records* (*History*, VI, 173, 240, 1922). The earlier site-factors are treated in Dr Ormsby's *London on the Thames* (1924). Recent work on the Restoration period includes N. G. Brett-James's *Growth of Stuart London* (1935), while the relevant portions of Defoe's *Tour* have been edited, with a wealth of annotation and of contemporary engravings, by Sir M. M. Beeton and E. B. Chancellor as *A Tour through London* (1929). M. D. George's "Increase of Population in the 18th Century as illustrated by London" (*Econ. Journal*, XXXII, 325, 1922) and her *London Life in the 18th Century* (1925) are indispensable. See also P. E. Jones and A. V. Judges, "London Population in the Late Seventeenth Century" *Econ. Hist. Rev.* VI, 45 (1935); this discussion deals with the city alone.

A Collection of the Yearly Bills of Mortality from 1657 to 1758 (1759) contains interesting essays by Graunt, Petty, and Corbyn Morris, while points may be gleaned from the controversy on depopulation and luxury started by Richard Price in 1780.

Parliamentary reports containing much information are those of the Select Committees on *Water Supply of the Metropolis* (1821, v) and on the *State of the Port of London* (1836). On this subject, W. Vaughan's *Collections of Tracts...on Wet Docks* (1797), J. Elmes's *Scientific Survey of the Port of London* (1838), and Sir J. G. Broodbank's *History of the Port of London* (1921), with Professor Rodwell Jones's admirable *Geography of London River* (1931) are all of value.

The most useful contemporary maps are by Faithorne and Newcourt (1658) and the splendid productions of J. Rocque (1741–46) and R. Horwood (1799). The 18th century charts of Whitworth (1775) and Heather (1794) are useful as giving the general aspect of the river, but the first really detailed charts were published by the Admiralty in 1812 and 1830. The Ordnance Survey 3-inch map and Stanford's Contour Map of London are indispensable modern maps; Stanford's have also a reprint of Rocque. Sir Laurence Gomme's paper on London maps (*Geog. Journ.* XXX, 489, 616 (1908)) is interesting.

INDEX

Cornwall, prehistoric, 11, 12, 16, 18, 23, 26 ff.; Roman, 52, 69 ff.; medieval, 176, 206, 217, 232, 242, 246, 256 n., 259, 272, 280 ff., 287, 291, 299; 16th cent. 346, 349, 350, 352, 354, 378; 17th cent. 389, 406, 417, 418, 421, 427 n. 440 n., 442; 18th cent. 469, 471, 473, 488, 503, 510, 522

Cotswold Hills, 12, 14, 16, 19, 37, 43, 53, 54, 56, 66, 68, 127, 254, 272, 275, 286, 287, 339, 340, 348, 350, 356, 360, 409, 487, 494, 507

Cotton manufacture, 351, 372, 379, 413 ff., 431 ff., 492, 501, 502, 508, 509, 547. *See also* Textiles

Coventry, medieval, 233, 249, 287; 16th cent. 352, 371; 17th cent. 411, 413, 420, 426; 18th cent. 495

Cowes, 441

Cranborne Chase, 66, 67

Crewkerne, 349

Cricklade, 425, 480

Cromer, 522

Cromwell, Oliver, 451

Crowland, 180, 368, 458

Culham Hythe, 340, 341

Cumberland, prehistoric, 22; Saxon, 120; Scandinavian, 134, 135, 144, 157, 158; medieval, 178 n., 187 n., 226, 227, 231, 256, 258, 297; 16th cent. 375; 17th cent. 403, 421; 18th cent. 469, 471, 473, 484, 510 ff., 518, 526

Cumbria, 11, 12, 16, 17

Cutlery, 349, 511, 512

Dagenham, 543

Dairy produce, medieval, 196, 204, 288, 293; 16th cent. 365, 366; 17th cent. 395, 409, 431; 18th cent. 468, 477, 479, 481, 482, 487, 503, 547

Damasks, 276, 412

Damnonia, 126, 128

Danish settlement, chap. IV *passim*

Darlington, 228

Dartford, 475

Dartmoor, prehistoric, 12; Roman, 52; Saxon, 128; medieval, 176, 281; 18th cent. 507

36

Dartmouth, 281, 282, 287, 289, 290, 296, 354 n., 381 n.

Davenant, C., 394, 399, 401, 408, 436

Deal, 522

Dean, Forest of, 28 n., 56, 70, 72, 226, 256 n., 259, 284, 285, 349, 355 n., 360, 395, 396, 416, 421, 424, 431, 491, 512

Debenham, 366

Dee, Dr John, 383, 386

Dee, R., 16, 26, 59, 76, 296, 352 n., 372, 389, 426, 430

Deer, 223, 345, 356, 369, 375, 408, 474

Defoe, Daniel, 443, chap. XIII *passim*, 529, 533, 541

Deira, 99, 106, 122, 127

Denbighshire, 402, 511

Deptford, 273 n., 358, 423 n., 491, 534, 544

Derby, Saxon, 125; Scandinavian, 144, 149; medieval, 172 n., 214, 218 ff.; 16th cent. 369, 417, 426, 441; 18th cent. 481, 495

Derbyshire, prehistoric, 26; Roman, 75; Scandinavian, 134, 149, 150, 158, 161, 163; medieval, 169, 173, 178 n., 179 n., 205, 208, 210, 226, 227, 256; 16th cent. 345 n., 349, 369, 370; 17th cent. 396, 402, 404, 406, 411, 416 ff., 423, 426, 440; 18th cent. 471, 473

Derwent, R., 106, 141, 338, 344, 369, 426

Devizes, 94, 411, 487

Devonshire, prehistoric, 28; Saxon, 128, 154; medieval, 170, 176, 178 n., 206, 232, 242, 246, 249, 256 n., 258, 259 n., 263, 272, 280 ff., 287, 291; 16th cent. 345, 349, 351, 354, 378, 379; 17th cent. 389, 404, 408, 410, 415 ff., 442; 18th cent. 469, 471, 473, 487, 488, 490, 507, 522

Dobuni, 21, 43, 54

Dodson, Colonel, 458

Dogs, Isle of, 547

Domesday Book, chap. V *passim*

Don, R., 141, 425, 512, 514

Doncaster, 79, 80, 369, 417, 425

Donnington, 462

KING ALFRED'S COLLEGE
LIBRARY